Signal and Image Representation in Combined Spaces

Wavelet Analysis and Its Applications

The subject of wavelet analysis has recently drawn a great deal of attention from mathematical scientists in various disciplines. It is creating a common link between mathematicians, physicists, and electrical engineers. This book series will consist of both monographs and edited volumes on the theory and applications of this rapidly developing subject. Its objective is to meet the needs of academic, industrial, and governmental researchers, as well as to provide instructional material for teaching at both the undergraduate and graduate levels.

This is the seventh book of the series. It is designed to explore the wide range of powerful tools provided by the wavelet approach to signal processing and image analysis. Wavelet analysis is considered in a very broad sense in this volume. It includes adaptive waveform representation, various time-frequency and windowed Fourier transform tools, subband and image coding, as well as efficient numerical algorithms. As such, both theoretical development and state-of-the-art applications to signal and image processing are investigated.

The series editor wishes to thank the editors of this volume for an outstanding contribution to this book series.

This is a volume in
WAVELET ANALYSIS AND ITS APPLICATIONS

CHARLES K. CHUI, SERIES EDITOR

A list of titles in this series appears at the end of this volume.

Signal and Image Representation in Combined Spaces

Edited by
Yehoshua Zeevi
Electrical Engineering Department
Technion-Israel Institute of Technology
Technion City
Haifa, Israel

Ronald Coifman
Department of Mathematics
Yale University
New Haven, Connecticut

ACADEMIC PRESS
San Diego   London   Boston
New York   Sydney   Tokyo   Toronto

ACADEMIC PRESS
525 B Street, Suite 1900, San Diego, CA 92101-4495, USA
1300 Boylston Street, Chestnut Hill, MA 02167, USA
http://www.apnet.com

ACADEMIC PRESS LIMITED
24–28 Oval Road, London NW1 7DX, UK
http://www.hbuk.co.uk/ap/

Library of Congress Cataloging-in-Publication Data

Signal and image representation in combined spaces / edited by
    Yehoshua Zeevi, Ronald Coifman.
      p.   cm. — (Wavelet analysis and its applications ; v. 7)
    Includes bibliographical references and index.
    ISBN 0-12-777830-6 (alk. paper)
    1. Signal processing — Mathematics.   2. Image processing —
Mathematics.   3. Wavelets (Mathematics)   I. Zeevi, Y. Y.
II. Coifman, Ronald R. (Ronald Raphaël)   III. Series.
TK5102.9.S534   1997
621.382'2'015152433—dc21   97-30324
                                                                    CIP

Printed and bound in the United Kingdom

Transferred to Digital Printing, 2011

Contents

III. Redundant Waveform Representations for Signal Processing and Image Analysis 257

IV. Numerical Compression and Applications 339

V. Analysis of Waveform Representations 393

VI. Filter Banks and Image Coding 511

Preface

Wavelets are a generic name for a collection of self similar localized waveforms suitable for signal and image processing. The first set of such functions that constituted an orthonormal basis for $L^2(\mathbb{R})$ was introduced in 1910 by Haar. However the Haar functions do not have good localization in the combined time-frequency space and, therefore, in many cases do not satisfy the properties required in signal and image processing and analysis. The problem of how to construct functions that are well localized in both time and frequency was confronted by communication engineers dealing with the analysis of speech in the 1920s and 1930s. About half a century ago Gabor introduced the optimally localized function, obtained by windowing a complex exponential with a Gaussian window. The main advantage of this localized waveform is in achieving the lowest bound on the joint entropy, defined as the product of effective temporal or spatial extent and frequency bandwidth. However, the Gabor elementary functions, which span $L^2(\mathbb{R})$, are not orthogonal.

The subject of representation in combined spaces refers to wavelet-type and Gabor-type expansions. Such expansions are more suitable for the analysis and processing of natural signals and images than expansion by the traditional application of Fourier series, polynomials, and other functions of infinite support, since the nonstationarity of natural signals calls for localization in both time (or spatial variables in the case of images) and frequency (or scale) in their representation. While global transforms such as the Fourier transform, which is the most widely used in engineering, describe the spectrum of the entire signal as a whole, the wavelet-type and Gabor-type transforms allow for extraction of the local signatures of the signal as they vary in time, or along the spatial coordinates in the case of images. By correlating signals with appropriately chosen wavelets, certain analysis tasks such as feature extraction, signal compression, and recognition can be facilitated. The ability of wavelets to localize signals in time, or spatial variables in the case of images, allows for a multiresolution approach in signal processing. In fact, since the wavelet transform is defined by either its basic time-scale, position-scale, or decomposition structure, it naturally lends itself to multiresolution analysis. Yet, a great deal of freedom is left for the exact choice of the transform's kernel and various parameters. Thus, the wavelet approach provides us with a wide range of powerful tools for signal processing and analysis. These are described in this volume.

The general interrelated topics involving multiscale analysis, wavelet and Gabor analysis, can all be viewed as enhancing the traditional Fourier analysis by enabling an adaptation of combined time and frequency localization procedures to various tasks. The simple and basic transition from the global Fourier transform to the localized (windowed) Fourier analysis, consists of segmenting the signal into windows of fixed length, each of which is expanded by a Fast Fourier Transform (FFT) or Discrete Cosine Transform (DCT). This type of procedure corresponds to spectrograms, to Gabor transform, as well as to localized trigonometric transforms. A dual version of this procedure corresponds to filtering the signal, or windowing its Fourier transform, usually referred to as wavelet, wavelet packets, or subband coding transforms.

Wavelet analysis and more generally adapted waveform analysis has provided a simple comprehensive mathematical and algorithmic infrastructure for the localized signal processing tools, as well as many new tools which evolved as a result of the cross-fertilization of ideas originated in many fields, such as the Calderon-Zygmund theory in mathematics, multiscale ideas from geophysical seismic prospecting, mathematical physics of coherent states and wave packets, pyramid structures in image processing, band and subband filtering in signal processing, music, numerical analysis, etc. In this volume we don't intend to elaborate on the origin of these ideas, but rather on the current state of this elaborate toolkit and the relative advantages it brings to the scene.

While to some extent most of the qualitative analytical aspects of wavelet analysis, and of the windowed Fourier transform, have been well understood by mathematicians for at least 30 years, the recent explosion of activity and algorithms is due to the discovery of the orthogonal wavelets by Stromberg and Meyer, and the connection to Quadrature Mirror Filter (QMF) by Mallat and Daubechies. More fundamental yet is our better understanding of structures permitting construction of a multitude of orthogonal and nonorthogonal expansions customized to tasks at hand, and enabling the introduction of fast computational methods and realtime processing. The role and usefulness of redundancy in providing stability in signal representation, as opposed to efficiency, has also become clear by means of the application of frame analysis and the Zak transform.

Some of the main tasks that can be accomplished by the application of wavelet-based tools are related to feature extraction and efficient description of large data sets for processing and computations. This is the point where, instead of using algebraic or analytic formulas, functions or measured data are described efficiently by adapted waveforms which are, in turn, described algorithmically and designed specifically to optimize various tasks. Perhaps the most natural analogy to the new modes of analysis (or signal transcription) is provided by musical scores and orchestration; an overlay of time frequency analysis. The musical score is somewhat more general and abstract than the alphabet and corresponds roughly to a description of a piece of music by specifying which notes are being played, i.e., the note's characteristic pitch, amplitude, duration, and location in time. While traditional windowed Fourier analysis considers a Fourier representation of the signal in each window of space (or time), wavelets, wavelet packets, and their variants provide a description in which notes of different duration (or resolution) are superimposed. For images, this corresponds to an overlay of patterns of different size and scale. This multiscale representation allows for a better separation of textures and structures, and of decomposition of the textures into their basic elements. The complementary procedure introduces a new approach to speech, music, and image synthesis, yet to be further explored.

Most of the chapters in this book are based on the lectures delivered at the Neaman Workshop on Signal and Image Representation in Combined Spaces, held at Technion. Additional chapters were contributed by invitees who could not attend the workshop. The material presented in this volume brings together a rich variety of ideas that blend most aspects of analysis mentioned above. These papers can be clustered into affinity groups as follows:

Variations on the windowed Fourier transform and its applications, relating Fourier analysis to analysis on the Heisenberg group, are provided in the following group of papers: M. An, A. Bordzik, I. Gertner, and R. Tolimieri: "Weyl-Heisenberg System and the Finite Zak Transform;" M. Bastiaans: "Gabor's Expansion and the Zak Transform for Continuous-Time and Discrete-Time Signals;" W. Schempp: "Non-Commutative Affine Geometry and Symbol Calculus: Fourier Transform Magnetic Resonance Imaging and Wavelets;" M. Zibulski and Y. Y. Zeevi: "The Generalized Gabor Scheme and Its Application in Signal and Image Representation."

Constructions of special waveforms suitable for specific tasks are given in: J. S. Byrnes: "A Low Complexity Energy Spreading Transform Coder;" A. Cohen and N. Dyn: "Nonstationary Subdivision Schemes, Multiresolution Analysis and Wavelet Packets."

The use of redundant representations in reconstruction and enhancement is provided in: J. J. Benedetto: "Noise Reduction in Terms of the Theory of Frames;" Z. Cvetković and M. Vetterli: "Overcomplete Expansions and Robustness;" F. Bergeaud and S. Mallat: "Matching Pursuit of Images."

Applications of efficient numerical compression as a tool for fast numerical analysis are described in: A. Averbuch, G. Beylkin, R. Coifman, and M. Israeli: "Multiscale Inversion of Elliptic Operators;" A. Harten: "Multiresolution Representation of Cell-Averaged Data: A Promotional Review."

Approximation properties of various waveforms in different contexts are described in the following series of papers: A. J. E. M. Janssen: "A Density Theorem for Time-Continuous Filter Banks;" V. E. Katsnelson: "Sampling and Interpolation for Functions with Multi-Band Spectrum: The Mean Periodic Continuation Method;" M. A. Kon and L. A. Raphael: "Characterizing Convergence Rates for Multiresolution Approximations;" C. Chui and Chun Li: "Characterization of Smoothness via Functional Wavelet Transforms;" R. Lenz and J. Svanberg: "Group Theoretical Transforms, Statistical Properties of Image Spaces and Image Coding;" J. Prestin and K. Selig: "Interpolatory and Orthonormal Trigonometric Wavelets;" B. Rubin: "On Calderón's Reproducing Formula;" and "Continuous Wavelet Transforms on a Sphere;" V. A. Zheludev: "Periodic Splines, Harmonic Analysis and Wavelets."

Acknowledgments

The Neaman Workshop was organized under the auspices of The Israel Academy of Sciences and Humanites and co-sponsored by The Neaman Institute for Advanced Studies in Science and Technology; The Institute of Advanced Studies in Mathematics; The Institute of Theoretical Physics; and The Ollendorff Center of the Department of Electrical Engineering, Technion—Israel Institute of Technology.

Several people helped in the preparation of this manuscript. We wish to thank in particular Ms. Lesley Price for her editorial assistance and word-processing of the manuscripts provided by the authors, Ms. Margaret Chui for her editing and overall guidance in the preparation of the book, and Ms. Katy Tynan of Academic Press for her communications assistance.

Haifa, Israel   Yehoshua Y. Zeevi
New Haven, Connecticut   Ronald Coifman
June 1997

Contributors

Numbers in parentheses indicate where the authors' contributions begin.

M. AN (1), *Prometheus Inc., 52 Ashford Street, Allston, MA 02134*
[myoung@ccs.neu.edu]

AMIR AVERBUCH (341), *School of Mathematical Sciences, Tel Aviv University, Tel Aviv 69978, Israel*
[amir@math.tau.ac.il]

MARTIN J. BASTIAANS (23), *Technische Universiteit Eindhoven, Faculteit Elektrotechniek, Postbus 513, 5600 MB Eindhoven, Netherlands*
[M.J.Bastiaans@ele.tue.nl]

JOHN J. BENEDETTO (259), *Department of Mathematics, University of Maryland, College Park, Maryland 29742*
[jjb@math.umd.edu]

FRANÇOIS BERGEAUD (285), *Ecole Centrale Paris, Applied Mathematics Laboratory, Grande Voie des Vignes, F-92290 Châtenay-Malabry, France*
[francois@mas.ecp.fr]

GREGORY BEYLKIN (341), *Program in Applied Mathematics, University of Colorado at Boulder, Boulder, CO 80309-0526*
[beylkin@julia.colorado.edu]

A. BRODZIK (1), *Rome Laboratory/EROI, Hanscom AFB, MA 01731-2909*

J. S. BYRNES (167), *Prometheus Inc., 52 Ashford St., Allston, MA 02134*
[jbyrnes@cs.umb.edu]

CHARLES K. CHUI (395), *Center for Approximation Theory, Texas A&M University, College Station, TX 77843*
[cchui@tamu.edu]

ALBERT COHEN (189), *Laboratoire d'Analyse Numérique, Université Pierre et Marie Curie, 4 Place Jussieu, 75005 Paris, France*
[cohen@ann.jussieu.fr]

RONALD COIFMAN (341), *Department of Mathematics, P.O. Box 208283, Yale University, New Haven, CT 06520-8283*
[coifman@jules.math.yale.edu]

ZORAN CVETKOVIĆ (301), *Department of Electrical Engineering and Computer Sciences, University of California at Berkeley, Berkeley, CA 94720-1772*
[zoran@eecs.berkeley.edu]

NIRA DYN (189), *School of Mathematical Sciences, Sackler Faculty of Exact Sciences, Tel Aviv University, Tel Aviv 69978, Israel*
[niradyn@math.tau.ac.il]

I. GERTNER (1), *Computer Science Department, The City College of New York, Convent Avenue at 138th Street, New York, NY 10031*
[csicg@csfaculty.engr.ccny.cuny.edu]

AMI HARTEN (361), *School of Mathematical Sciences, Tel-Aviv University, Tel-Aviv, 69978 Israel*

MOSHE ISRAELI (341), *Faculty of Computer Science, Technion-Israel Institute of Technology, Haifa 32000, Israel*
[israeli@cs.technion.ac.il]

A. J. E. M. JANSSEN (513), *Philips Research Laboratories, WL-01, 5656 AA Eindhoven, The Netherlands*

VICTOR E. KATSNELSON (525), *Department of Theoretical Mathematics, The Weizmann Institute of Science, Rehovot 76100, Israel*
[katze@wisdom.weizmann.ac.il]

MARK A. KON (415), *Department of Mathematics, Boston University, Boston, MA 02215*

REINER LENZ (553), *Department of Electrical Engineering, Linköping University, S-58183 Linköping, Sweden*
[reiner@isy.liu.se]

CHUN LI (395), *Institute of Mathematics, Academia Sinica, Beijing 100080, China.*

STÉPHANE MALLAT (285), *Ecole Polytechnique, CMAP, 91128 Palaiseau Cedex, France*
[mallat@cmapx.polytechnique.fr]

JÜRGEN PRESTIN (201), *FB Mathematik, Universität Rostock, D–18051 Rostock, Germany*
[prestin@mathematik.uni-rostock.d400.de]

LOUISE A. RAPHAEL (415), *Department of Mathematics, Howard University, Washington, DC 20059*

BORIS RUBIN (439, 457), *Department of Mathematics, Hebrew University of Jerusalem, Givat Ram 91904, Jerusalem, Israel*
[boris@math.huji.ac.il]

WALTER SCHEMPP (71), *Lehrstuhl fuer Mathematik I, University of Siegen, D-57068 Siegen, Germany*
[schempp@mathematik.uni-siegen.d400.de]

KATHI SELIG (201), *FB Mathematik, Universität Rostock, D–18051 Rostock, Germany*
[selig@mathematik.uni-rostock.d400.de]

JONAS SVANBERG (553), *Department of Electrical Engineering, Linköping University, S-58183 Linköping, Sweden*
[svan@isy.liu.se]

R. TOLIMIERI (1), *Electrical Engineering Department, The City College of New York, Convent Avenue at 138th Street, New York, NY 10031*

MARTIN VETTERLI (301), *Department d'Electricite EPFL, CH-1015 Lausanne, Switzerland*
[martin.vetterli@de.epfl.ch]

YEHOSHUA Y. ZEEVI (121), *Department of Electrical Engineering, Technion–Israel Institute of Technology, Haifa 32000, Israel*
[zeevi@ee.technion.ac.il]

VALERY A. ZHELUDEV (477), *School of Mathematical Sciences, Tel Aviv University, 69978 Tel Aviv, Israel*
[zhel@math.tau.ac.il]

MEIR ZIBULSKI (121), *Multimedia Department, IBM Science and Technology, MATAM, Haifa 31905, Israel*
[meirz@vnet.ibm.com]

I.

Variations of the Windowed Fourier Transform and Its Applications

Weyl-Heisenberg Systems and the
Finite Zak Transform

M. An, A. Brodzik, I. Gertner, and R. Tolimieri

Abstract. Previously, a theoretical foundation for designing algorithms for computing Weyl-Heisenberg (W-H) coefficients at critical sampling was established by applying the finite Zak transform. This theory established clear and easily computable conditions for the existence of W-H expansion and for stability of computations. The main computational task in the resulting algorithm was a 2-D finite Fourier transform.

In this work we extend the applicability of the approach to rationally over-sampled W-H systems by developing a deeper understanding of the relationship established by the finite Zak transform between linear algebra properties of W-H systems and function theory in Zak space. This relationship will impact on questions of existence, parameterization, and computation of W-H expansions.

Implementation results on single RISC processor of i860 and the PARAGON parallel multiprocessor system are given. The algorithms described in this paper possess highly parallel structure and are especially suited in a distributed memory, parallel-processing environment. Timing results show that real-time computation of W-H expansions is realizable.

§1   Introduction

During the last four years powerful new methods have been introduced for analyzing Wigner transforms of discrete and periodic signals [10, 11, 13] based on finite W-H expansions [2, 5, 6, 12]. A recent work [10] adapted these methods to gain control over the cross-term interference problem [9] by constructing signal systems in time frequency space for expanding Wigner transforms from W-H systems based on Gaussian-like signals.

The computational feasibility of the method in [10] depends strongly on the availability of efficient and stable algorithms for computing W-H expansion coefficients. Since W-H systems are not orthogonal, standard Hilbert space inner-product methods do not generally apply. Moreover,

Signal and Image Representation in Combined Spaces
Y. Y. Zeevi and R. R. Coifman (Eds.), pp. 3–21.

since critically sampled W-H systems may not form a basis, over-sampling in time frequency is necessary for the existence of arbitrary signal expansions. In fact, this is usually the case for systems based on the Gaussian. In [10, 11, 12, 13, 15], the concept of biorthogonals was applied to the problem of W-H coefficient computation. In [15], the Zak transform provided the framework for computing biorthogonals for rationally over-sampled W-H systems forming frames. A similar approach for critically and integer over-sampled W-H systems can be found in [3, 4]. The goal in this work is somewhat different in that major emphasis is placed on describing linear spans of W-H systems that are not necessarily complete and on establishing, in a form suitable for RISC and parallel processing, algorithms for computing W-H coefficients of signals in such linear spans. For the most part, our approach extends on that developed in [3, 14] and frame theory, although an important part in [15] plays no role in this work. However, as in these previous works [7, 8], the finite Zak transform will be established as a fundamental and powerful tool for studying critically sampled and rationally over-sampled W-H systems and for designing algorithms for computing W-H coefficients for discrete and periodic signals. The role of the finite Zak transform is analogous to that played by the Fourier transform in replacing complex convolution computations by simple pointwise multiplication. In this new setting, properties of W-H systems, such as their spanning space and dimension, can be determined by simple operations on functions in Zak space. This relationship will impact on questions of existence, parameterization, and computation of W-H expansions.

In the over-sampled case, both integer and rational over-sampling are investigated. Implementation results on single RISC processor of i860 and the PARAGON parallel multiprocessor system are given for sample sizes both of powers of 2 and mixed sizes with factors 2, 3, 4, 5, 6, 7, 8, and 9. The algorithms described in this paper possess highly parallel structure and are especially suited in a distributed memory, parallel-processing environment. Timing results on single i860 processor and on 4- and 8-node computing systems show that real-time computation of W-H expansions is realizable.

In Section 2, the basic preliminaries will be established. Algorithms will be described in Section 3 for critically sampled W-H systems, in Section 4 for integer over-sampled systems, and in Section 5 for rationally over-sampled systems. Implementation results will be given in Sections 6, 7, and 8.

§2   Preliminaries

2.1   Weyl-Heisenberg systems

Choose an integer $N > 0$. A discrete function $f(a)$, $a \in Z$ is called N-*periodic* if
$$f(a + N) = f(a), \qquad a \in \mathbf{Z}.$$
Denote by $L(N)$ the Hilbert space of all N-periodic functions with inner product
$$< f,g > = \sum_{a=0}^{N-1} f(a)g^*(a), \qquad f,g \in L(N).$$
For $g \in L(N)$ and $0 \le m, n < N$ define $g_{m,n} \in L(N)$ by
$$g_{m,n}(a) = g(a + m)e^{-2\pi i n a/N}, \qquad a \in \mathbf{Z}. \tag{2.1.1}$$
The functions in the set $\{g_{m,n} : 0 \le m, n < N\}$ are called *Weyl-Heisenberg wavelets* having generator g.

Suppose $N = KM$ with positive integers K and M. The collection of N functions
$$\{g_{kM,mK} : 0 \le k < K, \;\; 0 \le m < M\}$$
denoted by (g, M, K) is called a *critically sampled Weyl-Heisenberg* (W-H) system. Critically sampled W-H systems have been extensively studied in several works including [3] where the finite Zak transform was used to establish conditions for such systems to be a basis of $L(N)$. The extension to 2-D for applications to image representation and image analysis can be found in [14].

Suppose $N = K'M'$ is a second factorization of N into two positive integers. The collection of $K'M$ functions
$$\{g_{k'M',mK} : 0 \le k' < K', \;\; 0 \le m < M\}$$
denoted by (g, M', K) is called a general W-H system. The system (g, M', K) is called *over-sampled* if $M' < M$ and *under-sampled* if $M' > M$. Over-sampled systems are necessarily redundant having finer time-resolution as compared with associated critically sampled systems. A dual theory can easily be developed which introduces redundancies by having finer frequency-resolution.

An expansion of a signal $f \in L(N)$ as a linear combination over a W-H system is called a *W-H expansion*, with the corresponding coefficients called *W-H coefficients*. In general, except for critically sampled W-H systems forming a basis of $L(N)$, W-H coefficients are not uniquely determined. W-H basis were studied in [3, 14].

An over-sampled W-H system (g, M', K) is called an *integer* over-sampled system if $R = M/M'$ is an integer and a *rationally* over-sampled system otherwise.

2.2 Finite Zak transform (FZT)

Fundamental properties of FZT have been described in several works [3, 14] with applications of FZT to algorithms for computing W-H expansion coefficients for a critically sampled W-H basis. We will briefly outline the one-dimensional case.

Suppose $N = KM$. For $f \in L(N)$ define the *finite Zak Transform* (FZT), $Z(K)f(a,b)$, $a, b \in \mathbf{Z}$ by

$$Z(K)f(a,b) = \sum_{k=0}^{K-1} f(a + Mk)e^{2\pi i b k/K}, \qquad a, b \in \mathbf{Z}. \qquad (2.2.1)$$

The functional equations

$$Z(K)f(a + M, b) = e^{-2\pi i b/K} Z(K)f(a,b), \qquad a, b \in \mathbf{Z} \qquad (2.2.2)$$

$$Z(K)f(a, b + K) = Z(K)f(a,b), \qquad a, b \in \mathbf{Z} \qquad (2.2.3)$$

imply $Z(K)f$ is N-periodic in each variable and is completely determined by its values

$$Z(K)f(a,b), \qquad 0 \le a < M, \ \ 0 \le b < K. \qquad (2.2.4)$$

Denote by $L(M, K)$ the Hilbert space of all functions $F(a,b)$, $0 \le a < M$, $0 \le b < K$, with inner product

$$\langle F, G \rangle = \sum_{a=0}^{M-1} \sum_{b=0}^{K-1} F(a,b)G^*(a,b), \qquad F, G \in L(M, K). \qquad (2.2.5)$$

Define $Z_0(K)f \in L(M, K)$ by

$$Z_0(K)f(a,b) = Z(K)f(a,b), \qquad 0 \le a < M, \ \ 0 \le b < K. \qquad (2.2.6)$$

The mapping $K^{-1/2}Z_0(K)$ is an isometry from $L(N)$ onto $L(M, K)$. If $F \in L(M, K)$ and $f \in L(N)$ is defined by

$$f(a + Mk) = K^{-1} \sum_{b=0}^{K-1} F(a,b)e^{-2\pi i b k/K}, \qquad 0 \le a < K, \ \ 0 \le b < K. \qquad (2.2.7)$$

Then $F = Z_0(K)f$.

An important relationship exists between the FZT of W-H wavelets corresponding to a fixed generator g given by

$$Z(K)g_{m,n}(a,b) = e^{-2\pi i a n/N} Z(K)g(a + m, b - n), \qquad a, b \in \mathbf{Z}. \qquad (2.2.8)$$

In particular,

$$Z(K)g_{kM,mK}(a,b) = Z(K)g(a,b)e^{-2\pi i(ma/M+kb/K)}, \quad a,b \in \mathbf{Z},$$
$$0 \le k < K, \quad 0 \le m < M. \tag{2.2.9}$$

From these relationships we can prove two fundamental results which govern the application of FZT to the analysis of W-H expansions. Set $F = Z_0(K)f$ and $G = Z_0(K)g$.

First fundamental result

$$F(a,b)G^*(a,b) = \frac{1}{M} \sum_{k=0}^{K-1} \sum_{m=0}^{M-1} <f,g_{kM,mK}> e^{-2\pi i(ma/M+kb/K)}.$$
$$\tag{2.2.10}$$

The second result uses the FZT to unravel a W-H expansion as a product in Zak space.

Second fundamental result For $f \in L(N)$,

$$f = \sum_{k=0}^{K-1} \sum_{m=0}^{M-1} c(kM,mK)g_{kM,mK} \tag{2.2.11}$$

if and only if

$$F = GP, \quad P \in L(M,K),$$

where

$$P(a,b) = \sum_{k=0}^{K-1} \sum_{m=0}^{M-1} c(kM,mK)e^{-2\pi i(am/M+bk/K)}. \tag{2.2.12}$$

For critically sampled W-H systems, we have the following result:

Theorem 1. *The critically sampled W-H system*

$$(g,M,K) = \{g_{kM,mK} : 0 \le k < K, \quad 0 \le m < M\} \tag{2.2.13}$$

is a basis of $L(N)$ if and only if G never vanishes.

§3 Critically sampled W-H systems

Generalizations of the results in the previous section depend on an analysis of the zero-sets of FZT. Consider a critically sampled system (g,M,K) and set $G = Z_0(K)g$. Denote the zero-set of G by ζ.

Theorem 2. *A function $f \in L(N)$ is in the linear span of (g, M, K) if and only if $F = Z_0(K)f$ vanishes on ζ. The dimension of the linear span of (g, M, K) is $N - J$ where J is the number of points in ζ.*

Proof: By the second fundamental result we can identify the linear span of (g, M, K) with the space of all F of the form $F = GP$ with $P \in L(M, K)$. In particular, if f is in the linear span of (g, M, K), then F must vanish on ζ. Conversely, suppose F vanishes on ζ. Define $P \in L(M, K)$ by

$$P(a, b) = \begin{cases} F(a, b)/G(a, b), & (a, b) \notin \zeta \\ 0, & \text{otherwise.} \end{cases}$$

Then $F = GP$ and F is in the linear span of (g, M, K). Since we have shown that the linear span of (g, M, K) can be identified with the space of all $F \in L(M, K)$ which vanish on ζ, the theorem follows. ∎

Denote by $L(\zeta)$ the space of all functions $\alpha \in L(M, K)$ which vanish on the complement, ζ^c, of ζ. Theorem 2 leads to the following algorithm for computing expansion coefficients of f relative to (g, M, K). To each $\alpha \in L(\zeta)$, define the function $P^\alpha \in L(M, K)$ by

$$P^\alpha(a, b) = \begin{cases} F(a, b)/G(a, b), & (a, b) \notin \zeta \\ \alpha(a, b), & (a, b) \in \zeta. \end{cases}$$

By the second fundamental result, a collection of W-H coefficients is given by the 2-D $M \times K$ FT of $P^\alpha(a, b)$.

If ζ is not empty, the expansion coefficients are not uniquely determined. In fact, every f in the linear span of (g, M, K) has a J-dimensional space of W-H expansions over (g, M, K) parameterized by $L(\zeta)$.

§4   Integer over-sampled W-H systems

Consider an integer over-sampled W-H system $\mathbf{g} = (g, M', K)$. Since $K' = RK$ with R an integer, each $0 \le k' < K'$ can be written uniquely as

$$k' = r + kR, \quad 0 \le r < R, \quad 0 \le k < K.$$

Since

$$g_{k'M', mK} = (g_{rM', 0})_{kM, mK},$$

$\mathbf{g}$ is a union of critically sampled W-H systems

$$\mathbf{g} = \bigcup_{r=0}^{R-1} \mathbf{g}_r, \quad \mathbf{g}_r = (g_r, M, K), \quad g_r = g_{rM', 0}, \quad 0 \le r < R. \tag{4.1}$$

It is just as simple to consider the more general case where **g** is the union of arbitrary critically sampled W-H systems.

For any subset γ of $\{(a, b) : 0 \leq a < M, \ 0 \leq b < K\}$ define $F|\gamma \in L(M, K)$ by

$$F|\gamma(a, b) = \begin{cases} F(a, b), & (a, b) \in \gamma, \\ 0, & \text{otherwise.} \end{cases}$$

Theorem 3. *Suppose* **g** *is the union of critically sampled W-H systems,* $\mathbf{g}_r = (g_r, M, K), \ 0 \leq r < R$. *Denote the zero-set of* $G_r = Z_0(K)g_r$ *by* ζ_r *and set* $\zeta = \cap_{r=0}^{R-1}\zeta_r$. *Then* f *is in the linear span of* **g** *if and only if* $F = Z_0(K)f$ *vanishes on* ζ. *The dimension of the linear span of* **g** *is* $N - J$ *where* J *is the number of points in* ζ.

Proof: If f is in the linear span of **g**, then we can write $f = \sum_{r=0}^{R-1} f_r$ where f_r is in the linear span of $\mathbf{g}_r$. Theorem 2 implies $F_r = Z_0(K)f_r$ vanishes on ζ_r. Since $F = \sum_{r=0}^{R-1} F_r$, f vanishes on ζ.

Conversely, suppose F vanishes on ζ. If we can write $F = \sum_{r=0}^{R-1} F_r$ where F_r vanishes on ζ_r, then Theorem 2 implies that f is in the linear span of **g**. The following construction determines one such decomposition. Define $F_r \in L(M, K), 0 \leq r < R$ by

$$F_0 = F|\zeta_0^c$$

$$F_1 = F|\zeta_0 \cap \zeta_1^c$$

$$\vdots$$

$$F_{R-1} = F|\zeta_0 \cap \cdots \cap \zeta_{R-2}.$$

By definition, F_r vanishes on ζ_r and since F vanishes on ζ, $F = \sum_{r=0}^{R-1} F_r$. Since the linear span of **g** can be identified with the space of all $F \in L(M, K)$, which vanish on ζ, the theorem is proved. ∎

From the construction in the theorem we have the following:

Corollary 1. *If* f *is in the linear span of* **g**, *then we can write* $F = \sum_{r=0}^{R-1} F_r$, *where* $F_r F_s = 0$ *whenever* $r \neq s$, $0 \leq r, s < R$.

Choose $f \in L(N)$ in the linear span of **g**. An algorithm for computing a W-H expansion of f over **g** is given as follows.

- Decompose $F = Z_0(K)f$

$$F = \sum_{r=0}^{R-1} F_r, \qquad F_r \in L(M, K)$$

where F_r vanishes on the zero set ζ_r of G_r, $0 \leq r < R$.

- Compute the collection of 2D $M \times K$ FT of

$$P_r(a, b) = \frac{F_r(a, b)}{G_r(a, b)}, \qquad 0 \le r < R.$$

This stage is understood to be taken as in the critically sampled case, with arbitrary values assigned to the quotient at points where the functions G_r, $0 \le r < R$ vanish.

If we assume that $T \log T$ computations are needed for the T-point FT, then the complexity of one W-H expansion computation is

$$N \log K + R(N \log K + N \log M) + RN \tag{4.2}$$

but advantage can be taken of the large number of zero data values.

The coefficient set of W-H expansions of $f \in L(N)$ over **g** is parameterized by the collection of decompositions of F and by the arbitrarily assigned values to the quotients at the points ζ_r, $0 \le r < R$.

§5   Rationally over-sampled W-H systems

Consider the rationally over-sampled W-H system $\mathbf{g}' = (g, M', K)$ where $N = MK = M'K'$. $R = M/M'$ is no longer an integer. Denote the least common multiple of M and M' by $\overline{M}$ and set $\overline{M} = MS = M'S'$. S and S' are positive integers such that S divides K, S' divides K', and $N = \overline{M}\frac{K}{S} = \overline{M}\frac{K'}{S'}$.

Arguing as in the integer over-sampled case, we have that $\mathbf{g}'$ is the union of the under-sampled W-H systems

$$\mathbf{g}'_{s'} = (g_{s'}, \overline{M}, K), \qquad g_{s'} = g_{s'M', 0}, \ \ 0 \le s' < S'.$$

Since $\overline{M} = MS$ with S a positive integer, the under-sampled W-H system $\mathbf{g}'_{s'}$ is contained in the critically sampled W-H system $\mathbf{g}_{s'} = (g_{s'}, M, K)$. Denote the union of these critically sampled W-H systems by **g** and set $G_{s'} = Z_0(K)g_{s'}$.

Theorem 4. *A function $f \in L(N)$ is in the linear span of $\mathbf{g}'$ if and only if $F = Z_0(K)f$ has the form*

$$F = \sum_{s'=0}^{S'-1} G_{s'} P_{s'}, \tag{5.1}$$

where $P_{s'} \in L(M, K)$ satisfies

$$P_{s'}\left(a, b + \frac{K}{S}\right) = P_{s'}(a, b), \qquad 0 \le s' < S', \quad 0 \le a < M, \quad 0 \le b < \frac{K}{S}. \tag{5.2}$$

Proof: Since

$$Z(K)(g_{\overline{k}\,\overline{M},mK}(a,b) = G(a,b)e^{-2\pi i(\frac{m}{M}a+\overline{k}\frac{S}{K}b)}$$

the theorem follows from the second fundamental result. ∎

If an expansion of the form given in the theorem can be found, then arguing as before, a collection of W-H expansion coefficients of f over $\mathbf{g}'$ is given by the collection of 2-D $M\frac{K}{S}$ FTs of

$$P_{s'}(a,b), \quad 0 \le a < M, \ 0 \le b < \frac{K}{S}, \ 0 \le s' < S'.$$

In [1], an algorithm was given for computing W-H coefficients for rationally over-sampled W-H systems based on pseudo-matrix inversion of the matrix function

$$\mathbf{G}(a,b) = \left[G_{s'}\left(a,b+s\frac{K}{S}\right) \right]_{0 \le s < S, 0 \le s' < S'}.$$

Implementation results for this case will be given below. An alternate approach will be taken in this work which presents an *iterative* algorithm more in line with the philosophy of the preceding sections. We will describe an algorithm which for any $f \in L(N)$ computes a W-H expansion for the orthogonal projection of f onto the linear span of $\mathbf{g}'$.

Denote by $L(M,\frac{K}{S})$ the subspace of all $P \in L(M,K)$ satisfying $P(a,b+\frac{K}{S}) = P(a,b)$ $0 \le a < M$, $0 \le b < K$. The following result describes an algorithm for computing orthogonal projections onto the subspace $G \cdot L(M,\frac{K}{S}) = \{GP : P \in L(M,\frac{K}{S})\}$.

Theorem 5. *Suppose $F \in L(M,K)$ has the form $F = GP$ with $P \in L(M,K)$. Then there exists $P' \in L(M,\frac{K}{S})$ satisfying the condition*

$$\sum_{s=0}^{S-1} \left| G\left(a,b+s\frac{K}{S}\right) \right|^2 P'(a,b) = \sum_{s=0}^{S-1} \left| G\left(a,b+s\frac{K}{S}\right) \right|^2 P\left(a,b+s\frac{K}{S}\right),$$

$$0 \le a < M, 0 \le b < \frac{K}{S}, \qquad (5.3)$$

and $F' = GP'$ is the orthogonal projection of F onto $G \cdot L(M,\frac{K}{S})$.

Proof: Since the right-hand side of (5.3) vanishes at any point (a,b), $0 \le a < M$, $0 \le b < \frac{K}{S}$, at which

$$\sum_{s=0}^{S-1} \left| G\left(a,b+s\frac{K}{S}\right) \right|^2 = 0$$

and we can solve (5.3) for some $P' \in L(M, \frac{K}{S})$. Define $Q \in L(M, K)$ by

$$Q = |G|^2 (P - P'). \tag{5.4}$$

By (5.3) we have

$$\sum_{s=0}^{S-1} Q\left(a, b + s\frac{K}{S}\right) = 0, \quad 0 \le a < M, \quad 0 \le b < \frac{K}{S}$$

which, since $P' \in L(M, \frac{K}{S})$, implies

$$< G(P - P'), GP' > \; = \; < Q, P' > \; = 0.$$

Defining $P'' \in L(M, K)$ by $P = P' + P''$, we have that

$$GP = GP' + GP''$$

is an orthogonal decomposition in $L(M, K)$, completing the proof of the theorem. ∎

The computation of P' requires N additions and multiplications.

Algorithm for computing W-H coefficients

- For each $0 \le s' < S'$, compute $P_{s'} \in L(M, K)$ such that

$$F|\zeta_{s'}^c = G_{s'} P_{s'}, \quad P_{s'} \in L(M, K).$$

- Compute the orthogonal decomposition

$$G_{s'} P_{s'} = G_{s'} P'_{s'} + G_{s'} P''_{s'}.$$

- If $P'_{s'} = 0$ for all $0 \le s' < S'$, then f is orthogonal to the linear span of $\mathbf{g}'$, and we are done.

- Otherwise, choose $0 \le s_0 < S'$ such that

$$\|G_{s_0} P'_{s_0}\| \ge \|G_{s'} P'_{s'}\|, \quad 0 \le s' < S',$$

and iterate the previous steps with $F - G_{s_0} P'_{s_0}$ replacing F.

Since

$$F = G_{s_0} P'_{s_0} + (G_{s_0} P''_{s_0} + F|\zeta_{s_0})$$

is an orthogonal decomposition, we have

$$\|F - G_{s_0} P'_{s_0}\| < \|F\|,$$

and at some point of the iteration, we will arrive at $F = F' + F''$, with $F' = Z_0(K)f'$ with f' in the linear span of $\mathbf{g}'$ and $F'' = Z_0(K)f''$, f'' orthogonal to $\mathbf{g}'$. A W-H expansion of f' over $\mathbf{g}'$ can be given by a collection of 2-D $M \times \frac{K}{S}$ FTs as before.

§6   Implementation results

In this section we describe implementation issues and present timing results for the implementation of the algorithms presented in the previous sections. Implementations on a single Intel i860 RISC microprocessor as well as on the Paragon multi-processor parallel platform are reported.

6.1   Critical sampling (C.S.)

We have tested three basic analysis functions:

- Gaussian function

  When K and M are both even integers, the FZT of Gaussian window function has a zero at $(K/2, M/2)$. Set $Q(K/2, M/2) = 0.0$. The total energy of Gabor coefficients will be minimum.

  When either K or M is an odd integer, or both of them are odd integers, the FZT of Gaussian window function has no zeros.

- Rectangular function

  A small-size rectangular window will result in FZT with no zeros. For example, $N = K \times M = 1200$, a window of width 90 centered at 600, has no zeros in Zak space.

  A rectangular window of width 150 centered at 600 has zeros in Zak space located at: (j,8), (j,16), (j,24), (j,32), where j=0 to 39.

- Triangular function

  When either K or M is an odd integer, or both of them are odd integers, there are no zeros in Zak space.

  A relatively small triangular window will result in a single zero at the center of Zak space. For example, $N = 40 \times 30 = 1200$, a window of 61 non-zero values centered at 600, has one zero in Zak space at (20, 15).

We have implemented the computation for Critical Sampling case: the main program is in FORTRAN and the FFT modules are fine-tuned i860 assembly with mixed sizes. Timing results are given in Tables 1 and 2.

Complexity

For a real input signal f, the FZT of f is Hermitian symmetric along K-dimension. If the analysis signal is also real, then the 2-D $M \times K$ $Q(a,b)$ has the same symmetry. The inverses of the FZT of $g(a,b)$ are pre-computed and stored in memory. The complexity of the computation ($F(n)$ denotes the complexity of n-point FFT):

$Z(K)f$ (FZT of f)	$M \times$ real $F(K)$
$Z(K)f/Z(K)g$	$K/2 \times M$ multiplications
2-D FT of Q	$M \times$ Herm. $F(K)$
Herm. Symm. along K	$K \times$ real $F(M)$

Size N	2-D $K \times M$	Time
256	16×16	0.67
512	16×32	1.20
1024	32×32	2.02
2048	32×64	3.98
4096	64×64	7.41
8192	64×128	14.96
16384	128×128	29.82
32768	128×256	60.89
65536	256×256	125.55
131072	256×512	264.60
262144	512×512	566.99

Table 1. Timing results (in milliseconds) on the Intel i860 RISC microprocessor (critical sampling – 2^k).

§7   Integer over-sampling

We choose the decomposition $F = Z(K)f = \sum_{r=0}^{R-1} F_r$ such that $F_1, \ldots,$ F_{R-1} each has only one non-zero point, so that the computation of the 2-D FT of $Q_1(a,b), \ldots Q_{R-1}(a,b)$ is trivial. The codes are similar to a critically sampled case with data rearrangement at the end.

7.1   Rational over-sampling

In [12], the authors point out that a Gaussian window function over-sampled by more than 20 percent (5/4), does not have significant influence. We have implemented the computation for over-sampling rates 3/2 and 5/4. Again, the main routine is coded in FORTRAN, and the DFT

Size N	2-D $K \times M$	Time
384	8×48	1.47
768	16×48	1.99
1536	32×48	3.12
3072	64×48	5.91
3072	128×24	6.15
6144	128×48	12.07
6144	64×96	12.48
12288	512×24	26.07
12288	128×96	24.05
24576	256×96	48.70
49152	256×192	98.71
98304	256×384	203.52
98304	512×192	209.12
196608	512×384	433.41
393216	1024×384	1011.61

Table 2. Timing results (in milliseconds) on the Intel i860 RISC Microprocessor (critical sampling – mixed sizes).

routines are fine-tuned i860 assembly codes for mixed sizes. For the complex singular value decomposition (SVD), we used the LINPACK routine. We have tested three basis functions:

- Gaussian basis function

  Rational over-sampling of 3/2 and 5/4 were tested. If the *rank* $(G(a, b))$ equals to 2 or 4 correspondingly, then **g** is complete and every f has a W-H expansion over **g**.

- Rectangular basis function

  Rational over-sampling by 3/2 and 5/4 are tested. Rectangular window sizes have to be chosen such that it is not a factor of K along K-dimension to have every f expandable in the W-H system.

- Triangular basis function

  An example of size $N = 40 \times 30 = 1200$ has been tested with rational over-sampling by 3/2. The experimental results are:

  A window of size 101 centered at 600 results in an expandable W-H system.

A window of size 151 centered at 600 results in an expandable W-H system.

A window of size 201 results in point (20,10) being a zero singular value in Zak transform space.

Complexity

In the case of real input and real analysis signals, the FZT is Hermitian symmetric along K-dimension. We can show that the S' 2-D $M\frac{K}{S} P_s(a,b)$ has Hermitian symmetry along $\frac{K}{S}$-dimension. The complexity of real-time computation is:

FZT of f	$M\times$ real $F(K)$
$G^+(a,b)F(a,b)$	$M \times \frac{K}{S}$ matrix $S' \times S$ multiply a vector S
S' 2-D FT of P_s with Hermitian Symmetry along $\frac{K}{S}$	$S' \times M\times$ Hermitian $F(\frac{K}{S})$, $S' \times \frac{K}{S}$ real $F(M)$

Timing results of various sizes are given in Tables 3 and 4.

Size N	2-D $K \times M$	Time
384	16×24	2.06
768	32×24	2.97
1536	64×24	5.31
3072	64×48	10.79
3072	128×24	10.05
6144	128×48	20.85
6144	64×96	22.86
12288	128×96	43.15
24576	256×96	84.71
49152	256×192	171.39
98304	256×384	412.12
98304	512×192	413.50
196608	512×384	840.02

Table 3. Timing results (in milliseconds) in the Intel i860 RISC microprocessor (rational over-sampling (3/2)).

Size N	2-D $K \times M$	Time
320	8×40	2.82
640	16×40	3.85
1280	32×40	5.66
2560	64×40	9.65
5120	128×40	16.42
5120	64×80	18.32
10240	128×80	32.09
10240	64×160	37.99
20480	128×160	67.65
40960	128×320	134.08
81920	256×320	258.40
163840	512×320	522.19
327680	512×640	1149.76

Table 4. Timing results (in milliseconds) on the Intel i860 microprocessor (rational over-sampling (5/4)).

§8   Parallel implementation

Assume that a distributed memory parallel computer has p $(< \min(K, M))$ processors. Set

$$P = K/K_1 = M/K_2. \tag{8.1}$$

The algorithms described in Sections 3, 4 and 5 possess highly parallel structure. They are particularly suitable in a distributed memory multiprocessor system. For example, in the critically sampled case, the algorithm can be implemented as follows:

- Each processor receives K_1 K-point input data

- Compute K_1 K-point real FFT

- Point-wise multiplication of the pre-calculated Zak transform of the basis function $1/Z(K)g(a, b)$

- Compute K_1 K-point Hermitian FFT

- Data permutation between processors (matrix transpose)

- Compute K_2 M-point real FFT

Implementation of an integer over-sampled case has a similar structure to the critically sampled case, and the rationally over-sampled case has a better parallel structure, since it has S' relatively small 2-D $\frac{K}{S} \times M$

FFT's, and they might be carried out locally in each processor without interprocessor data permutation. Timing results of critical sampling on the Intel 4-nodes and 8-nodes Paragon are given in Tables 5 and 6. The parallel flow diagram is given in Figure 1.

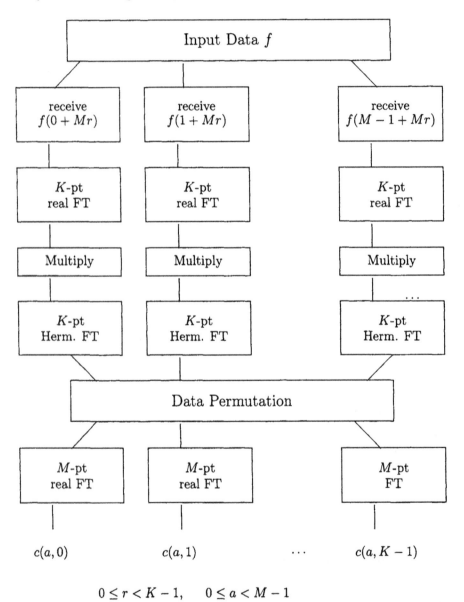

$$0 \le r < K - 1, \quad 0 \le a < M - 1$$

Figure 1. Parallel implementation flow diagram.

Size N	2-D $K \times M$	Time
16384	128×128	10.06
32768	128×256	19.66
65536	256×256	39.31
131072	256×512	80.24
262144	512×512	163.10
524288	512×1024	368.99
1048576	1024×1024	801.82
2097152	1024×2048	1661.96

Table 5. Timing results (in milliseconds) on the Intel Paragon (4-nodes).

Size N	2-D $K \times M$	Time
65536	256×256	22.18
131072	256×512	42.45
262144	512×512	86.32
524288	512×1024	189.54
1048576	1024×1024	404.32
2097152	1024×2048	840.17
8388608	2048×2048	1716.03

Table 6. Timing results (in milliseconds) on the Intel Paragon (8-nodes).

§9   Conclusions

Algorithms for the computation of Weyl-Heisenberg (W-H) coefficients for the cases of critical sampling, integer over-sampling, and rational over-sampling have been presented, and easily computable conditions for the existence of W-H expansions have been derived in terms of the Zak transform of the signal and the analysis function. We have shown that the algorithms described lead to very efficient FFT-based implementations both for single DSP processor systems as well as for parallel multi-processor configurations.

Acknowledgments.   The research of M. An was supported by ARPA F49620-C-91-0098, and research of R. Tolimieri is supported by AFOSR RF#447323.

References

[1] An, A., G. Kechriotis, C. Lu, and R. Tolimieri, The computation of Weyl-Heisenberg coefficients for critically sampled and over-sampled signals, *Proceedings of the Int. Conf. SPAT ICSPAT94* **2**, October 1994, pp. 824–829.

[2] Auslander, L. and R. Tolimieri, On finite Gabor expansion of signals, *IMA Proc. Signal Processing*, Springer-Verlag, New York, 1988.

[3] Auslander, L., I. Gertner, and R. Tolimieri, Finite Zak transforms and the finite Fourier transforms, *IMA on Radar and Sonar, Part II* **39** Springer-Verlag, New York, 1991, 21–36.

[4] Auslander, L., I. Gertner, and R. Tolimieri, The discrete Zak transform application to time-frequency analysis and synthesis of nonstationary signals, *IEEE Trans. Signal Processing* **39**(4) (1991), 825–835.

[5] Bastiaans, M. J., Gabor signal expansion and degree of freedom of a signal, *Optica Acta* **29** (1982), 1223–1229.

[6] Gabor, D., Theory of communications, *Proceedings IEE* III **93** (1946), 429–457.

[7] Gertner, I. and G. A. Geri, Image representation using Hermite functions, *Biological Cybernetics* **71** (1994), 147–151.

[8] Gertner, I. and G. A. Geri, *J. Optical Soc. Amer.* **11** (8) (1994), 2215-2219.

[9] Hlawatsch, F., Interference terms in the Wigner distribution, in *Proc. 1984 Int. Conf. on DSP*, Florence, Italy, 1984, pp. 363–367.

[10] Qian, S. and J. M. Morris, Wigner distribution decomposition and cross-term deleted representation, *Signal Processing* **27** (1992), 125–144.

[11] Raz, S., Synthesis of signals from Wigner distributions: Representation on biorthogonal basis, *Signal Processing* **20**(4) (1990), 303–314.

[12] Wexler, J. and S. Raz, Discrete Gabor expansions, *Signal Processing* **21**(3) (1990), 207–220.

[13] Wexler, J. and S. Raz, Wigner-space synthesis of discrete-time periodic signals, *IEEE Trans. Signal Processing* **40**(8) (1990).

[14] Zeevi, Y. Y. and I. Gertner, The finite Zak transform: An efficient tool for image representation and analysis, *J. Visual Comm. and Image Representation* **3**(1) (1992), 13–23.

[15] Zibulski, M. and Y. Y. Zeevi, Oversampling in the Gabor scheme, *IEEE Trans. Signal Processing* **41**(8) (1993).

M. An
Prometheus Inc.
52 Ashford Street
Allston, MA 02134
MYOUNG@CCS.neu.edu

A. Brodzik
Rome Laboratory/EROI
Hanscom AFB, MA 01731-2909

I. Gertner
Computer Science Department
The City College of New York
Convent Avenue at 138th Street
New York, NY 10031
csicg@csfaculty.engr.ccny.cuny.edu

R. Tolimieri
Electrical Engineering Department
The City College of New York
Convent Avenue at 138th Street
New York, NY 10031

Gabor's Expansion and the Zak Transform for Continuous-Time and Discrete-Time Signals

Martin J. Bastiaans

Abstract. Gabor's expansion of a signal into a discrete set of shifted and modulated versions of an elementary signal is introduced and its relation to sampling of the sliding-window spectrum is shown. It is shown how Gabor's expansion coefficients can be found as samples of the sliding-window spectrum, where, at least in the case of critical sampling, the window function is related to the elementary signal in such a way that the set of shifted and modulated elementary signals is bi-orthonormal to the corresponding set of window functions.

The Zak transform is introduced and its intimate relationship to Gabor's signal expansion is demonstrated. It is shown how the Zak transform can be helpful in determining the window function that corresponds to a given elementary signal and how it can be used to find Gabor's expansion coefficients.

The continuous-time as well as the discrete-time case are considered, and, by sampling the continuous frequency variable that still occurs in the discrete-time case, the discrete Zak transform and the discrete Gabor transform are introduced. It is shown how the discrete transforms enable us to determine Gabor's expansion coefficients via a fast computer algorithm, analogous to the well-known fast Fourier transform algorithm.

Not only Gabor's critical sampling is considered, but also, for continuous-time signals, the case of oversampling by an integer factor. It is shown again how, in this case, the Zak transform can be helpful in determining a (no longer unique) window function corresponding to a given elementary signal. An arrangement is described which is able to generate Gabor's expansion coefficients of a rastered, one-dimensional signal by coherent-optical means.

§1   Introduction

It is sometimes convenient to describe a time signal $\varphi(t)$, say, not in the time domain, but in the frequency domain by means of its *frequency spectrum,*

Signal and Image Representation in Combined Spaces
Y. Y. Zeevi and R. R. Coifman (Eds.), pp. 23–69.

23

i.e. the *Fourier transform* $\bar{\varphi}(\omega)$ of the function $\varphi(t)$, which is defined by

$$\bar{\varphi}(\omega) = \int \varphi(t)e^{-j\omega t}dt; \qquad (1.1)$$

a bar on top of a symbol will mean throughout that we are dealing with a function in the frequency domain. (Unless otherwise stated, all integrations and summations in this paper extend from $-\infty$ to $+\infty$.) The *inverse* Fourier transformation takes the form

$$\varphi(t) = \frac{1}{2\pi} \int \bar{\varphi}(\omega)e^{j\omega t}d\omega. \qquad (1.2)$$

The frequency spectrum shows us the *global* distribution of the energy of the signal as a function of frequency. However, one is often more interested in the momentary or *local* distribution of the energy as a function of frequency.

The need for a *local frequency spectrum* arises in several disciplines. It arises in music, for instance, where a signal is usually described not by a time function nor by the Fourier transform of that function, but by its *musical score*; indeed, when a composer writes a score, he prescribes the frequencies of the tones that should be present at a certain moment. It arises in optics: geometrical optics is usually treated in terms of *rays*, and the signal is described by giving the directions (see frequencies) of the rays (see tones) that should be present at a certain position (see time moment). It arises also in mechanics, where the position and the momentum of a particle are given simultaneously, leading to a description of mechanical phenomena in a *phase space*.

A candidate for a local frequency spectrum is *Gabor's signal expansion*. In 1946 Gabor [14] suggested the expansion of a signal into a discrete set of properly shifted and modulated Gaussian elementary signals [5, 6, 7, 14, 15, 16]. A quotation from Gabor's original paper might be useful. Gabor writes in the summary:

> Hitherto communication theory was based on two alternative methods of signal analysis. One is the description of the signal as a function of time; the other is Fourier analysis. ... But our everyday experiences ... insist on a description in terms of *both* time and frequency. ... Signals are represented in two dimensions, with time and frequency as co-ordinates. Such two-dimensional representations can be called 'information diagrams,' as areas in them are proportional to the number of independent data which they can convey. ... There are certain 'elementary signals' which occupy the smallest possible area in the information diagram. They are harmonic oscillations modulated by a probability pulse. Each elementary signal can be

considered as conveying exactly one datum, or one 'quantum of information.' Any signal can be expanded in terms of these by a process which includes time analysis and Fourier analysis as extreme cases.

Although Gabor restricted himself to an elementary signal that had a Gaussian shape, his signal expansion holds for rather arbitrarily shaped elementary signals [5, 6, 7].

We will restrict ourselves to one-dimensional time signals; the extension to two or more dimensions, however, is rather straightforward. Most of the results can be applied to continuous-time as well as discrete-time signals. We will treat continuous-time signals in Sections 2, 3, and 4 (and also in Sections 7 and 8), and we will transfer the concepts to the discrete-time case in Sections 5 and 6. To distinguish continuous-time from discrete-time signals, we will denote the former with curved brackets and the latter with square brackets; thus $\varphi(t)$ is a continuous-time and $\varphi[n]$ a discrete-time signal. We will use the variables in a consistent manner: in the continuous-time case, the variables t, m and T have something to do with time, the variables ω, k and Ω have something to do with frequency, and the relation $\Omega T = 2\pi$ holds throughout; in the discrete-time case, the variables n, m and N have something to do with time, the variables θ, k and Θ have something to do with frequency, and the relation $\Theta N = 2\pi$ holds throughout.

In his original paper, Gabor restricted himself to a *critical* sampling of the time-frequency domain; this is the case that we consider in Sections 2–6. In Section 2 we introduce Gabor's signal expansion, we introduce a window function with the help of which the expansion coefficients can be found, and we show a way in which—at least in principle—this window function can be determined. In Section 3 we introduce the Zak transform and we use this transform to determine the window function that corresponds to a given elementary signal in a mathematically more attractive way. A more general application of the Zak transform to Gabor's signal expansion is described in Section 4. We translate the concepts of Gabor's signal expansion and the Zak transform to discrete-time signals in Section 5. Finally, in Section 6, we introduce, on the analogy of the well-known discrete Fourier transform, a discrete version of the Zak transform; the discrete versions of the Fourier and the Zak transform enable us to determine Gabor's expansion coefficients by computer via a fast computer algorithm.

In Section 7 we extend Gabor's concepts to the case of *oversampling*, in particular to oversampling by an *integer* factor. We use the Fourier transform and the Zak transform again to transform Gabor's signal expansion into a mathematically more attractive form and we show how a (no longer unique) window function can be determined in this case of integer oversampling. Finally, in Section 8, we introduce an optical arrangement which is able to generate Gabor's expansion coefficients of a rastered, one-dimensional signal by coherent-optical means.

§2   Gabor's signal expansion

Let us consider an *elementary signal* $g(t)$, which may or may not have a Gaussian shape. Gabor's original choice was a Gaussian [14],

$$g(t) = 2^{1/4}e^{-\pi(t/T)^2}, \tag{2.1}$$

where we have added the factor $2^{\frac{1}{4}}$ to normalize $(1/T) \int |g(t)|^2 dt$ to unity, but in this paper the elementary signal may have a rather arbitrary shape; we will use Gabor's choice of a Gaussian-shaped elementary signal as an example only. From the elementary signal $g(t)$, we construct a discrete set of *shifted and modulated versions* $g_{mk}(t)$ defined by

$$g_{mk}(t) = g(t - mT)e^{jk\Omega t}, \tag{2.2}$$

where the time shift T and the frequency shift Ω satisfy the relationship $\Omega T = 2\pi$, and where m and k may take all integer values. Gabor stated in 1946 that any reasonably well-behaved signal $\varphi(t)$ can be expressed in the form

$$\varphi(t) = \sum_m \sum_k a_{mk} g_{mk}(t), \tag{2.3}$$

with properly chosen coefficients a_{mk}. Thus *Gabor's signal expansion* represents a signal $\varphi(t)$ as a superposition of properly shifted (over discrete distances mT) and modulated (with discrete frequencies $k\Omega$) versions of an elementary signal $g(t)$. We note that there exists a completely *dual* expression in the frequency domain; in this paper, however, we will concentrate on the time-domain description.

Gabor's signal expansion is related to the *degrees of freedom* of a signal: each expansion coefficient a_{mk} represents one complex degree of freedom [8, 14]. If a signal is, roughly, limited to the space interval $|t| < \frac{1}{2}a$ and to the frequency interval $|\omega| < \frac{1}{2}b$, the number of complex degrees of freedom equals the number of Gabor coefficients in the time-frequency rectangle with area ab, this number being about equal to the *time-bandwidth product* $ab/2\pi$. The reason for Gabor to choose a Gaussian-shaped elementary signal was that for such a signal, each shifted and modulated version, which conveys exactly one degree of freedom, occupies the smallest possible area in the time-frequency domain. Indeed, if we choose the elementary signal according to (2.1), the 'duration' of such a signal and the 'duration' of its Fourier transform—defined as the square roots of their normalized second-order moments (see [23, Sect. 8-2])—read $T/2\sqrt{\pi}$ and $\Omega/2\sqrt{\pi}$, respectively, and their product takes the minimum value $\frac{1}{2}$.

Two special choices of the elementary signal might be instructive. If we choose a *rectangular-shaped* elementary signal such that $g(t) = 1$ for $-\frac{1}{2}T < t \le \frac{1}{2}T$ and $g(t) = 0$ outside that time interval, then Gabor's signal expansion has an easy interpretation: we simply consider the signal $\varphi(t)$ in successive time intervals of length T and describe the signal in each time interval by means of a Fourier series. In the case of a *sinc-shaped* elementary signal $g(t) = \sin(\pi t/T)/(\pi t/T)$—and hence $\bar{g}(\omega) = T$ for $-\frac{1}{2}\Omega < \omega \le \frac{1}{2}\Omega$ and $\bar{g}(\omega) = 0$ outside that frequency interval—Gabor's signal expansion has again an easy interpretation: we simply consider the signal in successive frequency intervals of length Ω and describe the signal in each frequency interval by means of the well-known sampling theorem for band-limited signals.

For the rectangular- or sinc-shaped elementary signals considered in the previous paragraph, the discrete set of shifted and modulated versions of the elementary signal $g_{mk}(t)$ is *orthonormal*; in general, however, this need not be the case, which implies that Gabor's expansion coefficients a_{mk} cannot be determined in the usual way. Let us consider two elements $g_{mk}(t)$ and $g_{nl}(t)$ from the (possibly non-orthonormal) set of shifted and modulated versions of the elementary signal, and let their inner product be denoted by $d_{n-m,l-k}$; hence

$$\int g_{nl}^*(t)g_{mk}(t)dt = d_{n-m,l-k} = d_{m-n,k-l}^* . \qquad (2.4)$$

It is easy to see that for Gabor's choice of a Gaussian elementary signal, the array d_{mk} takes the form

$$d_{mk} = T(-1)^{mk}e^{-\frac{1}{2}\pi(m^2+k^2)}, \qquad (2.5)$$

which does not have the form of a product of two Kronecker deltas $T\delta_m\delta_k$; therefore, the set of shifted and modulated versions of a *Gaussian* elementary signal is *not* orthonormal.

Gabor's expansion coefficients can easily be found, even in the case of a non-orthonormal set $g_{mk}(t)$, if we could find a *window function* $w(t)$ such that

$$a_{mk} = \int \varphi(t)w_{mk}^*(t)dt, \qquad (2.6)$$

where we have used, again, the short-hand notation (see (2.2))

$$w_{mk}(t) = w(t - mT)e^{jk\Omega t}.$$

Such a window function should satisfy the two *bi-orthonormality conditions* [5, 6, 7]

$$\int w_{nl}^*(t)g_{mk}(t)dt = \delta_{n-m}\delta_{l-k} \qquad (2.7)$$

and

$$\sum_m \sum_k w_{mk}^*(t_1) g_{mk}(t_2) = \delta(t_1 - t_2); \qquad (2.8)$$

we will show later that the first bi-orthonormality condition implies the second one, so we can concentrate on the first one. The first bi-orthonormality condition guarantees that if we start with an array of coefficients a_{mk}, construct a signal $\varphi(t)$ via (2.3) and subsequently substitute this signal into (2.6), we end up with the original coefficients array; the second bi-orthonormality condition guarantees that if we start with a certain signal $\varphi(t)$, construct its Gabor coefficients a_{mk} via (2.6) and subsequently substitute these coefficients into (2.3), we end up with the original signal. We thus conclude that the two equations (2.3) and (2.6) form a *transform pair*.

We remark that (2.6), with the help of which we can determine Gabor's expansion coefficients, is, in fact, a sampled version of the *sliding-window spectrum* [7, 10] (or complex spectrogram, or windowed Fourier transform, or short-time Fourier transform), where the sampling appears on the time-frequency lattice $(mT, k\Omega)$ with $\Omega T = 2\pi$. In quantum mechanics this lattice is known as the Von Neumann lattice [4, 20], but for obvious reasons we prefer to call it the *Gabor lattice* in the context of this paper. Hence, whereas sampling the sliding-window spectrum yields the Gabor coefficients, Gabor's signal expansion itself can be considered as a way to reconstruct a signal from its sampled sliding-window spectrum. The name *window function* for the function $w(t)$ that corresponds to a given elementary signal $g(t)$ will thus be clear.

It is easy to see that the window function $w(t)$ is proportional to the elementary signal $g(t)$ if the set $g_{mk}(t)$ is orthonormal. In the remainder of this section we show a first way in which a window function can be found if the set $g_{mk}(t)$ is non-orthonormal. We therefore express the window function by means of its Gabor expansion (2.3) with expansion coefficients c_{mk}, say [5],

$$w(t) = \sum_m \sum_k c_{mk} g_{mk}(t) \qquad (2.9)$$

and try to find the array of coefficients c_{mk}. We therefore consider the first bi-orthonormality condition (2.7)

$$\delta_m \delta_k = \int w^*(t) g_{mk}(t) dt$$

and substitute from the Gabor expansion (2.9) for the window function, yielding

$$\delta_m \delta_k = \int \left[\sum_n \sum_l c_{nl}^* g_{nl}^*(t) \right] g_{mk}(t) dt.$$

We rearrange factors

$$\delta_m \delta_k = \sum_n \sum_l c_{nl}^* \int g_{nl}^*(t) g_{mk}(t) dt$$

and substitute from (2.4)

$$\delta_m \delta_k = \sum_n \sum_l c_{nl}^* d_{n-m,l-k} = \sum_n \sum_l c_{nl}^* d_{m-n,k-l}^* \, .$$

The first bi-orthonormality relation thus leads to the condition

$$\sum_n \sum_l c_{nl} d_{m-n,k-l} = \delta_m \delta_k, \tag{2.10}$$

in which the left-hand side has the form of a *convolution* of the given array d_{mk} with the array c_{mk} that we have to determine. Equation (2.10) can be solved, in principle, when we introduce the *Fourier transform* of the arrays according to

$$\bar{d}(t, \omega; T) = \sum_m \sum_k d_{mk} e^{-j(m\omega T - k\Omega t)} \quad \text{(with } \Omega T = 2\pi) \tag{2.11}$$

and a similar expression for $\bar{c}(t, \omega; T)$. Note that these Fourier transforms are periodic in the time variable t and the frequency variable ω with periods T and Ω, respectively:

$$\bar{d}(t + mT, \omega + k\Omega; T) = \bar{d}(t, \omega; T). \tag{2.12}$$

Hence, in considering such Fourier transforms we can restrict ourselves to the *fundamental Fourier interval* $(-\frac{1}{2}T < t \le \frac{1}{2}T, -\frac{1}{2}\Omega < \omega \le \frac{1}{2}\Omega)$. The *inverse* Fourier transformation reads

$$d_{mk} = \frac{1}{2\pi} \int_T \int_\Omega \bar{d}(t, \omega; T) e^{j(m\omega T - k\Omega t)} dt d\omega \tag{2.13}$$

and a similar expression for c_{mk}; $\int_T \cdot dt$ and $\int_\Omega \cdot d\omega$ denote integrations over one period T and Ω, respectively. After Fourier transforming both sides of (2.10), the convolution transforms into a *product*, and (2.10) takes the form

$$\bar{c}(t, \omega; T) \bar{d}(t, \omega; T) = 1. \tag{2.14}$$

The function $\bar{c}(t, \omega; T)$ can easily be found from the latter relationship, provided that the inverse of $\bar{d}(t, \omega; T)$ exists, and inverse Fourier transforming $\bar{c}(t, \omega; T)$ (see (2.13)) then results in the array c_{mk} that we are looking for.

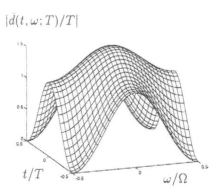

$|\bar{d}(t,\omega;T)/T|$

t/T ω/Ω

Figure 1. The Fourier transform $d(t,\omega;T)/T$ in the case of a Gaussian elementary signal.

Let us consider how things work out for Gabor's choice of a Gaussian elementary signal. After Fourier transforming the array (2.5) we get

$$\frac{1}{T}\bar{d}(t,\omega;T) = \theta_3(\Omega t)\theta_3(\omega T)+\theta_3(\Omega t)\theta_2(\omega T)+\theta_2(\Omega t)\theta_3(\omega T)-\theta_2(\Omega t)\theta_2(\omega T),$$

(2.15)

with

$$\theta_3(x) = \theta_3(x;e^{-2\pi}) = \sum_m e^{-2\pi m^2}e^{j2mx}$$

(2.16)

and

$$\theta_2(x) = \theta_2(x;e^{-2\pi}) = \sum_m e^{-2\pi(m+\frac{1}{2})^2}e^{j(2m+1)x}.$$

(2.17)

The functions $\theta(x;e^{-2\pi})$ are known as *theta functions* [1, 30] with *nome* $e^{-2\pi}$. The Fourier transform $(1/T)\bar{d}(t,\omega;T)$ has been depicted in Figure 1, where we have restricted ourselves to the fundamental Fourier interval.

Note that the values of $(1/T)\bar{d}(t,\omega;T)$ for $(t=\frac{1}{2}T+mT, \omega=\frac{1}{2}\Omega+k\Omega)$ read

$$\frac{1}{T}\bar{d}(\tfrac{1}{2}T+mT,\tfrac{1}{2}\Omega+k\Omega;T) = \theta_3^2(0)\left[1-2\frac{\theta_2(0)}{\theta_3(0)}-\frac{\theta_2^2(0)}{\theta_3^2(0)}\right].$$

Since the nome equals $e^{-2\pi}$, we have the relation $\theta_2(0)/\theta_3(0) = \sqrt{2}-1 = \tan(\pi/8)$ (see [30, p. 525]), and we conclude that $\bar{d}(t,\omega;T)$ has (double) zeros for $(t = \frac{1}{2}T + mT, \omega = \frac{1}{2}\Omega + k\Omega)$. Inversion of $\bar{d}(t, \omega; T)$ in order to find $\bar{c}(t, \omega; T)$ may thus be difficult.

Zeros of $\bar{d}(t, \omega; T)$ not only prohibit an easy determination of the window function $w(t)$, but they lead to another unwanted property: they enable us to construct a (not identically zero) function $\bar{z}(t, \omega; T)$ such that the product $\bar{z}(t, \omega; T)\bar{d}(t, \omega; T)$ (see (2.14)) vanishes. For a Gaussian elementary signal, with zeros for $(t = \frac{1}{2}T + mT, \omega = \frac{1}{2}\Omega + k\Omega)$, we might choose

$$\bar{z}(t, \omega; T) = \sum_m \sum_k \delta(t - \tfrac{1}{2}T - mT) 2\pi\delta(\omega - \tfrac{1}{2}\Omega - k\Omega)z. \qquad (2.18)$$

Inverse Fourier transforming this function yields the array

$$z_{mk} = (-1)^{m+k}z, \qquad (2.19)$$

which is a *homogeneous solution* of (2.10). Hence, Gabor coefficients might *not* be *unique*: if c_{mk} are Gabor coefficients that determine the window function $w(t)$, then $c_{mk} + z_{mk}$ are valid Gabor coefficients, as well!

In the next section we present a different and mathematically more attractive way to find the window function $w(t)$ that corresponds to a given elementary signal $g(t)$.

§3 Zak transform

In this section we introduce the *Zak transform* [31, 32, 33] and we show its intimate relationship to Gabor's signal expansion. The Zak transform $\tilde{\varphi}(t, \omega; \tau)$ [31, 32, 33] of a signal $\varphi(t)$ is defined as a one-dimensional Fourier transformation of the sequence $\varphi(t + m\tau)$ (with m taking on all integer values and t being a mere parameter), hence

$$\tilde{\varphi}(t, \omega; \tau) = \sum_m \varphi(t + m\tau)e^{-jm\omega\tau}; \qquad (3.1)$$

throughout we will denote the Zak transform of a signal by the same symbol as the signal itself, but marked by a tilde on top of it. We remark that the Zak transform $\tilde{\varphi}(t, \omega; \tau)$ is *periodic* in the frequency variable ω with period $2\pi/\tau$ and *quasi-periodic* in the time variable t with quasi-period τ:

$$\tilde{\varphi}\left(t + m\tau, \omega + k\frac{2\pi}{\tau}; \tau\right) = \tilde{\varphi}(t, \omega; \tau)e^{jm\omega\tau}. \qquad (3.2)$$

Hence, in considering the Zak transform we can restrict ourselves to the *fundamental Zak interval* $(-\frac{1}{2}\tau < t \leq \frac{1}{2}\tau, -\pi/\tau < \omega \leq \pi/\tau)$. The *inverse* relationship of the Zak transform has the form

$$\varphi(t + m\tau) = \frac{\tau}{2\pi} \int_{2\pi/\tau} \tilde{\varphi}(t, \omega; \tau)e^{jm\omega\tau} d\omega; \qquad (3.3)$$

it will be clear that the variable t in the latter equation can be restricted to an interval of length τ, with m taking on all integer values. From the properties of the Zak transform we mention Parseval's energy theorem, which leads to the relationship

$$\frac{1}{2\pi} \int_\tau \int_{2\pi/\tau} |\tilde{\varphi}(t,\omega;\tau)|^2 dt d\omega = \frac{1}{\tau} \int |\varphi(t)|^2 dt. \qquad (3.4)$$

The Zak transform $\tilde{\varphi}(t,\omega;\tau)$ provides a means to represent an arbitrarily long one-dimensional time function (or one-dimensional frequency function) by a two-dimensional time-frequency function on a rectangle with *finite area* 2π. This two-dimensional function $\tilde{\varphi}(t,\omega;\tau)$ is known as the Zak transform, because Zak was the first who systematically studied this transformation in connection with solid state physics [31, 32, 33]. Some of its properties were known long before Zak's work, however. The same transform is called Weil-Brezin map and it is claimed that the transform was already known to Gauss [28]. It was also used by Gel'fand (see, for instance, [27, Chap. XIII]); Zak seems, however, to have been the first to recognize it as the versatile tool it is. The Zak transform has many interesting properties and also interesting applications to signal analysis, for which we refer to [17, 18]. In this section we will show how the Zak transform can be applied to Gabor's signal expansion.

We want to make an observation to which we will return later on in this paper. Suppose that, for small τ for instance, we can approximate a function $g(t)$ by the piecewise constant function

$$g(t) = \sum_n g_n \text{rect}\left(\frac{t - n\tau}{\tau}\right), \qquad (3.5)$$

where $\text{rect}(x) = 1$ for $-\frac{1}{2} < x \le \frac{1}{2}$ and $\text{rect}(x) = 0$ outside that interval. In the time interval $-\frac{1}{2}\tau < t \le \frac{1}{2}\tau$, the Zak transform $\tilde{g}(t,\omega;\tau)$ then takes the form

$$\tilde{g}(t,\omega;\tau) = \sum_n g_n e^{-jn\omega\tau} = \bar{g}(\omega\tau); \qquad (3.6)$$

note that this Zak transform does not depend on the time variable t, and that the one-dimensional Fourier transform $\bar{g}(\omega\tau)$ of the sequence g_n arises. We remark that Parseval's energy theorem (3.4) now leads to the relation

$$\frac{1}{2\pi} \int_\tau \int_{2\pi/\tau} |\tilde{g}(t,\omega;\tau)|^2 dt d\omega = \frac{\tau}{2\pi} \int_{2\pi/\tau} |\bar{g}(\omega\tau)|^2 d\omega = \sum_n |g_n|^2. \qquad (3.7)$$

We still have to solve the problem of finding the window function $w(t)$ that corresponds to a given elementary signal $g(t)$ such that the biorthonormality conditions (2.7) and (2.8) are satisfied. We consider again the first bi-orthonormality condition (2.7)

$$\delta_m \delta_k = \int g(\tau) w^*_{mk}(\tau) d\tau$$

and apply a Fourier transformation (see (2.11)) to both sides of this condition, yielding

$$1 = \sum_m \sum_k \left[\int g(\tau) w^*(\tau - mT) e^{-jk\Omega\tau} d\tau \right] e^{-j(m\omega T - k\Omega t)}.$$

We rearrange factors

$$1 = \sum_m \left[\int g(\tau) w^*(\tau - mT) \left(\sum_k e^{-jk\Omega(\tau - t)} \right) d\tau \right] e^{-jm\omega T}$$

and replace the sum of exponentials by a sum of Dirac functions

$$1 = \sum_m \left[\int g(\tau) w^*(\tau - mT) \left(T \sum_n \delta(\tau - t - nT) \right) d\tau \right] e^{-jm\omega T}.$$

We rearrange factors again

$$1 = T \sum_m \sum_n \left[\int g(\tau) w^*(\tau - mT) \delta(\tau - t - nT) d\tau \right] e^{-jm\omega T}$$

and evaluate the integral

$$1 = T \sum_m \sum_n g(t + nT) w^*(t + [n - m]T) e^{-jm\omega T}.$$

After a final rearranging of factors we find

$$
\begin{aligned}
1 &= T \sum_n g(t + nT) e^{-jn\omega T} \left[\sum_m w^*(t + [n - m]T) e^{j(n-m)\omega T} \right] \\
&= T \left[\sum_n g(t + nT) e^{-jn\omega T} \right] \left[\sum_m w(t + mT) e^{-jm\omega T} \right]^*,
\end{aligned}
$$

in which expression we recognize (see (3.1)) the definitions for the Zak transforms $\tilde{g}(t, \omega; T)$ and $\tilde{w}(t, \omega; T)$ of the two functions $g(t)$ and $w(t)$, respectively; hence

$$T \tilde{g}(t, \omega; T) \tilde{w}^*(t, \omega; T) = 1. \tag{3.8}$$

The first bi-orthonormality condition (2.7) thus transforms into a *product*, enabling us to find the window function $w(t)$ that corresponds to a given elementary signal $g(t)$ in an easy way:

- from the elementary signal $g(t)$ we derive its Zak transform $\tilde{g}(t, \omega; T)$ via definition (3.1);

- under the assumption that division by $\tilde{g}(t, \omega; T)$ is allowed, the function $\tilde{w}(t, \omega; T)$ can be found with the help of relation (3.8);

- finally, the window function $w(t)$ follows from its Zak transform $\tilde{w}(t, \omega; T)$ by means of the inversion formula (3.3).

It is shown in Appendix A that the window function $w(t)$ found in this way also satisfies the second bi-orthonormality condition (2.8).

Let us consider Gabor's original choice of a Gaussian elementary signal again. The Zak transform $\tilde{g}(t, \omega; \alpha T)$ of the Gaussian signal (2.1) reads

$$\tilde{g}(t, \omega; \alpha T) = 2^{\frac{1}{4}} e^{-\pi(t/T)^2} \theta_3 \left(\alpha\pi \left[\frac{\omega}{\Omega} - j\frac{t}{T} \right]; e^{-\pi\alpha^2} \right), \qquad (3.9)$$

where

$$\theta_3 \left(z; e^{-\pi\alpha^2} \right) = \sum_m e^{-\pi\alpha^2 m^2} e^{j2mz} \qquad (3.10)$$

is a theta function again, in this case with nome $e^{-\pi\alpha^2}$. This Zak transform has been depicted in Figure 2 for several values of the parameter $\tau = \alpha T$, where we have restricted ourselves to the fundamental Zak interval; note that for $\alpha \leq \frac{1}{3}$, the Zak transform becomes almost independent of t, as we have mentioned before. We remark that the Zak transform of a Gaussian signal has zeros for $(t = \frac{1}{2}\alpha T + m\alpha T, \omega = \frac{1}{2}\Omega/\alpha + k\Omega/\alpha)$.

In the case of a Gaussian elementary signal and choosing $\tau = T$ (Gabor's original choice), the Zak transform of the window function takes the form

$$T\tilde{w}(t, \omega; T) = \frac{1}{\tilde{g}^*(t, \omega; T)} = 2^{-\frac{1}{4}} e^{\pi(t/T)^2} \frac{1}{\theta_3(\pi\zeta; e^{-\pi})}, \qquad (3.11)$$

in which expression we have set, for convenience, $\zeta = \omega/\Omega + jt/T$. In the fundamental Zak interval, the function $1/\theta_3(\pi\zeta; e^{-\pi})$ can be expressed as

$$\frac{1}{\theta_3(\pi\zeta; e^{-\pi})} = \left(\frac{K_o}{\pi} \right)^{-3/2} \left[c_0 + 2 \sum_{m=1}^{\infty} (-1)^m c_m \cos(2\pi m\zeta) \right], \qquad (3.12)$$

where the coefficients c_m are defined by

$$c_m = \sum_{n=0}^{\infty} (-1)^n e^{-\pi(n+\frac{1}{2})(2m+n+\frac{1}{2})} \qquad (3.13)$$

(see, for instance, [30, p. 489, Example 14]); the constant K_o is the complete elliptic integral for the modulus $\frac{1}{2}\sqrt{2}$: $K_o = 1.85407468$ (see, for instance,

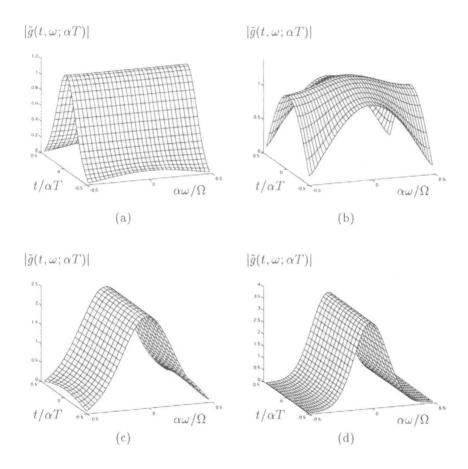

Figure 2. The Zak transform $\widetilde{g}(t, \omega; \alpha T)$ in the case of a Gaussian elementary signal for different values of α: (a) $\alpha = 2$, (b) $\alpha = 1$, (c) $\alpha = \frac{1}{2}$, and (d) $\alpha = \frac{1}{3}$.

$g(t), Tw(t)$

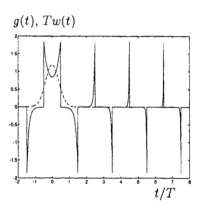

t/T

Figure 3. A Gaussian elementary signal $g(t)$ (dashed line) and its corresponding window function $Tw(t)$ (solid line).

[30, p. 524]). It is now easy to determine the window function $w(t)$ via the inversion formula (3.3), yielding

$$Tw(t + mT) = 2^{-\frac{1}{4}} e^{\pi(t/T)^2} \left(\frac{K_o}{\pi}\right)^{-3/2} (-1)^m c_m e^{2\pi m(t/T)} \qquad (3.14)$$

with $-\frac{1}{2}T < t \le \frac{1}{2}T$, and hence

$$Tw(t) = 2^{-\frac{1}{4}} e^{\pi(t/T)^2} \left(\frac{K_o}{\pi}\right)^{-3/2} \sum_{p+\frac{1}{2} \ge |t/T|}^{\infty} (-1)^p e^{-\pi(p+\frac{1}{2})^2}. \qquad (3.15)$$

The Gaussian elementary signal $g(t)$ and its corresponding window function $Tw(t)$ have been depicted in Figure 3.

A practical way to represent this particular window function is in the form [19]

$$Tw(t) = 2^{-\frac{1}{4}} \left(\frac{K_o}{\pi}\right)^{-3/2} (-1)^m e^{\pi[(t/T)^2 - (m+\frac{1}{2})^2]}$$

$$\sum_{p=m}^{\infty} (-1)^{p-m} e^{-\pi[(p+\frac{1}{2})^2 - (m+\frac{1}{2})^2]}, \qquad (3.16)$$

where m is the nonnegative integer defined by $(m - \frac{1}{2})T < |t| \le (m + \frac{1}{2})T$. Since the summation in the latter expression yields a result which is close to unity for any value of m (0.998133 for $m = 0$, 0.999997 for $m = 1, \ldots,$ 1 for $m = \infty$), this representation leads to the approximation

$$Tw(t) \simeq 2^{-\frac{1}{4}} \left(\frac{K_o}{\pi}\right)^{-3/2} (-1)^m e^{\pi[(t/T)^2 - (m+\frac{1}{2})^2]} \tag{3.17}$$

with m defined by $(m - \frac{1}{2})T < |t| \le (m + \frac{1}{2})T$.

We mention the property that in the case of a Gaussian elementary signal, whose Fourier transform has the same form as the elementary signal, the Fourier transform of the corresponding window function has the same form as the window function itself. Moreover, we note that for large positive r, the extrema of the window function read

$$Tw(\pm[r + \frac{1}{2}]T) \simeq 2^{-\frac{1}{4}} \left(\frac{K_o}{\pi}\right)^{-3/2} (-1)^r, \tag{3.18}$$

which implies that $|w(t)|$ does not decrease with increasing value of $|t|$. More properties of this particular window function can be found elsewhere [17].

Since the Zak transform of the Gaussian elementary signal has (simple) zeros for $(t = \frac{1}{2}T + mT, \omega = \frac{1}{2}\Omega + k\Omega)$, we can again construct a (not identically zero) function

$$\tilde{z}(t, \omega; T) = \sum_m \sum_k (-1)^m \delta(t - \frac{1}{2}T - mT) 2\pi \delta(\omega - \frac{1}{2}\Omega - k\Omega) z \tag{3.19}$$

for which the product $T\tilde{g}(t, \omega; T)\tilde{z}^*(t, \omega; T)$ (see (3.8)) vanishes. Thus, with the help of the inversion formula (3.3), a function

$$z(t) = zT \sum_m (-1)^m \delta(t - \frac{1}{2}T - mT) \tag{3.20}$$

occurs, which is a *homogeneous solution* of the bi-orthonormality relation (2.7). We conclude that if the Zak transform $\tilde{g}(t, \omega; T)$ of an elementary signal $g(t)$ has zeros, the corresponding window function may not be unique: if $w(t)$ is a window function, then $w(t) + z(t)$ is a proper window function, too.

Zeros in $\tilde{g}(t, \omega; T)$ may be even worse. When we apply Parseval's energy theorem (3.4) to $w(t)$ and substitute from relation (3.8), we get

$$\frac{1}{T} \int |w(t)|^2 dt = \frac{1}{2\pi} \int_T \int_\Omega |\tilde{w}(t, \omega; T)|^2 dt d\omega$$
$$= \frac{1}{2\pi} \int_T \int_\Omega \frac{1}{|T\tilde{g}(t, \omega; T)|^2} dt d\omega. \tag{3.21}$$

From this relationship we conclude that in the case of zeros in $\tilde{g}(t, \omega; T)$, the corresponding window function $w(t)$ may not be quadratically integrable. This consequence of zeros in $\tilde{g}(t, \omega; T)$ is even worse than the fact that the window function is not unique; it may cause very bad convergence properties in the determination of Gabor's expansion coefficients.

§4   Gabor and Zak transforms for continuous-time signals

Now that we have shown that, at least in principle, a window function $w(t)$ can be found corresponding to a given elementary signal $g(t)$, we will focus again on the two relations (2.3) and (2.6). These two relations form a *transform pair*, as has been remarked before, and we will therefore associate more appropriate names for these relations. We define the *Gabor transform* by means of relation (2.6)

$$a_{mk} = \int \varphi(t) w^*_{mk}(t) dt;$$

the Gabor transform thus yields the array of Gabor coefficients a_{mk} that corresponds to a signal $\varphi(t)$. The *inverse* Gabor transform will then be defined by relation (2.3)

$$\varphi(t) = \sum_m \sum_k a_{mk} g_{mk}(t);$$

the inverse Gabor transform reconstructs the signal $\varphi(t)$ from its array of Gabor coefficients a_{mk}.

In Section 3 we have seen that the Zak transform can be helpful in determining the window function $w(t)$ for a given elementary signal $g(t)$. In this section we will study the possibility of applying the Zak transform to the Gabor transform and its inverse.

Let us first apply a Fourier transformation (see (2.11)) to the array of Gabor coefficients a_{mk},

$$\bar{a}(t, \omega; T) = \sum_m \sum_k a_{mk} e^{-j(m\omega T - k\Omega t)},$$

and let us substitute from the Gabor transform (2.6):

$$\bar{a}(t, \omega; T) = \sum_m \sum_k \left[\int \varphi(\tau) w^*(\tau - mT) e^{-jk\Omega\tau} d\tau \right] e^{-j(m\omega T - k\Omega t)}.$$

Along the same lines as the ones that we followed in deriving (3.8), we can now proceed to express the latter relationship in the form

$$\bar{a}(t, \omega; T) = T\tilde{\varphi}(t, \omega; T)\bar{w}^*(t, \omega; T). \tag{4.1}$$

Hence, the Gabor transform (2.6) can be transformed into the *product form* (4.1).

A product form can also be found for the inverse Gabor transform. If we apply a Zak transformation (see (3.1)) to both sides of (2.3), we get

$$\tilde{\varphi}(t, \omega; T) = \sum_{n} \left[\sum_{m} \sum_{k} a_{mk} g(t + nT - mT) e^{jk\Omega t} \right] e^{-jn\omega T}.$$

After rearranging factors

$$\tilde{\varphi}(t, \omega; T) = \sum_{m} \sum_{k} a_{mk} e^{-j(m\omega T - k\Omega t)}$$

$$\times \left[\sum_{n} g(t + [n - m]T) e^{-j(n-m)\omega T} \right]$$

$$= \left[\sum_{m} \sum_{k} a_{mk} e^{-j(m\omega T - k\Omega t)} \right] \left[\sum_{n} g(t + nT) e^{-jn\omega T} \right],$$

we immediately get the product relation

$$\tilde{\varphi}(t, \omega; T) = \bar{a}(t, \omega; T) \tilde{g}(t, \omega; T). \tag{4.2}$$

Now that we have found product forms for the Gabor transform and its inverse, we have also found a different way of determining Gabor's expansion coefficients a_{mk}, without explicitly determining a window function $w(t)$:

- from the signal $\varphi(t)$ and the elementary signal $g(t)$ we derive their Zak transforms $\tilde{\varphi}(t, \omega; T)$ and $\tilde{g}(t, \omega; T)$, respectively, according to the definition (3.1);

- under the assumption that division by $\tilde{g}(t, \omega; T)$ is allowed, the function $\bar{a}(t, \omega; T)$ can be found by means of the product relation (4.2);

- finally, the expansion coefficients a_{mk} follow from the function $\bar{a}(t, \omega; T)$ with the help of the inverse Fourier transformation (2.13).

Again, we conclude that Gabor's expansion coefficients may be non-unique in the case that $\tilde{g}(t, \omega; T)$ has zeros. In that case, homogeneous solutions z_{mk} may occur for which the Fourier transform $\bar{z}(t, \omega; T)$ satisfies the relation

$$\bar{z}(t, \omega; T) \tilde{g}(t, \omega; T) = 0. \tag{4.3}$$

Relation (4.3), which is similar to the product relation (4.2) with $\tilde{\varphi}(t, \omega; T) = 0$, can be transformed into the relation

$$\sum_m \sum_k z_{mk} g_{mk}(t) = 0, \tag{4.4}$$

which is similar to relation (2.3) with $\varphi(t) = 0$. Relation (4.4) shows that certain arrays of nonzero coefficients in Gabor's signal expansion may yield a zero result. We thus conclude that Gabor's signal expansion may be non-unique: if the array of coefficients a_{mk} yields the signal $\varphi(t)$, then the array $a_{mk} + z_{mk}$ yields the same signal.

§5   Gabor and Zak transforms for discrete-time signals

Until now we have only considered continuous-time signals. In this and the following section we will extend the concepts of the Gabor transform and the Zak transform to discrete-time signals [9].

Let us consider a discrete-time signal $\varphi[n]$; to distinguish the discrete-time case from the continuous-time case, we will use square brackets [ ] to denote a discrete-time signal, whereas we used curved brackets ( ) to denote a continuous-time signal. The Fourier transform of a discrete-time signal is defined by

$$\bar{\varphi}(\theta) = \sum_n \varphi[n] e^{-j\theta n}; \tag{5.1}$$

note that this Fourier transform is *periodic* in the frequency variable θ with period 2π. The inverse Fourier transformation reads

$$\varphi[n] = \frac{1}{2\pi} \int_{2\pi} \bar{\varphi}(\theta) e^{j\theta n} d\theta, \tag{5.2}$$

where the integration extends over one period 2π. As already expressed in the introductory Section 1, we will consistently use n, m, and N as time variables, θ, k and Θ as frequency variables, and the relation $\Theta N = 2\pi$ holds throughout.

On the analogy of the Gabor transform (2.6) for continuous-time signals, we introduce the Gabor transform for discrete-time signals

$$a_{mk} = \sum_n \varphi[n] w_{mk}^*[n] \tag{5.3}$$

with the short-hand notation

$$w_{mk}[n] = w[n - mN] e^{jk\Theta n} \tag{5.4}$$

again (see (2.2)); in the discrete-time case, the positive integer N and the parameter $\Theta = 2\pi/N$ are the respective counterparts of T and $\Omega = 2\pi/T$

in the continuous-time case. We remark that the Gabor transform for discrete-time signals is an array that is *periodic* in the frequency variable k with period N; this periodicity in the Gabor transform, like the periodicity in the Fourier transform, results from the discrete nature of the signal. On the analogy of the corresponding relation (2.3) for continuous-time signals, the inverse Gabor transform for discrete-time signals takes the form

$$\varphi[n] = \sum_{m} \sum_{k=<N>} a_{mk} g_{mk}[n], \tag{5.5}$$

where the summation over k extends over one period N of the periodic array a_{mk}.

On the analogy of the continuous-time case, we need the Fourier transform of the array a_{mk}, which will be defined by (see (2.11))

$$\bar{a}(n, \theta; N) = \sum_{m} \sum_{k=<N>} a_{mk} e^{-j(m\theta N - k\Theta n)} \qquad \text{(with } \Theta N = 2\pi\text{)}, \tag{5.6}$$

where the summation over k extends again over one period N. We remark that this Fourier transform is periodic in the (discrete) time index n with period N and periodic in the (continuous) frequency variable θ with period Θ. The inverse Fourier transform reads (see (2.13))

$$a_{mk} = \frac{1}{2\pi} \sum_{n=<N>} \int_{\Theta} \bar{a}(n, \theta; N) e^{j(m\theta N - k\Theta n)} d\theta, \tag{5.7}$$

where the summation over n extends over one period N, and the integration over θ extends over one period Θ.

Furthermore, we need the Zak transform for discrete-time signals, which will be defined by (see (3.1))

$$\tilde{\varphi}(n, \theta; N) = \sum_{m} \varphi[n + mN] e^{-jm\theta N}. \tag{5.8}$$

The Zak transform in the discrete-time case is periodic in the (continuous) frequency variable θ with period Θ and quasi-periodic in the (discrete) time index n with quasi-period N (see (3.2)):

$$\tilde{\varphi}(n + mN, \theta + k\Theta; N) = \tilde{\varphi}(n, \theta; N) e^{jm\theta N}. \tag{5.9}$$

The inverse Zak transform now reads (see (3.3))

$$\varphi[n + mN] = \frac{1}{\Theta} \int_{\Theta} \tilde{\varphi}(n, \theta; N) e^{jm\theta N} d\theta, \tag{5.10}$$

where the time index n can be restricted to an interval of length N, with m taking on all integer values.

Using the discrete-time equivalents of the Gabor transform, the Fourier transform, and the Zak transform, it is not difficult to show that the Gabor transform (5.3) can be transformed into the product form (see (4.1))

$$\bar{a}(n, \theta; N) = N\tilde{\varphi}(n, \theta; N)\tilde{w}^*(n, \theta; N) \qquad (5.11)$$

and the inverse Gabor transform (5.5) into the product form (see (4.2))

$$\tilde{\varphi}(n, \theta; N) = \bar{a}(n, \theta; N)\tilde{g}(n, \theta; N). \qquad (5.12)$$

Let us now consider Gabor's choice of a Gaussian elementary signal $g(t)$ again (see (2.1)), and let us consider a discrete-time version $g[n]$ of this elementary signal, *symmetrically* positioned with respect to the sampling grid on the t-axis with a sampling distance T/N:

$$g[n] \;=\; g\left(n\frac{T}{N}\right) = 2^{\frac{1}{4}}e^{-\pi(n/N)^2} \qquad (N \text{ odd}); \qquad (5.13)$$

$$g[n] \;=\; g\left([n + \tfrac{1}{2}]\frac{T}{N}\right) = 2^{\frac{1}{4}}e^{-\pi([n+\frac{1}{2}]/N)^2} \qquad (N \text{ even}). \qquad (5.14)$$

The corresponding Zak transforms $\tilde{g}(n, \theta; N)$ follow from substituting from (5.13) and (5.14) into (5.8), and read (see (3.9))

$$\tilde{g}(n, \theta; N) = 2^{\frac{1}{4}}e^{-\pi(n/N)^2}\theta_3(\pi\zeta^*; e^{-\pi}) \qquad (\text{with } \zeta = \theta/\Theta + jn/N) \quad (5.15)$$

$$\tilde{g}(n, \theta; N) = 2^{\frac{1}{4}}e^{-\pi([n+\frac{1}{2}]/N)^2}\theta_3(\pi\zeta^*; e^{-\pi}) \qquad (\text{with } \zeta = \theta/\Theta + j[n + \tfrac{1}{2}]/N) \qquad (5.16)$$

for odd and even N, respectively. It is important to note that in the discrete-time case the zeros of the theta function do not occur on a raster point! Hence, the Zak transform $\tilde{g}(n, \theta; N)$ of a Gaussian elementary signal has no zeros.

Let us, for the sake of simplicity, take N odd, which implies that there is a sampling point at $t = 0$. The corresponding window function $w[n]$ then reads (see (3.15))

$$Nw[n] = 2^{-\frac{1}{4}}e^{\pi(n/N)^2}\left(\frac{K_o}{\pi}\right)^{-3/2}\sum_{p+\frac{1}{2}\geq|n/N|}^{\infty}(-1)^pe^{-\pi(p+\frac{1}{2})^2}, \qquad (5.17)$$

or, in a different form (see (3.16)),

$$Nw[n] \;=\; 2^{-\frac{1}{4}}\left(\frac{K_o}{\pi}\right)^{-3/2}(-1)^me^{\pi[(n/N)^2-(m+\frac{1}{2})^2]}$$

$$\sum_{p=m}^{\infty}(-1)^{p-m}e^{-\pi[(p+\frac{1}{2})^2-(m+\frac{1}{2})^2]}, \qquad (5.18)$$

which form can then be approximated by (see (3.17))

$$Nw[n] \simeq 2^{-\frac{1}{4}} \left(\frac{K_o}{\pi} \right)^{-3/2} (-1)^m e^{\pi[(n/N)^2 - (m+\frac{1}{2})^2]}, \qquad (5.19)$$

with m defined by $(m - \frac{1}{2})N < |n| \leq (m + \frac{1}{2})N$. We remark that for large positive r, the extrema of the window function read

$$Nw \left[\pm \left(\left[r + \tfrac{1}{2} \right] N - \tfrac{1}{2} \right) \right] \simeq 2^{-\frac{1}{4}} \left(\frac{K_o}{\pi} \right)^{-3/2}$$

$$\times (-1)^r e^{\pi[(r+\frac{1}{2}-1/2N)^2 - (r+\frac{1}{2})^2]}$$

$$= 2^{-\frac{1}{4}} \left(\frac{K_o}{\pi} \right)^{-3/2} (-1)^r e^{\pi[(1/2N)^2 - (r+\frac{1}{2})/N]}$$

$$\simeq 2^{-\frac{1}{4}} \left(\frac{K_o}{\pi} \right)^{-3/2} (-1)^r e^{-\pi(r/N)}. \qquad (5.20)$$

Unlike in the continuous-time case, where $|w(t)|$ did not decrease with increasing value of $|t|$ (see (3.18)), the function $|w[n]|$ does eventually decrease exponentially with increasing value of $|n|$. A similar result holds when N is chosen even.

§6   Discrete Gabor transform and discrete Zak transform

In this section we will introduce *discrete versions* of the Gabor and the Zak transform [2, 3, 21, 22, 24, 25, 26, 29] by *sampling* the *continuous* variable θ that arises both in the Fourier transform (5.6) and in the Zak transform (5.8) of discrete-time signals. We start with the Fourier transform (5.6)

$$\bar{a}(n, \theta; N) = \sum_m \sum_{k=<N>} a_{mk} e^{-j(m\theta N - k\Theta n)},$$

which is periodic in the *discrete* time index n with period N, and periodic in the *continuous* frequency variable θ with period Θ. We define the array $\bar{a}[n, l; N, M]$ as *samples* of this Fourier transform $\bar{a}(n, \theta; N)$

$$\begin{aligned} \bar{a}[n, l; N, M] &= \bar{a} \left(n, \frac{\Theta}{M} l; N \right) \\ &= \sum_m \sum_{k=<N>} a_{mk} e^{-j[m(2\pi/M)l - k(2\pi/N)n]}; \qquad (6.1) \end{aligned}$$

thus we have sampled the frequency axis such that in each period of length Θ there appear M equally-spaced sampling points a distance Θ/M apart.

We then define the array A_{mk} as a kind of *inverse* Fourier transform (see (5.7)) of the array $\bar{a}[n, l; N, M]$ by

$$A_{mk} = \frac{1}{MN} \sum_{n=<N>} \sum_{l=<M>} \bar{a}[n, l; N, M] e^{j[m(2\pi/M)l - k(2\pi/N)n]}. \qquad (6.2)$$

We remark that, whereas the array a_{mk} is periodic in the frequency variable k with period N but does not have a periodicity with respect to the time index m, the array A_{mk} is periodic in both the frequency variable k *and* the time index m with periods N and M, respectively. It can easily be shown that the relation between A_{mk} and a_{mk} reads

$$A_{mk} = \sum_{r} a_{m+rM,k}, \qquad (6.3)$$

from which relation we conclude that A_{mk} is a *summation of the array a_{mk} and all its replicas shifted along the m-direction over distances rM* ($r = -\infty, \ldots, +\infty$). Of course, the array $\bar{a}[n, l; N, M]$ can directly be expressed in the array A_{mk} through the relationship (see (6.1))

$$\bar{a}[n, l; N, M] = \sum_{m=<M>} \sum_{k=<N>} A_{mk} e^{-j[m(2\pi/M)l - k(2\pi/N)n]}. \qquad (6.4)$$

The latter relationship (6.4) between the arrays $\bar{a}[n, l; N, M]$ and A_{mk} is known as the *discrete Fourier transform*, whereas the relation (6.2) is consequently called the *inverse* discrete Fourier transform. It is important to note that if $a_{mk} \neq 0$ in an m-interval of length M and vanishes outside that interval, the array a_{mk} is just one period of the periodic array A_{mk} and can thus be reconstructed from A_{mk} and, hence, from $\bar{a}[n, l; N, M]$.

We will now perform an analogous procedure for the Zak transform (5.8)

$$\tilde{\varphi}(n, \theta; N) = \sum_{m} \varphi[n + mN] e^{-jm\theta N},$$

which, like the Fourier transform (5.6), is periodic in the *continuous* frequency variable θ with period Θ, as well. We define the array $\tilde{\varphi}[n, l; N, M]$ as *samples* of this Zak transform $\tilde{\varphi}(n, \theta; N)$

$$\tilde{\varphi}[n, l; N, M] = \tilde{\varphi}(n, \frac{\Theta}{M}l; N) = \sum_{m} \varphi[n + mN] e^{-jm(2\pi/M)l}; \qquad (6.5)$$

thus we have again sampled the frequency axis such that in each period of length Θ there appear M equally-spaced sampling points a distance Θ/M apart. We then define the function $\Phi[n + mN]$ as a kind of *inverse* Zak transform (see (5.10)) of the array $\tilde{\varphi}[n, l; N, M]$ by

$$\Phi[n + mN] = \frac{1}{M} \sum_{l=<M>} \tilde{\varphi}[n, l; N, M]e^{jm(2\pi/M)l}. \tag{6.6}$$

It can easily be shown that the relationship

$$\Phi[n] = \sum_{r} \varphi[n + rMN] \tag{6.7}$$

holds; thus, $\Phi[n]$ is a *summation of the signal* $\varphi[n]$ *and all its replicas shifted over distances* rMN $(r = -\infty, \ldots, +\infty)$. The sequence $\Phi[n]$ is thus periodic in the time index n with period MN. Of course, the array $\tilde{\varphi}[n, l; N, M]$ can directly be expressed in the sequence $\Phi[n]$ through the relationship (see (6.5))

$$\tilde{\varphi}[n, l; N, M] = \sum_{m=<M>} \Phi[n + mN]e^{-jm(2\pi/M)l}. \tag{6.8}$$

On the analogy of the discrete Fourier transform (6.4), we shall call the latter relationship (6.8) between the array $\tilde{\varphi}[n, l; N, M]$ and the sequence $\Phi[n]$ the *discrete Zak transform*, whereas the relation (6.6) will consequently be called the *inverse* discrete Zak transform. It is important again to note that if $\varphi[n] \neq 0$ in an interval of length MN and vanishes outside that interval, the signal $\varphi[n]$ is just one period of the periodic sequence $\Phi[n]$ and can thus be reconstructed from $\Phi[n]$ and, hence, from $\tilde{\varphi}[n, l; N, M]$.

We are now prepared to sample the product form (5.11) of the Gabor transform for discrete-time signals, leading to

$$\bar{a}[n, l; N, M] = N\tilde{\varphi}[n, l; N, M]\tilde{w}^*[n, l; N, M]. \tag{6.9}$$

Upon substituting the latter expression into the inverse discrete Fourier transform (6.2), we get

$$A_{mk} = \frac{1}{MN} \sum_{n=<N>} \sum_{l=<M>} N\tilde{\varphi}[n, l; N, M]\tilde{w}^*[n, l; N, M]$$
$$e^{j[m(2\pi/M)l-k(2\pi/N)n]},$$

in which relation we substitute from the discrete Zak transform (6.8) for $\Phi[n]$

$$A_{mk} = \frac{1}{M} \sum_{n=<N>} \sum_{l=<M>} \left[\sum_{r=<M>} \Phi[n + rN]e^{-jr(2\pi/M)l} \right]$$
$$\tilde{w}^*[n, l; N, M]e^{j[m(2\pi/M)l-k(2\pi/N)n]}.$$

We rearrange factors

$$A_{mk} = \sum_{n=<N>} \sum_{r=<M>} \Phi[n+rN]$$

$$\left[\frac{1}{M} \sum_{l=<M>} \tilde{w}[n,l;N,M]e^{j(r-m)(2\pi/M)l}\right]^* e^{-jk(2\pi/N)n}$$

and recognize (see (6.6)) the inverse discrete Zak transform of $W[n]$

$$A_{mk} = \sum_{n=<N>} \sum_{r=<M>} \Phi[n+rN]W^*[n+rN-mN]e^{-jk(2\pi/N)n}.$$

If we introduce, on the analogy of (5.4), the short-hand notation

$$W_{mk}[n] = W[n-mN]e^{jk\Theta n} = W[n-mN]e^{jk(2\pi/N)n}, \qquad (6.10)$$

the coefficients A_{mk} take the form

$$A_{mk} = \sum_{n=<N>} \sum_{r=<M>} \Phi[n+rN]W_{mk}^*[n+rN],$$

which finally leads to

$$A_{mk} = \sum_{n=<MN>} \Phi[n]W_{mk}^*[n]. \qquad (6.11)$$

The latter relationship will be called the *discrete Gabor transform*. As we remarked before, the discrete Gabor transform A_{mk} shows, apart from the usual periodicity with period N in the k-direction, due to the discrete nature of the signal, a *periodicity with period M in the m-direction*.

The *inverse* of the discrete Gabor transform results from sampling the product form (5.12) of the inverse of the Gabor transform for discrete-time signals, leading to

$$\tilde{\varphi}[n,l;N,M] = \bar{a}[n,l;N,M]\tilde{g}[n,l;N,M]. \qquad (6.12)$$

Upon substituting the latter expression into the inverse discrete Zak transform (6.6), we get

$$\Phi[n] = \frac{1}{M} \sum_{l=<M>} \bar{a}[n,l;N,M]\tilde{g}[n,l;N,M],$$

in which relation we substitute from the discrete Fourier transformation (6.4)

$$\Phi[n] = \frac{1}{M} \sum_{l=<M>} \left[\sum_{m=<M>} \sum_{k=<N>} A_{mk}e^{-j[m(2\pi/M)l-k(2\pi/N)n]}\right]$$

$$\tilde{g}[n,l;N,M].$$

We rearrange factors

$$\Phi[n] = \sum_{m=<M>} \sum_{k=<N>} A_{mk} \left[\frac{1}{M} \sum_{l=<M>} \tilde{g}[n,l;N,M]e^{-jm(2\pi/M)l} \right]$$
$$e^{jk(2\pi/N)n}$$

and recognize (see (6.6)) the inverse discrete Zak transform of $G[n]$

$$\Phi[n] = \sum_{m=<M>} \sum_{k=<N>} A_{mk}G[n-mN]e^{jk(2\pi/N)n}.$$

With the short-hand notation (see (6.10))

$$G_{mk}[n] = G[n-mN]e^{jk(2\pi/N)n},$$

the inverse discrete Gabor transform thus reads

$$\Phi[n] = \sum_{m=<M>} \sum_{k=<N>} A_{mk}G_{mk}[n]. \qquad (6.13)$$

What is the importance of having a *discrete* Gabor transform (6.11), whereas we are in fact only interested in the coefficients of Gabor's signal expansion and thus in the *normal* Gabor transform (5.3)? Assume that the signal $\varphi[n] \neq 0$ in an interval of length N_φ and vanishes outside that interval, and that the window function $w[n] \neq 0$ in an interval of length N_w and vanishes outside that interval; then the coefficients of the (normal) Gabor expansion (5.3)

$$a_{mk} = \sum_n \varphi[n]w_{mk}^*[n]$$

can only be $\neq 0$ in an m-interval of length M, say, where M is the smallest integer for which the relation $MN \geq N_\varphi + N_w - 1$ holds. Now take M such that $MN \geq N_\varphi + N_w - 1$ and construct the periodic signal sequence $\Phi[n] = \sum_r \varphi[n+rMN]$ and the periodic window sequence $W[n] = \sum_r w[n+rMN]$ according to (6.7). In that case, the array a_{mk} of the (normal) Gabor transform (5.3) can be identified with one period of the array A_{mk} of the discrete Gabor transform (6.11)

$$A_{mk} = \sum_{n=<MN>} \Phi[n]W_{mk}^*[n].$$

The array A_{mk} (and thus a_{mk}) can be computed via the discrete Zak transform and the inverse discrete Fourier transform, and in computing these transforms we can use a fast computer algorithm known as the *fast Fourier transform* (FFT). Calculating the discrete Gabor transform A_{mk} in such a way could be called the *fast discrete Gabor transform* and is equivalent to the *fast convolution* well-known in digital signal processing. In detail we proceed as follows:

- from the signal $\varphi[n]$ and the window function $w[n]$ we determine, with the use of a fast computer algorithm, their discrete Zak transforms $\tilde{\varphi}[n, l; N, M]$ and $\tilde{w}[n, l; N, M]$, respectively, via (6.5);

- the discrete Fourier transform $\bar{a}[n, l; N, M]$ follows from the product form (6.9) of the discrete Gabor transform;

- from the array $\bar{a}[n, l; N, M]$ we determine, with the use of a fast computer algorithm again, the inverse discrete Fourier transform A_{mk}, according to (6.2);

- the array of Gabor expansion coefficients a_{mk} then follows as one period of the periodic array A_{mk}.

In general, the signal $\varphi[n]$ does not vanish outside a certain interval or, if it does, the interval can be too large. In that case we can apply *overlap-add* techniques by splitting the signal $\varphi[n]$ in parts and treating all parts separately. In detail, we proceed as follows. We represent the signal $\varphi[n]$ as a sequence of partial signals $\varphi^{(r)}[n]$, where each partial signal vanishes outside an interval N_φ; hence,

$$\varphi[n] = \sum_r \varphi^{(r)}[n] \tag{6.14}$$

with

$$\varphi^{(r)}[n] = \begin{cases} \varphi[n] & \text{for } rN_\varphi \le n \le (r+1)N_\varphi - 1 \\ 0 & \text{elsewhere.} \end{cases} \tag{6.15}$$

Upon substituting the expansion (6.14) into the Gabor transform (5.3) we get

$$a_{mk} = \sum_n \left[\sum_r \varphi^{(r)}[n] \right] w_{mk}^*[n] = \sum_r \left[\sum_n \varphi^{(r)}[n] w_{mk}^*[n] \right] = \sum_r a_{mk}^{(r)}, \tag{6.16}$$

where each partial Gabor transform

$$a_{mk}^{(r)} = \sum_n \varphi^{(r)}[n] w_{mk}^*[n]$$

can be evaluated along the lines described in the previous paragraph. The last summation in (6.16) must take into account, of course, the overlap between the partial Gabor transforms.

§7 Gabor's expansion in the case of integer oversampling

In his original paper, Gabor restricted himself to a *critical* sampling of the time-frequency domain, where the expansion coefficients can be interpreted as independent data, *i.e.* degrees of freedom of a signal. It is the aim of this section to extend, for continuous-time signals, Gabor's concepts to the case of *oversampling*, in which case the expansion coefficients are no longer independent.

Let us consider an elementary signal $g(t)$ again, but let us now construct a discrete set of shifted and modulated versions defined as (see (2.2))

$$g(t - maT)e^{jk\beta\Omega t}, \tag{7.1}$$

where the time shift αT and the frequency shift $\beta\Omega$ satisfy the relationships $\Omega T = 2\pi$ and $\alpha\beta \leq 1$, and where m and k may take all integer values. Gabor's signal expression would then read (see (2.3))

$$\varphi(t) = \sum_m \sum_k a_{mk} g(t - maT)e^{jk\beta\Omega t}. \tag{7.2}$$

Gabor's original signal expansion was restricted to the special case $\alpha\beta = 1$ (and more particular $\alpha = \beta = 1$), in which case the expansion coefficients a_{mk} can be identified as degrees of freedom of the signal. For $\alpha\beta > 1$, the set of shifted and modulated versions of the elementary signal is not complete and thus cannot represent any arbitrary signal, while for $\alpha\beta < 1$, the set is overcomplete, which implies that Gabor's expansion coefficients become dependent and can no longer be identified as degrees of freedom. In the special case $\alpha\beta = 1$, it has been shown in Section 3 how a window function $w(t)$ can be found such that the expansion coefficients can be determined via the so-called Gabor transform, which would now take the form (see (2.6))

$$a_{mk} = \int \varphi(t)w^*(t - maT)e^{-jk\beta\Omega t}dt. \tag{7.3}$$

It is the aim of this section to show how a window function can be found when the parameters α and β satisfy the relation $\alpha\beta = 1/p < 1$ in the special case that p is a positive *integer*.

That a window function can be found in the case of oversampling (*i.e.* $\alpha\beta < 1$) is not surprising. To see this, let us consider the *continuous* analogues of Gabor's signal expansion and the Gabor transform. The Gabor transform (7.3) can be considered as a sampled version of the sliding-window spectrum $s(t, \omega)$ [7, 10] of the signal $\varphi(t)$, defined as

$$s(t, \omega) = \int \varphi(\tau)w^*(\tau - t)e^{-j\omega\tau}d\tau, \tag{7.4}$$

where the sampling appears on the time-frequency lattice $(t = m\alpha T, \omega = k\beta\Omega)$. Gabor's expansion coefficients follow from the sliding-window spectrum through the relation $a_{mk} = s(m\alpha T, k\beta\Omega)$. It is well known that the signal $\varphi(t)$ can be reconstructed from its sliding-window spectrum $s(t, \omega)$ in many different ways, one of them reading

$$\varphi(\tau) \int |w(t)|^2 dt = \frac{1}{2\pi} \int\int s(t, \omega) w(\tau - t) e^{j\omega\tau} dt d\omega. \qquad (7.5)$$

It is not difficult to see that the latter signal representation is a continuous analogue of Gabor's signal expansion (7.2), and that it can be derived from this expansion by letting the time step αT and the frequency step $\beta\Omega$ tend to zero. In fact, the signal representation (7.5) is identical to Gabor's signal expansion (7.2) with an *infinitely dense* sampling lattice. We conclude that in that limiting case, the window function $w(t)$ may be chosen such that it is proportional to the elementary signal $g(t)$. Later on in this section we will derive a detailed transition from Gabor's signal expansion to its continuous analogue, by letting $\alpha\beta \downarrow 0$.

Using the Fourier transform (see (2.11)) and the Zak transform (see (3.1)), it can be shown (see Appendix B) that the Gabor transform (7.3) can be transformed into the *product form*

$$\bar{a}(t, \omega; T) = \alpha p T \tilde{\varphi}\left(\alpha p t, \frac{\omega}{\alpha}; \alpha p T\right) \tilde{w}^*\left(\alpha p t, \frac{\omega}{\alpha}; \alpha T\right). \qquad (7.6)$$

If we consider the domains of the functions $\bar{a}(t, \omega; T)$, $\tilde{\varphi}(\alpha p t, \omega/\alpha; \alpha p T)$, and $\tilde{w}(\alpha p t, \omega/\alpha; \alpha T)$ in the fundamental Fourier interval $(-\frac{1}{2}T < t \leq \frac{1}{2}T, -\frac{1}{2}\Omega < \omega \leq \frac{1}{2}\Omega)$, we note that, whereas the Fourier transform $\bar{a}(t, \omega; T)$ appears only *once* in the fundamental Fourier interval, the Zak transforms $\tilde{\varphi}(\alpha p t, \omega/\alpha; \alpha p T)$ and $\tilde{w}(\alpha p t, \omega/\alpha; \alpha T)$ appear *p-fold*: $\tilde{\varphi}(\alpha p t, \omega/\alpha; \alpha p T)$ as p identical stripes with height Ω/p and width T, and $\tilde{w}(\alpha p t, \omega/\alpha; \alpha T)$ as p stripes with width T/p and height Ω, which stripes are identical to each other apart from the factor $e^{j\omega T}$ (see the periodicity property (3.2) of the Zak transform).

Note that the product form (7.6) of the Gabor transform enables us to determine Gabor's expansion coefficients in an easy way:

- we first determine the Zak transform $\tilde{\varphi}(\alpha p t, \omega/\alpha; \alpha p T)$ of the signal $\varphi(t)$ and the Zak transform $\tilde{w}(\alpha p t, \omega/\alpha; \alpha T)$ of the window function $w(t)$ by means of definition (3.1);

- we then find the Fourier transform $\bar{a}(t, \omega; T)$ by means of the product rule (7.6);

- we finally determine Gabor's expansion coefficients a_{mk} via the inverse Fourier transformation (see (2.13)).

Using the Fourier transform and the Zak transform, it can also be shown (see Appendix C) that Gabor's signal expansion (7.2) can be transformed into the *sum-of-products form*

$$\tilde{\varphi}\left(\alpha pt, \frac{\omega}{\alpha}; \alpha pT\right) = \frac{1}{p} \sum_{r=\langle p \rangle} \bar{a}\left(t, \omega + r\frac{\Omega}{p}; T\right) \tilde{g}\left(\alpha pt, \frac{\omega + r\Omega/p}{\alpha}; \alpha T\right),$$

(7.7)

where the expression $r = \langle p \rangle$ is used throughout as a short-hand notation for an interval of p successive integers ($r = 0, 1, 2, \ldots, p - 1$, for instance; due to the periodicity of the Fourier transform and the Zak transform, however, any sequence of p successive integers can be chosen). If we consider the domains of the three functions $\tilde{\varphi}(\alpha pt, \omega/\alpha; \alpha pT)$, $\bar{a}(t, \omega; T)$, and $\tilde{g}(\alpha pt, \omega/\alpha; \alpha T)$ in the fundamental Fourier interval ($-\frac{1}{2}T < t \leq \frac{1}{2}T, -\frac{1}{2}\Omega < \omega \leq \frac{1}{2}\Omega$), we note as before that, whereas the Fourier transform $\bar{a}(t, \omega; T)$ appears only *once* in the fundamental Fourier interval, the Zak transforms $\tilde{\varphi}(\alpha pt, \omega/\alpha; \alpha pT)$ and $\tilde{g}(\alpha pt, \omega/\alpha; \alpha T)$ appear *p-fold*: $\tilde{\varphi}(\alpha pt, \omega/\alpha; \alpha pT)$ as p identical stripes with height Ω/p and width T, and $\tilde{g}(\alpha pt, \omega/\alpha; \alpha T)$ as p stripes with width T/p and height Ω, which stripes are identical to each other apart from the factor $e^{j\omega T}$. If we would use the sum-of-products form (7.7) to reconstruct the signal from its Gabor transform, it is important to note that, since the Zak transform $\tilde{\varphi}(\alpha pt, \omega/\alpha; \alpha pT)$ is periodic in ω with period Ω/p and quasi-periodic in t with quasi-period T, we can restrict ourselves to the interval ($-\frac{1}{2}T < t \leq \frac{1}{2}T, -\frac{1}{2}\Omega/p < \omega \leq \frac{1}{2}\Omega/p$), which is smaller than the fundamental Fourier interval.

We will now prove that Gabor's signal expansion (7.2) and the Gabor transform (7.3) form a transform pair, by showing that for any elementary signal $g(t)$, a window function $w(t)$ can be constructed. If we substitute from the product form of the Gabor transform (7.6) into the sum-of-products form of Gabor's signal expansion (7.7), we get

$$\tilde{\varphi}\left(\alpha pt, \frac{\omega}{\alpha}; \alpha pT\right) = \frac{1}{p} \sum_{r=\langle p \rangle}$$

$$\left[\alpha pT\tilde{\varphi}\left(\alpha pt, \frac{\omega + r\Omega/p}{\alpha}; \alpha pT\right) \tilde{w}^*\left(\alpha pt, \frac{\omega + r\Omega/p}{\alpha}; \alpha T\right)\right]$$

$$\tilde{g}\left(\alpha pt, \frac{\omega + r\Omega/p}{\alpha}; \alpha T\right).$$

After rearranging,

$$\tilde{\varphi}\left(\alpha pt, \frac{\omega}{\alpha}; \alpha pT\right) = \alpha T \sum_{r=\langle p \rangle} \tilde{\varphi}\left(\alpha pt, \frac{\omega + r\Omega/p}{\alpha}; \alpha pT\right)$$

$$\tilde{w}^*\left(\alpha pt, \frac{\omega + r\Omega/p}{\alpha}; \alpha T\right) \tilde{g}\left(\alpha pt, \frac{\omega + r\Omega/p}{\alpha}; \alpha T\right)$$

and using the periodicity property (3.2) of $\tilde{\varphi}(\alpha pt, \omega/\alpha; \alpha pT)$, we get

$$\tilde{\varphi}\left(\alpha pt, \frac{\omega}{\alpha}; \alpha pT\right) = \tilde{\varphi}\left(\alpha pt, \frac{\omega}{\alpha}; \alpha pT\right)$$

$$\left[\alpha T \sum_{r=\langle p\rangle} \tilde{w}^*\left(\alpha pt, \frac{\omega + r\Omega/p}{\alpha}; \alpha T\right) \tilde{g}\left(\alpha pt, \frac{\omega + r\Omega/p}{\alpha}; \alpha T\right)\right].$$

From the latter equality, which should hold for any signal $\varphi(t)$, we conclude that Gabor's signal expansion (7.2) and the Gabor transform (7.3) form a transform pair if the elementary signal $g(t)$ and the window function $w(t)$ satisfy the condition

$$\alpha T \sum_{r=\langle p\rangle} \tilde{w}^*\left(\alpha pt, \frac{\omega + r\Omega/p}{\alpha}; \alpha T\right) \tilde{g}\left(\alpha pt, \frac{\omega + r\Omega/p}{\alpha}; \alpha T\right) = 1. \quad (7.8)$$

Note that since the Zak transform $\tilde{\varphi}(\alpha pt, \omega/\alpha; \alpha pT)$ is periodic in ω with period Ω/p and quasi-periodic in t with quasi-period T, we can restrict ourselves again to the interval $(-\frac{1}{2}T < t \leq \frac{1}{2}T, -\frac{1}{2}\Omega/p < \omega \leq \frac{1}{2}\Omega/p)$.

In Gabor's original case, *i.e.* for $p = 1$ (and $\alpha = 1$), the Gabor transform and Gabor's signal expansion form a transform pair, if the window function $w(t)$ and the elementary signal $g(t)$ satisfy the relation $T\tilde{w}^*(t, \omega; T)$ $\tilde{g}(t, \omega; T) = 1$ (see (3.8)). This relation enables us to determine a window function for a given elementary signal. However, finding the corresponding window function in this way can be difficult if the Zak transform of the elementary signal has zeros, which is very often the case [17] if the elementary signal is continuous and square integrable. Moreover, from Parseval's energy theorem, we already concluded (see (3.21)) that the window function $w(t)$ may not be quadratically integrable when the Zak transform of the elementary signal $\tilde{g}(t, \omega; T)$ has zeros. The difficulties that we encounter for critical sampling ($p = 1$, Gabor's original case) have in fact led to this study of oversampling ($p > 1$).

For $p > 1$, the window function that corresponds to a given elementary signal is *not unique*. This is in accordance with the fact that in the case of oversampling, the set of shifted and modulated versions of the elementary signal is overcomplete, and that Gabor's expansion coefficients are dependent and can no longer be considered as degrees of freedom, as we have mentioned before. One way to find a window function in the case of oversampling by an *integer* factor p is the following:

With the short-hand notations

$$w_r(t, \omega) = \tilde{w}\left(\alpha pt, \frac{\omega + r\Omega/p}{\alpha}; \alpha T\right) \quad (7.9)$$

and

$$g_r(t,\omega) = \tilde{g}\left(\alpha pt, \frac{\omega + r\Omega/p}{\alpha}; \alpha T\right), \qquad (7.10)$$

and choosing, for convenience, $r = 0, 1, \ldots, p-1$, we can construct the p-dimensional row vectors of functions

$$\mathbf{w} = [w_0(t,\omega) \quad w_1(t,\omega) \quad \cdots \quad w_{p-1}(t,\omega)] \qquad (7.11)$$

and

$$\mathbf{g} = [g_0(t,\omega) \quad g_1(t,\omega) \quad \cdots \quad g_{p-1}(t,\omega)]. \qquad (7.12)$$

With the help of these row vectors, (7.8) can be expressed in the elegant inner product form

$$\alpha T \mathbf{g} \mathbf{w}^* = 1, \qquad (7.13)$$

where, as usual, the asterisk in connection with a vector denotes complex conjugation *and* transposition.

In the case of oversampling, the conditions (7.8) and (7.13) do not lead to a unique solution for the window function $w(t)$, which allows us to impose additional constraints on the solution. Let us, for instance, impose the condition of minimum L_2-norm. It is well known that the *optimum solution* in the sense of *minimum L_2-norm* can be found with the help of the so-called *generalized (Moore-Penrose) inverse* [11] $\mathbf{g}^\dagger$ of $\mathbf{g}$, defined by

$$\mathbf{g}^\dagger = \mathbf{g}^*(\mathbf{g}\mathbf{g}^*)^{-1}; \qquad (7.14)$$

note that $\mathbf{g}\mathbf{g}^\dagger = 1$ and that $\mathbf{g}^\dagger \mathbf{g}\mathbf{g}^* = \mathbf{g}^*$. The optimum solution $\mathbf{w}_{opt}$ then reads

$$\mathbf{w}_{opt} = \frac{1}{\alpha T}(\mathbf{g}^\dagger)^* = \frac{1}{\alpha T}(\mathbf{g}\mathbf{g}^*)^{-1}\mathbf{g}. \qquad (7.15)$$

Of course, if we proceed in this way, we will find, for any t and ω, the minimum L_2-norm solution for the vector $\mathbf{w}$. It is not difficult to show, however, that minimum L_2-norm of the vector $\mathbf{w}$ corresponds to minimum L_2-norm of the Zak transform $\tilde{w}(\alpha pT, \omega/\alpha; \alpha T)$ and thus, with the help of Parseval's energy theorem (3.4), to minimum L_2-norm of the window function $w(t)$ itself. In Figure 4 we have depicted the Zak transforms of the optimum window functions that correspond to the Gaussian elementary signal (2.1) for different values of α and β, resulting in different values of oversampling p, while in Figure 5 we have depicted these optimum window functions themselves. We remark that the resemblance between the window function and the elementary signal increases with increasing values of p.

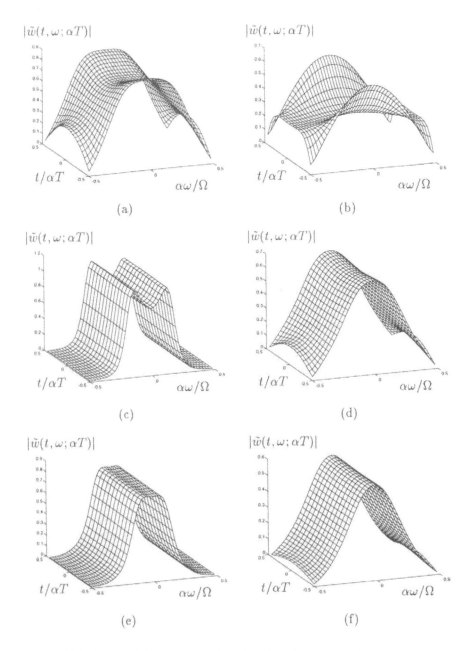

Figure 4. The Zak transform $\widetilde{w}(t, \omega; \alpha T)$ in the case of a Gaussian elementary signal for different values of oversampling: (a) $\alpha = \beta = 1/\sqrt{2}, p = 2$, (b) $\alpha = 1, \beta = 1/3, p = 3$, (c) $\alpha = 1/3, \beta = 1, p = 3$, (d) $\alpha = \beta = 1/\sqrt{3}, p = 3$, (e) $\alpha = 1/3, \beta = 3/4, p = 4$, and (f) $\alpha = \beta = 1/2, p = 4$.

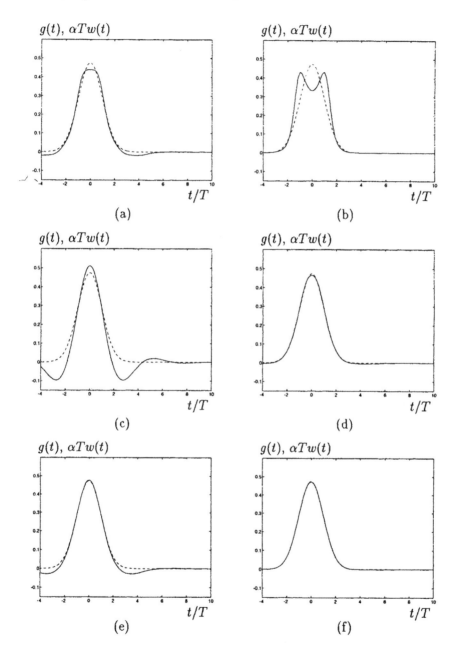

Figure 5. A Gaussian elementary signal $g(t)$ (dashed line) and its corresponding window function $\alpha T w(t)$ (solid line) for different values of oversampling: (a) $\alpha = \beta = 1/\sqrt{2}, p = 2$, (b) $\alpha = 1, \beta = 1/3, p = 3$, (c) $\alpha = 1/3, \beta = 1, p = 3$, (d) $\alpha = \beta = 1/\sqrt{3}, p = 3$, (e) $\alpha = 1/3, \beta = 3/4, p = 4$, and (f) $\alpha = \beta = 1/2, p = 4$.

Let us finally consider the L_2-norm $\mathbf{gg^*}$ in the limiting case that $\alpha T \downarrow 0$ and $p \to \infty$:

$$\mathbf{gg^*} = \sum_{r=0}^{p-1} |g_r(t,\omega)|^2 = \sum_{r=0}^{p-1} \left| \tilde{g} \left(\alpha pt, \frac{\omega + r\Omega/p}{\alpha}; \alpha T \right) \right|^2.$$

Since $|\tilde{g}(t,\omega;\alpha T)|$ is almost independent of t for small values of αT (see (3.6)), we might as well write

$$\mathbf{gg^*} \simeq \sum_{r=0}^{p-1} \left| \tilde{g} \left(t, \frac{\omega + r\Omega/p}{\alpha}; \alpha T \right) \right|^2,$$

and for large values of p—remembering that ω is restricted to a (small) interval of length Ω/p—this might as well be written as

$$\mathbf{gg^*} \simeq \sum_{r=0}^{p-1} \left| \tilde{g} \left(t, \frac{r\Omega}{p\alpha}; \alpha T \right) \right|^2.$$

In the limit of $\alpha T \downarrow 0$ and $p \to \infty$, we might as well write the latter expression in the form

$$\mathbf{gg^*} \simeq \frac{p}{2\pi} \int_{\alpha T} \int_{\Omega/\alpha} |\tilde{g}(t,\omega;\alpha T)|^2 dt d\omega,$$

which, with the help of Parseval's energy theorem (3.4), can be expressed in the form

$$\mathbf{gg^*} \simeq \frac{p}{\alpha T} \int |g(t)|^2 dt.$$

Since $\mathbf{g} = \alpha T(\mathbf{gg^*})\mathbf{w}_{opt}$, we finally conclude that

$$g(t) \simeq \left[p \int |g(t)|^2 dt \right] w_{opt}(t),$$

or

$$g(t) \simeq \frac{w_{opt}(t)}{p \int |w_{opt}(t)|^2 dt}. \qquad (7.16)$$

We can link this result to the continuous analogue (7.5) of Gabor's signal expansion. Approximating the double integration in this continuous analogue by a double summation we get

$$\varphi(\tau) \simeq \frac{\alpha\beta}{\int |w(t)|^2 dt} \sum_m \sum_k s(m\alpha T, k\beta\Omega) w(\tau - m\alpha T) e^{jk\beta\Omega\tau},$$

which expression has indeed the form of a Gabor expansion (see (7.2)) with expansion coefficients $a_{mk} = s(m\alpha T, k\beta\Omega)$ and with an elementary signal $g(t)$ that is proportional to the window function (see (7.16)).

§8 Coherent-optical generation of the Gabor coefficients

In this section we describe a coherent-optical set-up from which Gabor's expansion coefficients of a rastered, one-dimensional signal can be generated.

We already noted that (7.6) allows an easy determination of the array of Gabor coefficients a_{mk} via the Zak transform. Since we are in fact only dealing with Fourier transformations, (7.6) enables a coherent-optical implementation. Let us therefore consider the optical arrangement depicted in Figure 6, and let us identify the two variables t and ω in the Zak transforms and the Fourier transform, as the x- and y-coordinates. Moreover, let us, for the sake of convenience, take $\alpha = 1/p$, with which (7.6) reduces to

$$\bar{a}(t,\omega;T) = T\tilde{\varphi}(t,p\omega;T)\tilde{w}^*\left(t,p\omega;\frac{T}{p}\right). \tag{8.1}$$

A plane wave of monochromatic laser light is normally incident upon a transparency situated in the input plane. The transparency contains the signal $\varphi(x)$ in a rastered format. With X being the width of this raster and $p\mu X$ (with $\mu > 0$) being the spacing between the raster lines, the light amplitude $\varphi_i(x_i, y_i)$ just behind the transparency reads

$$\varphi_i(x_i, y_i) = \text{rect}\left(\frac{x_i}{X}\right)\sum_n \varphi(x_i + nX)\delta(y_i - np\mu X). \tag{8.2}$$

An anamorphic optical system between the input plane and the middle plane performs a Fourier transformation in the y-direction and an ideal imaging (with inversion) in the x-direction. Such an anamorphic system can be realized as shown, for instance, using a combination of a spherical and a cylindrical lens. The anamorphic operation results in the light amplitude

$$\begin{aligned}
\varphi_1(x,y) &= \iint \varphi_i(x_i, y_i)e^{-j\gamma_i y y_i}\delta(x - x_i)dx_i dy_i \\
&= \text{rect}\left(\frac{x}{X}\right)\tilde{\varphi}(x, p\mu\gamma_i y; X) \tag{8.3}
\end{aligned}$$

just in front of the middle plane; the parameter γ_i contains the effect of the wave length λ of the laser light and the focal length f_i of the spherical lens: $\gamma_i = 2\pi/\lambda f_i$. Note that in (8.3) we have introduced the Zak transform of $\varphi(x)$, defined by (3.1) with t replaced by x, ω replaced by $p\mu\gamma_i y$, and τ replaced by X.

A transparency with amplitude transmittance

$$m(x,y) = \text{rect}\left(\frac{x}{X}\right)\text{rect}\left(\frac{y}{Y}\right)X\tilde{w}^*\left(x, p\mu\gamma_i y; \frac{X}{p}\right), \tag{8.4}$$

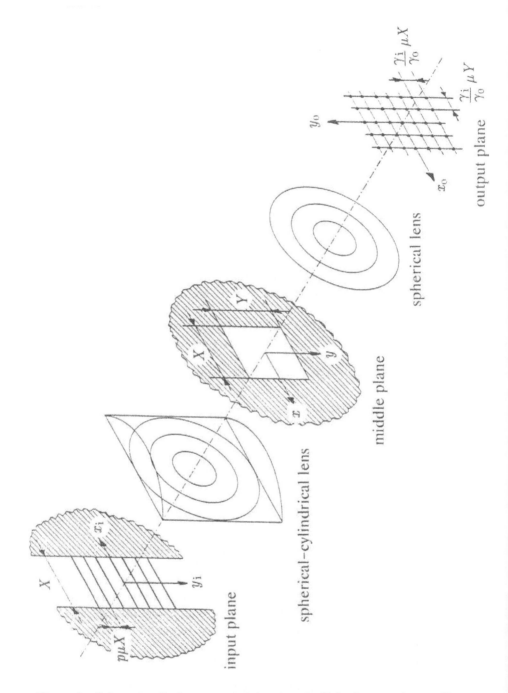

Figure 6. Coherent-optical arrangement to generate Gabor's expansion coefficients of a rastered, one-dimensional signal.

where $\mu\gamma_i Y = 2\pi/X$, is situated in the middle plane. Just behind this transparency, the light amplitude takes the form

$$\varphi_2(x,y) = m(x,y)\varphi_1(x,y) = \text{rect}\left(\frac{x}{X}\right)\text{rect}\left(\frac{y}{Y}\right)\bar{a}(x,\mu\gamma_i y; X), \quad (8.5)$$

where use has been made of (8.1), with t replaced by x, ω replaced by $\mu\gamma_i y$, and T replaced by X. Note that the aperture $\text{rect}(x/X)\text{rect}(y/Y)$ contains *one* period of the periodic Fourier transform $\bar{a}(x,\mu\gamma_i y; X)$, p periods of the (periodic) Zak transform $\bar{\varphi}(x,p\mu\gamma_i y; X)$, and p quasi-periods of the (quasi-periodic) Zak transform $\tilde{w}^*(x,p\mu\gamma_i y; X/p)$.

One of the reasons for oversampling is the additional freedom in choosing the window function $w(t)$. In particular, a window function can be chosen so that it is mathematically well-behaved; this is usually not the case for the original Gabor expansion with critical sampling ($p = 1$). Indeed, when the Zak transform $\tilde{g}$ of the elementary signal $g(t)$ has zeros, the Zak transform $\tilde{w} = 1/T\tilde{g}^*$ (see (3.8)) of the corresponding window function $w(t)$ has poles, which reflects the bad behavior of the window function. We remark that in the case of oversampling ($p > 1$) the Zak transform $\tilde{w}(x,p\mu\gamma_i y; X/p)$ can be constructed so that it has no poles, and, hence, a practical transparency can indeed be fabricated!

Finally, a two-dimensional Fourier transformation is performed between the middle plane and the output plane. Such a Fourier transformation can be realized as shown, for instance, using a spherical lens. The light amplitude in the output plane then takes the form

$$\begin{aligned}\varphi_o(x_o, y_o) &= \frac{1}{XY}\int\int \varphi_2(x,y)e^{-j\gamma_o(x_o x - y_o y)}dxdy \\ &= \sum_m\sum_k a_{mk}\,\text{sinc}\left(\frac{\gamma_o}{\gamma_i}\frac{x_o}{\mu Y} - k\right)\text{sinc}\left(\frac{\gamma_o}{\gamma_i}\frac{y_o}{\mu X} - m\right)\end{aligned}\quad (8.6)$$

where the sinc-function $\text{sinc}(z) = \sin(\pi z)/(\pi z)$ has been introduced; the parameter γ_o, again, contains the effects of the wave length λ of the laser light and the focal length f_o of the spherical lens: $\gamma_o = 2\pi/\lambda f_o$. We conclude that Gabor's expansion coefficients appear on a rectangular lattice of points

$$a_{mk} = \varphi_o\left(k\frac{\gamma_i}{\gamma_o}\mu Y, m\frac{\gamma_i}{\gamma_o}\mu X\right) \quad (8.7)$$

in the output plane.

We remark that it is not an essential requirement that the input transparency consists of Dirac functions. When we replace the practically unrealizable Dirac functions $\delta(y - np\mu X)$ by realizable functions $d(y - np\mu X)$, say, then (8.2) reads

$$\varphi_i(x_i, y_i) = \text{rect}\left(\frac{x_i}{X}\right)\sum_n \varphi(x_i + nX)d(y_i - np\mu X) \quad (8.8)$$

and the light amplitude $\varphi_1(x, y)$ just in front of the middle plane takes the form (see (8.3))

$$\varphi_1(x, y) = \text{rect}\left(\frac{x}{X}\right)\tilde{\varphi}(x, p\mu\gamma_i y; X)\bar{d}(\gamma_i y). \tag{8.9}$$

The additional factor $\bar{d}(\gamma_i y)$, the Fourier transform of $d(y)$, can easily be compensated for by means of a transparency in the middle plane.

The technique described in this section to generate Gabor's expansion coefficients fully utilizes the two-dimensional nature of the optical system, its parallel processing features, and the large space-bandwidth product possible in optical processing. The technique exhibits a resemblance to *folded spectrum* techniques, where space-bandwidth products in the order of 300,000 are reported [12]. In the case of speech processing, where speech recognition and speaker identification are important problems, such a space-bandwidth product would allow us to process speech fragments of about one minute.

§9   Conclusion

In this paper we have presented Gabor's expansion of a signal into a discrete set of properly shifted and modulated versions of an elementary signal. We have also described the inverse operation and the Gabor transform (with which Gabor's expansion coefficients can be determined), and we have shown how the expansion coefficients can be determined, even in the case that the set of elementary signals is not orthonormal. The key solution was the construction of a window function such that the discrete set of shifted and modulated versions of the window function is bi-orthonormal to the corresponding set of elementary signals. Thus, we have shown a strong relationship between Gabor's expansion of a signal on the one hand and sampling of the sliding-window spectrum of the signal on the other hand.

The Gabor lattice played a key role in the first part of this paper. It is the regular lattice ($t = mT, \omega = k\Omega$) with $\Omega T = 2\pi$ in the time-frequency domain, in which each cell occupies an area of 2π. The density of the Gabor lattice is thus equal to the *Nyquist density* $1/2\pi$, which, as is well known in information theory, is the minimum time-frequency density needed for full transmission of information. Gabor's expansion coefficients can then be interpreted as degrees of freedom of the signal.

We have introduced the Zak transform and we have shown its intimate relation to the Gabor transform. Not only did we consider the Gabor transform and the Zak transform in the continuous-time case, but we have also considered the discrete-time case. Furthermore, we have introduced the *discrete* Gabor transform and the *discrete* Zak transform, by sampling the continuous frequency variable that still occurred in the discrete-time case. The discrete transforms enable us to determine Gabor's expansion coefficients via a fast computer algorithm, analogous to the fast Fourier transform algorithm well known in digital signal processing.

Using the Zak transform, we have seen that—at least for critically sampling the time-frequency domain on the lattice $(t = mT, \omega = k\Omega)$—the Gabor transform, as well as Gabor's signal expansion itself, can be transformed into a product form. Determination of the expansion coefficients via the product forms may be difficult, however, because of the occurrence of zeros in the Zak transform. One way of avoiding the problems that arise from these zeros is to sample the time-frequency domain on a denser lattice $(t = m\alpha T, \omega = k\beta\Omega)$, with the product $\alpha\beta$ smaller than 1. In this paper we have considered the special case $\alpha\beta = 1/p$, where p is a positive integer. We have shown that in this case of oversampling by an integer factor, the Gabor transform can again be transformed into a product form; furthermore, Gabor's signal expansion itself can likewise be transformed into a sum-of-products form. Using these product forms, it was possible to show that the Gabor transform and Gabor's signal expansion form a transform pair. The properties of denser lattices have already been studied in many other papers [2, 3, 13, 21, 22, 24, 25, 26, 29, 34].

The process of oversampling introduces dependence between the Gabor coefficients; whereas these coefficients can be considered as degrees of freedom in the case of critical sampling, they can no longer be given such an interpretation in the case of oversampling. In controlling the dependence between the Gabor coefficients, we were able to avoid the problems that arise from the occurrence of zeros in the Zak transform. In particular, it was shown how the window function that appears in the Gabor transform, can be constructed from the elementary signal that is used in Gabor's signal expansion. The additional freedom caused by oversampling allowed us to construct the window function in such a way that it is mathematically well-behaved; in particular, we showed a way to find an (optimum) window function that has minimum L_2-norm. Moreover, it was shown that for large oversampling the optimum window function becomes proportional to the elementary signal; this result is in accordance with the continuous analogue of Gabor's signal expansion.

Finally, a coherent-optical arrangement was described which is able to generate Gabor's expansion coefficients of a rastered, one-dimensional signal via the Zak transform. The technique described there, which resembles folded spectrum techniques, fully utilizes the two-dimensional nature of the optical system, its parallel processing features, and the large space-bandwidth product possible in optical processing. Due to the possibility of avoiding the problems that arise from the occurrence of zeros in the Zak transform, the required optical transparency can indeed be fabricated.

We conclude this paper by drawing attention to some related topics: the *wavelet transform* of a signal and the way of representing a signal as a discrete set of *wavelets*. There is some resemblance between these topics and the ones that were presented in this paper. But, whereas the Gabor transform and the sliding-window spectrum lead to a *time-frequency* representation of the signal, the wavelet transform leads to a *time-scale* representation. And whereas the Gabor lattice is linear both in the time and the frequency variable, the lattice that is used in the wavelet representation is nonlinear. An excellent review on the wavelet transform can be found in [13].

Appendix A. The second bi-orthonormality condition (2.8)

We will show that the first bi-orthonormality condition (2.7), which in product form reads (see (3.8))

$$T\tilde{w}^*(t,\omega;T)\tilde{g}(t,\omega;T) = 1,$$

implies the second bi-orthonormality condition (2.8)

$$\sum_m \sum_k w_{mk}^*(t_1)g_{mk}(t_2) = \delta(t_2 - t_1).$$

We start with the left-hand side of the second bi-orthonormality condition (2.8)

$$\sum_m \sum_k w_{mk}^*(t_1)g_{mk}(t_2) = \sum_m \sum_k w^*(t_1 - mT)e^{-jk\Omega t_1}g(t_2 - mT)e^{jk\Omega t_2}.$$

After expressing the window function $w(t_1 - mT)$ by means of its Zak transform $\tilde{w}(t_1,\omega;T)$ (see (3.3)), the latter expression takes the form

$$\sum_m \sum_k \left[\frac{1}{\Omega}\int_\Omega \tilde{w}(t_1,\omega;T)e^{-jm\omega T}d\omega\right]^* e^{-jk\Omega t_1}g(t_2 - mT)e^{jk\Omega t_2}.$$

We rearrange factors

$$\sum_k e^{jk\Omega(t_2-t_1)}\frac{1}{\Omega}\int_\Omega \tilde{w}^*(t_1,\omega;T)\left[\sum_m g(t_2 - mT)e^{jm\omega T}\right]d\omega$$

and recognize (see (3.1)) the Zak transform $\tilde{g}(t_2,\omega;T)$ of the function $g(t_2)$

$$\sum_k e^{jk\Omega(t_2-t_1)}\frac{1}{\Omega}\int_\Omega \tilde{w}^*(t_1,\omega;T)\tilde{g}(t_2,\omega;T)d\omega.$$

We replace the sum of exponentials by a sum of Dirac functions

$$T \sum_n \delta(t_2 - t_1 - nT) \frac{1}{\Omega} \int_\Omega \tilde{w}^*(t_1, \omega; T) \tilde{g}(t_2, \omega; T) d\omega$$

and replace the variable t_2 in the integral by $t_1 + nT$

$$T \sum_n \delta(t_2 - t_1 - nT) \frac{1}{\Omega} \int_\Omega \tilde{w}^*(t_1, \omega; T) \tilde{g}(t_1 + nT, \omega; T) d\omega.$$

We use the quasi-periodicity property (see (3.2)) of the Zak transform $\tilde{g}(t_1, \omega; T)$

$$\sum_n \delta(t_2 - t_1 - nT) \frac{1}{\Omega} \int_\Omega T \tilde{w}^*(t_1, \omega; T) \tilde{g}(t_1, \omega; T) e^{jn\omega T} d\omega$$

and substitute from the product form (3.8) of the first bi-orthonormality condition

$$\sum_n \delta(t_2 - t_1 - nT) \frac{1}{\Omega} \int_\Omega e^{jn\omega T} d\omega.$$

We evaluate the integral

$$\sum_n \delta(t_2 - t_1 - nT) \delta_n$$

and conclude that the expression reduces to the required Dirac function

$$\delta(t_2 - t_1).$$

Appendix B. Derivation of the product form (7.6)

In the Fourier transform (see (2.11)) of the array a_{mk}, we substitute from the Gabor transform (7.3)

$$\bar{a}(t, \omega; T) = \sum_m \sum_k \left[\int \varphi(\tau) w^*(\tau - m\alpha T) e^{jk\beta\Omega\tau} d\tau \right] e^{-j(m\omega T - k\Omega t)}.$$

In the right-hand side of the latter equation, we rearrange factors

$$\sum_m \left[\int \varphi(\tau) w^*(\tau - m\alpha T) \left\{ \sum_k e^{-jk\Omega(\beta\tau - t)} \right\} d\tau \right] e^{-jm\omega T}$$

and replace the sum of exponentials by a sum of Dirac functions

$$\sum_m \left[\int \varphi(\tau) w^*(\tau - m\alpha T) \left\{ \frac{T}{\beta} \sum_k \delta \left(\tau - \frac{t + kT}{\beta} \right) \right\} d\tau \right] e^{-jm\omega T}.$$

We rearrange factors again

$$\frac{T}{\beta}\sum_m\left[\sum_k\int \varphi(\tau)w^*(\tau-m\alpha T)\delta\left(\tau-\frac{t+kT}{\beta}\right)d\tau\right]e^{-jm\omega T}$$

and evaluate the integral

$$\frac{T}{\beta}\sum_m\left[\sum_k\varphi\left(\frac{t+kT}{\beta}\right)w^*\left(\frac{t+kT}{\beta}-m\alpha T\right)\right]e^{-jm\omega T}.$$

After a final rearrangement of factors, we get

$$\frac{T}{\beta}\sum_k\varphi\left(\frac{t}{\beta}+k\frac{T}{\beta}\right)e^{-jk(\omega/\alpha)(T/\beta)}$$

$$\left[\sum_m w^*\left(\frac{t}{\beta}-[m-\frac{k}{\alpha\beta}]\alpha T\right)e^{-j(m-k/\alpha\beta)(\omega/\alpha)\alpha T}\right]$$

$$=\frac{T}{\beta}\sum_k\varphi\left(\frac{t}{\beta}+k\frac{T}{\beta}\right)e^{-jk(\omega/\alpha)(T/\beta)}$$

$$\left[\sum_m w\left(\frac{t}{\beta}-[m-pk]\alpha T\right)e^{j(m-pk)(\omega/\alpha)\alpha T}\right]^*.$$

In the last expression, we recognize the definitions (see (3.1)) for the Zak transforms $\tilde{\varphi}(t/\beta,\omega/\alpha;T/\beta)$ and $\tilde{w}(t/\beta,\omega/\alpha;\alpha T)$ of the signal $\varphi(t)$ and the window function $w(t)$, respectively, which leads to the result

$$\bar{a}(t,\omega;T)=\frac{T}{\beta}\tilde{\varphi}\left(\frac{t}{\beta},\frac{\omega}{\alpha};\frac{T}{\beta}\right)\tilde{w}^*\left(\frac{t}{\beta},\frac{\omega}{\alpha};\alpha T\right)$$

or, with $\beta=1/\alpha p$,

$$\bar{a}(t,\omega;T)=\alpha pT\tilde{\varphi}\left(\alpha pt,\frac{\omega}{\alpha};\alpha pT\right)\tilde{w}^*\left(\alpha pt,\frac{\omega}{\alpha};\alpha T\right).$$

Appendix C. Derivation of the sum-of-products form (7.7)

In Gabor's signal expansion (7.2), we substitute from the inverse Fourier transform (see (2.13))

$$\varphi(\tau)=\sum_m\sum_k\left[\frac{1}{2\pi}\int_T\int_\Omega \bar{a}(t,\omega;T)e^{j(m\omega T-k\Omega t)}dtd\omega\right]$$

$$\times\, g(\tau-m\alpha T)e^{jk\beta\Omega\tau}.$$

In the right-hand side of the latter equation, we rearrange factors

$$\frac{1}{2\pi} \int_T \int_\Omega \bar{a}(t,\omega;T) \left[\sum_m g(\tau - m\alpha T)e^{jm\omega T} \right] \left[\sum_k e^{-jk\Omega(t-\beta\tau)} \right] dt d\omega.$$

We replace the sum of exponentials by a sum of Dirac functions and recognize the Zak transform of the elementary signal $g(t)$ (see (3.1)):

$$\frac{1}{2\pi} \int_T \int_\Omega \bar{a}(t,\omega;T)\tilde{g}\left(\tau,\frac{\omega}{\alpha};\alpha T\right) \left[T \sum_k \delta(t - \beta\tau + kT) \right] dt d\omega.$$

We rearrange factors again and substitute from the periodicity property (see (2.12)) of $\bar{a}(t,\omega;T)$

$$\frac{1}{\Omega} \int_\Omega \left[\sum_k \int_T \bar{a}(t+kT,\omega;T)\delta(t+kT-\beta\tau)dt \right] \tilde{g}\left(\tau,\frac{\omega}{\alpha};\alpha T\right) d\omega,$$

and we replace the summation over k together with the time integral over the finite interval T by an integral over the entire time axis

$$\frac{1}{\Omega} \int_\Omega \left[\int \bar{a}(t,\omega;T)\delta(t-\beta\tau)dt \right] \tilde{g}\left(\tau,\frac{\omega}{\alpha};\alpha T\right) d\omega.$$

Evaluation of the resulting time integral yields the intermediate result

$$\varphi(\tau) = \frac{1}{\Omega} \int_\Omega \bar{a}(\beta\tau,\omega;T)\tilde{g}\left(\tau,\frac{\omega}{\alpha};\alpha T\right) d\omega.$$

We now write down the definition of the Zak transform (see (3.1))

$$\tilde{\varphi}\left(\frac{t}{\beta},\frac{\omega_o}{\alpha};\frac{T}{\beta}\right) = \sum_n \varphi\left(\frac{t}{\beta}+n\frac{T}{\beta}\right) e^{-jn(\omega_o/\alpha)(T/\beta)}.$$

In the right-hand side of the latter equation, we substitute from the intermediate result above

$$\sum_n \frac{1}{\Omega} \left[\int_\Omega \bar{a}\left(\beta\left[\frac{t}{\beta}+n\frac{T}{\beta}\right],\omega;T\right) \tilde{g}\left(\frac{t}{\beta}+n\frac{T}{\beta},\frac{\omega}{\alpha};\alpha T\right) d\omega \right]$$

$$\times e^{-jn(\omega_o/\alpha)(T/\beta)}.$$

We rearrange

$$\sum_n \frac{1}{\Omega} \left[\int_\Omega \bar{a}(t+nT,\omega;T)\tilde{g}\left(\frac{t}{\beta}+pn\alpha T,\frac{\omega}{\alpha};\alpha T\right) d\omega \right] e^{jpn\omega_o T}$$

and use the periodicity property (see (2.12)) of $\bar{a}(t, \omega; T)$ and the (quasi-)
periodicity property (see (3.2)) of $\tilde{g}(t/\beta, \omega/\alpha; \alpha T)$

$$\sum_n \frac{1}{\Omega} \left[\int_\Omega \bar{a}(t, \omega; T)\tilde{g}\left(\frac{t}{\beta}, \frac{\omega}{\alpha}; \alpha T\right) e^{jpn(\omega/\alpha)\alpha T} d\omega \right] e^{-jpn\omega_o T}.$$

We rearrange factors

$$\frac{1}{\Omega} \int_\Omega \bar{a}(t, \omega; T)\tilde{g}\left(\frac{t}{\beta}, \frac{\omega}{\alpha}; \alpha T\right) \left[\sum_n e^{jnp(\omega-\omega_o)T}\right] d\omega$$

and replace the sum of exponentials by a sum of Dirac functions

$$\frac{1}{\Omega} \int_\Omega \bar{a}(t, \omega; T)\tilde{g}\left(\frac{t}{\beta}, \frac{\omega}{\alpha}; \alpha T\right) \left[\frac{\Omega}{p} \sum_n \delta\left(\omega - \omega_o - n\frac{\Omega}{p}\right)\right] d\omega.$$

We rearrange factors again

$$\frac{1}{p} \sum_n \int_\Omega \bar{a}(t, \omega; T)\tilde{g}\left(\frac{t}{\beta}, \frac{\omega}{\alpha}; \alpha T\right) \delta\left(\omega - \omega_o - n\frac{\Omega}{p}\right) d\omega$$

and replace the summation over n by a double summation over r and m
through the substitution $-n = mp - r$, where r extends over an interval of
length p

$$\frac{1}{p} \sum_{r=\langle p \rangle} \sum_m \int_\Omega \bar{a}(t, \omega; T)\tilde{g}\left(\frac{t}{\beta}, \frac{\omega}{\alpha}; \alpha T\right) \delta\left(\omega - \omega_o + [mp - r]\frac{\Omega}{p}\right) d\omega.$$

We substitute from the periodicity properties of the Fourier transform and
the Zak transform (see (2.12) and (3.2))

$$\frac{1}{p} \sum_{r=\langle p \rangle}$$

$$\left[\sum_m \int_\Omega \bar{a}(t, \omega + m\Omega; T)\tilde{g}\left(\frac{t}{\beta}, \frac{\omega + m\Omega}{\alpha}; \alpha T\right) \delta\left(\omega + m\Omega - \omega_o - r\frac{\Omega}{p}\right) d\omega\right]$$

and replace the summation over m together with the frequency integral
over the finite interval Ω by an integral over the entire frequency axis

$$\frac{1}{p} \sum_{r=\langle p \rangle} \int \bar{a}(t, \omega; T)\tilde{g}\left(\frac{t}{\beta}, \frac{\omega}{\alpha}; \alpha T\right) \delta\left(\omega - \omega_o - r\frac{\Omega}{p}\right) d\omega.$$

Evaluation of the integral and replacing ω_o by ω results in

$$\tilde{\varphi}\left(\frac{t}{\beta}, \frac{\omega}{\alpha}; \frac{T}{\beta}\right) = \frac{1}{p}\sum_{r=\langle p\rangle} \bar{a}\left(t, \omega + r\frac{\Omega}{p}; T\right) \tilde{g}\left(\frac{t}{\beta}, \frac{\omega + r\Omega/p}{\alpha}; \alpha T\right)$$

or, with $\beta = 1/\alpha p$

$$\tilde{\varphi}\left(\alpha pt, \frac{\omega}{\alpha}; \alpha pT\right) = \frac{1}{p}\sum_{r=\langle p\rangle} \bar{a}\left(t, \omega + r\frac{\Omega}{p}; T\right) \tilde{g}\left(\alpha pt, \frac{\omega + r\Omega/p}{\alpha}; \alpha T\right).$$

References

[1] Abramowitz, M. and I. A. Stegun (eds.), *Handbook of Mathematical Functions*, Dover, New York, 1970.

[2] Auslander, L., I. C. Gertner, and R. Tolimieri, The discrete Zak transform application to time-frequency analysis and synthesis of nonstationary signals, *IEEE Trans. Signal Process.* **39** (1991), 825–835.

[3] Balart, R. and R. S. Orr, Computational accuracy and stability issues for the finite, discrete Gabor transform, in *Proc. IEEE-SP International Symposium on Time-Frequency and Time-Scale Analysis*, Victoria, BC, Canada, October 1992, 403–406.

[4] Bargmann, V., P. Butera, L. Girardello, and J. R. Klauder, On the completeness of the coherent states, *Rep. Math. Phys.* **2** (1971), 221–228.

[5] Bastiaans, M. J., Gabor's expansion of a signal into Gaussian elementary signals, *Proc. IEEE* **68** (1980), 538–539.

[6] Bastiaans, M. J., The expansion of an optical signal into a discrete set of Gaussian beams, *Optik* **57** (1980), 95–102.

[7] Bastiaans, M. J., Sampling theorem for the complex spectrogram, and Gabor's expansion of a signal in Gaussian elementary signals, *Opt. Eng.* **20** (1981), 594–598.

[8] Bastiaans, M. J., Gabor's signal expansion and degrees of freedom of a signal, *Opt. Acta* **29** (1982), 1223–1229.

[9] Bastiaans, M. J., On the sliding-window representation in digital signal processing, *IEEE Trans. Acoust. Speech Signal Processing* **ASSP-33** (1985), 868–873.

[10] Bastiaans, M. J., Gabor's signal expansion and its relation to sampling of the sliding-window spectrum, in *Advanced Topics in Shannon Sampling and Interpolation Theory*, R.J. Marks II (ed.), Springer, New York, 1993, 1–35.

[11] Campbell, S. L. and C. D. Meyer, Jr., *Generalized Inverses of Linear Transformations*, Piman, London, 1979.

[12] Casasent, D., Optical signal processing, in *Optical Data Processing; Applications*, Topics in Applied Physics, Vol. 23, D. Casasent (ed.), Springer, Berlin, 1978, Chap. 8.

[13] Daubechies, I., The wavelet transform, time-frequency localization and signal analysis, *IEEE Trans. Inform. Theory* **36** (1990), 961–1005.

[14] Gabor, D., Theory of communication, *Proc. Inst. Electr. Eng.* **93 (III)** (1946), 429–457.

[15] Helstrom, C. W., An expansion of a signal in Gaussian elementary signals, *IEEE Trans. Inform. Theory* **IT-12** (1966), 81–82.

[16] Janssen, A. J. E. M., Gabor representation of generalized functions, *J. Math. Anal. Appl.* **83** (1981), 377–396.

[17] Janssen, A. J. E. M., Bargmann transform, Zak transform and coherent states, *J. Math. Phys.* **23** (1982), 720–731.

[18] Janssen, A. J. E. M., The Zak transform: a signal transform for sampled time-continuous signals, *Philips J. Res.* **43** (1988), 23–69.

[19] Kritikos, H. N. and P. T. Farnum, Approximate evaluation of Gabor expansions, *IEEE Trans. Syst., Man Cybern.* **SMC-17** (1987), 978–981.

[20] Neumann von, J., *Mathematical Foundations of Quantum Mechanics*, Princeton University Press, 1955.

[21] Orr, R. S., Finite discrete Gabor transform relations under periodization and sampling, in *Proc. IEEE-SP International Symposium on Time-Frequency and Time-Scale Analysis*, Victoria, BC, Canada, October 1992, 395–398.

[22] Orr, R. S., The order of computation for finite discrete Gabor transforms, *IEEE Trans. Signal Process.* **41** (1993), 122–130.

[23] Papoulis, A., *Signal Analysis*, McGraw-Hill, New York, 1977.

[24] Qian, S., K. Chen, and S. Li, Optimal biorthogonal sequence for finite discrete-time Gabor expansion, *Signal Process.* **27** (1992), 177–185.

[25] Qian S. and D. Chen, A general solution of biorthogonal analysis window functions for orthogonal-like discrete Gabor transform, in *Proc. IEEE-SP International Symposium on Time-Frequency and Time-Scale Analysis*, Victoria, BC, Canada, October 1992, 387–389.

[26] Qian, S. and D. Chen, Discrete Gabor transform, *IEEE Trans. Signal Process.* **41** (1993), 2429–2438.

[27] Reed, M. and B. Simon, *Methods and Modern Mathematical Physics. IV Analysis of Operators*, Academic Press, New York, 1978.

[28] Schempp, W., Radar ambiguity functions, the Heisenberg group, and holomorphic theta series, *Proc. Am. Math. Soc.* **92** (1984), 103–110.

[29] Wexler, J. and S. Raz, Discrete Gabor expansions, *Signal Process.* **21** (1990), 207–221.

[30] Whittaker, E. T. and G. N. Watson, *A Course of Modern Analysis*, Cambridge University Press, 1927.

[31] Zak, J., Finite translations in solid-state physics, *Phys. Rev. Lett.* **19** (1967), 1385–1387.

[32] Zak, J., Dynamics of electrons in solids in external fields, *Phys. Rev.* **168** (1968), 686–695.

[33] Zak, J., The kq-representation in the dynamics of electrons in solids, *Solid State Phys.* **27** (1972), 1–62.

[34] Zibulski, M. and Y. Y. Zeevi, Oversampling in the Gabor scheme, *IEEE Trans. Signal Process.* **41** (1993), 2679–2687.

Martin J. Bastiaans
Technische Universiteit Eindhoven
Faculteit Elektrotechniek
Postbus 513
5600 MB Eindhoven, Netherlands
M.J.Bastiaans@ele.tue.nl

Non-Commutative Affine Geometry and Symbol Calculus: Fourier Transform Magnetic Resonance Imaging and Wavelets

Walter Schempp

Dedicated to Professor Satoru Igari, Tohoku University,
on the occasion of his sixtieth birthday

Die Welt als Phantom und Matrize.
Günther Anders, 1956

Theorien kommen zustande durch ein vom empirischen Material inspiriertes Verstehen, welches am besten im Anschluß an Plato als Zur-Deckung-Kommen von inneren Bildern mit äußeren Objekten und ihrem Verhalten zu deuten ist. Die Möglichkeit des Verstehens zeigt aufs Neue das Vorhandensein regulierender typischer Anordungen, denen sowohl das Innen wie das Außen des Menschen unterworfen sind.
Wolfgang Pauli, 1961

Timing is everything ...
Robert B. Lufkin, The MRI Manual, 1990

Variations on a theme by Kepler.
Victor W. Guillemin and Shlomo Sternberg, 1990

Cherchez la lésion!
Robert I. Grossman, 1994

Signal and Image Representation in Combined Spaces
Y. Y. Zeevi and R. R. Coifman (Eds.), pp. 71-119.
ISBN 0-12-777830-6

Abstract. Magnetic Resonance Imaging (MRI) is one of the most significant advances in medical imaging in this century. The multi-planar capability and high-resolution imaging are unmatched by any other current clinical imaging technique. The physical and mathematical principles on which this non-invasive imaging technology is based are as complex as the computer-controlled synergy of the MRI scanner organization. The MRI scanner is a quantum electrodynamical (QED) device, the components of which implement and detect the wavelet interference and resonance phenomena which underly the symplectic MRI filter-bank processing. Notice the fact that QED has approved the semi-classical approach to quantum holography. In this paper, a semi-classical QED treatment of the basic MRI system organization is presented, with deep roots in the Keppler phase triangulation procedure of physical astronomy and in the symplectically invariant symbol calculus of pseudodifferential operators. It is based on non-commutative geometry of affine transvections and non-commutative Fourier analysis. This allows for modeling the interference patterns of phase coherent wavelets by distributional harmonic analysis on the Heisenberg nilpotent Lie group G of quantum mechanics. Geometric quantization yields the tomographic slices by the planar coadjoint orbit stratification $\mathcal{O}_\nu, \nu \neq 0$, of the unitary dual $\hat{G}$ of the Heisenberg group G. The resonance of affine wavelets is treated by harmonic analysis on the affine solvable Lie group $\mathbf{GA}(\mathbb{R})$, which is intrinsically operating on non-trivial blocks of G. It allows the acquisition of the coordinates of high-resolution tomographic scans within the tomographic slices by the Lauterbur spatial encoding technique of quantum holography.

§1 General introduction

The Radon transform $\mathcal{R}$ maps a function f in the Schwartz space $\mathcal{S}(\mathbb{R}^n)$ into the set of its integrals over hyperplanes ($n \geq 2$):

$$f \rightsquigarrow \mathcal{R}f(\theta, s) = \int_{<x|\theta>=s} f(x)\,\mathrm{d}x \quad \left(\theta \in \mathbf{S}_{n-1}, s \in \mathbb{R}\right).$$

Because $\mathcal{R}$ solves the reconstruction problem of X-ray computed tomography (CT) by means of line integrals, it opened up a window for geometric analysis on medical imaging and visualization. Several of the long-standing problems with standard dynamic X-ray CT were overcome with the introduction of spiral X-ray CT scanning in 1990. However, the limited intrinsic spatial resolution and, of course, the invasiveness of the ionizing radiation remained. As a result of experience with ionizing radiation, the absorption rate of a spinal examination with a modern X-ray CT scanner is about twenty times the absorption rate of a plain radiograph, and a cranial neuroanatomic X-ray CT examination needs about thirty times the absorption rate of a highly sensitive X-ray film. Because the genotoxic and oncogenic effects of the new generation of X-ray CT scanners are not negligible, cranial X-ray CT examinations should be restricted to the radiological evaluation of emergency patients with acute head trauma.

The introduction and vertiginous development of new imaging techniques have revolutionized modern radiology and neuroradiology during the last 15 years. Among these techniques, X-ray CT and magnetic resonance imaging (MRI) are without doubt the ones that have had the most important impact on patient evaluation protocols. MRI is a radically different, non-invasive radiodiagnostic imaging technique based on QED, with superior soft tissue contrast resolution over X-ray CT. The QED organization of a MRI scanner resembles a pumped laser more than a X-ray scanner. The extraordinarily large innate contrast (which can be on the order of several hundred percent for two soft tissues) is the single most distinguishing feature of MRI when compared to X-ray-based modalities like CT ([14]). In X-ray CT neuroimgaing, for example, the difference in electron density (and therefore in X-ray scattering) between gray and white matter is less than one percent. Because MRI has a strong perspective to image-guided therapy, it forms the most significant imaging advance in medical technology since the introduction of X-rays by Wilhelm Conrad Röntgen in 1895. For X-ray examinations 100 years ago, the exposure to radiation took between 10 to 120 minutes. In less than 20 years, MRI, with its high sensitivity to pathological processes, excellent anatomical detail, multiplanar imaging capability, and lack of ionizing radiation has had a major impact on routine clinical diagnoses ([4]). Frequently, MRI is the definitive examination, providing invaluable information to help the surgeon not only to understand the underlying pathoanatomy but also to make the critical decision regarding surgical intervention without planar restriction. Cranial MRI should be performed to assess the full extent of intra-axial damage whenever X-ray CT does not adequately explain the patient's neurological status, specifically in all patients presenting persistent impairment or delayed improvement of consciousness.

Essential to diagnosis by any medical imaging modality is detectability and locability. The medical discipline of radiodiagnostics only exists today because of the discovery of bands of radiation in the electromagnetic spectrum which can penetrate human tissue, and thus can be exploited by computerized Fourier analysis for medical imaging and visualization purposes. The MRI modality employs radiation in the radiofrequency band, and therefore represents the final such window in the electromagnetic spectrum for medical imaging and visualization. Since its first introduction, it has fired the imagination of MRI scientists and clinicians. MRI has revolutionized both the understanding and the radiodiagnostics of pathology of the central nervous system, the musculoskeletal system, and the vascular system. There is now general agreement that MRI has advanced to a degree that if access to it were readily available, it would be the preferred initial investigation technique for almost all types of intracranial pathology and for any conditions of the spinal column, joints, and muscles in the body. Its ability to accurately depict the anatomy of the central nervous system ([2]), the morphology of the musculoskeletal system ([4]), the extent of breast cancer in images of infiltrating ductal carcinoma ([4]), and in most of the potential applications now equals or exceeds that of X-ray CT. Many clinical studies have shown evidence of a clear superiority of MRI over X-ray CT in the detection of alterations in soft tissue composition produced by disease ([3, 14]). Due to the abundance of information, MRI is a lot harder to read than X-ray CT. MRI is not, however, always more specific.

Despite its widespread use in all modern medical centers as a preeminent diagnostic imaging modality, MRI appears as a remarkably recent technologic development that is in a constant state of flux, and therefore is as far from maturation as it ever was. In all advanced medical imaging centers, MRI has established itself as the most versatile and precise approach for tomographic imaging of much of the human body that has displaced plain radiographs of roentgenography, X-ray CT, diagnostic arthrography, myelography, and even angiography, as the imaging study of choice for a growing number of diseases. As a result of the recent incursion of MRI protocols into the field of functional brain imaging, brain function can be mapped with MRI sensitized to regional blood-oxygen-level-dependent (BOLD) changes that originate from the cortical parenchyma due to neural activation ([23]). Functional MRI (fMRI) is the only tomographic imaging modality presently available for accomplishing brain activity visualization during mental processes and the performance of cognitive tasks in vivo, and non-invasively without the use of exogenous contrast agents ([20]). Because there is nothing as compelling to the understanding of a neurological disease as being able to directly visualize the neuropathological state of the brain, the fMRI techniques have significantly expanded the potential clinical role of MRI, which includes the localization of epileptic foci, presurgical treatment planning, intraoperative mapping of the diseased brain function, and elucidation of normal brain function.

Tomography, from the Greek $\tau \acute{o} \mu o \varsigma$, the slice, and $\gamma \rho \alpha \varphi \acute{\iota} \varsigma$, the crayon, refers to cross-sectional imaging of the internal anatomy of the human body from either transmission or scattering quantum data. The cross-sections of the main directions are the central, coronal, and sagittal tomographic slices.

- In contrast to the X-ray CT modality of radiodiagnostic imaging, nuclear magnetic resonance preserves the differential phase as well as the frequency according to the semi-classical QED description of response data phase holograms. MRI is not a transmission imaging technique, but a phase coherent radar-type tomographic imaging modality of computer-controlled, actively backscattering cross-sectional targets. The most notable distinguishing feature of MRI, when compared to X-ray CT, is the absence of a universal gray scale.

Crystallography, from the Greek $\kappa\rho\upsilon\sigma\tau\alpha\lambda\lambda\sigma\varsigma$, *the ice*, may be viewed as a mathematical extension of the theory of array antennas. Therefore, the problem of probing crystals and determining molecular structures by the techniques of X-ray crystallography introduced by Max von Laue in 1912, had a strong impact on mathematics. It is the problem of determining the positions of the atoms in the crystal when only the amplitudes of the diffraction maxima are available from diffraction experiments. However, the associated phases which are also needed if one is to determine the equal-atom crystal and molecular structures from the experimental observations, are lost in the diffraction experiment. By exploiting a priori structural information of the sample, it can be established that the lost phase information is contained in the measured diffraction intensities and can be recovered provided that the molecular structure is not too large ([16]). For proteins and virus', however, these methods of solving the phase problem appear to be unsuccessful. Hence, phase coherent imaging techniques such as electron holography ([40]) reveal to be more appropriate for determining the position of single atoms in crystal structures. In medical imaging, some of the MRI findings have resulted in deepening and fortifying the knowledge of the underlying pathomechanism of disease processes such as multiple sclerosis (MS) and degenerative disc disease of the spine. In MS the principal constituent of the white matter of the central nervous system, the myeline, is attacked, thus impairing neural function. The first MRI scans of MS lesions were displayed in 1981. MRI has now completely superseded X-ray CT as a tool for imaging MS plaques, whether for the purposes of diagnosis or for monitoring the course of the disease. It has provided a unique opportunity to study the evolving pathologic process of MS in vivo ([31]) and the problems of epileptic processes as they relate to anatomically and functional abnormalities of the human brain ([22]).

Holography from the Greek $\dot{o}\lambda\dot{o}\varsigma$, *the whole*, is the signal record of both the amplitude and phase information content of coherent wavelets. The quadrature phased-array multicoil receiver technique of MRI, for long spinal cord imagery say, is based on the holographic principles of phase preservation as used in phase coherent data storage of radar-array processing in synthetic aperture radar (SAR) imagery ([26]). In SAR imaging, the aperture is synthesized by recording the radio signal received at two or more array antenna elements and later processing these phases coherently within a digital processor or an optical processor, which acts as a massively parallel multichannel cross-correlator ([24, 25]). Synthesizing an aperture by allowing the earth's rotation to sweep a phased-array of fixed antenna elements through space also was at the basis of the development of synthetic aperture radio telescopes by Martin Ryle in 1952. It follows that the ideas of tomography are closely related to the coherent data-processing methods of synthetic antenna developed in radar imaging and radio astronomy. The transmission systems, antennae, and receiver coils of MRI scanner organizations should be considered from this point of view. In particular, this aspect holds for quadrature phased-array multicoils as used for long spinal cord imagery.

In X-ray CT, the Radon transform $\mathcal{R}$ provides a decomposition into plane waves whereas the MRI modality is based on response data phase holograms generated by computer-controlled wavelet interference. The echo wavelet elicited is generated by the relaxation-weighted spin isochromat densities (literally, 'spins of the same color') themselves in response to a computer-controlled external perturbation. According to the semi-classical approach to quantum holography, which has been approved by QED, the spin precession under the strong external magnetic flux density serves as a natural reference for the arrays of phase-locked Bloch vectors. It generates a natural symplectic structure on the planar tomographic slices, which is basic for the FTMRI modality. The ensuing phase shift of period 4 has become routine in the implementation of various multiple spin echo trains by pulse transmission gates with a quartz-controlled stable local oscillator ([30]) similar to the STALO technique of SAR transmission (see Figure 1).

A factor that sets clinical imaging experts apart from specialists in other medical disciplines who consider MRI scanners as black boxes ([8]), and forms the strongest explanation for the imaging expert's claim to an indispensable role in medicine, is an understanding of the sciences of imaging and visualization. Both are based on mathematical methods. From the mathematical point of view, the FTMRI modality depends on

- Non-commutative affine geometry
- Non-commutative Fourier analysis

Tomographic imaging means, in the context of radiodiagnostic medicine, displaying internal anatomy in cross-sections by non-commutative geometry of mappings. Non-commutative Fourier analysis allows for translation of the geometry of affine mappings into the frequency domain of phase coherent wavelets to make the geometry accessible to computer-controlled wavelet backscattering, interference, and phase conjugation. Interference and phase conjugation, however, are resonance phenomena both of which require parallel synchronization. In this sense, computer-controlled synchronized timing is everything in FTMRI.

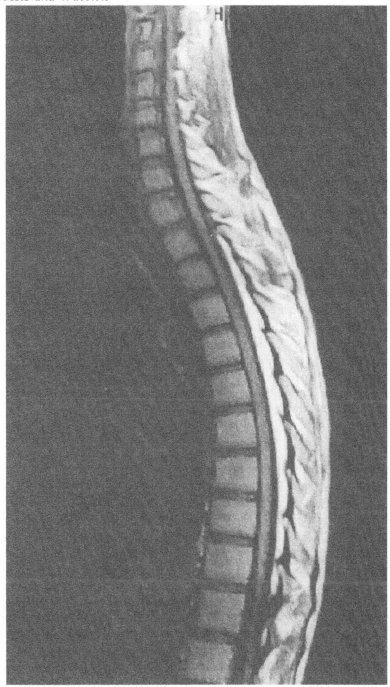

Figure 1. FTMRI using a quadrature phased-array spin multicoil.

In the case of the FTMRI modality, the appropriate choice of mappings are the affine transvections. A transvection is an endomorphism of a real vector space which leaves pointwise fixed every element of a homogeneous hyperplane ([10]). Every transvection is an affine bijection of determinant 1 whose set of fixed points is exactly the given hyperplane. Transvections generate the unimodular groups and play with respect to the symplectic group an analog role, as do symmetries with respect to the orthogonal group. Their tensor powers are called tranvectants.

In FTMRI, there are basically three standard types of affine transvections: central, coronal, and sagittal transvections which are implemented together with their affine bijections by the Heisenberg nilpotent Lie group G. It is a surprising fact that despite the complexity of the MRI system organization, the structure of this specific nilpotent Lie group is rich enough to allow for a study of the non-invasive diagnostic FTMRI modality. The key to this semi-classical QED approach is an embedding of Fourier analysis into the symplectically invariant symbol calculus of pseudodifferential operators in conjunction with the planar coadjoint orbit stratification $\mathcal{O}_\nu, \nu \neq 0$, of the unitary dual $\hat{G}$ of G.

The mathematical methods utilized in this paper are not restricted to the treatment of FTMRI, and apply to the semi-classical approach to other coherent imaging techniques as well. They allow, for instance, for a description of the resonance phenomena occurring in a bubble flow which intercepts an acoustic beam. Ultrasonic tomography allows one to obtain a two-dimensional cross-sectional image of the bubble flux density ([44]).

§2   Kernel distributions

Distribution theory, usually formulated as a local extension theory in terms of open subsets of the Euclidean vector space $\mathbb{R}^n$ or a C^∞-differential manifold, can be thought of as the completion of differential calculus just as Lebesgue integration theory can be thought of as the completion of integral calculus. Distribution theory, when applied to unitary linear representations of Lie groups G, provides the basis for non-commutative Fourier analysis and symbol calculus. Specifically in FTMRI, distributional harmonic analysis on the Heisenberg nilpotent Lie group G is suitable to describe relaxation-weighted spin isochromat densities excited in the planar coadjoint orbits $\mathcal{O}_\nu, \nu \neq 0$, of the stratification of $\hat{G}$ in terms of the symplectically invariant symbol calculus. The symbol calculus allows us to embed the von Neumann approach to quantum mechanics, which is based on the category of Hilbert spaces into the Dirac approach (which is based on complex locally convex topological vector spaces of tempered distributions). For the preparation of the symbol calculus, some generalities of the calculus of Schwartz kernels are needed ([37]). The full QED meaning of the kernel distributions for the semi-classical approach to FTMRI can be appreciated in the context of the Heisenberg nilpotent Lie group G, which provides two-dimensional Fourier analysis with an extra symplectic structure.

For simplicity, let G denote a unimodular Lie group and dg a Haar measure of G. Then dg forms the element 1_G of the complex vector space $\mathcal{D}'(G)$ of distributions on G. The topological antidual of the complex vector space $\mathcal{D}(G)$ of infinitely differentiable, compactly supported, complex-valued functions on G under its natural anti-involution of complex conjugation

$$\psi \rightsquigarrow \bar{\psi}$$

is the vector space $\mathcal{D}'(G)$ of those antilinear forms on $\mathcal{D}(G)$, which are continuous with respect to the canonical inductive limit topology of $\mathcal{D}(G)$. As a realization of the Wigner invariance theorem, the involutory anti-automorphism of $\mathcal{D}'(G)$, which is contragredient to the natural anti-involution of $\mathcal{D}(G)$, plays a crucial role in FTMRI in modeling the refocusing phenomenon of phase conjugation. Note that the complex vector spaces $\mathcal{D}(G)$ and $\mathcal{D}'(G)$ form their own anti-spaces.

Let U denote a linear representation of G acting continuously on the Hilbert subspace $\mathcal{H}$ of $\mathcal{D}'(G)$ by Hilbert space automorphisms of $\mathcal{H}$. Then $\mathcal{H}$ forms a closed vector subspace of $\mathcal{D}'(G)$ under its weak dual topology in the sense that the linear injection

$$\mathcal{H} \hookrightarrow \mathcal{D}'(G)$$

forms a continuous mapping. The norm topology of $\mathcal{H}$ is finer than the topology induced on $\mathcal{H}$ by the weak dual topology of $\mathcal{D}'(G)$, the mapping

$$G \times \mathcal{H} \ni (g, \psi) \rightsquigarrow U(g)\psi \in \mathcal{H}$$

is simultaneously continuous, and the vector subspace $\mathcal{H}^{+\infty}$ of smooth vectors for the unitary linear representation U is everywhere dense in $\mathcal{H}$ with respect to the norm topology. Thus, for every element of the complex vector space $\mathcal{H}^{+\infty}$, the trajectory through $\psi \in \mathcal{H}^{+\infty}$, which is defined by

$$\boxed{\tilde{\psi} : g \rightsquigarrow U(g)\psi}$$

forms an infinitely differentiable mapping of G in $\mathcal{H}$ ([38]). The mapping

$$\psi \rightsquigarrow \tilde{\psi}$$

defines a continuous linear embedding of $\mathcal{H}$ into the topological vector space

$$\mathcal{C}^0(G; \mathcal{H}) \cong \mathcal{C}^0(G) \hat{\otimes} \mathcal{H}$$

of continuous functions of G with values in the complex Hilbert space $\mathcal{H}$ under the locally convex vector space topology of compact convergence. For the case of the Heisenberg group G with center C, and $\mathcal{H} = L^2(\mathbb{R})$, the standard complex Hilbert space over the bi-infinite time scale $\mathbb{R}$, the continuous linear embedding

$$\mathcal{H} \hookrightarrow \mathcal{C}^0(G/C; \mathcal{H})$$

performed at resonance frequency $\nu \neq 0$ with the linear Schrödinger representation U^ν of G in $\mathcal{H}$ with projective kernel C, is at the basis of the symplectically invariant symbol calculus approach to FTMRI. According to the semi-classical QED description of response data phase holograms, the canonical projection

$$G \longrightarrow G/C$$

allows by passing to the quotient mod C for implementation of the phase dispersion and its compensation by phase conjugation in the laboratory coordinate frame attached to the cross-section G/C to C in G.

The continuous injection $\mathcal{H} \hookrightarrow \mathcal{C}^0(G; \mathcal{H})$ identifies $\mathcal{H}$ with a closed vector subspace $\tilde{\mathcal{H}}^0$ of $\mathcal{C}^0(G; \mathcal{H})$, and $\mathcal{H}^{+\infty}$ with the closed vector subspace

$$\tilde{\mathcal{H}}^{+\infty} = \left\{ \tilde{\psi} \,\middle|\, U(g')\tilde{\psi}(g) = \tilde{\psi}(g'.g), \quad g', g \in G \right\}$$

of $\mathcal{C}^\infty(G; \mathcal{H})$. Notice that G acts infinitely differentiable on the vector space $\tilde{\mathcal{H}}^{+\infty}$ of smooth trajectories via left translations by the reflected group elements $g'^{-1} \in G$. Of course, each smooth trajectory $\tilde{\psi} \in \tilde{\mathcal{H}}^{+\infty}$ through $\psi \in \mathcal{H}^{+\infty}$ defines a distribution on G with values in $\mathcal{H}$ by the standard prescription

$$\tilde{\psi}(f) = \int_G \tilde{\psi}(g) f(g) \mathrm{d}g,$$

where $f \in \mathcal{D}(G)$ denotes an arbitrary test function on G. It is called the trajectory distribution associated to $\psi \in \mathcal{H}^{+\infty}$.

Let μ denote a compactly supported scalar measure on G and $\check{\mu}$ the image of μ under the involutory homeomorphism $g \rightsquigarrow g^{-1}$ of G onto itself. Then, as observed above,

$$U(g')\tilde{\psi} = \check{\varepsilon}_{g'} \star \tilde{\psi} \qquad (g' \in G)$$

for $\tilde{\psi} \in \tilde{\mathcal{H}}^{+\infty}$ and the Dirac measure $\varepsilon_{g'}$ located at $g' \in G$. The weak integral taken in $\mathcal{H}$

$$U(\mu) = \int_G U(g) \mathrm{d}\mu(g)$$

extends U by averaging over G. If μ is absolutely continuous with respect to $\mathrm{d}g$ and admits a density f, which is integrable over G with respect to

dg, and vanishes in a neighborhood of the point at infinity, then $U(f) = U(f.dg)$ is given by the prescription

$$U(f) = \int_G U(g)f(g)\mathrm{d}g.$$

The extension $U(\mu)$ satisfies the distributional convolution identity

$$U(\mu)\tilde{\psi} = \check{\mu} \star \tilde{\psi}$$

for all trajectory distributions $\tilde{\psi} \in \tilde{\mathcal{H}}^{+\infty}$. The elements of the complex vector space $\tilde{\mathcal{H}}^{-\infty}$ of distributions $K \in \mathcal{D}'(G; \mathcal{H})$ on G with values in $\mathcal{H}$ satisfying the convolution identity

$$U(\mu)K = \check{\mu} \star K$$

for all compactly supported scalar measure μ on G are called kernel distributions of $\mathcal{H}$ for the given unitary linear representation U of G in $\mathcal{H}$. Equivalently,

$$U(\mu)K(f) = K(\mu \star f)$$

holds for all test functions $f \in \mathcal{D}(G)$. If the complex vector space $\tilde{\mathcal{H}}^{-\infty}$ is endowed with the locally convex vector space topology induced by $\mathcal{D}'(G; \mathcal{H})$, the continuous inclusions

$$\tilde{\mathcal{H}}^{+\infty} \hookrightarrow \tilde{\mathcal{H}}^0 \hookrightarrow \tilde{\mathcal{H}}^{-\infty}$$

hold. In this sequence, the vector space of trajectory distributions $\tilde{\mathcal{H}}^{+\infty}$ is isomorphic to the space $\mathcal{H}^{+\infty}$ of smooth vectors for U, and the vector space $\tilde{\mathcal{H}}^0$ is isomorphic to the representation space $\mathcal{H}$ of the unitary linear representation U of G.

The important step is to realize that the unitary representation U of G in $\mathcal{H}$ gives rise to a continuous representation $\tilde{U}$ of G acting in the trajectory space

$$\tilde{\mathcal{H}}^0 \cong \mathcal{H}$$

according to the prescription

$$\boxed{\tilde{U}(g') : \psi \rightsquigarrow (U(g')\psi)^{\sim} \qquad (g' \in G).}$$

The polarized symbol map $\tilde{U}$ extends to an infinitely differentiable linear representation $g' \rightsquigarrow \tilde{U}(g')$ of G in $\tilde{\mathcal{H}}^{-\infty}$ by right translations. Thus,

$$\tilde{U}(g')K = K \star \check{\varepsilon}_{g'}$$

for $K \in \tilde{\mathcal{H}}^{-\infty}$. In particular, it follows

$$\tilde{U}(\mu)\tilde{\psi} = \tilde{\psi} \star \check{\mu}$$

by convolving the trajectories $\tilde{\psi} \in \tilde{\mathcal{H}}^{+\infty}$ from the right with the reflected compactly supported scalar measure μ on G. Moreover, the filter cascade identity

$$\boxed{\tilde{U}(S \star T)\tilde{\psi} = \tilde{U}(S) \circ \tilde{U}(T)\tilde{\psi}}$$

holds for all compactly supported complex distributions $S, T \in \mathcal{D}'(G)$.

The kernel distribution $K \in \tilde{\mathcal{H}}^{-\infty}$ of $\mathcal{H}$ for the representation U of G is a called generating kernel distribution ([38]) provided the image space $K(\mathcal{D}(G))$ is an everywhere dense vector subspace of $\mathcal{H}$. Let U be a unitary left regular representation of G on the Hilbert subspace $\mathcal{H}$ of $\mathcal{D}'(G)$ in the usual sense that the unitary group action $U(g)\psi$ of G on $\psi \in \mathcal{H}$ is performed via left translations $\varepsilon_g \star \psi$ by elements $g \in G$. Then,

$$U(\mu)\psi = \mu \star \psi$$

holds for each compactly supported scalar measure μ on G and each element $\psi \in \mathcal{H}$. It follows that the Schwartz kernel H of $\mathcal{H}$ in $\mathcal{D}'(G)$ is defined by convolution from the left with a uniquely defined positive definite distribution $H^\bullet$ on G. Therefore, the canonical reproducing kernel of the representation U of G is defined by $\star H^\bullet$. The existence of the mapping

$$\boxed{H : \psi \rightsquigarrow \psi \star H^\bullet}$$

is a consequence of the Schwartz kernel theorem ([37, 38]). The mapping $K \rightsquigarrow K^\bullet$ of $\tilde{\mathcal{H}}^{-\infty}$ onto

$$\mathcal{H}^{-\infty} = \{T \in \mathcal{D}'(G) \,|\, f \star T \in \mathcal{H}, \quad f \in \mathcal{D}(G)\}$$

forms a linear bijection which uniquely extends the inverse bijection

$$\tilde{\psi} \rightsquigarrow \psi$$

of $\tilde{\mathcal{H}}^0$ onto $\mathcal{H}$ and identifies the representation $\tilde{U}$ with the representation $g \rightsquigarrow U(g)$ of left translations $\varepsilon_g \star$. In particular,

$$(U(S)K)^\bullet = U(S)K^\bullet$$

holds for all kernel distributions $K \in \tilde{\mathcal{H}}^{-\infty}$ and all compactly supported complex distributions $S \in \mathcal{D}'(G)$. Moreover,

$$U(g)T = \varepsilon_g \star T,$$

and

$$U(S)T = S \star T$$

holds for all elements $g \in G$ and distributions $T \in \mathcal{H}^{-\infty}$ on G.

As a consequence of the preceding reasonings, it follows

- There exists a bijective correspondence between the unitary regular representations of G and the positive definite distributions on G.

The contragredient representation $\check{U}$ of U is defined by the inverse of the transposed operators according to the rule

$$\check{U} : g \rightsquigarrow {}^t U(g^{-1}).$$

Therefore, G acts continuously by $\check{U}$ on the topological antidual $\bar{\mathcal{H}}'$ of $\mathcal{H}$ under its weak dual topology. The coefficients of $\check{U}$ are the complex conjugates of the continuous coefficient functions of U and therefore encode important properties of the unitary linear representation U of G.

§3 The synchronization

The Heisenberg group G as a maximal nilpotent subgroup

$$G \hookrightarrow \mathbf{SL}(3, \mathbb{R})$$

forms a connected and simply connected three-dimensional nilpotent Lie group of level 2 generated by transvections ([36]). It is unimodular and represents a non-split central group extension

$$\mathbb{R} \lhd G \longrightarrow \mathbb{R} \oplus \mathbb{R}$$

where the invariant subgroup is isomorphic to the center C of G, and defines a smooth action on the cross-section G/C. The one-parameter action $\mathbb{R} \lhd G \longrightarrow \mathbb{R} \oplus \mathbb{R}$ defines a smooth dynamical system that has an infinitesimal generator operating on the complex vector space $\tilde{\mathcal{H}}^{-\infty}$ associated to $\mathcal{H} = L^2(\mathbb{R})$, the standard complex Hilbert space over the bi-infinite time scale $\mathbb{R}$. This infinitesimal generator is basic for the encryption and decryption by the FTMRI modality.

Notice also that the mapping

$$z \rightsquigarrow \begin{pmatrix} 1 & z \\ 0 & 1 \end{pmatrix}$$

defines a faithful representation of the additive group $\mathbb{R}$ in the plane $\mathbb{R} \oplus \mathbb{R}$ realized by transvections. Note also that the mapping

$$\dot{\exp}_G : \mathrm{Lie}(G) \longrightarrow G$$

of the Heisenberg Lie algebra $\mathrm{Lie}(G)$, and its inverse

$$\log_{\mathrm{Lie}(G)} : G \longrightarrow \mathrm{Lie}(G)$$

are diffeomorphisms: As a connected and simply connected nilpotent Lie group, G is of exponential type.

The central transvections of G associated with a homogeneous line L inside the plane $\mathbb{R} \oplus \mathbb{R}$ of fixed points admit a coordinatization with respect to the laboratory coordinate frame attached to the cross-section G/C. In terms of elementary matrices it reads

$$C = \left\{ \begin{pmatrix} 1 & 0 & z \\ 0 & 1 & 0 \\ 0 & 0 & 1 \end{pmatrix} \middle| z \in \mathbb{R} \right\}.$$

The faithful block representation of C by central transvections in the plane $\mathbb{R} \oplus \mathbb{R}$, defined by deletion of the middle row and column of the elementary matrices belonging to the line C,

$$\begin{pmatrix} 1 & 0 & z \\ 0 & 1 & 0 \\ 0 & 0 & 1 \end{pmatrix} \rightsquigarrow \begin{pmatrix} 1 & z \\ 0 & 1 \end{pmatrix}$$

allows for a linear ramp action of the disconnected affine Lie group

$$\mathbf{GA}(\mathbb{R}) = \left\{ \begin{pmatrix} \alpha & \beta \\ 0 & 1 \end{pmatrix} \middle| \alpha \neq 0, \beta \in \mathbb{R} \right\}$$

on the homogeneous plane spanned by the lines L and C. The affine bijections represented by the elements of $\mathbf{GA}(\mathbb{R})$ are transvections for $\alpha = 1$, dilations of ratio $\alpha \neq 0$ for $\beta = 0$ including the phase conjugating symmetries for $\alpha = -1$ and $\beta = 0$, and linear gradients for $\alpha \neq 0$ and $\beta \in \mathbb{R}$, which generate the superposition of a frequency modulating linear ramp of slope α. Therefore, $\mathbf{GA}(\mathbb{R})$ is called the frequency modulation group of the Lauterbur spatial encoding technique of quantum holography ([21]).

Because a dilation never commutes with a transvection unless one or both is the identity map, the commutator subgroup of $\mathbf{GA}(\mathbb{R})$ is given by the set of transvections in $\mathbb{R} \oplus \mathbb{R}$

$$[\mathbf{GA}(\mathbb{R}), \mathbf{GA}(\mathbb{R})] = \left\{ \begin{pmatrix} 1 & \beta \\ 0 & 1 \end{pmatrix} \middle| \beta \in \mathbb{R} \right\},$$

and the center of $\mathbf{GA}(\mathbb{R})$ is trivial. The linear ramp action of $\mathbf{GA}(\mathbb{R})$ on the homogeneous plane spanned by the lines L and C can be faithfully pulled back to the affine Larmor frequency scale on the dual of the transversal line C of the non-split central group extension $\mathbb{R} \triangleleft G \longrightarrow \mathbb{R} \oplus \mathbb{R}$. The affine Larmor frequency scale itself allows for slice selection by resonance with the frequency modulating linear ramp.

Let (x, y) denote the coordinates with respect to the laboratory coordinate frame of the homogeneous plane $\mathbb{R} \oplus \mathbb{R} \cong G/C$ of fixed points
associated with the central transvections. Then the unipotent matrices

$$G = \left\{ \begin{pmatrix} 1 & x & z \\ 0 & 1 & y \\ 0 & 0 & 1 \end{pmatrix} \mid x, y, z \in \mathbb{R} \right\}$$

present a coordinatization of the closed subgroup G of the unimodular
group $\mathbf{SL}(3, \mathbb{R})$. Indeed, G is a closed subgroup of the standard Borel
subgroup of the linear group $\mathbf{GL}(3, \mathbb{R})$ consisting of the upper triangular
matrices

$$\begin{pmatrix} x_{11} & x_{12} & x_{13} \\ 0 & x_{22} & x_{23} \\ 0 & 0 & x_{33} \end{pmatrix}$$

with real entries such that

$$\det \begin{pmatrix} x_{11} & x_{12} & x_{13} \\ 0 & x_{22} & x_{23} \\ 0 & 0 & x_{33} \end{pmatrix} = x_{11}.x_{22}.x_{33} \neq 0.$$

The Borel subgroup leaves a flag consisting of a homogeneous line L and a
plane containing L globally invariant. The condition

$$x_{11} = x_{22} = x_{33} = 1$$

ensures the aforementioned embedding

$$G \hookrightarrow \mathbf{SL}(3, \mathbb{R})$$

and achieves that the line L is kept pointwise fixed under the action of the
Borel subgroup. With respect to this coordinatization, the multiplication
law of G reads

$$\begin{pmatrix} 1 & x & z \\ 0 & 1 & y \\ 0 & 0 & 1 \end{pmatrix} \cdot \begin{pmatrix} 1 & x' & z' \\ 0 & 1 & y' \\ 0 & 0 & 1 \end{pmatrix} = \begin{pmatrix} 1 & x + x' & z + z' + x.y' \\ 0 & 1 & y + y' \\ 0 & 0 & 1 \end{pmatrix}$$

so that a left and right Haar measure of G is given by Lebesgue measure

$$dx \otimes dy \otimes dz$$

of the space $\mathbb{R}^3$. Then the center C of G carries the Haar measure dz.
    The preceding polarized presentation of G gives rise to the matrix

$$\begin{pmatrix} 0 & 1 \\ 0 & 0 \end{pmatrix}$$

of the central bilinear form. On the other hand, the Heisenberg Lie algebra of nilpotent matrices

$$\mathrm{Lie}(G) = \left\{ \begin{pmatrix} 0 & a & c \\ 0 & 0 & b \\ 0 & 0 & 0 \end{pmatrix} \,\bigg|\, a, b, c \in \mathbb{R} \right\}$$

as a maximal nilpotent Lie subalgebra

$$\mathrm{Lie}(G) \hookrightarrow \mathrm{Lie}\big(\mathbf{SL}(3, \mathbb{R})\big)$$

admits the one-dimensional center

$$\log_{\mathrm{Lie}(G)} C = \left\{ \begin{pmatrix} 0 & 0 & c \\ 0 & 0 & 0 \\ 0 & 0 & 0 \end{pmatrix} \,\bigg|\, c \in \mathbb{R} \right\}.$$

The Lie bracket of $\mathrm{Lie}(G)$ gives rise to the symplectic matrix

$$J = \begin{pmatrix} 0 & 1 \\ -1 & 0 \end{pmatrix}$$

of the central transvection. The symplectic matrix J of period 4 can be unitarized by inserting the factor $\frac{1}{2}$ so that the symplectic presentation of G reads

$$\begin{pmatrix} 1 & x & z \\ 0 & 1 & y \\ 0 & 0 & 1 \end{pmatrix} \cdot \begin{pmatrix} 1 & x' & z' \\ 0 & 1 & y' \\ 0 & 0 & 1 \end{pmatrix} = \begin{pmatrix} 1 & x + x' & z + z' + \frac{1}{2}\det\left(\begin{smallmatrix} x & y \\ x' & y' \end{smallmatrix}\right) \\ 0 & 1 & y + y' \\ 0 & 0 & 1 \end{pmatrix}.$$

Under the symplectic presentation of G, the restriction of the homeomorphism $\log_{\mathrm{Lie}(G)} : G \longrightarrow \mathrm{Lie}(G)$ to the center C of G identifies with the identity map of the central factor in $\mathrm{Lie}(G)$

$$\log_{\mathrm{Lie}(G)} | C = \mathrm{id}_{\mathbb{R}}.$$

In terms of the MRI technology, $\frac{1}{2}J$ is called the swapping matrix. It forms the infinitesimal generator at the angle $\frac{\pi}{2}$ of rotations in a symplectic affine plane and acts by doubling the number of variables. It turns out that the swapping matrix allows to detect the spectral information content of phase coherent wavelets on the bi-infinite time scale $\mathbb{R}$.

For the purposes of MRI it is appropriate to use the row vectors of upper triangular matrices for the coordinatization of the non-split central group extension $\mathbb{R} \lhd G \longrightarrow \mathbb{R} \oplus \mathbb{R}$ and the associated smooth dynamical system, because the row vectors of the matrices represent the coordinates with respect to a basis of the dual of the underlying three-dimensional real vector space (whose coordinates represent the column vectors of the transposed matrices). Apart from the faithful block representation

$$z \rightsquigarrow \begin{pmatrix} 1 & z \\ 0 & 1 \end{pmatrix}$$

of the line C by central transvections, the coordinatization also gives rise to the faithful block representation

$$y \rightsquigarrow \begin{pmatrix} 1 & y \\ 0 & 1 \end{pmatrix}$$

of G by coronal transvections. Swapping then yields the faithful block representation of G by sagittal transvections

$$x \rightsquigarrow \begin{pmatrix} 1 & x \\ 0 & 1 \end{pmatrix}.$$

The frequency modulation group $\mathbf{GA}(\mathbb{R})$ acts from the left on the overlapping blocks of G. By transition to the frequency domain, it is this intrinsic linear ramp action of the group $\mathbf{GA}(\mathbb{R})$ on the blocks inside G which mathematically describes the resonance phenomena on which the MRI modality is based. Therefore, the non-invasive diagnostic FTMRI modality can be studied by means of Fourier analysis on this additional Lie group, which act on the blocks of G. The synergy of this block organization inside G is realized by the MRI scanner organization.

The canonical commutation relations of quantum mechanics that underlie the Heisenberg uncertainty principle, pushed down by the diffeomorphism $\exp_G$ from $\mathrm{Lie}(G)$ to the Lie group G of central, sagittal, and coronal transvections, provide by passing to the quotient mod C the cross-section G/C to the center C in G with the structure of a symplectic affine plane. Thus, the symplectic cross-section to C in G carries the structure of a plane consistently endowed with both the structure of an affine plane and a symplectic plane. By passing to the quotient mod C and complexification, the vertical projection G/C of G carries the structure of a complex plane

$$W \cong (\mathbb{R} \oplus \mathbb{R}, J) \cong \mathbb{C},$$

and any transvection having G/C as its plane of fixed points actually forms a symplectic transvection. The tensor powers of the symplectic transvections inside the tensor algebra of W give rise to expansions of holomorphic functions on W in terms of transvectants.

The Keppler phase triangulation procedure (Albert Einstein, 1951: ein wahr haft genialer Einfall ...), which allowed Johann Keppler to establish his three fundamental laws of physical astronomy, is based on the ingenious idea of semi-classically interpreting (x, y) as differential phase-local frequency coordinates with respect to a coordinate frame rotating with frequency $\nu \neq 0$. According to the semi-classical QED description of response data phase holograms, the differential phases and local frequencies represent the spectral information content of phase coherent wavelets on the bi-infinite time scale $\mathbb{R}$. Notice that the problem of measuring phase angles is far from being as evident as the classical expositions like to make them believe.

The implementation of the parallel synchronization of the Keppler phase triangulation procedure in the rotating coordinate frame implies a transition from the cross-section G/C to the planar coadjoint orbit stratification of the unitary dual $\hat{G}$ of the Heisenberg group G. Due to the coadjoint action of G on the dual $\mathrm{Lie}(G)^\star$ of the real vector space $\mathrm{Lie}(G)$ which is given by ([33])

$$\mathrm{CoAd}_G \begin{pmatrix} 1 & x & z \\ 0 & 1 & y \\ 0 & 0 & 1 \end{pmatrix} = \begin{pmatrix} 1 & 0 & -y \\ 0 & 1 & x \\ 0 & 0 & 1 \end{pmatrix},$$

the basic identification reads

$$\boxed{\mathrm{Lie}(G)^\star/\mathrm{CoAd}_G(G) \cong \hat{G}.}$$

The elements of the symplectic quotient $\mathrm{Lie}(G)^\star/\mathrm{CoAd}_G(G)$ are identified as follows:

(i) single points in the singular plane $\nu = 0$, and

(ii) symplectic affine planes $\{\mathcal{O}_\nu | \nu \neq 0\}$,

respectively. Hence, the spectrum of G provides a continuous open mapping onto the central factor $\mathbb{R} \lhd G \longrightarrow \mathbb{R} \oplus \mathbb{R}$, taking a copy of $\mathbb{R} \oplus \mathbb{R}$ onto 0, and a single point onto each non-zero resonance frequency $\nu \in \mathbb{R}$. In terms of the $C^\star$-algebra of G, considered as a bundle of $C^\star$-algebras, the fibers are labeled by the vertical coordinate, the frequency scale $\nu \in \mathbb{R}$. The fibers are

(i) the singular $C^\star$-algebra $\mathcal{C}_0(\mathbb{R} \oplus \mathbb{R})$ of continuous complex valued functions on the plane $\mathbb{R} \oplus \mathbb{R}$ vanishing at infinity, when $\nu = 0$, and

(ii) the standard $C^\star$-algebra $\mathcal{LC}(\mathcal{H})$ of compact operators on the complex Hilbert space $\mathcal{H} = L^2(\mathbb{R})$, when $\nu \neq 0$,

with respect to their natural anti-involutions.

In FTMRI, the basic observation is that the planar coadjoint orbits $\mathcal{O}_\nu, \nu \neq 0$, of the stratification of the dual $\hat{G}$ of G form the central tomographic slices (ii) having attached to them a coordinate frame rotating at the frequency $\nu \neq 0$ of excitation.

- The FTMRI modality allows for visualization of the planar coadjoint orbits $\mathcal{O}_\nu \in \mathrm{Lie}(G)^\star/\mathrm{CoAd}_G(G)$ admitting the resonance frequency $\nu \neq 0$ as their vertical coordinate.

The central tomographic slices, with their rotating coordinate frames attached, can be selected by resonance with the unitary central character

$$
\chi_\nu : C \ni \begin{pmatrix} 1 & 0 & z \\ 0 & 1 & 0 \\ 0 & 0 & 1 \end{pmatrix} \rightsquigarrow \begin{pmatrix} 1 & z \\ 0 & 1 \end{pmatrix} \rightsquigarrow e^{2\pi\nu i z} \in \mathbf{S}_1
$$

of frequency $\nu \neq 0$ associated to the central transvections of C. Specifically, the tomographic slices of scout-view scans are selected as single, whole planes by standard clinical MRI protocols in a plane orthogonal to the desired image plane by resonance with χ_ν to allow more accurate tomographic slice positioning. The basic resonance processing of single, whole planes is a fundamental consequence of the Stone-von Neumann theorem of quantum mechanics ([33]) in conjunction with the linear ramp action of the frequency modulation group $\mathbf{GA}(\mathbb{R})$ on the homogeneous plane spanned by the lines L and C. The linear ramp action of the Lauterbur spatial encoding technique of quantum holography generates the affine Larmor frequency scale on the dual of C. Due to the gyromagnetic ratio of protons

$$
\frac{\gamma}{2\pi} = 42.5743 \,\mathrm{MHz/T},
$$

the angular resonance frequency at a standard external magnetic flux density B_0 of 1.5 T is given by

$$
2\pi\nu = 63.86 \,\mathrm{MHz}
$$

For comparison: The earth's average magnetic field has a flux density of about 0.5 Gauß (1 T $= 10^4$ Gauß), which induces protons in the human body to process at an angular frequency of about 2.1 kHz. The external magnetic flux densities in clinical MRI use vary from approximately 0.5 T to 4.0 T and are strong when compared to the earth's magnetic flux density. Therefore, it is important to remember that metal objects taken into MRI scan rooms can become lethal projectiles.

The coadjoint orbit picture of the unitary dual of the affine solvable Lie group

$$
\mathbf{GA}(\mathbb{R}) \hookrightarrow \mathbf{GL}(2, \mathbb{R})
$$

of affine bijections associated to the central, coronal, and sagittal transvections reads

$$
\mathbb{R} \cup \mathcal{O}_+ \cup \mathcal{O}_-
$$

where $\mathcal{O}_\pm$ denotes the open upper/lower complex half-plane. Its coadjoint orbit symmetries

$$(\mathcal{O}_+, \mathcal{O}_-)$$

indicate how to compensate the phase dispersion induced by the computer-controlled wavelet distortion of central tomographic slice selection by the phase conjugation symmetry

$$\boxed{\begin{pmatrix} -1 & 0 \\ 0 & 1 \end{pmatrix} \in \mathbf{GA}(\mathbb{R})}$$

For

$$\psi \in L^2(\mathbb{R}_+^\times)$$

and

$$w = \beta + i\alpha \in \mathcal{O}_\pm,$$

define the function

$$g(w) = \int_{\mathbb{R}_+^\times} e^{2\pi i w t} t^n \psi(t)\, dt.$$

By an application of the Fourier co-transform at spin level

$$n \in \frac{1}{2}\mathbb{N}^\times.$$

The integral converges, as follows from the Cauchy-Schwarz-Bunjakowski inequality and the Laplace transform:

$$|g(w)| \le \left(\int_{\mathbb{R}_+^\times} t^{2n} e^{-2\alpha t} dt \right) \cdot \left(\int_{\mathbb{R}_+^\times} |\psi(t)|^2 dt \right) = \Gamma(2n+1)^{\frac{1}{2}} (2\alpha)^{-\left(n+\frac{1}{2}\right)} \|\psi\|_2$$

The function g is holomorphic/anti-holomorphic on the open upper/lower complex half-plane $\mathcal{O}_\pm$. An application of the Plancherel theorem yields the identity

$$\int_{\mathbb{R}} |g(\beta + i\alpha)|^2 d\beta = 2\pi \int_{\mathbb{R}_+^\times} e^{-2\alpha t} t^{2n} |\psi(t)|^2 dt,$$

so that g has finite norm $\|g\|_2$ on the Poincaré half-plane $\mathcal{O}_+$, carrying the standard measure

$$\frac{d\alpha.d\beta}{\alpha^2},$$

which is invariant under the canonical linear fractional action of the unimodular group $\mathbf{SL}_2(\mathbb{R})$ defined by

$$\begin{pmatrix} a & b \\ c & d \end{pmatrix}.w = \frac{aw+b}{cw+d}, \qquad \det \begin{pmatrix} a & b \\ c & d \end{pmatrix} = 1.$$

Then it follows the isomorphy

$$\mathcal{O}_+ \cong \mathbf{SL}_2(\mathbb{R})/\mathbf{SO}(2, \mathbb{R}),$$

and

$$\|g\|_2^2 = \int \int_{\mathcal{O}_\pm} |g(w)|^2 \alpha^{(2n+1)} \frac{\mathrm{d}\alpha.\mathrm{d}\beta}{\alpha^2},$$

or explicitly,

$$\|g\|_2^2 = 2\pi \int_{\mathbf{R}_+^\times} \alpha^{(2n-1)} \int_{\mathbf{R}_+^\times} e^{-2\alpha t} t^{2n} |\psi(t)|^2 \mathrm{d}\alpha \mathrm{d}t = 2\pi 2^{-2n} \Gamma(2n) \|\psi\|_2^2.$$

The affine wavelet transform $\mathcal{W}_n$ of spin level $n \in \frac{1}{2}\mathbb{N}^\times$ associated to the matrix

$$\begin{pmatrix} \alpha & \beta \\ 0 & 1 \end{pmatrix} \in \mathbf{GA}(\mathbb{R})$$

in the orbit of the natural action of the linear group $\mathbf{GL}(2, \mathbb{R})$ by the holomorphic/anti-holomorphic discrete series representations of $\mathbf{SL}_2(\mathbb{R})$ on the open upper/lower complex half-plane $\mathcal{O}_\pm$, with stabilizer at $i \in \mathcal{O}_+$ and $-i \in \mathcal{O}_-$, respectively, reads

$$\boxed{\mathcal{W}_n : \psi \rightsquigarrow \left(w \rightsquigarrow \beta^{2n+1} \int_{\mathbf{R}_+^\times} e^{(2\pi i \alpha \beta - \beta^2)t} t^n \psi(t)\mathrm{d}t \right).}$$

The attenuation factor of the affine wavelet transform $\mathcal{W}_n$ indicates the scattering effect of phase dispersion in response to an external perturbation. For echo wavelet detection, the phase dispersion must be corrected by phase conjugation because the external distortion destroys phase coherence and degrades the quantum holograms. The distortion correction is performed by the inverse wavelet transform produced by an application of the symmetry $\begin{pmatrix} -1 & 0 \\ 0 & 1 \end{pmatrix}$ in the frequency modulation group $\mathbf{GA}(\mathbb{R})$. As a realization of the Wigner invariance theorem, the symmetry inverts the wavelet transform by recalling in 'inverse order shape'. The correction processing illuminates the importance of the affine wavelet transform $\mathcal{W}_n$ for the Lauterbur spatial encoding technique of quantum holography.

Gradient echo techniques are an essential part of the modern MRI examination and are offered as standard software by every major vendor. They have proved useful in a wide variety of applications, including spinal imagery, cardiac imaging, magnetic resonance angiography (MRA), three-dimensional imaging, echo planar imaging (EPI), and dynamic contrast techniques. An important feature of both gradient recalled echo (GRE) imaging and spoiled GRE imaging (SPGR) is that the gradient slope reversal refocuses only those spins that have been dephased by the linear ramp action itself. Specifically, phase shifts resulting from magnetic field inhomogeneities, static tissue susceptibility gradients, or chemical shifts are not compensated by the GRE and SPGR techniques. Therefore, it is important to notice that the principle of rephasing by computer-controlled gradient slope reversal of GRE and SPGR imaging does not correspond completely to the refocusing technique of spin echo (SE) imaging. In SE protocols, the local frequency fan narrowing performed by $\tilde{\psi} \in \tilde{\mathcal{H}}^{+\infty}$ is implemented by the symplectic form of the planar coadjoint orbits $\mathcal{O}_\nu, \nu \neq 0$, of the stratification of $\hat{G}$. The natural action of $\mathbf{SU}(2,\mathbb{C}) \hookrightarrow \mathbf{SO}(4,\mathbb{R})$ on the compact unit sphere $\mathbf{S}_2$ of $\mathbb{R}^3$ with stabilizer at the north-pole

$$\begin{pmatrix} i & 0 \\ 0 & -i \end{pmatrix}$$

provides the Hopf projector $\mathbf{S}_3 \longrightarrow \mathbf{S}_2$. It provides by structure transport from the tangent plane of $\mathbf{S}_2$ at the north-pole the complexified coadjoint orbit $\mathcal{O}_\nu \in \mathrm{Lie}(G)^\star/\mathrm{CoAd}_G(G)$ with the parameterization

$$\begin{pmatrix} 0 & x + iy \\ x - iy & 0 \end{pmatrix}$$

of two contragredient rotating coordinate frames of frequency $\nu \neq 0$. The complex parameterization is consistent with the structure of

$$\mathcal{O}_\nu \in \mathrm{Lie}(G)^\star/\mathrm{CoAd}_G(G)$$

as a symplectic affine plane. Parallel synchronization of the Keppler phase triangulation procedure by computer-controlled radiofrequency pulses performing phase conjugation with respect to the rotating coordinate frame attached to the planar coadjoint orbit $\mathcal{O}_\nu$ of the stratification of $\hat{G}$ to compensate the dephasing effect of transverse magnetization decay within the planar coadjoint orbit $\mathcal{O}_\nu \in \mathrm{Lie}(G)^\star/\mathrm{CoAd}_G(G)$; and the switching of a linear gradient pulse that performs phase conjugation with respect to the affine Larmor frequency scale along the dual of the transversal line C to correct the phase dispersion induced by the computer-controlled wavelet perturbation of central tomographic slice selection, generates the computer-controlled timing diagram of a specific exciting pulse sequence at resonance frequency $\nu \neq 0$. The parallel synchronization parameters, which are central to the understanding of tissue contrast, are the waiting time intervals: the time to echo T_E and the time to repeat T_R. In this way, the computer-controlled score of the standard SE choreography is generated by implementation of the waiting times. The phase of an SE modulation function is a linear ramp. The strength and duration of the pulse determines the flip angle θ of the Bloch vector. Because the implementation of choreographic scores that perform parallel synchronizations of Keppler configurations is at the heart of MRI protocols, computer-controlled synchronized timing is everything in FTMRI.

Soon after the initial discovery of the phenomenon of nuclear magnetic resonance (NMR), Erwin Louis Hahn, a former radar technician, performed an experiment that, in conceptual originality and elegance, transcended all that had preceded it—the SE experiment. The SE imaging technique has long been the workhorse of clinical MRI because of its generally high signal intensity and contrast compared with other imaging techniques. In joint imaging, for instance, the SE technique is most commonly used so that even today, the SE protocol remains the gold standard for diagnostic MRI due to its inherent contrast properties, its robust signal strength, its flexibility for contrast adaption, and its relative insensitivity to artifacts that plague other techniques such as GRE imaging.

The SE imaging technique is characterized by a $\frac{\pi}{2}$ pulse that flips the Bloch vector into the selected planar coadjoint orbit of G followed by a contragredient π pulse performing phase conjugation, and generates a refocused echo wavelet by local frequency fan narrowing at the time T_E. This computer-controlled synchronized pulsing is repeated at time intervals of length T_R to record the response data phase holograms by switching the linear frequency gradient pulses and by stepwise increasing the slope $\alpha > 0$ of the frequency modulating linear ramps $\left(\begin{smallmatrix} \alpha & \beta \\ 0 & 1 \end{smallmatrix}\right) \in \mathbf{GA}(\mathbb{R})$ in the score, as displayed in Figure 2 supra.

- The echo time T_E is the time the MRI scanner waits after the applied $\frac{\pi}{2}$ pulse to receive the echo signal.
- The repetition time T_R represents the time that elapses from the $\frac{\pi}{2}$ pulse of one repetition to the same pulse in the next repetition.

Advanced methods of sampling the acquisition data, such as the fast SE imaging technique (FSE) or turbo SE acquisition, significantly decrease the scan times by acquiring trains of multiple spin echoes instead of one in each repetition interval of length T_R. The number of phase encoding steps that are simultaneously acquired is termed the echo train length (ETL). Typically, an ETL ≤ 16 can be used in clinical neuroimaging ([20]). STALO-controlled transmission gates for sequences of coherent pulses and substantial computer memory, however, are required to handle a long ETL. For high-resolution FSE studies even an ETL scan parameter of 64 may be chosen. After each echo, the phase encoding is rewound to preserve phase coherence and allow formation of the subsequent spin echoes. In the sense of parallel data acquisition, computer-controlled synchronized timing is everything in FTMRI. Figure 3 shows a high-resolution FSE study of the cranium.

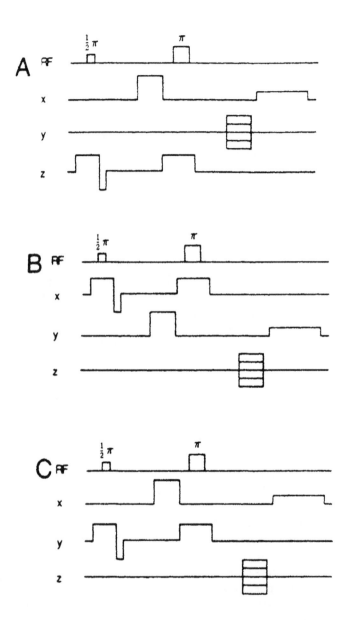

Figure 2. Timing diagrams of synchronized pulse sequences for central (A), sagittal (B), and coronal (C) tomographic slices.

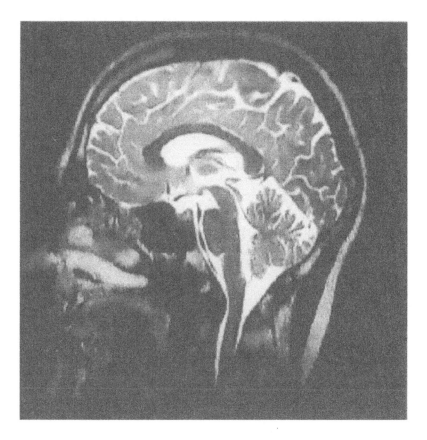

Figure 3. High-resolution midsagittal tomographic slice FSE study of the cranium.

Recall that an irreducible unitary linear representation U of G with projective kernel C is square integrable mod C if and only if one of the non-trivial coefficient functions of U, hence all the coefficient functions of U, belong to the complex Hilbert space $L^2(G/C)$. By Godement's theorem ([39]), an equivalent condition is the isomorphy of U to a subrepresentation of the regular representation acting on $L^2(G/C)$. Equivalently, U is a discrete summand of the linear representation

$$U^\chi = \operatorname{ind}_C^G(\chi)$$

of G, which is unitarily induced by the unitary central character

$$\chi = U|C$$

of G, hence the term discrete series representation of G ([32]). Two isomorphic irreducible unitary linear representations of G are either simultaneously square integrable mod C or not ([33]). The strong geometric condition of flatness of the symplectic affine planar coadjoint orbits
$\mathcal{O}_\nu \in \mathrm{Lie}(G)^\star / \mathrm{CoAd}_G(G)$, admitting the vertical coordinate $\nu \neq 0$ implies
the square integrability mod C of the associated isomorphy classes in $\hat{G}$
of irreducible unitary linear representations U^ν of G ([6]). It follows the
fundamental identity of central tomographic slice selection at resonance
frequency ν:

$$\boxed{U^\nu | C = \chi_\nu \qquad (\nu \neq 0)}$$

and each irreducible unitary linear representation V^ν of G satisfying the
resonance condition of the unitary central character

$$V^\nu | C = \chi_\nu \qquad (\nu \neq 0)$$

is unitarily isomorphic to U^ν, hence attached to the same planar coadjoint
orbit $\mathcal{O}_\nu$ of the stratification of $\hat{G}$, and therefore a discrete summand of
the unitary linear representation $U^{\chi_\nu} = \mathrm{ind}_C^G(\chi_\nu)$ of G. According to
the Stone-von Neumann theorem of quantum mechanics, the isomorphism
of U^ν onto V^ν is given by the unitary action σ of the metaplectic group
$\mathbf{Mp}(2, \mathbb{R})$ on $\mathcal{H} = L^2(\mathbb{R})$ ([19]). Specifically,

$$\boxed{\sigma(J) = \bar{\mathcal{F}}_{\mathbf{R}}}$$

is the Fourier cotransform of the bi-infinite time scale $\mathbb{R}$.

- For $\nu \neq 0$, the unitary action σ of the metaplectic group $\mathbf{Mp}(2, \mathbb{R})$
  links the symplectic structure J of the planar coadjoint orbit $\mathcal{O}_\nu \in$
  $\mathrm{Lie}(G)^\star / \mathrm{CoAd}_G(G)$ to the spectral information content of phase
  coherent wavelets on the bi-infinite time scale $\mathbb{R}$.

The semidirect product of $\mathbf{Mp}(2, \mathbb{R})$ and G is the Jacobi group $G^J =$
$\mathbf{Mp}(2, \mathbb{R}) \propto G$. Its Lie algebra $\mathrm{Lie}(G^J)$ reflects the canonical commutation
relations of quantum mechanics

$$[X, Y] = X.Y - Y.X = 1$$

in the real Weyl algebra of non-commutative polynomials in two variables
$\{X, Y\}$ of degree ≤ 2.

The Schwartz kernel of $\mathcal{H}$ in $\mathcal{H}$ is the canonical isomorphism

$$j : \bar{\mathcal{H}}' \longrightarrow \mathcal{H}$$

admitting the inverse kernel of $\bar{\mathcal{H}}'$ in $\bar{\mathcal{H}}'$, which is given by

$$j^{-1} : \mathcal{H} \longrightarrow \bar{\mathcal{H}}'.$$

The identity

$$\check{U}^\nu | C = \bar{\chi}_\nu \qquad (\nu \neq 0)$$

for the unitary central character of the contragredient representation $\check{U}^\nu$ acting on $\bar{\mathcal{H}}'$ shows that $\check{U}^\nu$ is also square integrable mod C for $\nu \neq 0$, that it is isomorphic to $U^{-\nu}$, that its isomorphy class as an element of $\hat{G}$ is associated to the conjugate planar coadjoint orbit $\mathcal{O}_{-\nu} \in \text{Lie}(G)^*/\text{CoAd}_G(G)$, and that it forms a discrete summand of the unitary linear representation $U^{\chi_{-\nu}} = \text{ind}_C^G(\chi_{-\nu})$ of G. The covariant representation

$$\check{U}^\nu \hat{\otimes} U^\nu$$

of G acts unitarily on the complex Hilbert space of Hilbert-Schmidt integral operators

$$\mathcal{L}_2(\mathcal{H}) \hookrightarrow \mathcal{LC}(\mathcal{H})$$

of $\mathcal{H}$, which is endowed with its covariant tracial scalar product. It weakly contains the one-dimensional identity representation of G having as its associated coadjoint orbit the single point set

$$\{0\} \in \text{Lie}(G)^*/\text{CoAd}_G(G),$$

which allows to map $\mathcal{O}_\nu$ onto $\mathcal{O}_{-\nu}$ by central reflection:

$$\mathcal{O}_\nu \longleftrightarrow \mathcal{O}_{-\nu} \qquad (\nu \neq 0).$$

The action of the linear Schrödinger representation U^ν of G at the matrix $\begin{pmatrix} 1 & x & z \\ 0 & 1 & y \\ 0 & 0 & 1 \end{pmatrix} \in G$ on the phase coherent wavelet ψ in the complex Schwartz space $\mathcal{S}(\mathbb{R}) \hookrightarrow L^2(\mathbb{R})$ on the bi-infinite time scale $\mathbb{R}$ reads

$$\boxed{U^\nu\left(\begin{pmatrix} 1 & x & z \\ 0 & 1 & y \\ 0 & 0 & 1 \end{pmatrix}\right) : \psi \rightsquigarrow \left(t \rightsquigarrow e^{2\pi i \nu(z+yt)}\psi(t+x)\right).}$$

The action of the linear Schrödinger representation U^ν implements the iteration of the smooth one-parameter action of $\mathbb{R}$ on $\mathcal{S}(\mathbb{R})$ by translation of the bi-infinite time scale $\mathbb{R}$, and the dual action of $\mathbb{R}$ on $\mathcal{S}(\mathbb{R})$, which correspond to the faithful block representations of G by sagittal and coronal transvections, respectively. The infinitesimal generator of the associated smooth dynamical system $\mathbb{R} \lhd G \longrightarrow \mathbb{R} \oplus \mathbb{R}$, which is operating on the complex vector space $\tilde{\mathcal{H}}^{-\infty}$ of kernel distributions of $\mathcal{H} = L^2(\mathbb{R})$, is given by the formula

$$\omega_\nu = \tilde{U}^\nu(-\dot{\varepsilon}_0).$$

Using the Newtonian notation,

$$\dot{\varepsilon}_0 = \frac{d\varepsilon_0}{dt}$$

denotes the temporal derivative of the Dirac measure ε_0 at the origin of the bi-infinite time scale $\mathbb{R}$ associated with the smooth dynamical system $\mathbb{R} \lhd G \longrightarrow \mathbb{R} \oplus \mathbb{R}$. According to the semi-classical QED description of response data phase holograms excited by $\tilde{\psi}$, parameter x denotes the differential phase coordinate, and the parameter y denotes the local frequency coordinate with respect to a coordinate frame rotating with frequency $\nu \neq 0$. Finally, z, the vertical coordinate of the central tomographic slice $\mathcal{O}_\nu \in \text{Lie}(G)^\star/\text{CoAd}_G(G)$, is dual to the overall resonance frequency $\nu \neq 0$. All three coordinates are introduced by computer-controlled wavelet perturbations that require distortion correction by phase conjugation within the score.

To complement the semi-classical QED approach to quantum holography, it should be noticed that the linear Schrödinger representation U^ν of G admits an alternate realization by the Bargmann-Fock model of lowering and raising operators of quantum mechanics operating on holomorphic functions on W ([33]). The standard L^2 Sobolev inequality establishes that the complex vector space of smooth vectors for U^ν acting on $\mathcal{H} = L^2(\mathbb{R})$ is formed by the Schwartz space

$$\mathcal{H}^{+\infty} = \mathcal{S}(\mathbb{R})$$

of complex-valued smooth functions on the bi-infinite time scale $\mathbb{R}$, rapidly decreasing at infinity such that all their derivatives are also rapidly decreasing at infinity. It is well known that $\mathcal{S}(\mathbb{R})$ is a complex nuclear locally convex topological vector space under its canonical Fréchet space topology, and that $\mathcal{H}^{+\infty}$ is a normal vector space of complex distributions on $\mathbb{R}$ in the sense that the canonical injection

$$\mathcal{D}(\mathbb{R}) \hookrightarrow \mathcal{H}^{+\infty}$$

is continuous and admits an everywhere dense image ([37]). By extension, it provides $\mathcal{H}^{+\infty}$ with its natural anti-involution. The complex vector space $\tilde{\mathcal{H}}^{-\infty}$, which is isomorphic to $\mathcal{H}^{-\infty}$ under the locally convex vector space topology induced by $\mathcal{D}'(G; \mathcal{H})$, forms the Sobolev space consisting of all tempered distributions $T \in \mathcal{S}'(\mathbb{R} \oplus \mathbb{R})$ such that their symplectic convolution products satisfy

$$f \star_\nu T \in L^2(\mathbb{R} \oplus \mathbb{R})$$

for all functions $f \in \mathcal{S}(\mathbb{R} \oplus \mathbb{R})$. It contains the irreducible $\mathcal{S}(\mathbb{R} \oplus \mathbb{R})$ - module $\mathcal{S}(\mathbb{R} \oplus \mathbb{R})$ as well as the irreducible $L^2(\mathbb{R} \oplus \mathbb{R})$ - module $L^2(\mathbb{R} \oplus \mathbb{R})$ in the sense of the symplectic convolution product $\star_\nu$ of $\tilde{\mathcal{H}}^{-\infty}$. Thus, the continuous inclusions

$$\mathcal{S}(\mathbb{R} \oplus \mathbb{R}) \hookrightarrow L^2(\mathbb{R} \oplus \mathbb{R}) \hookrightarrow \tilde{\mathcal{H}}^{-\infty} \hookrightarrow \mathcal{S}'(\mathbb{R} \oplus \mathbb{R})$$

hold. Due to the swapping symplectic matrix $\frac{1}{2}J$ associated to the Lie bracket of $\mathrm{Lie}(G)$, the symplectic convolution product induced by the linear Schrödinger representation U^ν on the planar coadjoint orbit $\mathcal{O}_\nu \in \mathrm{Lie}(G)^\star/\mathrm{CoAd}_G(G)$ takes the explicit form

$$f \star_\nu g(x, y; \nu) = \frac{1}{2} \int_{\mathbb{R} \oplus \mathbb{R}} f(x', y'; \nu).g(x - x', y - y'; \nu) e^{\pi i \nu \det \left(\begin{smallmatrix} x & y \\ x' & y' \end{smallmatrix}\right)} \mathrm{d}x' \mathrm{d}y'.$$

The normalization factor in front of the integral comes from the formal degree of the linear Schrödinger representation U^ν of G. Taking into account the iteration of smooth actions implemented by U^ν, the symplectic convolution product $f \star_\nu g$ reveals to be jointly continuous for

$$f, g \in \mathcal{S}(\mathbb{R} \oplus \mathbb{R}) \cong \mathcal{S}(\mathbb{R}) \hat{\otimes} \mathcal{S}(\mathbb{R}) \cong \mathcal{S}(\mathbb{R}; \mathcal{S}(\mathbb{R}))$$

and $\nu \neq 0$. Hence, the irreducible $\mathcal{S}(\mathbb{R} \oplus \mathbb{R})$ - module $\mathcal{S}(\mathbb{R} \oplus \mathbb{R})$ is a complex Fréchet algebra under the symplectic convolution. The symplectic Fourier transform is defined by the involutory isomorphism

$$\mathcal{S}(\mathbb{R} \oplus \mathbb{R}) \ni f \rightsquigarrow \hat{f} \in \mathcal{S}(\mathbb{R} \oplus \mathbb{R}),$$

where

$$\boxed{\hat{f} = f \star_\nu (1_\nu \otimes 1_\nu).}$$

The symplectic convolution with the constant distribution $1_\nu \otimes 1_\nu$ on the symplectic affine plane $\mathcal{O}_\nu \in \mathrm{Lie}(G)^\star/\mathrm{CoAd}_G(G)$ explicitly reads

$$\hat{f}(x, y; \nu) = \frac{1}{2} \int_{\mathbb{R} \oplus \mathbb{R}} f(x', y'; \nu) e^{\pi i \nu . \det \left(\begin{smallmatrix} x & y \\ x' & y' \end{smallmatrix}\right)} \mathrm{d}x' \mathrm{d}y' \qquad ((x, y) \in \mathbb{R} \oplus \mathbb{R}).$$

It follows from the Stone-von Neumann theorem of quantum mechanics ([33]) or from a smooth version of the Takesaki-Takai duality theorem ([12])

applied to the dynamical system $\mathbb{R} \lhd G \longrightarrow \mathbb{R} \oplus \mathbb{R}$, that phase averaging of the proton-weighted spin isochromat density

$$f \in L^2(\mathbb{R} \oplus \mathbb{R}) \hookrightarrow S'(\mathbb{R} \oplus \mathbb{R})$$

by the mod C square integrable linear Schrödinger representation U^ν of G, leads to the Hilbert-Schmidt integral operator

$$U^\nu(f) \in \mathcal{L}_2(\mathcal{H}),$$

realized by an integral operator of kernel in $L^2(\mathbb{R} \oplus \mathbb{R})$. The integrated form $U^\nu(f)$ of U^ν extends the evaluation at the Dirac measure $\varepsilon_{\begin{pmatrix} 1 & x & 0 \\ 0 & 1 & y \\ 0 & 0 & 1 \end{pmatrix}}$

by an average over the cross-section G/C to the center C with respect to the measure

$$f(x, y; \nu) \, . \, \mathrm{d}x \otimes \mathrm{d}y.$$

- The square integrability mod C of the linear Schrödinger representation U^ν of G for $\nu \neq 0$ allows to embed by means of the symbol calculus the von Neumann approach to quantum mechanics. This is based on the category of Hilbert spaces into the Dirac approach, which is based on complex locally convex topological vector spaces of tempered distributions.

Let $< . , . >$ denote the bracket which defines the topological vector space antiduality

$$(S(\mathbb{R}), S'(\mathbb{R})).$$

It implements the involutory anti-automorphism of the complex vector space $S'(\mathbb{R})$ of tempered distributions on the bi-infinite time scale $\mathbb{R}$, which is contragredient to the natural anti-involution of $S(\mathbb{R})$. Moreover, it provides $S'(\mathbb{R})$ with its weak dual topology under which $S'(\mathbb{R})$ forms a complex nuclear locally convex topological vector space ([37]).

- The complex vector spaces $S(\mathbb{R})$ and $S'(\mathbb{R})$ form their own antispaces, and the sesquilinear form $< . , . >$ is consistent with the internal scalar product of the standard complex Hilbert space $L^2(\mathbb{R}) \hookrightarrow S'(\mathbb{R})$.

The holographic transform attached to the symplectic affine plane $\mathcal{O}_\nu \in \mathrm{Lie}(G)^*/\mathrm{CoAd}_G(G)$ at resonance frequency $\nu \neq 0$ is defined by

$$\mathcal{H}_\nu : S(\mathbb{R}) \ni \psi \rightsquigarrow \left((x, y) \rightsquigarrow < U^\nu \left(\begin{pmatrix} 1 & x & 0 \\ 0 & 1 & y \\ 0 & 0 & 1 \end{pmatrix} \right) \psi, 1_\nu > . \, 1_\nu \otimes 1_\nu \right).$$

According to the semi-classical QED description of response data phase holograms, the tempered distribution $1_\nu \in \mathcal{S}'(\mathbb{R})$ represents a rotation at the frequency $\nu \neq 0$ of the unitary central character χ_ν associated to the planar coadjoint orbit $\mathcal{O}_\nu$, and $\mathcal{H}_\nu(\psi)$ performs the temporal cross-correlation of $1_\nu \in \mathcal{S}'(\mathbb{R})$ with the echo wavelet $\psi \in \mathcal{S}(\mathbb{R})$ elicited at resonance frequency $\nu \neq 0$. The echo wavelet $\psi \in \mathcal{S}(\mathbb{R})$ represents the non-linear spin isochromat's response to the exciting pulse sequence at resonance frequency $\nu \neq 0$. An application of the Paley-Wiener-Schwartz theorem allows to smooth out on the Fourier transform side the Poisson bracket by an expansion in terms of transvectants of the symplectic affine plane $\mathcal{O}_\nu \in \mathrm{Lie}(G)^\star/\mathrm{CoAd}_G(G)$.

- For $\nu \neq 0$, the holographic transform $\mathcal{H}_\nu$ attached to the symplectic affine plane $\mathcal{O}_\nu \in \mathrm{Lie}(G)^\star/\mathrm{CoAd}_G(G)$ forms a quantum analog of the Liouville density in phase space. The FTMRI modality is a quantum holographic process.

Because the continuous linear mapping $\mathcal{H}_{-\nu}$ defined by the coefficient of the contragredient representation

$$\check{U}^\nu \cong U^{-\nu}$$

of G corresponding to $1_\nu \in \mathcal{S}'(\mathbb{R})$ commutes with the left regular action of G/C on $L^2(G/C)$, an application of Schur's theorem ([33]) to the von Neumann algebra formed by the weakly closed commutant of $U(G)$ in the group of automorphisms of $\mathcal{H}$ yields

$$\mathcal{H}_{-\nu} \circ j \circ \mathcal{H}_\nu = \mathrm{id}_{\mathcal{S}'(\mathbb{R}\oplus\mathbb{R})},$$

and similarly,

$$\mathcal{H}_\nu \circ j^{-1} \circ \mathcal{H}_{-\nu} = \mathrm{id}_{\mathcal{S}'(\mathbb{R}\oplus\mathbb{R})}.$$

Thus, the identity

$$\bar{\mathcal{H}}'_\nu = \mathcal{H}_{-\nu}$$

holds for the inverse kernel of $\mathcal{H}_\nu$. In consistency with the Bochner-Plancherel-Schwartz characterization of positive definite distributions, the tempered distribution

$$\boxed{\mathcal{H}^\bullet_\nu = 1_\nu \otimes 1_\nu}$$

on the planar coadjoint orbit $\mathcal{O}_\nu \in \mathrm{Lie}(G)^\star/\mathrm{CoAd}_G(G)$ of unitary central character χ_ν, provides the uniquely defined canonical reproducing kernel $\star_\nu \mathcal{H}^\bullet_\nu$ of the linear Schrödinger representation U^ν, $\nu \neq 0$, of G. Notice that $\mathcal{H}^\bullet_\nu \in \tilde{\mathcal{H}}^{-\infty}$ represents the orientation class of $\mathcal{S}'(\mathbb{R} \oplus \mathbb{R})$. Conversely, the

preceding identity and its infinitesimally generating equivalent imply the
square integrability mod C of U^ν and the flatness of the planar coadjoint
orbit $\mathcal{O}_\nu \in \mathrm{Lie}(G)^*/\mathrm{CoAd}_G(G)$ associated to the isomorphy class of U^ν in
the unitary dual $\hat{G}$ ([13]).

- The flatness of the symplectic affine planes $\mathcal{O}_\nu, \nu \neq 0$, allows to identify the trace of the holographic transform $\mathcal{H}_\nu$ attached to the planar
  coadjoint orbit $\mathcal{O}_\nu \in \mathrm{Lie}(G)^*/\mathrm{CoAd}_G(G)$ of the Heisenberg group G
  at resonance frequency ν as follows:

$$\mathrm{tr}\,\mathcal{H}_\nu = \mathcal{H}_\nu^\bullet.$$

In terms of the exterior differential 2-form representing the swapping
matrix $\frac{1}{2}J$, the infinitesimal generator ω_ν of the smooth dynamical
system $\mathbb{R} \lhd G \longrightarrow \mathbb{R} \oplus \mathbb{R}$ is the rotational curvature form

$$\omega_\nu = \pi i \nu \,.\, dx \wedge dy = -\frac{\pi}{2}\nu \,.\, dw \wedge d\bar{w}\,.$$

The symplectic character formula holds in the density space $\tilde{\mathcal{H}}^{-\infty}$,
where (x, y) denote the differential phase-local frequency coordinates
with respect to a coordinate frame rotating with frequency $\nu \neq 0$,
and $w = x + iy \in W$. The upsampling Pfaffian of the canonical
symplectic form $\frac{1}{\pi i}\omega_\nu$ reads

$$\mathrm{Pf}\left(\frac{1}{\pi i}\omega_\nu\right) = \nu.$$

The identity

$$U^\nu(\mu)\mathcal{H}_\nu = \mu \star_\nu \mathcal{H}_\nu,$$

which holds for all compactly supported scalar measures μ on the symplectic affine plane $\mathcal{O}_\nu \in \mathrm{Lie}(G)^*/\mathrm{CoAd}_G(G)$, implies

$$\mathcal{H}_\nu(\psi) = \tilde{\psi} \star_\nu \mathcal{H}_\nu^\bullet.$$

Hence,

$$< \mathcal{H}_\nu(\psi), \mathcal{H}_\nu^\bullet > = \tilde{\psi}$$

for all elements $\psi \in \mathcal{H}^{-\infty}$. It follows that $\mathcal{H}_\nu(\psi) \in \mathcal{S}'(\mathbb{R} \oplus \mathbb{R})$ represents a
reproducing kernel of $\mathcal{H} = L^2(\mathbb{R})$ in $\mathcal{H}^{-\infty}$. The uniquely defined canonical
reproducing kernel $\star_\nu \mathcal{H}^\bullet$ where $\mathcal{H}_\nu^\bullet \in \tilde{\mathcal{H}}^{-\infty}$ acts as the differential phase-
local frequency reference of the rotating coordinate frame of frequency $\nu \neq$
0, which is attached to the planar coadjoint orbit $\mathcal{O}_\nu \in \mathrm{Lie}(G)^*/\mathrm{CoAd}_G(G)$.
According to the semi-classical QED description of response data phase
holograms, it can be thought of as the reference of a Keppler configuration

consisting of precession arrays of phase-locked Bloch vectors inside the symplectic affine plane $\mathcal{O}_\nu$.

According to Harish-Chandra's philosophy, the correct way to think of characters of Lie groups is as distributions. In terms of the symplectic presentation of G, the symplectic character formula displayed supra can be looked at as a generalization of the relation

$$\exp \circ \mathrm{tr} = \det \circ \exp$$

to Schwartz kernels. From the calculus of Schwartz kernels in the locally convex topological vector space $\mathcal{S}'(\mathbb{R} \oplus \mathbb{R})$ of tempered distributions, follow the symplectic filter-bank identities

$$\sqrt{\nu} \cdot \mathcal{H} = \sqrt{\star_\nu \mathcal{H}_\nu^\bullet} \, \mathcal{L}_2(\mathcal{H}) \qquad (\nu > 0),$$

and

$$\sqrt{-\nu} \cdot \mathcal{H} = \sqrt{\star_{-\nu} \mathcal{H}_{-\nu}^\bullet} \, \mathcal{L}_2(\mathcal{H}) \qquad (\nu < 0).$$

The kernel K_f^ν associated to the Hilbert-Schmidt integral operator $U^\nu(f) \in \mathcal{L}_2(\mathcal{H})$ for $f \in L^2(\mathbb{R} \oplus \mathbb{R})$ extends from $f \in \mathcal{S}(\mathbb{R} \oplus \mathbb{R})$ to its antidual $\mathcal{S}'(\mathbb{R} \oplus \mathbb{R})$ by the rule

$$K_f^\nu(x, y) = 2e^{-\pi i \nu x y} \cdot f \star_\nu \mathcal{H}_\nu^\bullet(x, y) \qquad ((x, y) \in \mathbb{R} \oplus \mathbb{R}).$$

The preceding identity leads to the following result, which explains the important role played by the symplectic filter-bank processing in FTMRI. Several methods of MRI, such as the spin presaturation technique by gradient spoiling for motion artifact reduction in variable-thickness slabs, are based on symplectic filter-bank processing.

- The generating kernel distribution K_f^ν with respect to the rotating coordinate frame attached to the symplectic affine plane $\mathcal{O}_\nu \in \mathrm{Lie}(G)^\star/\mathrm{CoAd}_G(G)$, obtains by symplectic filtering of the proton-weighted spin isochromat density f with the uniquely defined canonical reproducing kernel $\star_\nu \mathcal{H}_\nu^\bullet$, associated to the holographic transform $\mathcal{H}_\nu$ at resonance frequency $\nu \neq 0$.

The irreducible unitary linear representation V^ν of G satisfying the resonance condition of the unitary central character, and therefore attached to the same planar coadjoint orbit $\mathcal{O}_\nu \in \mathrm{Lie}(G)^\star/\mathrm{CoAd}_G(G)$, defines a holographic transform isomorphic to $\mathcal{H}_\nu$ by the unitary action σ of the metaplectic group $\mathbf{Mp}(2, \mathbb{R})$ on the Hilbert space $\mathcal{H} = L^2(\mathbb{R})$, and therefore a generating kernel distribution isomorphic to K_f^ν.

§4   Relaxation filtering

A standard clinical SE protocol consists of a computer-controlled $\frac{\pi}{2}$ pulse that flips by resonance the Bloch vector into the selected central tomographic slice. Because there are also phased-arrays of contragredient Bloch vectors rotating in the opposite sense to the spin precession at resonance frequency ν, the following contragredient π pulse at $\frac{1}{2}T_E$ performs a local frequency fan narrowing by phase conjugation (see Figure 2). According to the (T_1, T_2) relaxation decomposition of G, the tissue specific spin relaxation rates

$$\frac{1}{T_k} \qquad (k \in \{1, 2\})$$

and the flip angle θ of the Bloch vector provides a nonlinear amplitude modulation

$$\boxed{w_{T_1, T_2, \theta} : t \rightsquigarrow \beta_{T_1, \theta}(t) . e^{-|t|/T_2} \sin \theta}$$

of the exciting pulse sequence $\psi \in \mathcal{S}'(\mathbb{R})$ at resonance frequency $\nu \neq 0$. Due to the Bloch gyroscopic equation that is rotationally invariant, $w_{T_1, T_2, \theta}$ includes the central spin relaxation weight ([36])

$$\beta_{T_1, \theta} : t \rightsquigarrow e^{-|t|/T_1} \cos \theta + \left(1 - e^{-|t|/T_1}\right)$$

for normalized equilibrium magnetization.

The modulation of the proton-weighted spin isochromat density f by the relaxation weight $\tilde{w}_{T_1, T_2, \theta}$ allows for control of the contrast resolution by synchronization of the temporal filter-bank design parameters T_E and T_R of the exciting pulse sequence $\psi \in \mathcal{S}'(\mathbb{R})$ at resonance frequency $\nu \neq 0$, with the dephasing signal

$$K^{\nu}_{\tilde{w}_{T_1, T_2, \theta} \star_{\nu} f} \in \mathcal{S}'(\mathbb{R} \oplus \mathbb{R}).$$

The preceding symplectic convolution represents the dephasing effect of transverse magnetization decay within the planar coadjoint orbit $\mathcal{O}_{\nu} \in \text{Lie}(G)^{\star}/\text{CoAd}_G(G)$.

- Spin relaxation performs a symplectic filtering of the proton-weighted spin isochromat density f with temporal filter-bank design parameters T_E and T_R.

In standard clinical MRI protocols, short T_E and short T_R provide T_1-weighting, long T_E and long T_R provide T_2-weighting, short T_E and long T_R provide proton density-weighting. T_1-weighted scans are most useful for analyzing morphologic detail and are employed in conjunction with the gadolinium contrast agents because enhancing lesions are bright under T_1-weighting. In GRE imaging, the SPGR modality provides T_1-weighting because it eliminates any residual transverse magnetization or T_2 effect due to the phase dispersion that is caused by the application of additional spoiler gradients. T_2-weighted scans are very sensitive to increased water content and can visualize edema to great advantage. T_2-weighted scans are also most sensitive to differences in susceptibility. Proton density-weighted scans are helpful from the standpoint of diagnostic specificity because they give far better tissue differentiation than does X-ray CT.

Beyond these general guidelines, the imaging strategies differ depending upon their goals. In joint imaging, for instance, T_1-weighted scans better display the hyperintense bone marrow that contrasts with the signal void cortex. The intra-articular fluid collection and the hyaline cartilage are clearly delineated on T_2-weighted tomographic slices ([4]). With the introduction of FSE techniques, the information obtained by SE scans in the relatively long MRI examination time of 9 to 10 minutes has been reduced to 5 minutes. FSE protocols have brought a revolution to body, musculoskeletal, and spinal MRI, but has been slower to find a role in cerebral MRI. In breast imaging, the current method of choice is the rotating delivery of excitation off-resonance method (RODEO). This specialized GRE imaging technique uses jump-return radiofrequency pulse excitations that restore the magnetization of the on-resonance species but creates residual transverse magnetization of the off-resonance species. Typically, the transmitter is set at the major fat frequency, so that imaging by the RODEO pulse sequence achieves fat suppression and water excitation with a single pulse, eliminating the need for an additional saturation pulse for fat. Those protons with relatively long T_2 relaxation times can be spoiled to give low signal intensity in the resulting image. This radio frequency spoiling is particularly useful for the determination of wavelets originating from cysts or fluid in the MRI scans of the breast ([4]).

In MRI protocols, the relaxation times satisfy

$$T_2 < T_1,$$

whereas the synchronization parameters of echo and repetition times satisfy

$$T_E < T_R.$$

In standard MRI protocols of clinical neuroimaging and spinal imaging ([46]), short T_E typically means 20 ms, and long T_E means 160 ms, whereas short T_R typically means 600 ms and long T_R means 3000 ms. Due to the low signal-to-noise ratio, long T_E and short T_R scans are not used in clinical MRI protocols. In this sense again, computer-controlled synchronized timing is everything in FTMRI.

§5   FTMRI reconstruction

The MRI scan generated in the rotating coordinate frame, which is attached to the symplectic affine plane $\mathcal{O}_\nu \in \mathrm{Lie}(G)^\star/\mathrm{CoAd}_G(G)$, can be thought of as the result of a cascade of symplectic filtering processes ([21]). In consistency with the Heisenberg uncertainty principle, transition from the rotating coordinate frame to the laboratory coordinate frame attached to G/C is performed by the partial Fourier cotransform

$$\bar{\mathcal{F}}^2_{\mathbf{R}\oplus\mathbf{R}}\mathcal{H}^\bullet_\nu(x,y) = 1_\nu(x) \otimes \varepsilon_{-\nu y}.$$

Freezing of the local frequency coordinate y with respect to the rotating coordinate frame attached to the symplectic affine plane $\mathcal{O}_\nu$ at the overall resonance frequency $\nu \neq 0$, is performed by resonance with the affine Larmor frequency scale of the faithful block representation

$$y \rightsquigarrow \begin{pmatrix} 1 & y \\ 0 & 1 \end{pmatrix}.$$

For the purpose of multiple data line collecting filter-bank, organization of the response data phase holograms in the laboratory frame, adjusting the spectra of single whole resonance lines ([21]) of image data

$$\left\{ 1_\nu(x) \otimes \varepsilon_{-\nu y} \big| (x,y) \in \mathbb{R} \oplus \mathbb{R} \right\}$$

of $\mathcal{H}^\bullet_\nu \in \tilde{\mathcal{H}}^{-\infty}$ to the symplectic structure J of the planar coadjoint orbit $\mathcal{O}_\nu \in \mathrm{Lie}(G)^\star/\mathrm{CoAd}_G(G)$, is achieved by the affine Larmor frequency scale

$$\left\{ \varepsilon_{\nu x} \otimes 1_\nu(y) \big| (x,y) \in \mathbb{R} \oplus \mathbb{R} \right\}$$

swapped by σ and associated to the faithful block representation

$$x \rightsquigarrow \begin{pmatrix} 1 & x \\ 0 & 1 \end{pmatrix}$$

introduced earlier in the context of the Lauterbur spatial encoding technique of quantum holography.

The partial Fourier cotransform $\bar{\mathcal{F}}^2_{\mathbf{R}\oplus\mathbf{R}}$ is encapsulated in the symplectically invariant symbol map $\tilde{U}^\nu(K^\nu_f)$ in the sense of the theory of pseudodifferential operators ([17])

$$\tilde{U}^\nu(K^\nu_f) : (x,y) \rightsquigarrow \tilde{U}^\nu\left(\begin{pmatrix} 1 & x & 0 \\ 0 & 1 & y \\ 0 & 0 & 1 \end{pmatrix} \right)(K^\nu_f)$$

of the kernel

$$K^\nu_f \in L^2(\mathbb{R} \oplus \mathbb{R})$$

associated to the Hilbert-Schmidt integral operator $U^\nu(f) \in \mathcal{L}_2(\mathcal{H})$ for the proton-weighted spin isochromat density $f \in L^2(\mathbb{R} \oplus \mathbb{R})$. Indeed, the kernel function $\tilde{U}^\nu(K_f^\nu) \in L^2(\mathbb{R} \oplus \mathbb{R})$ takes the form

$$\tilde{U}^\nu(K_f^\nu)(x,y) = e^{-2\pi i \nu x y} \cdot \bar{\mathcal{F}}_{\mathbb{R} \oplus \mathbb{R}}^2(K_f^\nu)(x,y) \quad ((x,y) \in \mathbb{R} \oplus \mathbb{R}).$$

It represents the response data phase hologram, quantum electrodynamically performed by the MRI scanner organization from the excited proton-weighted spin isochromat density f in the laboratory coordinate frame. This is attached to the cross-section G/C by interference with the echo wavelet $\psi \in S(\mathbb{R})$ elicited at resonance frequency $\nu \neq 0$.

According to the semi-classical QED description of response data phase holograms, the central aspect of MRI is the spatial encoding of differential phase-local frequency coordinates. The phase holograms form symplectic spinors ([19]) which act as learning matrices consisting of matched filterbanks. The matched filter-bank processing to increase the signal-to-noise ratio ([27]) forms the link between MRI and coherent optical holography ([15, 34]). The two main advantages offered by the holographic storage technology are high information capacity in small volumes and massive parallelism in data access. Considered as learning matrices ([7]), the phase holograms form the link between MRI and neural networks. In the sense of photonics ([35]), phase coherence is everything in MRI.

The FTMRI modality is based on the marriage of the computer to quantum mechanics. Readout of the response data phase holograms $\tilde{U}^\nu(K_f^\nu)$ stored in the laboratory coordinate frame of G/C after transition from the rotating coordinate frame attached to the planar coadjoint orbit $\mathcal{O}_\nu, \nu \neq 0$ of the stratification of $\hat{G}$, follows by the symplectic Fourier transform

$$\tilde{U}^\nu(K_f^\nu)(x,y) = 2\big(e^{-\pi i \nu x' y'} f(x',y';\nu)\big)\widehat{\ }(2x, 2y) \quad ((x,y) \in \mathbb{R} \oplus \mathbb{R})$$

of pseudodifferential operators. Indeed, because $\tilde{U}^\nu(K_f^\nu)$ averages the phases of the proton-weighted spin isochromat densities in the symplectic affine plane $\mathcal{O}_\nu \in \mathrm{Lie}(G)^*/\mathrm{CoAd}_G(G)$, the symplectically invariant symbol calculus provides the reconstruction formula in the laboratory coordinate frame of G/C for the proton density-weighted spin isochromat density $f \in L^2(\mathbb{R} \oplus \mathbb{R})$

$$\boxed{f(x,y;\nu) = \frac{1}{2} e^{\pi i \nu x y} \cdot \widehat{\tilde{U}^\nu(K_f^\nu)}\left(\frac{1}{2}x, \frac{1}{2}y\right) \quad ((x,y) \in \mathbb{R} \oplus \mathbb{R})).}$$

Thus, the canonical reproducing kernel

$$\star_\nu \mathcal{H}_\nu^\bullet = \star_\nu \big(1_\nu \otimes 1_\nu\big)$$

forms the cryptographic key for the decoding of the quantum holographically encoded MRI encryption.

- The symplectic Fourier transform $f \rightsquigarrow \hat{f}$ allows for a decryption of the response data phase holograms spatially encoded by the MRI scanner organization as a FTMRI cipher text.

Due to the amplitude modulation of the exciting pulse sequences $\psi \in S'(\mathbb{R})$ at resonance frequency $\nu \neq 0$ by the spin relaxation weight $w_{T_1,T_2,\theta}$, the excited proton-weighted spin isochromat density $f \in L^2(\mathbb{R} \oplus \mathbb{R})$ sensitively depends upon the dynamical data

$$T_1, T_2, \theta$$

of the symplectic relaxation filter

$$\tilde{w}_{T_1,T_2,\theta} \star_\nu f \in S'(\mathbb{R} \oplus \mathbb{R})$$

and therefore, upon the waiting times T_E and T_R of the pulse sequence ψ, which is used for exciting the proton-weighted spin isochromat density f in the symplectic affine plane $\mathcal{O}_\nu \in \text{Lie}(G)^*/\text{CoAd}_G(G)$. It also trains the learning matrix of the neural network that underlies the phase memory of spin precession at resonance frequency $\nu \neq 0$. Due to the contrast processing by the symplectic relaxation filtering, there exists no universal gray scale in MRI, unlike scans performed by the X-ray CT modality.

The dependence of the symplectic relaxation filter from the flip angle θ of the Bloch vector in conjunction with compensation of the phase dispersion by gradient inversion is used in fast low-angle shot (FLASH) protocols or partial flip angle imaging. FLASH plays an important role in magnetic resonance mammography (MRM), for instance, because it allows one to identify even a carcinoma of the breast as small as 3 mm. For the standard SE pulse sequence that remedies the dephasing by phase conjugation, the amplitude modulation is performed by the spin relaxation weight factor. As a result of the symplectically invariant symbol calculus, the relaxation-weighted amplitude modulation arises

$$e^{-m T_E/T_2} \cdot \left(1 - e^{-T_R/T_1}\right) \cdot f,$$

where $m \geq 1$ denotes the echo number. In the sense of the temporal filter-bank design parameters T_E, T_R of spin relaxation, timing is everything in FTMRI.

The idea of filtering is also extremely valuable in three-dimensional MRI processing, specifically in MRA. From the symplectic character formula it follows that the Plancherel measure μ of G is concentrated on the dual of the line C. It is dual to the downsampled measure

$$\frac{1}{\text{Pf}(\mathcal{H}_\nu^\bullet)} \, \mathrm{d}z$$

on C, and therefore μ implements the affine Larmor frequency scale by

$$\mu = \nu.d\nu.$$

It has been less than a decade since the realization that blood flow could be imaged by MRI techniques. In MRA, tomographic slices $\mathcal{O}_\nu$ are integrated along C against μ in order to receive three-dimensional slabs by the Cavalieri fibration principle of the integration theory of homogeneous spaces. Image postprocessing techniques such as anisotropic filtering by convolution with a second-difference kernel can be designed to enhance small vessel details and to suppress noise and stationary tissue detail ([11]) so that MRA forms the most sensitive technique for the visualization of flow phenomena ([1]), eventually replacing even the recently developed magnetic transfer contrast techniques ([43]). Advances in filter-bank strategies rather than acquisition techniques in conjunction with visualization by the maximum-intensity projection (MIP) algorithm, may be the key toward optimization of MRA as a screening technique for intracranial aneurysms or vascular malformations and atherosclerotic disease of the carotid arteries ([42]).

§6 Fast MRI techniques

Conventional MRI protocols, specifically conventional SE protocols, are an intrinsically slow technique ([2]). It typically takes several minutes to acquire each set of scans so that a complete MRI examination that may require the application of several pulse sequences and assessment by scout-view scans in more than one planar coadjoint orbit $\mathcal{O}_\nu \in \mathrm{Lie}(G)^*/\mathrm{CoAd}_G$ (G), can take as along as an hour to complete. Therefore, it was the factor speed that has limited more widespread use and dissemination of MRI in the decade since the clinical MRI modality has been routinely available. Fast readout algorithms are based on symmetry reasonings (Quarter FTMRI) and FFT techniques of computerized Fourier analysis. To reduce signal-to-noise, conjugate response data synthesis is applied in the digital evaluation processing of the response data phase holograms performed by the MRI scanner organization. In addition, correction algorithms for spatially dependent phase shifts are implemented ([28]).

An alternative approach refers to the discrete non-split central group extension

$$\mathbf{Z} \lhd \Gamma \longrightarrow \mathbf{Z} \oplus \mathbf{Z},$$

which implements the J-invariant Gaussian lattice

$$\Gamma \hookrightarrow W$$

as stroboscopic lattice of the Keppler phase triangulation procedure by integral transvections. Notice that the stroboscopic lattice Γ immobilizes the rotating coordinate frame and implies the symplectic filtering version of the Poisson summation formula

$$e^{\pi i xy} \cdot \sum_{n \in \mathbf{Z}} e^{2\pi i(z+ny)} \psi(n+x) = e^{-\pi i xy} \cdot \sum_{n \in \mathbf{Z}} e^{2\pi i(z-nx)} \bar{\mathcal{F}}_{\mathbf{R}} \psi(n+y).$$

On the right hand side, the Poisson summation formula includes the one-dimensional Fourier cotransform $\bar{\mathcal{F}}_{\mathbf{R}}$ acting on $\psi \in \mathcal{S}(\mathbb{R})$. This formula gives rise to an intertwining operator of the translations of the compact nilmanifold G/Γ and the linear Schrödinger representation U^ν of G, which is termed system transfer function in photonics. Its inverse is nothing else than the periodization by Γ of the polarized symbol isomorphism

$$\tilde{U}^\nu : \mathcal{S}'(\mathbb{R} \oplus \mathbb{R}) \longrightarrow \mathcal{S}'(\mathbb{R} \oplus \mathbb{R}).$$

Notice that $\tilde{U}^\nu$ has been used to make the response data phase hologram, recorded with respect to the rotating coordinate frame attached to the symplectic affine plane $\mathcal{O}_\nu \in \mathrm{Lie}(G)^\star/\mathrm{CoAd}_G(G)$, stationary. Thus, the Keppler phase triangulation procedure in the rotating coordinate frame finds its natural formulation in the setting of the compact Heisenberg nilmanifold G/Γ. In addition, an application of the theory of currents on W ([37]) to the symplectic character formula projected to G/Γ, yields by passing to the quotient mod Γ the Pick area formula ([9]) via the J-invariant Gaussian lattice Γ. The Pick approach to the spatial resolution of FTMRI by the area A under the frequency or readout gradient is given by

$$A = \#\mathring{A} + \tfrac{1}{2}(\#\partial A) - 1$$

where $\#\mathring{A}$ denotes the number of interior sampling points, and $\#\partial A$ the number of ramp sampling points. The proof exploits the connection of the Bohr-Sommerfeld quantization rules with cyclic cohomology ([12]) in the plane W.

The stroboscopic lattice Γ gives rise to the ultra-fast EPI procedure of FTMRI, which collects all the response data in a single-shot or snapshot protocol ([29]). The first half of the exciting pulse sequence is essentially identical to a standard clinical SE protocol in that there is a $\frac{\pi}{2}$ pulse that flips the Bloch vector into the selected planar coadjoint orbit of G followed by a contragredient π pulse performing phase conjugation, and generates a refocused echo wavelet by local frequency fan narrowing at the time $T_\mathbf{E}$. The second half of the exciting pulse sequence is where EPI differs significantly from a standard SE acquisition. The readout gradient oscillates rapidly from positive to negative amplitude to form a train of gradient echoes. Each echo in the EPI echo train is phase encoded differently by the phase encode blips. Each oscillation of the readout gradient corresponds to one resonance line of image data and each phase encode blip corresponds to a transition from one line to the next. Thus, the most significant difference between the SE acquisition and the EPI acquisition is that the EPI pulse sequence acquires multiple resonance lines of image data during one interval of length T_R.

The EPI technique has many similarities to FSE. Both techniques collect multiple resonance lines of image data during each interval of length T_R, both techniques have an associated ETL that forms an approximate measure of the extent to which the data acquisition will be faster than a conventional SE technique. There is, however, also a significant difference between EPI and FSE protocols in that the EPI echo train does not contain any refocusing pulse. The lack of refocusing pulses in EPI protocols is responsible for the unique capability of EPI for snapshot tomographic imagery. Each line of the response data phase holograms is collected by reversing the readout gradient slope to form another gradient echo. The realization of the ideal Dirac comb kernel associated to the stroboscopic lattice Γ, however, involves the need for rapidly switching high gradient slope systems. Increasing the gradient slew rate in combination with increasing the receiver bandwidth of the imager, represent significant hardware requirements for performing high quality snapshot EPI protocols. Gradient rise time, which is half of the gradient ramp time, and maximum slope α must each be improved by at least a factor of two over today's standard. EPI protocols that use more than one shot to complete the image acquisition are referred to as multi-shot EPI protocols, and those that use only one shot to acquire the image are referred as single-shot or snapshot EPI protocols. Recently constructed EPI scanners, however, allow for producing snapshot tomographic images of the cranium with 0.75 mm in-plane resolution and 3 mm to 5 mm tomographic slice thickness at acquisition rates of 10–50 scans per second. At these rates, MRI applications enter an entirely new domain such as whole-heart perfusion studies in a single breath hold, and whole-brain functional neuroimaging. Though not considered to be a safety issue, peripheral nerve stimulation imposes limitations on the gradient ramp time. Therefore, it is quite understandable that in the last couple of years, EPI protocols have arguably become the most talked about of MRI acquisition capabilities. From completing the entire MRI examination in a matter of seconds to expanding the applications realm of FTMRI into radiodiagnostic territories traditionally claimed by other imaging modalities, the perceived clinical potential of EPI processing is indeed great ([45]). Figure 4 shows the comparison of oblique-central imgaes of the cranium from three clinical cases.

In MRI, a highly homogeneous magnetic flux density is required to achieve phase coherence over the desired imaging volume; typically, better

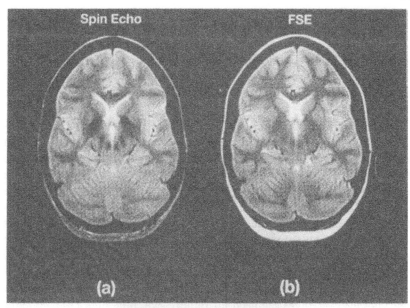

Figure 4. Comparison of oblique-central images of the cranium from clinical T_2-weighted ($T_R = 2500$ ms, $T_E = 80$ ms) SE protocol of 8:30 scan time (a), T_2-weighted ($T_R = 2500$ ms, $T_E = 80$ ms, ETL = 12) FSE protocol of 3:50 scan time (b), and T_2-weighted ($T_R = 2500$ ms, $T_E = 80$ ms, 16 shots) EPI protocol of 2:03 scan time (c).

than one part per million root-mean-square deviation over a 30 cm diameter of spherical volume is accepted. Using a linear magnetic field gradient imposed onto the homogeneous magnetic field density by the linear ramp action of the frequency modulation group $\mathbf{GA}(\mathbb{R})$, an accurate encoding of the in-plane positions of proton-weighted spin isochromat densities is made possible. In addition, the presence of a homogeneous magnetic flux density allows for the excitation of planar tomographic slices of even thickness and accurate flip angle in response to the spin manipulation pulses within the radiofrequency window. A special shielding is required to eliminate interference effects from external radiofrequency sources that distort and degrade the quantum holograms. The phase-constrained encoding (PACE) technique ([18]) is a technique of efficiently correcting inhomogeneity of the magnetic flux density induced by geometric distortion. It is particularly appropriate for the correction in open access magnet designs used in interventional MRI applications. Similar to the EPI technique, PACE also relies on the Dirac comb kernel associated to the stroboscopic lattice Γ of the Keppler phase triangulation procedure by integral transvections.

§7 Outlook

From the mathematical point of view, the disciplines of optical and digital signal processing overlap significantly. Over the years, however, the notational conventions and methodologies of the two areas have diverged to the point that engineers and scientists in one area often find similar or identical work in the other area nearly unrecognizable. The combination of mathematical physics, electrical engineering, and computerized Fourier analysis applied to FTMRI has had almost as big an impact on the medical discipline of radiodiagnostics as Röntgen's discovery of X-rays one hundred years ago. Because even highly respected clinical investigators of MRI are not familiar with the physics and mathematics underlying this sophisticated phase-holographic imaging modality, it is not surprising that the surface of knowledge actually obtainable via MRI techniques has been barely scratched. However, the future of the FTMRI modality continues to be bright, with the advent of routine second and subsecond imaging with current delivery of the next generation of MRI scanners. With further improvement in image quality, reduction in scan time, and substantial refinements, fMRI of cortical activation will have a major impact on cognitive neuroscience. The endogenous BOLD contrast technique to map the functional anatomy of the human brain will be useful for acquiring maps of specific eloquent cortical regions such as the visual cortex, the motor strip, and the silent speech area. BOLD-EPI is a particularly useful data acquisition modality for the localization of seizure foci in epileptic disorders.

In the past two years the application of fMRI has exploded. Specifically, the use of fMRI for the purposes of interventional medicine is highly promising. The most important interventional applications include image-guided frameless stereotaxy using multiplanar and three-dimensional presentation of MRI data sets, MRI-guided biopsies which provide histological diagnosis of primary breast lesions not visible by X-ray mammography, MRI-guided endoscopy, and real-time intraoperative imaging during minimally invasive procedures. Nevertheless, interventional MRI is still in its infancy. Like MRA, EPI, and MRM it has tremendous clinical potential, but this potential will take time to develop and mature. It is still too early to define its precise clinical range.

§8   Summary

MRI has developed into a major scientific field in the area of diagnostic imaging. A fascinating aspect of this powerful clinical imaging modality is that the contrast resolution admits a dynamic implementation. In MRI, the contrast is not fixed in advance such as in other radiodiagnostic imaging modalities, but can be adapted to the specific needs of a clinical examination. Clinicians who initially were pleasantly surprised by the imaging capabilities of MRI are now active participants in the gathering of data in this rapidly expanding field of non-invasive radiodiagnostics. The unique QED basis of NMR in conjunction with the imaging capabilities of current computer technology has made this imaging modality a target of interest for clinicians, physicists, biologists, engineers, and mathematicians. This basis is relatively new to radiodiagnostic imaging, particularly in comparison to the well-developed, long-standing familiarity with X-ray, ultrasound, and radionuclide imaging techniques. Even a decade after its first clinical applications appeared in the literature, applications of MRI continuously develop while software and hardware technology evolve. The more the MRI technique is used, the more new applications are discovered, and the higher are requirements of computer performance. Therefore, the future of MRI continues to be bright concerning its theoretical and clinical performance, with current delivery of the next generation of MRI scanners providing substantial further improvement in image quality and reduction in scan time.

The main contributions of MRI to medical science have not been discovering new diseases, but rather facilitating and improving the diagnostic accuracy of known disease processes by virtue of its unprecedented soft-tissue contrast resolution, tissue characterization, and multiplanar imaging capabilities of pathoanatomy ([41]). In addition, the possibility of non-invasive vascular imaging and functional imaging has been the focus of many research programs during the last few years, with dramatic progress ([23]). As a result of the increased contrast resolution, it brought neuroanatomists even further toward understanding the brain's complex architecture in adults ([2]), and the developing brain in pediatric subjects ([5]). MRI has exerted considerable impact on the practices of neuroradiology, neurology, and neurosurgery. The roads made accessible by clinical MRologists are in the direction of better medical care, the common goal of the medical community.

Acknowledgment. Parts of this paper have been prepared during a Visiting Fellowship at the University of California, San Francisco. The author is grateful to the MRI staff technologists of UCSF for their practical advice and help. Moreover, his special thanks go to the MRI neurologists of the Department of Radiology at Stanford University School of Medicine, for their expert clinical instructions during a Neuroimaging Visiting Fellowship at Stanford Medical Center, California, and to the team of physicists, neuroscientists, and staff MRologists of the Department of Radiology at Harvard Medical School for the competent advice given during a fMRI Visiting Fellowship at the NMR Center and the NMR Research Laboratories of Massachusetts General Hospital in Boston, Massachusetts. Finally, he gratefully acknowledges a series of valuable discussions on the calculus of Schwartz kernels with one of his former students at Ruhr-Universität Bochum, the late Dr. Ulrich Tippenhauer.

References

[1] Anderson, C. M., R. R. Edelman, and P. A. Turski, *Clinical Magnetic Resonance Angiography*, Raven Press, New York, 1993.

[2] Bauer, R., E. van de Flierdt, K. Mörike, and C. Wagner-Manslau, *MR Tomography of the Central Nervous System*, Second edition, Gustav Fischer Verlag, Stuttgart, Jena, New York, 1993.

[3] Barkovich, A. J., *Pediatric Neuroimgaing*, Second edition, Raven Press, New York, 1995.

[4] Beltran, J., (ed.), *Current Review of MRI*, First Edition, CM Current Medicine, Philadelphia, 1995.

[5] Bisese, J. H. and A.-M. Wang, *Pediatric Cranial MRI: An Atlas of Normal Development*, Springer-Verlag, Heidelberg, 1994.

[6] Brezin, J., Geometry and the method of Kirillov, in *Non-Commutative Harmonic Analysis*, J. Carmona, J. Dixmier, and M. Vergne (eds.), Lecture Notes in Mathematics, **466**, Spinger Verlag, Heidelberg, 1975, pp. 13–25.

[7] Carpenter, G. A., Neural network models for pattern recognition and associative memory, *Neural Networks* **2** (1989), 243–257.

[8] Crooks, L. E. and P. Rothschild, Clinical imaging, in *Pulsed Magnetic Resonance: NMR, ESR, and Optics, A Recognition of E.L. Hahn*, D. M. S. Bagguley (ed.), Clarendon Press, Oxford 1992, pp. 346–361.

[9] Diaz, R. and S. Robins, Pick's formula via the Weierstrass ℘-function, *Amer. Math. Monthly* **102** (1995), 431–437.

[10] Dieudonné, J. A., *La Géometrie des Groupes Classiques*, Troisième Edition, Ergebnisse der Mathematik und ihrer Grenzgebiete, Band 5, Springer-Verlag, Heidelberg, 1971.

[11] Du, Y. P., D. L. Parker, and W. L. Davis, Vessel enhancement filtering in three-dimensional MR angiography, *JMRI* **5** (1995), 353–359.

[12] Elliott, G. A., T. Natsume, and R. Nest, Cyclic cohomology for one-parameter smooth crossed products, *Acta Math.* **160** (1988), 285–305.

[13] Felix, R., When is a Kirillov coadjoint orbit a linear variety? *Proc. Amer. Math. Soc.* **86** (1982), 151–152.

[14] Grossman, C. B., *MRI and CT of the Head and the Spine*, Second edition, Williams & Wilkins, Baltimore, MD, 1995.

[15] Hahn, E. L., NMR and MRI in retrospect, *Phil. Trans. R. Soc. London* **A 333** (1990), 403–411. Also in *NMR Imaging*, P. Mansfield and E .L. Hahn (eds.), The Royal Society, London, 1990, pp. 1–9.

[16] Hauptman, H., A minimal principle in the direct methods of X-ray crystallography, in *Recent Advances in Fourier Analysis and Its Applications*, J. S. Byrnes and J. L. Byrnes (eds.), Kluwer Academic Publishers, Dordrecht, Boston, 1990, pp. 3–15.

[17] Hörmander, L., The Weyl calculus of pseudodifferential operators, *Comm. Pure Appl. Math.* **32** (1979), 359–343.

[18] Kim, J.K., D .B. Plewes, and R. M. Henkelman, Phase constrained encoding (PACE): A technique for MRI in large static field inhomogeneities, *Magn. Reson. Med.* **33** (1995), 497–505.

[19] Kostant, B., Symplectic spinors, in *Geometria Simplettica e Fisica Matematica, Symposia Mathematica* **14**, Istituto Nazionale di Alta Matematica Roma, Academic Press, New York 1974, pp. 139–152.

[20] Kucharczyk, J., M. Moseley, and A. J. Barkovich, (eds.), *Magnetic Resonance Neuroimaging*, CRC Press, Boca Raton, 1994

[21] Kumar, A., D. Welti, and R. R. Ernst, NMR Fourier zeugmatography, *J. Magn. Reson.* **18** (1975), 69–83.

[22] Kuzniecky, R. I. and G. D. Jackson, *Magnetic Resonance in Epilepsy*, Raven Press, New York 1995.

[23] Lee, A. T., G. H. Glover, and C. H. Meyer, Discrimination of large venous vessels in time-course spiral blood-oxygen-level-dependent magnetic-resonance functional neuroimaging, *Magn. Reson. Med.* **33** (1995), 745–754.

[24] Leith, E. N., Synthetic aperture radar, in *Optical Data Processing*, D. Casasent (ed.), Topics in Applied Physics **23**, Springer-Verlag, Heidelberg, 1978, pp. 89–117.

[25] Leith, E. N., Optical processing of synthetic aperture radar data, in *Photonic Aspects of Modern Radar*, H. Zmuda and E.N. Toughlian (eds.) Artech House, Boston, 1994, pp. 381–401.

[26] Leith, E. N. and A. L. Ingalls, Synthetic antenna data processing by wavefront reconstruction, *Appl. Opt.* **7** (1968), 539–544.

[27] Lu, D. and P. M. Joseph, A matched filter echo summation technique for MRI, *Magn. Reson. Imag.* **13** (1995), 241–249.

[28] MacFall, J. R., N. J. Pelc, and R. M. Vavrek, Correction of spatially dependent phase shifts for partial Fourier imaging, *Magn. Reson. Imag.* **6** (1988), 143–155.

[29] Mansfield, P., Imaging by nuclear magnetic resonance, in *Pulsed Magnetic Resonance: NMR, ESR, and Optics, A Recognition of E.L. Hahn*, D. M. S. Bagguley (ed.), Clarendon Press, Oxford 1992, pp. 317–345.

[30] Meiboom S. and D. Gill, Modified spin-echo method for measuring nuclear relaxation times, *Rev. Sci. Instrum.* **29** (1958), 668–691. Also in *NMR in Biomedicine: The Physical Basis*, E. Fukushima (ed.), Key papers in Physics **2**, American Institute of Physics, New York 1989, pp. 80-83.

[31] Miller, D. H. and W. I. McDonald, Neuroimaging in multiple sclerosis, *Clinical Neuroscience* **2** (1994), 215–224.

[32] Moore, C. C. and J. A. Wolf, Square integrable representations of nilpotent Lie groups, *Trans. Amer. Math. Soc.* **185** (1973), 445–462.

[33] Schempp, W., *Harmonic Analysis on the Heisenberg Nilpotent Lie Group, with Applications to Signal Theory* Pitman Research Notes in Mathematics Series **147**, Longman Scientific and Technical, London, 1986.

[34] Schempp, W., Phase coherent wavelets, Fourier transform magnetic resonance imaging, and synchronized time-domain neural networks, *Proc. V.A. Steklov Inst. Math.* **203** (1994), 389–428.

[35] Schempp, W., Quantum holography and magnetic resonance imaging, in *Circular-Grating, Light-Emitting Sources*, S.I. Najafi, N. Peyghambarian, and M. Fallahi (eds.), SPIE **2398**, Bellingham, WA 1995, pp. 34–40.

[36] Schempp, W., Wavelet interference, Fourier transform magnetic resonance imaging, and temporally encoded synchronized neural networks, to appear.

[37] Schwartz, L., Sous-espaces hilbertiens d'espaces vectoriels topologiques et noyaux associés (noyaux reproduisants), *J. Analyse Math.* **13** (1964), 115–256.

[38] Schwartz, L., Sous-espaces hilbertiens et noyaux associés; applications aux représentations des groupes de Lie. Deuxième Colloq. l'Anal. Fonct., *Centre Belge Recherches Mathématiques*, Librairie Universitaire, Louvain 1964, pp. 153–163.

[39] Shucker, D. S., Square integrable representations of unimodular groups, *Proc. Amer. Math. Soc.* **89** (1983), 169–172.

[40] Tonumura, A., *Electron Holography*, Springer Series in Optical Sciences **70**, Springer-Verlag, Heidelberg, 1993.

[41] Truwit, C. L. and T. E. Lempert, *High Resolution Atlas of Cranial Neuroanatomy*, Williams & Wilkins, Baltimore, MD, 1994.

[42] Vogl, T. J., *MR-Angiographie und MR-Tomographie des Gefäßsystems*, Springer-Verlag, Heidelberg, 1995.

[43] Wallner, B. and Herausgeber, *MR-Angiographie*, Georg Thieme Verlag, Stuttgart, New York, 1993.

[44] Wolf, J., Investigation of bubbly flow by ultrasonic tomography, *Part. Syst. Charact.* **5** (1988), 170–173.

[45] Worthington, B. S., J. L. Firth, G. K. Morris, I. R. Johnson, R. Coxon, A. M. Blamire, P. Gibbs, and P. Mansfield, The clinical potential of ultra-high-speed echo-planar imaging, *Phil. Trans. R. Soc. London A* **333** (1990), 507–514. Also in *NMR Imaging*, P. Mansfield and E.L. Hahn (eds.), The Royal Society, London 1990, pp. 105–112.

[46] Yock, D. H., Jr., *Magnetic Resonance Imaging of CNS Disease*, Mosby-Year Book, St. Louis, Baltimore, 1995.

Walter Schempp
Lehrstuhl fuer Mathematik I
University of Siegen
D-57068 Siegen, Germany
schempp@mathematik.uni-siegen.d400.de

The Generalized Gabor Scheme and Its Application in Signal and Image Representation

Meir Zibulski and Yehoshua Y. Zeevi

Abstract. The single-window Gabor scheme is generalized by incorporating multi-windows as well as kernels other than the exponential. Two types of schemes are considered, a continuous-variable scheme and a discrete-variable scheme. The properties of the sequence of representation functions are characterized by an approach which combines the concept of frames and the Zak transform. The frame operator associated with the multi-window Gabor-type frame is examined by representing the frame operator as a finite order matrix-valued function. Completeness and frame properties of the sequence of representation functions are examined in relation to the properties of the matrix-valued function. Calculation of the frame bounds and the dual frame, as well as the issue of tight frames, are considered. It is shown that the properties of the sequence of representation functions are essentially unchanged by replacing the widely-used exponential kernel with other kernels. For the continuous scheme, the Balian-Low theorem is generalized to consideration of a scheme of multi-windows. For the discrete scheme, generalized inverses and representations other than those based on the dual frame are considered. Examples illustrate the advantages of the multi-window scheme over the single-window scheme.

§1 Introduction

Many problems in engineering and science involve representation of a given signal by a sequence of functions, the so-called linear expansion of a signal. The two major issues involved in such an expansion are finding a suitable set of expansion coefficients and characterization of the sequence properties.

Two types of such sequences, extensively utilized and investigated in recent years, are the wavelets and the Gabor expansion. One of the advantages offered by using these types of sequences is their simple structure, which is based on a single-window (prototype) function translated and dilated in the wavelet case, or translated and modulated in the Gabor case. These two types of sequences are special cases of the following generalized sequence:

$$s_{r,m,n}(x) = g_r(x - na_r)\phi_{r,m}(x), \qquad (1.1)$$

Signal and Image Representation in Combined Spaces
Y. Y. Zeevi and R. R. Coifman (Eds.), pp. 121-164.

where $\{g_r(x)\}$ is a sequence of window functions, $\{a_r\}$ is a sequence of real numbers, and $\{\phi_{r,m}(x)\}$ is a sequence of kernel functions. For example, in the classical Gabor case [9], we use a single-window function $g(x)$, with $a_r = a$, b some positive constant, $\phi_{r,m}(x) = e^{j2\pi mbx}$, and $m, n \in \mathbf{Z}$. In the wavelet case, $\phi_{r,m}(x) = 1$, $g_r(x) = b^{-r/2}g(x/b^r)$, $a_r = ab^r$, $r, n \in \mathbf{Z}$, and $g(x)$ is a mother wavelet function.

In this paper we present some results which concern the properties of Gabor-type subsequences of the generalized sequence $\{s_{r,m,n}\}$. Analysis of a related subsequence can be found in [14].

§2   Preliminaries

2.1   Frames [6]

Definition 1. *A sequence $\{\psi_n\}$ in a Hilbert Space H constitutes a frame if there exists numbers $0 < A \leq B < \infty$ s.t. $\forall f \in H$ we have: $A\|f\|^2 \leq \sum_n |\langle f, \psi_n\rangle|^2 \leq B\|f\|^2$, where $\langle \cdot, \cdot \rangle$ denotes an inner-product in Hilbert space.*

Definition 2. *Given a frame $\{\psi_n\}$ in H, the frame operator $\mathcal{S} : H \to H$ is defined by : $\mathcal{S}f \triangleq \sum_n \langle f, \psi_n\rangle \psi_n$.*

Corollary 1. *Every $f \in H$ can be presented by means of the frame or the dual frame $\{\mathcal{S}^{-1}\psi_n\}$, as follows:*

$$f = \sum_n \langle f, \mathcal{S}^{-1}\psi_n\rangle \psi_n = \sum_n \langle f, \psi_n\rangle \mathcal{S}^{-1}\psi_n.$$

2.2   The Zak transform

The ZT [11] of a signal $f(x)$ is defined as follows:

$$(\mathcal{Z}f)(x, u) \triangleq \alpha^{1/2} \sum_{k \in \mathbf{Z}} f\Big[\alpha(x + k)\Big]e^{-j2\pi uk}, \quad -\infty < x, u < \infty, \quad (2.2.1)$$

with a fixed parameter $\alpha > 0$.

The ZT satisfies the following periodic and quasiperiodic relations:

$$(\mathcal{Z}f)(x, u + 1) = (\mathcal{Z}f)(x, u), \quad (\mathcal{Z}f)(x + 1, u) = e^{j2\pi u}(\mathcal{Z}f)(x, u). \quad (2.2.2)$$

As a consequence of these two relations, the ZT is completely determined by its values over the unit square $(x, u) \in ([0, 1)^2)$. This is the essence of this unitary mapping.

Based on the ZT defined by (2.2.1), we define the *Piecewise Zak Transform* (PZT) as a vector-valued function $F(x,u)$ of size p [27]:

$$F(x,u) = \left[F_0(x,u), F_1(x,u) \ldots, F_{p-1}(x,u) \right]^T, \qquad (2.2.3)$$

where

$$F_i(x,u) = (\mathcal{Z}f)\left(x, u + \frac{i}{p}\right), \quad 0 \leq i \leq p-1, \ i \in \mathbf{Z}. \qquad (2.2.4)$$

The vector-valued function $F(x,u)$ belongs to $L^2([0,1) \times [0,1/p); \mathbb{C}^p)$, which is a Hilbert space with the inner-product:

$$\langle F, G \rangle = \int_0^1 dx \int_0^{1/p} du \sum_{i=0}^{p-1} F_i(x,u) \overline{G_i(x,u)}.$$

Since the ZT is a unitary mapping from $L^2(\mathbb{R})$ to $L^2([0,1)^2)$, the PZT is a unitary mapping from $L^2(\mathbb{R})$ to $L^2([0,1) \times [0,1/p); \mathbb{C}^p)$. As a consequence, we obtain the following inner-product preserving property:

$$\int_{-\infty}^{\infty} f(x)\overline{g(x)}dx = \int_0^1 \int_0^1 (\mathcal{Z}f)(x,u)\overline{(\mathcal{Z}g)(x,u)}dxdu$$

$$= \int_0^1 dx \int_0^{1/p} du \sum_{i=0}^{p-1} F_i(x,u)\overline{G_i(x,u)}.$$

This unitary property of the PZT allows us the translation from $L^2(\mathbb{R})$ to $L^2([0,1) \times [0,1/p); \mathbb{C}^p)$, where issues regarding Gabor-type representations are often easier to deal with and understand.

§3   Multi-window Gabor-type expansions

3.1   Generalization of the Gabor scheme

Generalizing the Gabor scheme by using several window functions instead of a single-one, the representation of a given signal $f(x) \in L^2(\mathbb{R})$ is given by:

$$f(x) = \sum_{r=0}^{R-1} \sum_{m,n} c_{r,m,n} g_{r,m,n}(x), \qquad (3.1.1)$$

where

$$g_{r,m,n}(x) = g_r(x - na)e^{j2\pi mbx}, \qquad (3.1.2)$$

and $\{g_r(x)\}$ is a set of R distinct window functions. Such a set can incorporate, for example, Gaussian windows of various widths. In this case, with proper oversampling, one can overcome the uncertainty constraint expressed by the limitations of having either a high temporal (spatial) resolution and low frequency resolution or vice versa. This type of a richer Gabor-type representation can be instrumental in applications such as detection of transient signals [8], or in identification and recognition of various prototypic textures and other features in images [18]. Clearly, if $R = 1$ we obtain the single-window Gabor representation.

The characterization of the sequence $\{g_{r,m,n}\}$ can be divided into three categories according to the sampling density of the combined space (the so-called phase space density) defined by $d \triangleq R(ab)^{-1}$: undersampling— $d < 1$, critical sampling— $d = 1$, and oversampling— $d > 1$.

Detailed discussion of the results and proofs of theorems presented in this and the next sections can be found in [26, 30].

3.2 Matrix algebra approach for the analysis of frames' properties

In order to examine the properties of the sequence $\{g_{r,m,n}\}$, we consider the operator:

$$Sf = \sum_{r=0}^{R-1} \sum_{m,n} \langle f, g_{r,m,n} \rangle g_{r,m,n}. \tag{3.2.1}$$

This is clearly the frame operator if $\{g_{r,m,n}\}$ constitutes a frame. For the single-window Gabor scheme, this operator was examined by straightforward methods [4, 10], where application of the ZT was restricted to the case $ab = 1$. We show that, if the product ab is a rational number, it might be advantageous to examine this operator in $L^2([0,1) \times [0,1/p); \mathbb{C}^p)$ by using the PZT. We use the ZT with the parameter $\alpha = 1/b$. However, the same analysis can be applied with $\alpha = a$, but with a differently defined PZT, wherein the x-variable of the ZT is considered in a piecewise manner.

A major result of analysis in the PZT domain is the representation of the frame operator as a finite order matrix-valued function, as stated in the following theorem.

Theorem 1. *Let* $ab = p/q$, $p, q \in \mathbb{N}$, *and let* S_z *be the frame operator of the sequence, which is the PZT of* $\{g_{r,m,n}\}$. *The action of* S_z *in* $L^2([0,1) \times [0,1/p); \mathbb{C}^p)$ *is given by the following matrix algebra:*

$$(S_z F)(x, u) = \mathbf{S}(x, u) F(x, u), \tag{3.2.2}$$

where $\mathbf{S}(x, u)$ *is a* $p \times p$ *matrix-valued function whose entries are given by:*

$$S_{i,k}(x, u) = \frac{1}{p} \sum_{r=0}^{R-1} \sum_{l=0}^{q-1} Zg_r\left(x - l\frac{p}{q}, u + \frac{i}{p}\right) \overline{Zg_r\left(x - l\frac{p}{q}, u + \frac{k}{p}\right)};$$

$$i, k = 0, \cdots, p - 1, \qquad (3.2.3)$$

and the vector-valued function $F(x, u)$ is given by (2.2.3) and (2.2.4).

Since the PZT is a unitary transform, (3.2.2) is an isometrically isomorphic representation of $\mathcal{S}$ (3.2.1).

Using the PZT and the matrix representation of the operator $\mathcal{S}$, we examine the properties of the sequence $\{g_{r,m,n}\}$ for a rational ab. Since, in the case of undersampling the sequence $\{g_{r,m,n}\}$ is not complete, the results presented next are relevant in the cases of critical sampling and oversampling.

The following theorem examines the completeness of the sequence $\{g_{r,m,n}\}$ in relation to the structure of the matrix-valued function $\mathbf{S}(x, u)$.

Theorem 2. *Given $g_r \in L^2(\mathbb{R})$, $0 \le r \le R - 1$, and a matrix-valued function $\mathbf{S}(x, u)$, $(x, u) \in ([0, 1) \times [0, 1/p))$ as in (3.2.3), the sequence $\{g_{r,m,n}\}$ associated with $\{g_r\}$, $ab = p/q$, $p, q \in \mathbb{N}$ is complete if and only if $\det(\mathbf{S})(x, u) \ne 0$ a.e. on $[0, 1) \times [0, 1/p)$.*

We can also find the frame bounds of the sequence $\{g_{r,m,n}\}$. Let

$$\lambda_{\max}(\mathbf{S}) \stackrel{\triangle}{=} \text{ess sup}_{(x,u) \in ([0,1) \times [0,1/p))} \max_{1 \le i \le p} \lambda_i(\mathbf{S})(x, u) \qquad (3.2.4)$$

$$\lambda_{\min}(\mathbf{S}) \stackrel{\triangle}{=} \text{ess inf}_{(x,u) \in ([0,1) \times [0,1/p))} \min_{1 \le i \le p} \lambda_i(\mathbf{S})(x, u), \qquad (3.2.5)$$

where $\lambda_i(\mathbf{S})(x, u)$ are the eigenvalues of the matrix $\mathbf{S}(x, u)$. Then, the upper frame-bound $B = \lambda_{\max}(\mathbf{S})$, and the lower frame-bound $A = \lambda_{\min}(\mathbf{S})$. This result yields the following theorem.

Theorem 3. *The sequence $\{g_{r,m,n}\}$ associated with $\{g_r\}$, $g_r \in L^2(\mathbb{R})$, $0 \le r \le R - 1$, and $ab = p/q$, $p, q \in \mathbb{N}$ constitutes a frame if and only if $0 < \lambda_{\min}(\mathbf{S}) \le \lambda_{\max}(\mathbf{S}) < \infty$.*

Finding the eigenvalues of a matrix-valued function may be a difficult task. We therefore propose an alternative approach to determining whether $\{g_{r,m,n}\}$ constitutes a frame. First, we introduce a Lemma, which formulates a necessary and sufficient condition for the existence of an upper frame bound $B < \infty$.

Lemma 1. *The sequence $\{g_{r,m,n}\}$ associated with $\{g_r\}$, $g_r \in L^2(\mathbb{R})$, $0 \le r \le R - 1$ and $ab = p/q$, $p, q \in \mathbb{N}$, has an upper frame bound $B < \infty$ if and only if $(\mathcal{Z}g_r)(x, u)$ are all bounded a.e. on $(0, 1]^2$ $(\mathcal{Z}g_r \in L^\infty((0, 1]^2))$.*

Second, we present a theorem which determines whether the sequence $\{g_{r,m,n}\}$ constitutes a frame, when an upper frame bound exists, and which does not necessitate calculation of the eigenvalues of $\mathbf{S}(x, u)$.

Theorem 4. *Given $g_r \in L^2(\mathbb{R})$, $0 \le r \le R - 1$, such that there exists an upper frame bound $B < \infty$ for the sequence $\{g_{r,m,n}\}$ associated with $\{g_r\}$, and $ab = p/q$, $p, q \in \mathbb{N}$. The sequence $\{g_{r,m,n}\}$ constitutes a frame if and only if $0 < K \le \det(\mathbf{S})(x, u)$ a.e. on $[0, 1) \times [0, 1/p)$, where the matrix-valued function $\mathbf{S}(x, u)$ is as in (3.2.3).*

A sequence $\{\psi_n\}$ in a Hilbert space H constitutes a *tight frame* if

$$\sum_n |\langle f, \psi_n \rangle|^2 = A\|f\|^2$$

for all $f \in H$ (*i.e.* $A = B$). This is a special type of frame for which the upper bound equals the lower bound. For example, an orthonormal basis is a particular case of a tight frame with $A = 1$. In fact, a tight frame with $A = 1$ and $\|\psi_n\| = 1$ for all ψ_n constitutes an orthonormal basis [4]. Considering these properties of tight frames, we introduce the following theorem concerning tight frames and the matrix representation of the frame operator.

Theorem 5. *Given $g_r \in L^2(\mathbb{R})$, $0 \le r \le R - 1$, and a matrix-valued function $\mathbf{S}(x, u)$ as in (3.2.3), the sequence $\{g_{r,m,n}\}$ associated with $\{g_r\}$, $ab = p/q$, $p, q \in \mathbb{N}$ constitutes a tight frame if and only if $\mathbf{S}(x, u) = A\mathbf{I}$ a.e., where $\mathbf{I}$ is the identity matrix, and $A = \frac{q}{p} \sum_{r=0}^{R-1} \|g_r\|^2$.*

In order to use a representation such as suggested by corollary 1, we have to obtain the dual frame. In general, for this purpose one can use operator techniques [6, 4]. For the single-window scheme finding the dual frame $\{S^{-1}g_{m,n}\}$ is simplified since it is generated by a single dual frame window function [10, 4]. In fact, this is also the case for the multi-window scheme, *i.e.* let $\{\gamma_{r,m,n}\}$ denote the dual frame of $\{g_{r,m,n}\}$ then $\{\gamma_{r,m,n}\}$ is generated by a finite set of R dual frame window functions $\{\gamma_r\}$:

$$\gamma_{r,m,n}(x) = \gamma_r(x - na)e^{j2\pi mbx}, \quad 0 \le r \le R - 1,$$

where $\gamma_r = S^{-1}g_r$. Using the matrix representation (3.2.2) of the frame operator, the PZT of γ_r, is:

$$\Gamma_r(x, u) = \mathbf{S}^{-1}(x, u)G_r(x, u), \qquad (3.2.6)$$

that is, $\Gamma_r(x, u), G_r(x, u)$ are vector-valued functions in $L^2([0, 1) \times [0, 1/p);$ $\mathbb{C}^p)$, and $\mathbf{S}^{-1}(x, u)$ is the inverse matrix of $\mathbf{S}(x, u)$. For example, $\mathbf{S}^{-1}(x, u) = [\det(\mathbf{S})(x, u)]^{-1}\text{adj}(\mathbf{S})(x, u)$.

3.3 The Balian-Low theorem in the case of multi-windows

The case of a single-window Gabor scheme with critical sampling, *i.e.* $ab = 1$, was thoroughly examined in the literature [4, 10]. In short, by choosing an appropriate $g(x)$, the sequence $\{g_{r,m,n}\}$ can be complete and constitute a frame, albeit the problems of stability [4]. The sequence $\{g_{r,m,n}\}$ can, however, be complete and not constitute a frame, in which case the representation is unstable. A classical example of such an unstable scheme is the one with a Gaussian window function.

In the case of critical sampling, the following theorem of Balian and Low indicates that a wide range of "well behaved" – rapidly decaying and smooth functions – $g(x)$ are excluded from being proper candidates for generators of frames [4, 2, 5].

Theorem 6. *Given $g \in L^2(\mathbb{R})$, $a > 0$ and $ab = 1$, if the sequence $\{g_{r,m,n}\}$ constitutes a frame, then either $xg(x) \notin L^2(\mathbb{R})$ or $g'(x) \notin L^2(\mathbb{R})$.*

Note that $g'(x) \in L^2(\mathbb{R}) \Leftrightarrow \omega\hat{g}(\omega) \in L^2(\mathbb{R})$, where $\hat{g}$ is the Fourier transform of g.

One of the solutions for this problem is oversampling. In fact, it was recently proven that in the case of a Gaussian, the $\{g_{r,m,n}\}$ constitutes a frame for all $ab < 1$ [13, 22].

In the case of critical sampling of the multi-window scheme, an interesting question is whether we can "beat" the Balian-Low Theorem by utilizing several windows. If all the windows in the set $\{g_r\}$ are well-behaved functions, it is still not possible to beat the Balian-Low condition, as indicated by the following theorem.

Theorem 7. *Given $g_r \in L^2(\mathbb{R})$, $0 \le r \le R - 1$, $a > 0$ and $R(ab)^{-1} = 1$, if the sequence $\{g_{r,m,n}\}$, as in (3.1.2), constitutes a frame, then either $xg_r(x) \notin L^2(\mathbb{R})$ or $g'_r(x) \notin L^2(\mathbb{R})$ for some $0 \le r \le R - 1$.*

In fact, as we shall show later, Theorem 7 applies to a wider class of representation functions with a critical sampling of the combined space.

One of the advantages of using more than one window is the possibility of overcoming the constraint imposed by the Balian-Low Theorem on the choice of window functions, by adding an extra window function of proper nature such that the resultant scheme of critical sampling constitutes a frame. Whether one can find a non-well-behaved window function complementary to a set of well-behaved window functions (such that the inclusive set will generate a frame for critical sampling) depends on the nature of the set of the well-behaved window functions as indicated by the following proposition.

Proposition 1. *Let a set, $\{g_r\}$, $0 \leq r \leq R-2$, of $R-1$ window functions be given. Denote by $\mathbf{G}^0(x,u)$ the $R-1 \times R$ matrix-valued function with entries $G^0_{r,k}(x,u) = \overline{\mathcal{Z}g_r(x, u + \frac{k}{R})}$, and $\mathbf{P}(x,u) = \mathbf{G}^0(x,u)\mathbf{G}^{0^*}(x,u)$. There exists a window function $g_{R-1}(x)$ such that the inclusive set $\{g_r\}$, $0 \leq r \leq R-1$ generates a frame for the critical sampling case, if and only if $0 < K \leq \det(\mathbf{P})(x,u)$ a.e. on $[0,1) \times [0,1/R)$.*

An example of $R-1$ well-behaved window functions, which satisfy $0 < K \leq \det(\mathbf{P})(x,u)$ a.e. on $[0,1) \times [0,1/R)$, can be constructed in the following manner. Take a window function $g(x)$ such that the sequence $\{g(x - n/b)e^{j2\pi mx/a}\}$ constitutes a frame for $ab = R/(R-1)$. Note, that this is an oversampling scheme $(1/(ab) = (R-1)/R < 1)$ and that there exist, therefore, well-behaved window functions $g(x)$ such that $\{g(x - n/b)e^{j2\pi mx/a}\}$ constitutes a frame (for example the Gaussian function). Construct the following $R-1$ window functions:

$$g_r(x) = g\left(x - \frac{rR}{b(R-1)}\right).$$

Clearly these are well-behaved window functions. Moreover, we obtain $(\mathcal{Z}g_r)(x,u) = (\mathcal{Z}g)(x - r\frac{R}{R-1}, u)$, and the matrix-valued function $\mathbf{G}^0(x,u)$ equals the matrix-valued function $\mathbf{G}(x,u)$, which corresponds to the sequence $\{g(x - na)e^{j2\pi mbx}\}$ (denoted by $\{g_{r,m,n}\}$). In this case, $\{g_{r,m,n}\}$ corresponds to an undersampling scheme. By the duality principle, as presented in [21, Theorem 2.2(e)], since $\{g(x - n/b)e^{j2\pi mx/a}\}$ constitutes a frame, $\{g_{r,m,n}\}$ constitutes a Riesz basis for a sub-space of $L^2(\mathbb{R})$. It can therefore be shown that $0 < K \leq \det(\mathbf{P})(x,u)$ a.e. on $[0,1) \times [0,1/R)$.

§4 Further generalizations and applications of the multi-window scheme

4.1 A wavelet-type scheme based on a different sampling density for each window

Consider a generalization of the sequence $\{g_{r,m,n}\}$ where for each window function $g_r(x)$ there is a different set of parameters a_r, b_r. Explicitly, we have

$$g_{r,m,n}(x) = g_r(x - na_r)e^{j2\pi mb_r x}, \qquad (4.1.1)$$

and the sampling density of the combined space is:

$$d \stackrel{\triangle}{=} \sum_{r=0}^{R-1} (a_r b_r)^{-1}.$$

Similar to the analysis in Section 3, the characterization of the sequence $\{g_{r,m,n}\}$ can be divided into the categories of undersampling, critical sampling, and oversampling according to $d < 1$, $d = 1$, $d > 1$, respectively.

In order to analyze this kind of a scheme, we convert it into an equivalent one with $a_r = a$, $b_r = b$ for all r [26, 30] and utilize the tools presented in the previous section for the analysis of the sequence properties.

Utilizing the degrees of freedom of choosing a different set of parameters a_r, b_r for each window function g_r, we construct a wavelet-type scheme. Let α, β be positive, real numbers. Given a window function $g(x)$ let

$$g_r(x) = \alpha^{-r/2} g(\alpha^{-r} x).$$

Also, let $a_r = \beta \alpha^r$, and $a_r b_r = R/d$ for all r, where d is the sampling density of the combined space. We then have

$$g_{r,m,n}(x) = \alpha^{-r/2} g(\alpha^{-r} x - n\beta) e^{j2\pi \frac{mxR}{\beta \alpha^r d}}.$$

In this scheme, the width of the window is proportional to the translation step a_r, whereas, the product $a_r b_r$ is constant. As such, this scheme incorporates scaling, which is characteristic of wavelets and of the Gabor scheme with the logarithmically-distorted frequency axis [17]. However, in contrast with wavelets, this scheme has a finite number of window functions, *i.e.* resolution levels, and each of the windows is modulated by the infinite set of functions defined by the kernel. Each subset for fixed m can be considered as a finite (incomplete) wavelet-type set, in that it obeys the properties of scaling and translation of a complex prototypic "mother wavelet". For example, if R/d is an integer, the mother wavelet is defined by $g(x) e^{j2\pi \frac{mxR}{\beta d}}$, if not, this definition holds within a "complex phase". Thus, all the functions corresponding to each of these mother wavelets are self-similar. We illustrate in Figure 1 an example with a Gaussian window function and $\alpha = 2, \beta = 1, d = 1, R = 4$. The real part of $g_{r,m,n}(x)$ is depicted in Figure 1(a)-(c), for $m = 0, 1$ and 2, respectively. Each of the three parts of Figure 1 shows the four representation functions, $r = 0, 1, 2, 3$, $n = 1$, superimposed on each other.

Scaling has been realized as an important property of sets of representation functions which are used in the analysis of natural signals and images. Hence, the importance of this type of generalization of the Gabor scheme. The exponential kernel can be replaced by any non-exponential kernel which obeys some mild conditions.

4.2 Irregular periodic sampling of the combined space

As an application of the multi-window case we consider irregular periodic sampling of the combined space for the single window case. For such a

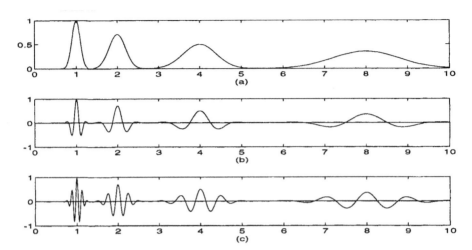

Figure 1. Example of wavelet-type Gaborian representation functions with a
Gaussian window. Each of the three displays of this figure, (a)-(c), illustrates
the real part of the four functions, $r = 0, 1, 2, 3$, $n = 1$, superimposed on each
other, with a different value for m: (a) $m = 0$ (b) $m = 1$ (c) $m = 2$. Note the
self-similarity of the functions within each subset of representation functions.

scheme, given two finite sets of length R, $\{a_n\}$, $\{b_n\}$, and a window function
g, we construct the R window functions

$$g_r(x) = g(x - a_r)e^{j2\pi b_r x}.$$

We can now view the representation functions

$$g_{r,m,n}(x) = e^{-j2\pi b_r na}g(x - na - a_r)e^{j2\pi x(mb+b_r)} \qquad (4.2.1)$$

as localized around the lattice $\cup_{r=0}^{R-1}(na + a_r, mb + b_r)$, $n, m \in \mathbf{Z}$ of the
combined (phase) space.

A particularly interesting case of such a scheme is $a_r = \frac{a}{R}r$, $b_r = \frac{b}{R}r$.
In this case, if $\{g_{r,m,n}\}$ constitutes a frame, the dual frame is generated by
a single-window function

$$\gamma_{r,m,n}(x) = e^{-j2\pi b_r na}\gamma(x - na - a_r)e^{j2\pi m(b+b_r)x},$$

where $\gamma = S^{-1}g$ [26, 30]. If we choose $R = 2$ we obtain the hexagonal
lattice (Figure 2(a)). An example of the $R = 3$ case is illustrated in Figure
2(b); The lattice points of the combined space and the two (not unique)
generating vectors which generate the lattice points are shown. Note that
for an arbitrary lattice, if the coordinates of the generating vectors are all
rational numbers, there exists a scheme, with irregular periodic sampling

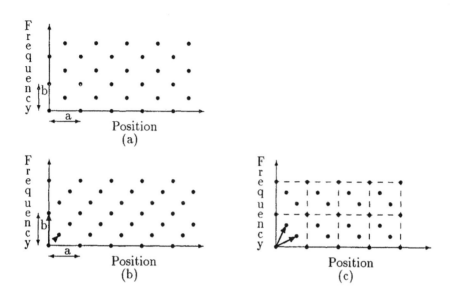

Figure 2. Irregular periodic sampling of the combined space. (a) Hexagonal lattice. (b) The case of $R = 3$ with $a_r = ar/3$, $b_r = br/3$. This is also an example of the Gabor representation on lattices, where the two generating vectors of the lattice points are shown. (c) An example of a lattice and of a rectangular sub-lattice (connected with dashed lines).

of the combined space, which generates the same lattice points of the combined space. This is true, since in such a case there exists a rectangular sub-lattice where each cell defined by the sub-lattice is replicated across the combined space. A simple example is illustrated in Figure 2(c), where the rectangular sub-lattice is connected with dashed lines.

4.3   Non-exponential kernels

The sequence of representation functions $\{g_{r,m,n}\}$ as defined in (3.1.2) according to the original scheme [9], is constructed by utilizing a sampled exponential kernel $e^{j2\pi mbx}$. In many applications of signal processing, other kernels, with a better performance in specific applications, are utilized.

Therefore, we further generalize the scheme of representation and consider
(3.1.1), with the sequence $\{g_{r,m,n}\}$ replaced by the sequence $\{g_{r,m,n}^{\phi}\}$:

$$g_{r,m,n}^{\phi}(x) = g_r(x - na)\phi_m(x). \tag{4.3.1}$$

We assume that $\{\phi_m\}$ constitutes a frame for $L^2([0, 1/b))$, and that each
$\phi_m(x)$ is a periodic function with a period of $1/b$. Note that this implies
that $\{\phi_m\}$ constitutes a frame for $L^2([y, y + 1/b))$, where $y \in \mathbb{R}$ is a parameter.

We first consider the special case where the sequence $\{\sqrt{b}\phi_m\}$ constitutes a tight frame. Clearly, the sequence $\{\sqrt{b}e^{j2\pi mbx}\}$ is a particular
example of such a sequence, since it constitutes an orthonormal basis. For
a single-window and a critical sampling case, a scheme involving orthogonal
bases was presented in [7]. The interesting (but not surprising) property of
such a generalization is that the frame property is not changed, as indicated
by the following theorem.

Theorem 8. *Let the sequence $\{\sqrt{b}\phi_m\}$, where each $\phi_m(x)$ is a $1/b$-periodic
function, constitute a tight frame for $L^2([0, 1/b))$, with a frame bound A_ϕ.
Also, let S' be the frame operator of the sequence $\{g_{r,m,n}^{\phi}\}$ as in (4.3.1),
and S be the frame operator of the sequence $\{g_{r,m,n}\}$ as in (3.1.2). We
then have $S' = A_\phi S$.*

The following corollary is a consequence of Theorem 8.

Corollary 2. *Let the sequences $\{g_{r,m,n}\}$, $\{g_{r,m,n}^{\phi}\}$, be as in Theorem 8,
then:*

1. *The sequence $\{g_{r,m,n}\}$ is complete/a frame/a tight frame if and only
   if the same is true for the sequence $\{g_{r,m,n}^{\phi}\}$.*

2. *Let A and B be the lower and upper frame bounds of the the sequence
   $\{g_{r,m,n}\}$ respectively, and let A' and B' be the lower and upper frame
   bounds of the sequence $\{g_{r,m,n}^{\phi}\}$ respectively, then, $A' = A_\phi A$, $B' =
   A_\phi B$.*

3. *Theorem 7 holds for the sequence $\{g_{r,m,n}^{\phi}\}$.*

4. *Let $\{g_{r,m,n}^{\phi}\}$ be a frame with the dual frame $\{\gamma_{r,m,n}^{\phi}\}$. Then, within a
   constant, $\{\gamma_{r,m,n}^{\phi}\}$ is generated by the same R functions which generate the dual frame of $\{g_{r,m,n}\}$, i.e. $\gamma_{r,m,n}^{\phi}(x) = A_\phi^{-1}\gamma_r(x - na)\phi_m(x)$,
   where $\gamma_r = S^{-1}g_r$.*

5. *If the sequence $\{\sqrt{b}\phi_m\}$ is an orthonormal basis, then the sequence
   $\{g_{r,m,n}\}$ is an exact frame (Riesz basis)/an orthonormal basis if and
   only if the same is true for the sequence $\{g_{r,m,n}^{\phi}\}$.*

Next, we consider the general case where $\{\phi_m\}$ constitutes a frame for $L^2([0,1/b))$ and each $\phi_m(x)$ is $1/b$-periodic. We then have the following theorem.

Theorem 9. *Let the sequence $\{\sqrt{b}\phi_m\}$ constitute a frame for $L^2([0,1/b))$, where each $\phi_m(x)$ is a $1/b$-periodic function. Then, the sequence $\{g^\phi_{r,m,n}\}$ as in (4.3.1), constitutes a frame if and only if the sequence $\{g_{r,m,n}\}$ as in (3.1.2), constitutes a frame. Moreover, let A_ϕ, B_ϕ, be the lower, upper frame bounds of $\{\sqrt{b}\phi_m\}$ respectively, and let A, B be the lower, upper frame bounds of $\{g_{r,m,n}\}$ respectively. Then, $A_\phi A \leq A' \leq B_\phi A$, $A_\phi B \leq B' \leq B_\phi B$, where A', B' are the lower, upper frame bounds of $\{g^\phi_{r,m,n}\}$ respectively.*

An arbitrary L^2 function can be represented by means of the sequence $\{g^\phi_{r,m,n}\}$ as follows [26, 30]:

$$f = \sum_{r=0}^{R-1} \sum_{m,n} \langle f, \gamma^\psi_{r,m,n} \rangle g^\phi_{r,m,n}.$$

where $\{\sqrt{b}\psi_m\}$ is the dual frame of $\{\sqrt{b}\phi_m\}$ in $L^2([0,1/b))$ and $\gamma_r = S^{-1}g_r$. Note that in general $\{\gamma^\psi_{r,m,n}\}$ is not the dual frame of $\{g^\phi_{r,m,n}\}$, nevertheless, if $\{g^\phi_{r,m,n}\}$ constitutes an exact frame (a Riesz basis), then, indeed, $\{\gamma^\psi_{r,m,n}\}$ is the dual frame (biorthonormal basis) of $\{g^\phi_{r,m,n}\}$, since, the expansion is unique.

§5   Discrete-variable (finite) multi-window Gabor-type schemes

5.1   Introduction

In the previous sections the continuous-variable Gabor scheme was generalized to incorporate several window functions as well as kernels other than the exponential. In this section we examine the discrete-finite case of the multi-window Gabor-type scheme. Moreover, we consider issues which are of interest only in the case of the discrete scheme (algorithms, complexity, etc.), and present some further analysis which was not considered in the previous section. Preliminary results were presented in [29].

We consider discrete-variable signals, which are L-periodic, that is signals which satisfy: $f(i + L) = f(i)$, $i \in \mathbf{Z}$. For such signals, given two divisors M, N of L, i.e. $L = N'M = M'N$, where M, N, M', N' are all positive integers, we propose the following scheme of representation:

$$f(i) = \sum_{r=0}^{R-1} \sum_{m=0}^{M-1} \sum_{n=0}^{N-1} c_{r,m,n} g_{r,m,n}(i), \tag{5.1.1}$$

where for a set of R window functions $\{g_r(i)\}$, we define

$$g_{r,m,n}(i) = g_r(i - mN') \exp\left(2\pi\imath\frac{in}{N}\right), \qquad (5.1.2)$$

and $\imath \triangleq \sqrt{-1}$. One can recognize two particular cases. First, if $R = 1$ we obtain the known discrete single-window (Gabor) scheme presented in [24]. Second, if $N = 1$ we obtain a finite discrete version of multirate filter banks, where each window function corresponds to a different filter. Thus, a scheme of type (5.1.1) incorporates the Gabor scheme (single-window) and multirate filter banks.

Let $d \triangleq RMN/L$ be the sampling density of the discrete combined space (so-called phase space). We then identify the three sampling categories as in the continuous case. (For further results and proofs see [26, 31].)

5.2   Preliminaries

The following notations are utilized in this section. As noted $\imath = \sqrt{-1}$. Lowercase boldfaced letters denote column vectors and vector-valued functions such as $\mathbf{f}$, $\mathbf{g}(i)$ respectively. Capital boldfaced letters denote matrices and matrix-valued functions such as $\mathbf{X}$, $\mathbf{G}(i)$, respectively. The notation $(\mathbf{G})_{i,k}$ stands for the entries of the matrix $\mathbf{G}$ (similar notation for vectors as well). The entries indices always start from zero. We use $\overline{f}$ to denote the complex conjugate of f, and $\mathbf{f}^*$ to denote the complex conjugate transpose of a vector $\mathbf{f}$ (similarly for matrices). $\mathbf{Z}$ denotes the integers and for a positive integer N, $i \in \underline{N}$ means $i = 0, 1, \cdots, N - 1$.

The concept of frames was presented in Section 2.1. Recall that given a frame $\{x_n\}$ in a Hilbert space H, every $f \in H$ can be represented by either the frame or the dual frame in the following manner:

$$f = \sum_n \langle f, \mathcal{S}^{-1}x_n \rangle x_n = \sum_n \langle f, x_n \rangle \mathcal{S}^{-1}x_n,$$

where $\mathcal{S}$ is the frame operator. Unless the frame is a basis, the representation coefficients $\langle f, \mathcal{S}^{-1}x_n \rangle$ are not unique. The choice of the dual frame for computing the representation coefficients yields the minimal energy solution of the representation coefficients [6]. For a finite-dimensional Hilbert space (as is our case), this solution corresponds to the so-called generalized inverse. Also, finite-dimensional spaces satisfying the frame property and completeness are identical. Therefore, if the sequence $\{x_n\}$ does not constitute a frame, it is impossible to expand any desired vector by means of $\{x_n\}$. We can, however, find the best approximation of the signal by means of the Moore-Penrose or generalized inverse. Moreover, since for finite-dimensional spaces, every sequence is a frame in its own span, the best approximation can also be found by means of the dual frame in the subspace spanned by the frame, and the dual frame can be found utilizing the Moore-Penrose inverse of the frame operator. For general finite dimensional spaces, these facts are summarized in the following proposition [26] (for infinite-dimensional spaces see [20]).

Proposition 2. *Let H be a finite-dimensional Hilbert space, and let $\{x_n\}$ be some finite sequence in H. Denote by $S^\dagger$ the Moore-Penrose inverse of the frame operator S. Then,*

(i) *$\{S^\dagger x_n\}$ is the dual frame of $\{x_n\}$ in the sub-space spanned by $\{x_n\}$.*

(ii) *For $f \in H$ define $f_{ap} = \sum_n \langle f, S^\dagger x_n \rangle x_n$, then, $\|f - f_{ap}\|$ is minimal, and the expansion coefficients $\langle f, S^\dagger x_n \rangle$ are of minimal norm. Moreover, $f_{ap} = \sum_n \langle f, x_n \rangle S^\dagger x_n$.*

The ZT is a fundamental tool in the analysis of the Gabor expansion. We utilize the discrete finite version of the transform [1, 25]. The Finite Zak Transform (FZT) of an L-periodic one-dimensional signal is defined by:

$$(\mathcal{Z}^a f)(i, v) \triangleq \sum_{k=0}^{b-1} f(i + ka) \exp\left(-2\pi i \frac{kv}{b}\right), \quad (i, v) \in \mathbf{Z}^2, \qquad (5.2.1)$$

where a, b are integers such that $L = ab$. The FZT is a DFT-based transform and therefore can be realized using fast algorithms [1, 25]. Unless explicitly stated otherwise, we will use the FZT with $a = N$, $b = M'$. We then omit the superscript a and write $(\mathcal{Z}f)(i, v) \triangleq (\mathcal{Z}^N f)(i, v)$.

Denote by $l^2(\mathbf{Z}/L)$ the Hilbert space of L-periodic square-summable one-dimensional signals with the following inner-product

$$\langle f, g \rangle = \sum_{i=0}^{L-1} f(i)\overline{g(i)}, \qquad (5.2.2)$$

where $f, g \in l^2(\mathbf{Z}/L)$. For the problem addressed in this paper it is natural to define vector-valued functions which express a sort of decimation. Given $f \in l^2(\mathbf{Z}/L)$, define a vector-valued function of size M' as:

$$\mathbf{f}(i) = [(\mathbf{f})_0(i), \ldots, (\mathbf{f})_{M'-1}(i)]^T, \qquad (5.2.3)$$

where

$$(\mathbf{f})_k(i) = f(i + kN), \ k \in \underline{M'}.$$

Note that it is sufficient to consider only $i \in \underline{N}$. The inner-product in the domain of vector-valued functions can be expressed as:

$$\langle \mathbf{f}, \mathbf{g} \rangle = \sum_{i=0}^{N-1} \mathbf{g}^*(i) \mathbf{f}(i),$$

Clearly $\langle f, g \rangle = \langle \mathbf{f}, \mathbf{g} \rangle$.

It was shown previously that, for the analysis of the Gabor scheme in the case of oversampling, the concept of vector-valued functions is also useful in the ZT domain. A vector-valued function obtained using the Piecewise Zak Transform (PZT) was introduced for such an analysis. For the discrete scheme, $L = MN' = NM'$. Let $L/(MN) = p/q$, where p, q are relatively prime integers, then, $M'/p = M/q$ is an integer. Based on the definition of the FZT, (5.2.1), we define the Piecewise Finite Zak Transform (PFZT) [29] as a vector-valued function of size p:

$$\tilde{\mathbf{f}}(i, v) = [(\tilde{\mathbf{f}})_0(i, v), \ldots, (\tilde{\mathbf{f}})_{p-1}(i, v)]^T, \tag{5.2.4}$$

where

$$(\tilde{\mathbf{f}})_k(i, v) \triangleq (\mathcal{Z}f)(i, v + k\frac{M'}{p}), \quad k \in \underline{p}. \tag{5.2.5}$$

Note that it is sufficient to consider $i \in \underline{N}, \ v \in \underline{M'/p}$. The vector-valued function $\tilde{\mathbf{f}}(i, v)$ belongs to a Hilbert space of vector-valued functions with the inner-product:

$$\langle \tilde{\mathbf{f}}, \tilde{\mathbf{g}} \rangle = \frac{1}{M'} \sum_{i=0}^{N-1} \sum_{v=0}^{\frac{M'}{p}-1} \tilde{\mathbf{g}}^*(i, v) \tilde{\mathbf{f}}(i, v). \tag{5.2.6}$$

We obtain here, too, the inner-product preserving property

$$\langle f, g \rangle = \langle \mathcal{Z}f, \mathcal{Z}g \rangle = \langle \tilde{\mathbf{f}}, \tilde{\mathbf{g}} \rangle. \tag{5.2.7}$$

5.3 Matrix algebra approach in the discrete case

As in the continuous case, we examine the operator:

$$\mathcal{S}f = \sum_{r=0}^{R-1} \sum_{m=0}^{M-1} \sum_{n=0}^{N-1} \langle f, g_{r,m,n} \rangle g_{r,m,n}. \tag{5.3.1}$$

If the sequence $\{g_{r,m,n}\}$ constitutes a frame, this is clearly the frame operator. We show in [26, 31] that in the signal domain of vector-valued functions, the action of the frame operator can be expressed in terms of matrix algebra as:

$$(\mathcal{S}\mathbf{f})(i) = \mathbf{S}(i)\mathbf{f}(i), \quad i \in \underline{N}, \tag{5.3.2}$$

where $\mathbf{S}(i)$ is a $M' \times M'$ matrix-valued function with elements given by

$$(\mathbf{S})_{k,l}(i) = N \sum_{r=0}^{R-1} \sum_{m=0}^{M-1} g_r(i + kN - mN') \overline{g_r(i + lN - mN')}, \qquad (5.3.3)$$

and $\mathbf{f}(i)$ is a vector-valued function of size M', defined by (5.2.3). The matrix $\mathbf{S}(i)$ is self-adjoint and positive semi-definite for each i, since by defining an $RM \times M'$ matrix-valued function $\mathbf{G}(i)$:

$$\mathbf{G}(i) = \begin{pmatrix} \mathbf{G}_0(i) \\ \vdots \\ \mathbf{G}_{R-1}(i) \end{pmatrix}, \qquad (5.3.4)$$

where each $\mathbf{G}_r(i)$ is a matrix-valued function of size $M \times M'$ with elements:

$$(\mathbf{G}_r)_{m,l}(i) = \overline{g_r(i + lN - mN')},$$

we obtain

$$\mathbf{S}(i) = N\mathbf{G}^*(i)\mathbf{G}(i). \qquad (5.3.5)$$

Note that $\mathbf{S}(i + N') = \mathbf{S}(i)$, *i.e.* the entries of $\mathbf{S}(i)$ are periodic with period N', when considering $i \in \mathbf{Z}$.

In [26, 31] we show that the action of the frame operator in the PFZT domain can be expressed in terms of the following matrix algebra:

$$(\mathcal{S}\tilde{f})(i, v) = \tilde{\mathbf{S}}(i, v)\tilde{\mathbf{f}}(i, v), \quad i \in \underline{N}, \ v \in \underline{M'/p}, \qquad (5.3.6)$$

where $\tilde{\mathbf{S}}(i, v)$ is a $p \times p$ matrix-valued function with elements given by

$$(\tilde{\mathbf{S}})_{k,j}(i, v) = \frac{N}{p} \sum_{r=0}^{R-1} \sum_{l=0}^{q-1} (\mathcal{Z}g_r)(i - lN', v + kM'/p) \overline{(\mathcal{Z}g_r)(i - lN', v + jM'/p)},$$

$$(5.3.7)$$

and $\tilde{\mathbf{f}}(i, v)$ is a vector-valued function of size p, defined by (5.2.4). As in the signal domain, the matrix $\tilde{\mathbf{S}}(i, v)$ is self-adjoint and positive semi-definite for each (i, v), since by defining a $Rq \times p$ matrix-valued function $\tilde{\mathbf{G}}(i, v)$:

$$\tilde{\mathbf{G}}(i, v) = \begin{pmatrix} \tilde{\mathbf{G}}_0(i, v) \\ \vdots \\ \tilde{\mathbf{G}}_{R-1}(i, v) \end{pmatrix}, \qquad (5.3.8)$$

where each $\tilde{\mathbf{G}}_r(i)$ is a matrix-valued function of size $q \times p$ with elements:

$$(\tilde{\mathbf{G}}_r)_{l,j}(i, v) = \overline{(\mathcal{Z}g_r)(i - lN', v + jM'/p)},$$

we obtain

$$\tilde{\mathbf{S}}(i, v) = \frac{N}{p}\tilde{\mathbf{G}}^*(i, v)\tilde{\mathbf{G}}(i, v). \tag{5.3.9}$$

Note that $\tilde{\mathbf{S}}(i + N', v) = \tilde{\mathbf{S}}(i, v)$, *i.e.* in the transform domain, too, the matrix $\tilde{\mathbf{S}}(i, v)$ is periodic.

5.4   Frame properties

The above analysis of the discrete case is similar conceptually to the analysis of the continuous case. For the discrete case, however, we presented also a signal domain analysis which does not appear in the continuous case (for such an analysis one has to employ infinite matrices [21]). The results obtained by such an analysis are similar to those of the continuous case.

The frame bounds A, B can be derived by calculating the eigenvalues of the matrix-valued function $\mathbf{S}(i)$, *i.e.*

$$A = \min_{i\in\underline{N},\ j\in\underline{M'}} \lambda_j(\mathbf{S})(i), \tag{5.4.1}$$

$$B = \max_{i\in\underline{N},\ j\in\underline{M'}} \lambda_j(\mathbf{S})(i), \tag{5.4.2}$$

where $\lambda_j(\mathbf{S})(i)$ are the eigenvalues of the matrix $\mathbf{S}(i)$. Also

$$A = \min_{i\in\underline{N},\ v\in\underline{M'}/p,\ j\in\underline{p}} \lambda_j(\tilde{\mathbf{S}})(i, v), \tag{5.4.3}$$

$$B = \max_{i\in\underline{N},\ v\in\underline{M'}/p,\ j\in\underline{p}} \lambda_j(\tilde{\mathbf{S}})(i, v), \tag{5.4.4}$$

where $\lambda_j(\tilde{\mathbf{S}})(i, v)$ are the eigenvalues of the matrix $\tilde{\mathbf{S}}(i, v)$. $B/A=1$.

The following theorems consider frame properties of the sequence $\{g_{r,m,n}\}$.

Theorem 10. *Given* $g_r \in l^2(\mathbf{Z}/L)$, $r \in \underline{R}$, *and a matrix-valued function* $\mathbf{S}(i)$ *as in (5.3.3) (or* $\tilde{\mathbf{S}}(i, v)$ *as in (5.3.7)), the sequence* $\{g_{r,m,n}\}$ *associated with* $\{g_r\}$ *constitutes a frame if and only if* $\det(\mathbf{S})(i) \neq 0$ *for all* $i \in \underline{N}$, *(or* $\det(\tilde{\mathbf{S}})(i, v) \neq 0$ *for all* $i \in \underline{N}$, $v \in \underline{M'}/p$).

Note that $\det(\mathbf{S})(i) \neq 0$ is equivalent to saying that $\mathbf{G}(i)$ is of full rank.

Theorem 11. *Given* $g_r \in l^2(\mathbf{Z}/L)$, $r \in \underline{R}$, *and a matrix-valued function* $\mathbf{S}(i)$ *as in (5.3.3) (or* $\tilde{\mathbf{S}}(i, v)$ *as in (5.3.7)), the sequence* $\{g_{r,m,n}\}$ *associated with* $\{g_r\}$ *constitutes a tight frame if and only if* $\mathbf{S}(i) = A\mathbf{I}$ *(or* $\tilde{\mathbf{S}}(i, v) = A\mathbf{I}$), *where* $\mathbf{I}$ *is the identity matrix, and* $A = \frac{MN}{L}\sum_{r=0}^{R-1}\|g_r\|^2$.

If $M = N = L$ (a maximal oversampling rate), $\mathbf{S}(i)$ is scalar-valued and, based on (5.3.2), (5.3.3), we have

$$(\mathcal{S}f)(i) = f(i)L \sum_{r=0}^{R-1} \|g_r\|^2,$$

i.e. the sequence $\{g_{r,m,n}\}$ is always a tight frame (for any set of window functions) with the bounds $A = B = L \sum_{r=0}^{R-1} \|g_r\|^2$. This property is called resolution of identity [29].

Note that, in general, if $\{x_n\}$ constitutes a frame, $\{\mathcal{S}^{-1/2}x_n\}$ constitutes a tight frame with $A = 1$ [23]. This fact can be utilized for the construction of tight frames of the form $\{\mathcal{S}^{-1/2}g_{r,m,n}\}$. Moreover, let $h_r = \mathcal{S}^{-1/2}g_r$, then it can be shown that $\mathcal{S}^{-1/2}g_{r,m,n} = h_{r,m,n}$, *i.e.* the tight frame is generated by the R window functions $h_r(i)$ (see [23, Proposition 4.6] for the single-window continuous-variable case and [12, Section 2.8] for the single-window discrete-variable case). Clearly, $h_r(x)$ can be found by utilizing $\mathbf{S}^{-1/2}(i)$ (or $\tilde{\mathbf{S}}^{-1/2}(i, v)$).

5.5   The dual frame

The structure of the dual frame is identical to that of the frame $\{g_{r,m,n}\}$, *i.e.* it is generated by R window functions $\gamma_r(i)$

$$\gamma_{r,m,n}(i) = \gamma_r(i - mN') \exp\left(2\pi i \frac{in}{N}\right). \tag{5.5.1}$$

In fact, $\gamma_r = \mathcal{S}^{-1}g_r$, and in the signal domain of vector-valued functions the R dual-frame window functions can be calculated by the inverse of the matrix $\mathbf{S}(i)$:

$$\gamma_r(i) = \mathbf{S}^{-1}(i)\mathbf{g}_r(i). \tag{5.5.2}$$

Note that $\mathbf{S}^{-1}(i)$ is in fact the inverse of the $M' \times M'$ matrix $\mathbf{S}(i)$ for each $i \in \underline{N}$, where i is a parameter.

Alternatively, the R windows of the dual frame can be calculated in the transform domain by the inverse of the matrix $\tilde{\mathbf{S}}(i, v)$:

$$\tilde{\gamma}_r(i, v) = \tilde{\mathbf{S}}^{-1}(i, v)\tilde{\mathbf{g}}_r(i, v), \tag{5.5.3}$$

where $\tilde{\gamma}_r(i, v), \tilde{\mathbf{g}}_r(i, v)$ are the PFZT of $\gamma_r(i), g_r(i)$.

Note that in the case of critical sampling the matrix-valued function $\mathbf{G}(i)$ (or $\tilde{\mathbf{G}}(i, v)$) is a square matrix. Therefore, $\mathbf{S}^{-1}(i) = \frac{1}{N}\mathbf{G}^{-1}(i)\mathbf{G}^{-1*}(i)$. Since $\mathbf{G}(i)$ can be constructed by utilizing the vector-valued functions $\mathbf{g}_r(i)$:

$$\mathbf{G}_r(i) = \begin{pmatrix} \mathbf{g}_r^*(i) \\ \vdots \\ \mathbf{g}_r^*(i - (M-1)N') \end{pmatrix}, \tag{5.5.4}$$

where $r \in \underline{R}$, and $\mathbf{G}(i)$ as in (5.3.4), we obtain for the dual-frame window functions

$$\gamma_r(i) = \frac{1}{N} \begin{pmatrix} (\mathbf{G}^{-1})_{0,rM}(i) \\ \vdots \\ (\mathbf{G}^{-1})_{M'-1,rM}(i) \end{pmatrix}. \tag{5.5.5}$$

That is, $\gamma_r(i)$ is equal to the appropriate column (the rM-th column) of the matrix-valued function $\frac{1}{N}\mathbf{G}^{-1}(i)$. Similarly, in the transform domain $\tilde{\gamma}_r(i)$ is equal to the appropriate column (the rq-th column) of the matrix-valued function $\frac{p}{N}\tilde{\mathbf{G}}^{-1}(i, v)$.

5.6   Generalized inverse and undersampling

Recall that by utilizing the Moore-Penrose or generalized inverse we obtain the minimal norm least-square solution to a representation problem in finite-dimensional spaces. In our case, if $\{g_{r,m,n}\}$ does not constitute a frame, we can use the Moore-Penrose inverse in order to find the minimal norm coefficients for which expansion of type (5.1.1) yields the best approximation of a given signal. As noted, the coefficients can be found by means of the Moore-Penrose inverse of the frame operator:

$$c_{r,m,n} = \langle f, \mathcal{S}^\dagger g_{r,m,n} \rangle.$$

The structure of the sequence $\{\mathcal{S}^\dagger g_{r,m,n}\}$, which is the dual frame of $\{g_{r,m,n}\}$ in its own span, is, as expected, identical to the structure of $\{g_{r,m,n}\}$. That is, let $\gamma_{r,m,n} = \mathcal{S}^\dagger g_{r,m,n}$, then (5.5.1) holds, where $\gamma_r = \mathcal{S}^\dagger g_r$. Moreover, the calculation of γ_r can be made similarly to (5.5.2) where $\mathbf{S}^{-1}(i)$ should be replaced by $\mathbf{S}^\dagger(i)$, or similarly to (5.5.3) where $\tilde{\mathbf{S}}^{-1}(i, v)$ should be replaced by $\tilde{\mathbf{S}}^\dagger(i, v)$. Furthermore, let $\boldsymbol{\Gamma}(i)$ ($\tilde{\boldsymbol{\Gamma}}(i, v)$), be a matrix-valued function associated with the set $\{\gamma_r\}$, in the same manner as $\mathbf{G}(i)$ ($\tilde{\mathbf{G}}(i, v)$) is associated with the set $\{g_r\}$. Based on the previous results and on the structure of these matrix-valued functions we have $\boldsymbol{\Gamma}^*(i) = \mathbf{S}^\dagger(i)\mathbf{G}^*(i)$. Since, $\mathbf{G}^\dagger = (\mathbf{G}^*\mathbf{G})^\dagger \mathbf{G}^*$ [3, Theorem 1.2.1], this implies

$$\boldsymbol{\Gamma}^*(i) = \frac{1}{N}\mathbf{G}^\dagger(i), \tag{5.6.1}$$

or in the transform domain

$$\tilde{\boldsymbol{\Gamma}}^*(i, v) = \frac{p}{N}\tilde{\mathbf{G}}^\dagger(i, v). \tag{5.6.2}$$

Equations (5.6.1) and (5.6.2) express an algorithm which can be applied for *all possible cases* of finding the dual-frame window functions in the sub-space spanned by the $\{g_{r,m,n}\}$.

As indicated by Theorem 10, the frame property of the sequence $\{g_{r,m,n}\}$ can be checked by examining $\det(\mathbf{S})(i)$ (or $\det(\tilde{\mathbf{S}})(i, v)$). If the sequence

$\{g_{r,m,n}\}$ does not constitute a frame, the dimension of the space spanned by the sequence $\{g_{r,m,n}\}$ can be calculated by examining the matrix-valued function $\mathbf{G}(i)$ (or $\tilde{\mathbf{G}}(i,v)$), as shown in the following theorem.

Theorem 12. *Given $g_r \in l^2(\mathbf{Z}/L)$, $r \in \underline{R}$, and a matrix-valued function $\mathbf{G}(i)$ as in (5.3.4) (or $\tilde{\mathbf{G}}(i,v)$ as in (5.3.8)), the dimension of the space spanned by the sequence $\{g_{r,m,n}\}$, associated with $\{g_r\}$, is equal to $\sum_{i=0}^{N-1} \text{rank}(\mathbf{G})(i)$ (or to $\sum_{i=0}^{N-1} \sum_{v=0}^{M'/p-1} \text{rank}(\tilde{\mathbf{G}})(i,v)$).*

As a matter of fact, in the cases of critical sampling and oversampling we would like to design the sequence $\{g_{r,m,n}\}$, i.e. choose the window functions g_r, such that the sequence constitutes a frame (or a basis) and the reconstruction be perfect. In the case of undersampling we have no other choice but to utilize the Moore-Penrose inverse in order to get the reconstruction error as small as possible. Moreover, given an undersampling scheme we would like the $\{g_{r,m,n}\}$ to constitute a basis to a sub-space of $l^2(\mathbf{Z}/L)$, i.e. choose the window functions g_r such that the $g_{r,m,n}$ are linearly independent. This basis property is examined in the next theorem through the introduction of the following matrix-valued functions

$$\mathbf{P}(i) = N\mathbf{G}(i)\mathbf{G}^*(i), \tag{5.6.3}$$

where $\mathbf{G}(i)$ is as in (5.3.4), and

$$\tilde{\mathbf{P}}(i,v) = \frac{N}{p}\tilde{\mathbf{G}}(i,v)\tilde{\mathbf{G}}^*(i,v), \tag{5.6.4}$$

where $\tilde{\mathbf{G}}(i,v)$ is as in (5.3.8).

Theorem 13. *Given $g_r \in l^2(\mathbf{Z}/L)$, $r \in \underline{R}$, and a matrix-valued function $\mathbf{P}(i)$ as in (5.6.3) (or $\tilde{\mathbf{P}}(i,v)$ as in (5.6.4)), in the case of undersampling, the sequence $\{g_{r,m,n}\}$ associated with $\{g_r\}$ constitutes a basis for a sub-space of $l^2(\mathbf{Z}/L)$, if and only if $\det(\mathbf{P})(i) \neq 0$ for all $i \in \underline{N}$ (or $\det(\tilde{\mathbf{P}})(i,v) \neq 0$ for all $i \in \underline{N}$, $v \in \underline{M'/p}$).*

Note that in case $\{g_{r,m,n}\}$ constitutes a basis in its own span, we have the following formulae for the calculation of the Moore-Penrose inverse [3, Theorem 1.3.2]

$$\mathbf{S}^\dagger(i) = N\mathbf{G}^*(i)\mathbf{P}^{-1}(i)\mathbf{P}^{-1}(i)\mathbf{G}(i),$$

and

$$\tilde{\mathbf{S}}^\dagger(i,v) = \frac{N}{p}\tilde{\mathbf{G}}^*(i,v)\tilde{\mathbf{P}}^{-1}(i,v)\tilde{\mathbf{P}}^{-1}(i,v)\tilde{\mathbf{G}}(i,v). \tag{5.6.5}$$

Moreover, since $\mathbf{G}(i)$ can be constructed utilizing the vector-valued functions $\mathbf{g}_r(i)$ as in (5.5.4), each dual-frame (basis) window function, $\gamma_r(i)$, is equal to the appropriate column (the rM-th column) of the matrix-valued function $\mathbf{G}^*(i)\mathbf{P}^{-1}(i)$. Similarly, in the transform domain, $\tilde{\gamma}_r(i)$ is equal to the appropriate column (the rq-th column) of the matrix-valued function $\tilde{\mathbf{G}}^*(i,v)\tilde{\mathbf{P}}^{-1}(i,v)$.

The case where $\{g_{r,m,n}\}$ constitutes an orthonormal basis in its own span can be characterized also by utilizing the matrix-valued functions $\mathbf{P}(i)$, $\tilde{\mathbf{P}}(i,v)$. The following theorem presents such a characterization.

Theorem 14. *Given $g_r \in l^2(\mathbf{Z}/L)$, $r \in \underline{R}$, and a matrix-valued function $\mathbf{P}(i)$ as in (5.6.3) (or $\tilde{\mathbf{P}}(i,v)$ as in (5.6.4)), in the case of undersampling, the sequence $\{g_{r,m,n}\}$ associated with $\{g_r\}$ constitutes an orthonormal basis for a sub-space of $l^2(\mathbf{Z}/L)$, if and only if $\mathbf{P}(i) = \mathbf{I}$ (or $\tilde{\mathbf{P}}(i,v) = \mathbf{I}$), where $\mathbf{I}$ is the identity matrix.*

5.7 Representations other than the dual frame

In the case of oversampling, the expansion coefficients are not unique. As noted, the dual frame provides the coefficients which are of minimal energy in l^2 sense. We can find, however, different coefficients which still satisfy (5.1.1) by calculating the inner products of the signal with sequences of the type $\{\gamma_{r,m,n}\}$, generated by different dual window functions, which are not the dual-frame window functions (see [19, 24] for an example in the single-window case).

Theorem 15. *Let $\{g_r\}$, $\{\gamma_r\}$ be two given sets of R functions each. Then,*

$$f = \sum_{r=0}^{R-1}\sum_{m=0}^{M-1}\sum_{n=0}^{N-1} \langle f, \gamma_{r,m,n}\rangle g_{r,m,n}, \qquad (5.7.1)$$

for all $f \in l^2(\mathbf{Z}/L)$, if and only if

$$N\mathbf{G}^*(i)\mathbf{\Gamma}(i) = \mathbf{I}, \quad \left(\frac{N}{p}\tilde{\mathbf{G}}^*(i,v)\tilde{\mathbf{\Gamma}}(i,v) = \mathbf{I}\right). \qquad (5.7.2)$$

Note that a necessary condition to satisfy (5.7.2) is that both $\{g_{r,m,n}\}$ and $\{\gamma_{r,m,n}\}$ constitute frames (but one is not necessarily the dual frame of the other), since both $\mathbf{G}(i)$, $\mathbf{\Gamma}(i)$, should be of full rank for each i [15, Section 12.6].

Given a sequence $\{g_{r,m,n}\}$ that constitutes a frame, *i.e.* $\mathbf{G}(i)$ is of full rank for each i, one of the possible solutions for $\mathbf{\Gamma}(i)$ that satisfies (5.7.2) is the Moore-Penrose inverse of $N\mathbf{G}^*(i)$. Clearly, this solution corresponds to the dual frame of $\{g_{r,m,n}\}$. It is well known that the Moore-Penrose inverse is of minimal Frobenius norm among all possible solutions. This implies that the sum

$$\sum_{r=0}^{R-1}\sum_{m=0}^{M-1}\sum_{l=0}^{M'-1} |\gamma_r(i + lN - mN')|^2,$$

attains its minimum for each i, if $\{\gamma_{r,m,n}\}$ is the dual frame of $\{g_{r,m,n}\}$. Therefore, among all possible sets $\{\gamma_r\}$ that satisfy (5.7.1), the window functions of the dual frame are of minimal norm. For the single-window scheme this fact was noticed in [12].

5.8 Single-window scheme

The representation scheme in the single-window case is:

$$f(i) = \sum_{m=0}^{M-1}\sum_{n=0}^{N-1} c_{m,n} g_{m,n}(i), \tag{5.8.1}$$

where for the window function $g(i)$ we define

$$g_{m,n}(i) = g(i - mN') \exp\left(2\pi i \frac{in}{N}\right), \quad L = N'M. \tag{5.8.2}$$

In the critical sampling case ($L = MN$), $\tilde{\mathbf{S}}(i,v) = N|\mathcal{Z}g(i,v)|^2$ is scalar-valued. Therefore, there is an advantage in using the FZT. For example, $\{g_{m,n}\}$ constitutes a frame (which is a basis in this case) if and only if $(\mathcal{Z}g)(i,v)$ does not vanish. Also, $\{g_{m,n}\}$ constitutes an orthonormal basis if and only if $N|\mathcal{Z}g(i,v)|^2 = 1$ for all $i \in \underline{N}$, $v \in \underline{M}$. The dual frame window is given by $(\mathcal{Z}\gamma)(i,v) = (N\overline{(\mathcal{Z}g)(i,v)})^{-1}$, and the expansion coefficients are

$$c_{m,n} = \frac{1}{M}\sum_{i=0}^{N-1}\sum_{v=0}^{M-1} \frac{(\mathcal{Z}f)(i,v)}{N(\mathcal{Z}g)(i,v)} \exp\left(-2\pi i \frac{in}{N}\right) \exp\left(2\pi i \frac{vm}{M}\right). \tag{5.8.3}$$

One of the disadvantages of the case of critical sampling is the instability that may occur because of a zero of $(\mathcal{Z}g)(i,v)$. One of the solutions to this problem is to consider an oversampling scheme ($MN > L$), in which case the FZT plays an important role as well [29]. The effect of oversampling is easily understood in the case of an integer oversampling rate, *i.e.* for $L/(MN) = 1/q$, where $q > 1$ is an integer. In such a case, $\tilde{\mathbf{S}}(i,v)$ is scalar-valued, $\tilde{\mathbf{S}}(i,v) = N\sum_{l=0}^{q-1} |(\mathcal{Z}g)(i - lN',v)|^2$. Therefore, if $\mathcal{Z}g$ has a single zero only (in a square of size $N \times M'$ in the (i,v) plane), such an oversampling will clearly stabilize the scheme. Note that if $\tilde{\mathbf{S}}(i,v)$ does not vanish, $i.e.$ $\{g_{m,n}\}$ constitutes a frame, the expansion coefficients are given by

$$c_{m,n} = \frac{1}{MN} \sum_{i=0}^{N-1} \sum_{v=0}^{M-1} \frac{(\mathcal{Z}f)(i,v)\overline{(\mathcal{Z}g)(i,v)}}{\sum_{l=0}^{q-1}|(\mathcal{Z}g)(i-lN',v)|^2}$$
$$\times \exp\left(-2\pi i \frac{in}{N}\right) \exp\left(2\pi i \frac{vm}{M}\right). \qquad (5.8.4)$$

Moreover, if $\tilde{S}(i,v)$ does vanish, the expansion coefficients which correspond to the generalized inverse solution can be found by summing in (5.8.4) only over the non-zero values of $\tilde{S}(i,v)$, similarly to the critical sampling case.

In some cases there is an advantage in performing the analysis in the signal domain. For example, consider $g(i)$ such that $g(i) = 0$ for $N \le i \le L-1$ ($g(i)$ is compactly supported). In such a case, $S(i)$ is diagonal with $(S)_{k,k}(i) = \sum_{m=0}^{M-1}|g(i+kN-mN')|^2$. Therefore, $\{g_{m,n}\}$ constitutes a frame if and only if $(S)_{0,0}(i)$ does not vanish for $i \in \underline{N'}$. It is a tight frame if and only if $(S)_{0,0}(i)$ is a constant function, and the dual frame is given by

$$\gamma(i) = \frac{g(i)}{\sum_{m=0}^{M-1}|g(i-mN')|^2}.$$

It is interesting to point out the particular case of an integer under-sampling rate, $i.e.$ $L/(MN) = p$, where p is an integer. In this case, $\tilde{G}(i,v)$ is of size $1 \times p$, $i.e.$ it is a vector-valued function, and $\tilde{P}(i,v) = N/p \sum_{k=0}^{p-1}|(\mathcal{Z}g)(i,v+kM'/p)|^2$ is scalar-valued. In fact, $\tilde{G}(i,v) = \tilde{g}^*(i,v)$. Therefore, if $\{g_{m,n}\}$ constitutes a basis in its own span, $i.e.$ $\tilde{P}(i,v)$ does not vanish, based on (5.6.5), the FZT of the window function that generates the dual (biorthonormal) basis is

$$(\mathcal{Z}\gamma)(i,v) = \frac{N(\mathcal{Z}g(i,v)}{p\sum_{k=0}^{p-1}|(\mathcal{Z}g)(i,v+kM'/p)|^2}. \qquad (5.8.5)$$

Note that if $\tilde{P}(i,v)$ does vanish we can still calculate $\tilde{S}^\dagger(i,v)$, in which case the dual frame (a frame in its own span) window function is given by (5.8.5), but is zero wherever $\tilde{P}(i,v)$ vanishes. This particular case was analyzed by a different method in [16], yielding a similar solution.

5.9 Nonrectangular sampling of the combined space

In the case of a single-window scheme, one may view the representation functions $g_{m,n}(i)$ as localized around the points (mN', nM') of the combined space of size $L \times L$. Clearly, the functions defined in (5.8.2) represent a rectangular sampling of the combined space (rectangular lattice). Utilizing the scheme of several window functions one may sample the combined space in a nonrectangular way.

For example, consider a two-window scheme such that $g_0(i) = g(i)$, $g_1(i) = g(i-a) \exp\left(2\pi i \frac{ib}{L}\right)$, where $a \in \underline{N'}$, $b \in \underline{M'}$. The combined space is sampled at the points $(mN', nM') \cup (mN' + a, nM' + b)$. We may construct a hexagonal lattice by taking $a = N'/2$, $b = M'/2$ (assuming N', M' is divisible by 2).

A general setting of a nonrectangular sampling, based on a single-window function $g(i)$, is based on (5.1.1), where the R-window functions g_r are defined by

$$g_r(i) = g(i-a_r) \exp\left(2\pi i \frac{ib_r}{L}\right), \tag{5.9.1}$$

and it is sufficient to consider $a_r \in \underline{N'}$, $b_r \in \underline{M'}$. We assume that $a_r = a_k$, $b_r = b_k$ only if $r = k$, and then have

$$g_{r,m,n}(i) = \exp\left(-2\pi i \frac{ib_r mN'}{L}\right) g(i - mN' - a_r) \exp\left(2\pi i \frac{i(nM' + b_r)}{L}\right).$$

In this setting the sequence $\{g_{r,m,n}\}$ is generated by a single-window function. This is not true for the dual frame $\{\gamma_{r,m,n}\}$, except for some particular cases. For example, if

$$a_r = \frac{N'}{R} r, \quad b_r = \frac{M'}{R} r$$

(assuming N', M' are divisible by R), the dual frame is generated by a single-window function $\gamma(i)$:

$$\gamma_{r,m,n}(i) = \exp\left(-2\pi i \frac{ib_r mN'}{L}\right) \gamma(i - a_r - mN') \exp\left(2\pi i \frac{i(nM' + b_r)}{L}\right),$$

and the frame window function can be found in the signal domain $\gamma(i) = \mathbf{S}^{-1}(i)\mathbf{g}(i)$ (or in the transform domain $\tilde{\gamma}(i, v) = \tilde{\mathbf{S}}^{-1}(i, v)\tilde{\mathbf{g}}(i, v)$) [26, 31].

5.10 Filter-bank schemes

For such a scheme $N = 1$, the scheme of representation becomes

$$f(i) = \sum_{r=0}^{R-1} \sum_{m=0}^{M-1} c_{r,m} g_{r,m}(i), \tag{5.10.1}$$

where

$$g_{r,m}(i) = g_r(i - mN'), \quad L = N'M. \tag{5.10.2}$$

The expansion coefficients in this case are given by

$$c_{r,m} = \sum_{i=0}^{L-1} f(i)\overline{\gamma_r(i - mN')}.$$

For each r, the expansion coefficients are obtained by a cyclic convolution of the signal with $\overline{\gamma_r(-i)}$ and decimation by the factor N'. Therefore, this is a filter-bank scheme, for L-periodic functions, consisting of R filters decimated by factor N'.

In the signal domain, the matrix $\mathbf{S}(i)$ is the largest possible of size $L \times L$. We prefer to make the analysis in the transform domain. In such a case, the FZT coincides with the DFT ($M' = L$), we therefore use the DFT directly. The DFT of an L-periodic signal is defined by

$$\hat{f}(v) = \sum_{i=0}^{L-1} f(i) \exp\left(-2\pi i \frac{iv}{L}\right), \quad i \in \mathbf{Z}, \tag{5.10.3}$$

and indeed $\hat{f}(v) = (\mathcal{Z}f)(0, v)$. The function $\hat{f}(v)$ is also L-periodic. The representation functions in the transform domain are

$$\hat{g}_{r,m}(v) = \hat{g}_r(v) \exp\left(-2\pi i \frac{vmN'}{L}\right). \tag{5.10.4}$$

Let

$$\hat{\mathbf{f}}(v) = [(\hat{\mathbf{f}})_0(v), \ldots, (\hat{\mathbf{f}})_{N'-1}(v)]^T, \tag{5.10.5}$$

where $(\hat{\mathbf{f}})_k(v) = \hat{f}(v + kM)$, $k \in \underline{N'}$ ($\hat{\mathbf{f}}(i) = \tilde{\mathbf{f}}(0, v)$). The action of the frame operator is then represented as

$$(\mathcal{S}\hat{\mathbf{f}})(v) = \hat{\mathbf{S}}(v)\hat{\mathbf{f}}(v), \quad v \in \underline{M}.$$

Since $M'/p = M$, $q = 1$, $p = N'$, based on (5.3.7), the matrix-valued function $\hat{\mathbf{S}}(v)$ is of size $N' \times N'$ with elements

$$(\hat{\mathbf{S}})_{k,l}(v) = \frac{1}{N'} \sum_{r=0}^{R-1} \hat{g}_r(v + kM)\overline{\hat{g}_r(v + lM)}. \tag{5.10.6}$$

Now, Theorems 10 and 11 can be used with the matrix-valued function $\hat{\mathbf{S}}(v)$.

5.11 Representation functions with non-exponential kernels

As in the continuous case, a general expansion scheme with kernel other than an exponential is based on (5.1.1), with the sequence $\{g_{r,m,n}\}$ replaced by the sequence $\{g^\phi_{r,m,n}\}$:

$$g^\phi_{r,m,n}(i) = g_r(i - mN')\phi_n(i). \tag{5.11.1}$$

Each of the functions $\phi_n(i)$ is N-periodic, and the sequence $\{\frac{1}{\sqrt{N}}\phi_n\}$ constitutes an orthonormal basis for $l^2(\mathbf{Z}/N)$.

This generalization does not change the frame properties of the sequence as shown by the following theorem.

Theorem 16. *Let the sequence $\{\frac{1}{\sqrt{N}}\phi_n\}_{n=0}^{N-1}$ be an N-periodic orthonormal basis for $l^2(\mathbf{Z}/N)$. Then, the sequence $\{g_{r,m,n}\}$ as in (5.1.2) and the sequence $\{g_{r,m,n}^{\phi}\}$ as in (5.11.1) have the same frame operator S.*

Note that the operator S is the same for the two sequences, even when $\{g_{r,m,n}\}$ does not constitute a frame.

As a consequence of Theorem 16, we obtain the following corollary.

Corollary 3. *Let the sequence $\{g_{r,m,n}\}$ be as in (5.1.2), and the sequence $\{g_{r,m,n}^{\phi}\}$ as in (5.11.1), where, $\{\frac{1}{\sqrt{N}}\phi_n\}_{n=0}^{N-1}$ constitutes an orthonormal basis for $l^2(\mathbf{Z}/N)$.*

1. *The sequence $\{g_{r,m,n}\}$ is a frame/a tight frame/a basis/an orthonormal basis if and only if the same is true for the sequence $\{g_{r,m,n}^{\phi}\}$.*

2. *If $\{g_{r,m,n}\}$ constitutes a frame, then $\{g_{r,m,n}^{\phi}\}$ constitutes a frame with the same frame bounds.*

3. *Let $\{g_{r,m,n}^{\phi}\}$ be a frame with the dual frame $\{\gamma_{r,m,n}^{\phi}\}$, then the dual frame is generated by the R functions $\gamma_r = S^{-1}g_r$, i.e. $\gamma_{r,m,n}^{\phi}(i) = \gamma_r(i - mN')\phi_n(i)$.*

In the case where $\{\frac{1}{\sqrt{N}}\phi_n\}$ constitutes a basis for $l^2(\mathbf{Z}/N)$, which is not necessary an orthonormal one, we obtain

$$f = \sum_{r=0}^{R-1}\sum_{m=0}^{M-1}\sum_{n=0}^{N-1}\langle f, \gamma_{r,m,n}^{\psi}\rangle g_{r,m,n}^{\phi}, \qquad (5.11.2)$$

where $\{\psi_n\}$ is the biorthonormal basis of $\{\phi_n\}$ in $l^2(\mathbf{Z}/N)$, and $\{\gamma_r\}$ are properly defined [26, 31].

As noted in the continuous case (where we considered a general case with $\{\psi_n\}$ a dual frame of $\{\phi_n\}$), $\{\gamma_{r,m,n}^{\psi}\}$ constitutes the dual frame of $\{g_{r,m,n}^{\phi}\}$ in the case of critical sampling, i.e. when $\{g_{r,m,n}^{\phi}\}$ constitutes a basis for $l^2(\mathbf{Z}/L)$. In the case of oversampling, $\{\gamma_{r,m,n}^{\psi}\}$ does not necessarily constitute the dual frame of $\{g_{r,m,n}^{\phi}\}$, as can be easily verified by considering some particular examples. Nevertheless, the nice structure of the expansion of type (5.11.2) might be attractive also in the oversampling case, especially when considering the computational cost of calculating the expansion coefficients.

§6   Examples of implementation

6.1   Single-window

Oversampling, in the case of either single-windows or multi-windows, is important since a scheme with oversampling solves a stability problem which exists in the case of critical sampling [4, 28]. Further, oversampling affords a better temporal (or spatial) segmentation than can be obtained by critical sampling when the signal is comprised of two temporally separated tones [29].

In the following example we illustrate the application of a discrete (finite) single-window scheme for image reconstruction from partial information. The image (Figure 3(a)) of size 240 × 256 is comprised of several texture patterns. The cases of critical sampling, where $M = 16$, $N = 15$, and oversampling with $p/q = 2/3$, where $M = 24$, $N = 15$ are compared. The window function is a Gaussian with effective width of 10 pixels. To illustrate the advantage offered by oversampling, we compare the reconstructions (approximations) of the image from partial information, in the cases of critical sampling and oversampling. Figures 3(b) and 3(c) show the reconstructed images in the cases of critical sampling and oversampling respectively, where each column is reconstructed as a 1-D signal using the 100 coefficients of the highest magnitude. An example of a 1-D signal taken along a column is shown in Figure 4, where the reconstructed signals (dashed lines) are superimposed on the original signal (solid line) in the cases of critical sampling (Figure 4(a)) and oversampling (Figure 4(b)). The reconstructed image and 1D signal appear to be better in the case of oversampling. This, however, may not be the case for an arbitrary selection of the scheme configuration and parameters, since the advantage of oversampling strongly depends on matching the width of the window to the properties of the image and on proper choice of the rate of oversampling.

6.2   Multi-windows – one-dimensional example

In the case of a multicomponent signal where the components are distinctly characterized in the time-frequency (position-frequency) space, it is not possible to select an optimal window for a single-window scheme. In such cases it is advantageous to use the multi-window scheme as is demonstrated by the following example. Consider the signal $f(i)$ (Figure 5(a)), of length $L = 480$, which is comprised of three different components: $f(i) = f_1(i) + f_2(i) + f_3(i)$. The signals $f_1(i), f_2(i)$ (Figure 5(b), (c)) are two time-limited tones overlapping in time. The signal $f_3(t)$ (Figure 5(d)) is a wide Gaussian envelope.

Three different schemes are utilized in order to analyze the structure of the signal, and their performances are compared. First, we consider a

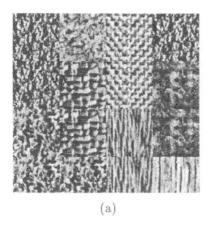

(a)

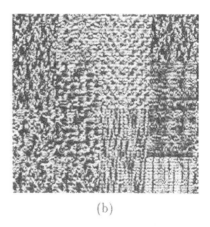

(b)

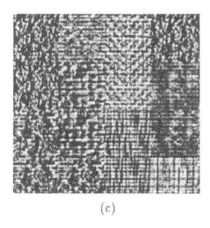

(c)

Figure 3. A mosaic made of eight natural textures reconstructed from partital information of the Gabor expansion coefficients. Compared are approximations of the image obtained by using the 100 coefficients of the highest magnitude in the reconstruction of each column. (a) Original image. (b) Reconstruction in the case of critical sampling. (c) Reconstruction in the case of oversampling. Note the better reconstruction obtained from the same number of coefficients in the case of oversampling.

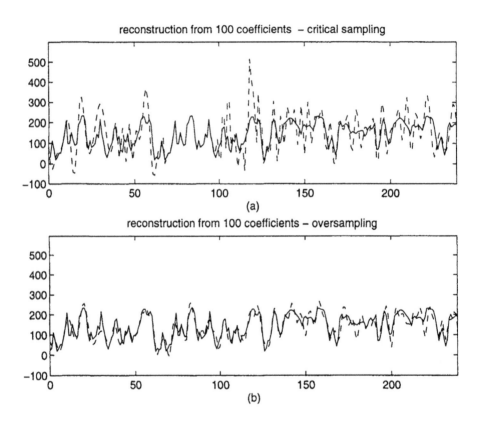

Figure 4. Signal reconstruction from partial information of the Gabor expansion coefficients. The effective width of the Gaussian window is ten pixels. Compared are the signal (solid line) and its reconstructions from 100 dominant coefficients (dashed lines). (a) Critical sampling. (b) Oversampling. Note the better reconstruction obtained from the same number of coefficients in the case of oversampling.

single-window scheme ($R = 1$) with a narrow Gaussian window (Figure 5(e)) and $M = 48, N = 30, p = 1, q = 3$. Second, we consider a single-window scheme with a wide Gaussian window (Figure 5(e)) and $M = 24, N = 60, p = 1, q = 3$. Finally, we consider a double-window scheme ($R = 2$) with both the narrow and wide windows and $M = 24, N = 30, p = 2, q = 3$.

Note that in all three cases we use the same number of representation functions, whereas, the tessellation of the combined space is different. Even in the case of a single-window there is a proper way or even an optimal one to tessellate the combined space once the structure (including width) of a window has been selected with reference to the signals to be analyzed. To be more specific, given a certain sampling density of the combined space, there is a degree of freedom of determining the distribution of the representation functions. However, if the criterion of optimal (minimal) condition number (recall that the condition number is equal to B/A) is to be satisfied, the distribution of the representation functions is dictated by the window. For this reason, in the two cases of the single-window scheme the distributions are different. Whereas, in the case of the wide window there is a high density of representation functions along the frequency axis and low density along the time (position) axis; in the case of a narrow window, the densities are the other way around. Consequently, in the case of the wide window there is a high resolution along the frequency axis on the expense of a wide spread along the time (position) axis, which eliminates the temporal (spatial) fine structure of the signal. In the case of the narrow window, the distributions of the functions and resultant resolutions are the other way around. In the case of a multi-window scheme, the distribution is also dictated by the windows but the scheme incorporates several degrees of freedom. Therefore, to simplify matters and reduce the number of degrees of freedom, we have limited the analysis to the case of identical overlaying sampling grids of the combined space for all windows. (Note that the distributions do not have to be identical and, in fact, should be different to satisfy some kind of optimality.) Based on these restrictions, and the fact that the number of representation functions is similar for all schemes, we obtained the distribution for the double-window scheme. (The distribution along the frequency axis is similar to the distribution for the narrow single-window scheme, and the distribution along the temporal (spatial) axis is similar to the distribution for the wide single-window scheme.)

The absolute values of the expansion coefficients corresponding to the narrow and wide single-window schemes are shown in Figure 6(c), (d) respectively, with corresponding grey-level plots shown in Figure 6(a), (b) respectively. Cross-sections at $m = 22$ ($mN' = 220$), for the narrow window, and at $m = 11$ ($mN' = 220$), for the wide window, are shown in Figure 6(e) and 6(f) respectively. The dual-frame windows are shown in Figure

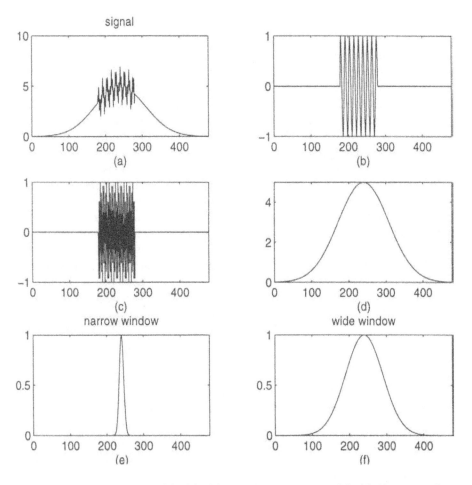

Figure 5. (a) The signal. (b), (c), (d) Signal components. (e), (f) Narrow and wide windows.

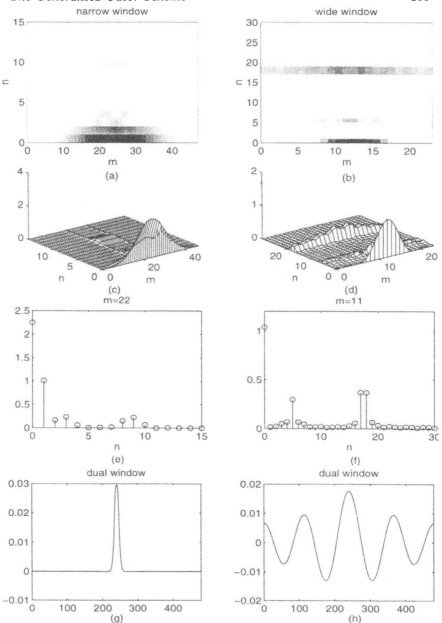

Figure 6. Single-window schemes. (a),(b) Grey-level plots of the expansion coefficients corresponding to the narrow and wide windows, respectively. (c), (d) The corresponding absolute values of the coefficients. (e), (f) A cross-section at $m = 22$ for the narrow window, and a cross-section at $m = 11$ for the wide window, respectively. (g), (h) Dual-frame window functions of the narrow and wide windows, respectively.

6(g),6(h), for the narrow and wide windows, respectively. Considering first the narrow window case, Figure 6 clearly depicts the Gaborian representation of the wide Gaussian envelope, along with some traces of the two tones. Note, however, that the finger print of the low frequency tone is not so clear, *i.e.* it merges with that of the wide Gaussian envelope (a spread over the frequency axis). In the case of the wide window, one can clearly see the three different frequencies corresponding to the Gaussian envelope and the two tones. However, the temporal (spatial) resolution of the two tones is very poor, *i.e.* there is a wide spread of the signal, in particular of the high frequency tone.

The absolute values of the expansion coefficients of the double-window scheme are shown in Figure 7(c) (with corresponding grey-level plot in Figure 7(a)) for the coefficients that correspond to the narrow window, and in Figure 7(d) and 7(b) for the coefficients that correspond to the wide window. A cross-section at $m = 11$ ($mN' = 220$) for the narrow and wide windows is shown in Figure 7(e) and 7(f), respectively. Also, the dual-frame windows are shown in Figure 7(g) and 7(h), for the narrow and wide windows, respectively. It appears as though the multi-window scheme can, with proper rate of oversampling, overcome in some way the limitations imposed by the uncertainty principle on the simultaneous resolution in time (position) and frequency. In the case of the multi-window scheme, each of the windows can lock on certain components of the signal if the windows are properly selected. By displaying the components corresponding to the two windows of the above (double-window) example separately, it is observed that this is indeed the case. The broadly tuned (in time) Gaussian component of the signal is clearly captured by the wide window, while the two tones are hardly represented (and are smeared in time) by this window. On the other hand, the two tones are clear and localized in the representation of the components corresponding to the narrow window, while the broad Gaussian component is hardly represented. To further stress this point, partial reconstruction by utilizing coefficients corresponding to each of the windows separately is shown in Figure 8. A signal reconstructed from functions that correspond to the narrow window only (dashed line), superimposed on a signal comprised of the two tones (solid line) is shown in Figure 8(a). A signal reconstructed from functions that correspond to the wide window only (dashed line), superimposed on the wide Gaussian component of the signal (solid line) is shown in Figure 8(b).

Note the different structure of the dual windows corresponding to the single-window and the double-window schemes. Clearly, in the double-window scheme one window is affected by the other (which is not so in the case of the single window). This in fact causes the advantages offered by a double-window (multi-window) scheme. Thus, the processing, signal component spearation, and identification obtained by the double-window

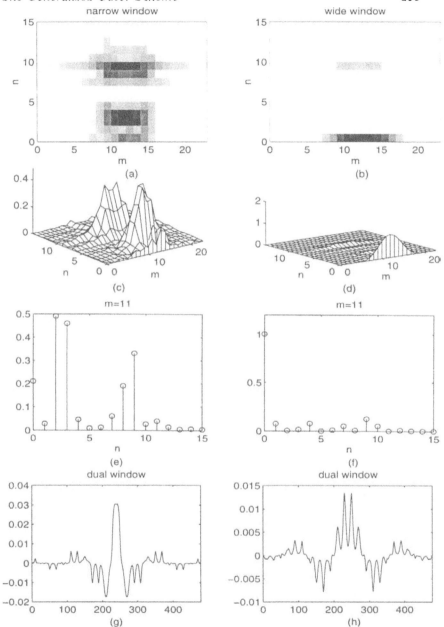

Figure 7. Double-window scheme. (a), (b) Grey-level plots of the expansion coefficients for the narrow and wide windows, respectively. (c), (d) The corresponding absolute values of the coefficients. (e), (f) A cross-section at $m = 11$ for the narrow and wide windows, respectively. (g), (h) Dual-frame window function for the narrow and wide windows, respectively.

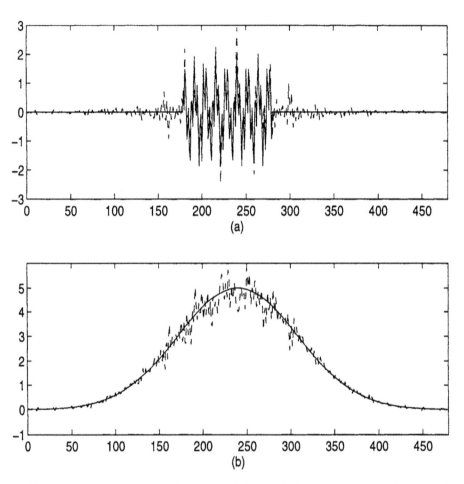

Figure 8. Reconstruction from a partial set of the coefficients in the case of the double-window scheme. (a) A signal reconstructed by utilizing coefficients corresponding to the narrow window only (dashed line), superimposed on the sum of the two tones (solid line). (b) A signal reconstructed by utilizing coefficients corresponding to the wide window only (dashed line), superimposed on the wide Gaussian envelope (solid line).

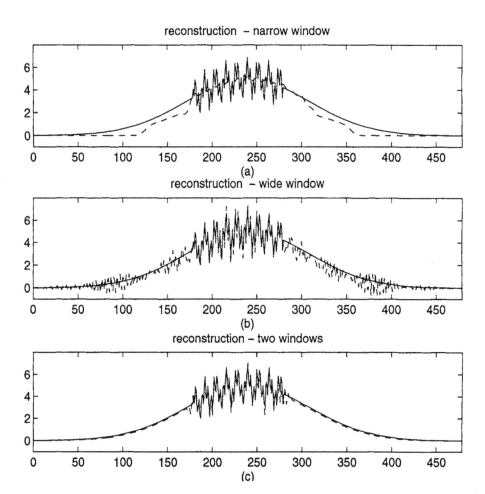

Figure 9. Reconstruction from partial information. (a) Single-window scheme – narrow window. (b) Single-window scheme – wide window. (c) Double-window scheme.

scheme, can not be obtained by a combination of processing by two single-window schemes.

Finally, we also compare the reconstruction of the signal from partial information. Figure 9(a), (b), and (c) shows the reconstructed signal (dashed line) superimposed on the original signal (continuous line) for the cases of single narrow window, single wide window and double window, respectively. The signal is reconstructed by using the 50 coefficients of the highest magnitude. The reconstruction in the double-window case is clearly better, which further stresses the advantage offered by the double-window scheme.

6.3   Multi-windows – two-dimensional example

In the following example we illustrate the application of the multi-window scheme in image reconstruction from partial information. The image (Figure 10 (a)) is of size 240×256 (the same image as in Figure 3(a)). Reconstructions (approximations) where each column is reconstructed as a 1-D signal using the 100 coefficients of the highest magnitude are compared for three different schemes. An image reconstructed by means of a single-window scheme, with a narrow Gaussian window and $M = 24, N = 15, p = 2, q = 3$, is shown in Figure 10(b). A similar reconstruction by means of a single-window scheme with a wide Gaussian window is shown in Figure 10(c). Finally, the application of a double-window scheme with both narrow and wide windows, and $M = 12, N = 15, p = 4, q = 3$, is shown in Figure 10(d). In all the above reconstructions, the same number of representation functions have been used. Reconstructions of a 1-D signal taken along a column for the single narrow window, single wide window and double-window schemes are shown in Figure 11(b), (c), (d), respectively. The reconstructed signal (dashed line) is superimposed on the original signal (continuous line). The window functions are shown in Figure 11(a). The reconstructed image appears to be better in the case of a double-window scheme. Note that the scheme which corresponds to the narrow window is the same as in Figure 3(c), which further stresses the advantage offered by multi-window schemes compared to schemes of a single window with oversampling.

§7   Conclusion

Wavelets and Gabor representation are widely used in the analysis and representation of nonstationary signals and images. However, the original uniform two-parameter wavelets and single Gaussian-window Gaborian scheme with critical sampling are not optimal for the analysis of all types of nonstationary signals. Various generalizations of these schemes

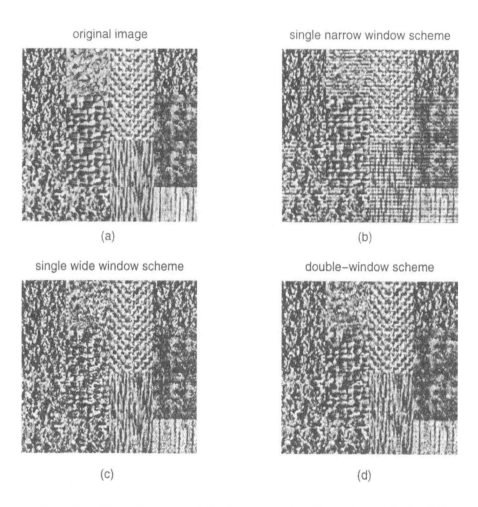

Figure 10. Reconstructions of the image utilizing the 100 coefficients of the highest magnitude. (a) Original image. (b) Single narrow-window scheme. (c) Single wide-window scheme. (c) Double-window scheme.

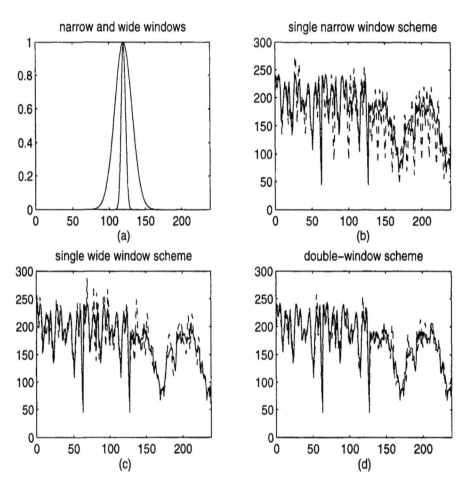

Figure 11. (a) Narrow and wide-window functions. (b), (c), (d) Reconstructions (dashed lines) of a 1-D signal taken along a column of the image (continuous line) by utilizing the 100 coefficients of the highest magnitude. (b) Single-window scheme – narrow window. (c) Single-window scheme – wide window. (d) Double-window scheme.

are therefore useful in the analysis and representation of certain types of nonstationary signals and images.

In this paper we considered various generalizations of the Gabor scheme, such as the one in which we used several windows instead of a single one and kernels other than the exponential. Two types of schemes were considered, a continuous-variable scheme and a discrete-variable scheme. The former is important from a theoretical viewpoint because it permits analytical examination, whereas the latter is important for computational implementation.

We presented a matrix-algebra approach for the analysis of the properties of the representation functions of the generalized Gabor scheme. By means of this new approach, we derived results concerning the completeness and frame properties of the sequence of representation functions. We also characterized tight frames and presented a method for finding the dual frame and therefore the expansion coefficients.

For the continuous scheme, the application of this new approach to the case of critical sampling led to the generalization of the so-called Balian-Low Theorem to the case of multi-window Gabor scheme. Also, a different sampling density for each window function and irregular periodic sampling of the combined space for a single-window scheme, was considered. The former led to a construction of a wavelet-type example with a finite number of resolution levels, whereas the latter generalized the concept of representation on lattices. By further generalizing the sequence of representation functions and utilizing nonexponential kernels, it was found that the properties of the generalized sequence are essentially similar to the properties of the original sequence with exponential kernel.

Utilizing a discrete version of the matrix approach, presented for the analysis of the continuous-variable scheme, we analyzed the properties of the discrete-variable scheme. Moreover, combining frames and generalized inverses, the approach was extended and the issues of undersampling as well as critical sampling and oversampling were considered. As a byproduct, a relation between the concept of frames and generalized inverses was introduced. The approach was further extended to the representation by means of matrix-valued functions in the signal domain. Particular cases of the multi-window scheme, such as the single-window scheme, nonrectangular sampling of the combined space, and filter-bank schemes were also examined. The case of representation functions with nonexponential kernel was analyzed.

Finally, some examples of implementation were presented, indicating the advantages offered by a multi-window scheme. It was illustrated that when component separation (clustering) of a multicomponent signal is considered, each window of a multi-window scheme can lock on certain components of the signal if the windows are properly selected, whereas it is not possible to select in such a common case of a multicomponent signal an optimal window for a single-window scheme. Furthermore, in such a case, reconstruction from partial information is also better in the case of a multi-window scheme than with a single-window scheme.

References

[1] Auslander, L., I. Gertner, and R. Tolimieri, The discrete Zak transform application to time-frequency analysis and synthesis of nonstationary signals, *IEEE Trans. Signal Processing* **39** (1991), 825–835.

[2] Benedetto, J. J., C. Heil, and D. F. Walnut, Differentiation and the Balian-Low theorem, *Journal of Fourier Analysis and Applications* **1** (1995), 355–402.

[3] Campbell, S. L. and C. D. Meyer, *Generalized Inverses of Linear Transformations*, Pitman, London, 1979.

[4] Daubechies, I., The wavelet transform, time-frequency localization and signal analysis, *IEEE Trans. Inform. Theory* **36** (1990), 961–1005.

[5] Daubechies, I. and A. J. E. M. Janssen, Two theorems on lattice expansions, *IEEE Trans. Inform. Theory* **39** (1993), 3–6.

[6] Duffin, R. J. and A. C. Schaeffer, A class of nonharmonic Fourier series. *Trans. Amer. Math. Soc.* **72** (1952), 341–366.

[7] Einziger, P. D., S. Raz, and S. Farkash, Gabor expansion on orthogonal bases, *Electron. Lett.* **25** (1989), 80–82.

[8] Friedlander, B. and B. Porat, Detection of transient signals by the Gabor representation, *IEEE Trans. Acoust., Speech, Signal Processing* **37** (1989), 169–180.

[9] Gabor, D., Theory of communication, *J. Inst. Elect. Eng. (London)* **93(III)** (1946), 429–457.

[10] Heil, C. and D. Walnut, Continuous and discrete wavelet transforms, *SIAM Rev.* **31** (1989), 628–666.

[11] Janssen, A. J. E. M., The Zak transform: A signal transform for sampled time-continuous signals, *Philips J. Res.* **43** (1988), 23–69.

[12] Janssen, A. J. E. M., Duality and biorthogonality for discrete-time Weyl-Heisenberg frames, 1994, preprint.

[13] Janssen, A. J. E. M., Signal analytic proofs of two basic results on lattice expansions, *Appl. Comp. Harmonic Anal.* (1994), 350–354.

[14] Janssen, A. J. E. M., A density theorem for time-continuous filter banks, in *Signal and Image Representation in Combined Spaces*, Y. Y. Zeevi and R. Coifman (eds.), Academic Press, Boston, 1997.

[15] Lancaster, P. and M. Tismenetsky, *The Theory of Matrices*, Academic Press, Orlando, 1985.

[16] Polyak, N., W. A. Pearlman, and Y. Y. Zeevi, Orthogonalization of circular stationary vector sequences and its application to the Gabor decomposition, *IEEE Trans. Signal Processing* **43** (1995), 1778–1789.

[17] Porat, M. and Y. Y. Zeevi, The generalized Gabor scheme of image representation in biological and machine vision, *IEEE Trans. Pattern Anal. Machine Intell.* **10** (1988), 452–468.

[18] Porat, M. and Y. Y. Zeevi, Localized texture processing in vision: Analysis and synthesis in the Gaborian space, *IEEE Trans. on Biomedical Engineering* **36** (1989), 115–129.

[19] Qian, S. and D. Chen, Discrete Gabor transform, *IEEE Trans. Signal Processing* **41** (1993), 2429–2438.

[20] Ron, A. and Z. Shen, Frames and stable bases for shift-invariant subspaces of $L_2(\mathbb{R}^d)$, preprint, *Canad. J. Math.* to appear.

[21] Ron, A. and Z. Shen, Weyl-Heisenberg frames and Riesz bases in $L_2(\mathbb{R}^d)$, preprint.

[22] Seip, K. and R. Wallstén, Density theorems for sampling and interpolation in the Bargmann-Fock space II, *J. Reine Angew. Math* **429** (1992), 107–113.

[23] Walnut, D. F., Continuity properties of the Gabor frame operator, *J. Math. Anal. Appl.* **165** (1992), 479–504.

[24] Wexler, J. and S Raz, Discrete Gabor expansions, *Signal Processing* **21** (1990), 207–220.

[25] Zeevi, Y. Y. and I. Gertner, The finite Zak transform: An efficient tool for image representation and analysis, *J. Visual Comm. and Image Represent.* **3** (1992), 13–23.

[26] Zibulski, M., *Gabor-type representations of signals and images*, PhD thesis, Technion–Israel Inst. of Tech., Israel, October 1995.

[27] Zibulski, M. and Y. Y. Zeevi, Gabor representation with oversampling, in *Proc. SPIE Conf. on Visual Communications and Image Processing'92*, P Maragos (ed.), Boston, 1992, **1818**, pp. 976–984.

[28] Zibulski, M. and Y. Y. Zeevi, Oversampling in the Gabor scheme, *IEEE Trans. Signal Processing* **41** (1993), 2679–2687.

[29] Zibulski, M. and Y. Y. Zeevi, Frame analysis of the discrete Gabor-scheme, *IEEE Trans. Signal Processing* **42** (1994), 942–945.

[30] Zibulski, M. and Y. Y. Zeevi, Analysis of multi-window Gabor-type schemes by frame methods, CC Pub. No. 101, Technion–Israel Inst. of Tech., Israel, April 1995.

[31] Zibulski, M. and Y. Y. Zeevi, Discrete multi-window Gabor-type transforms, CC Pub. No. 124, Technion–Israel Inst. of Tech., Israel, November 1995.

Meir Zibulski
Multimedia Department
IBM Science and Technology
MATAM, Haifa 31905
Israel
meirz@vnet.ibm.com

Yehoshua Y. Zeevi
Department of Electrical Engineering
Technion–Israel Institute of Technology
Haifa 32000, Israel
zeevi@ee.technion.ac.il

II.

Construction of Special Waveforms for Specific Tasks

A Low Complexity Energy Spreading Transform Coder

J. S. Byrnes

Abstract. We have shown how the Prometheus Orthonormal Set (PONS^{tm}), originally constructed to prove an uncertainty principle conjecture of H. S. Shapiro, can be effectively used to compress all common digital audio signals. This compression method is effective because of two fundamental properties: computational simplicity and energy spreading. Although there exist other transform coding methods, such as Walsh-Hadamard, which give compression while limiting the computational burden, we believe that the energy-spreading feature of PONS is unique. We discuss the various advantages that result from these properties, show how the multidimensional analogue of PONS is constructed, and present an algorithm to decompose multidimensional data sets into smaller blocks with uniformly bounded energy. We then indicate the application of PONS to image processing.

§1 Introduction

We defined in [4] a "Walsh-like" complete orthonormal sequence for $L^2(0, 2\pi)$ which satisfies several important properties. Each function in this *Prometheus Orthonormal Set* (PONS^{tm}) is piecewise ± 1, can change sign only at points of the form $\frac{k}{2^n}(2\pi), 1 \leq k \leq 2^n - 1$, and is easily computable using a straightforward and fast recursive algorithm. In addition to these features shared with the Walsh functions, PONS:

- Is optimal with respect to the global uncertainty principle described in [4]

- Yields the *uniform* crest factor $\sqrt{2}$

- Spreads energy almost equally among all transform domain bins.

We discuss the utilization of these properties below, beginning with applications to one-dimensional digital signals. Then the possible advantages over other signal processing methods are described. Finally, we consider the application of PONS to image processing, and give some examples of our preliminary results.

Signal and Image Representation in Combined Spaces
Y. Y. Zeevi and R. R. Coifman (Eds.), pp. 167–187.

§2 One-dimensional signals

2.1 Audio processing

Our first implementation of PONS [5] yields high quality data compression for audio. For 16-bit 44.1, 22.05, and 11.025 kHz monaural signals we achieve 4 to 1 compression with virtually no audible difference in sound. The compression and decompression algorithms are comparable in execution speed and can operate in real-time on all modern PCs (*e.g.* 386 or higher IBM compatible, current Macintosh, UNIX workstation, etc.). These results are to be contrasted with a current standard, ADPCM, where the same compression ratio is achieved, but at much higher computational cost and with much lower quality.

2.2 Spread spectrum communications

A second application of PONS, in the context of one-dimensional signals, is in multi-user spread spectrum communications. The recently developed IS-95 standard for commercial code-division multiple access (CDMA) communications involves a stage-direct sequence-spreading process, first with a Walsh function and then with a longer pseudonoise (PN) sequence. Our research has shown that PONS offers two important advantages over the Walsh functions in this application:

- The minimal crest factor property of the PONS sequences provides much more uniform spreading of the signal's energy across the frequency band. This increases robustness with respect to channel effects, such as narrowband fading and interference, and also reduces spectral features that can be exploited by an adversary in military scenarios.

- Spreading with PONS sequences rather than Walsh sequences yields signals requiring lower short-term ("peak") power to maintain a specified average transmission power. This is advantageous in view of the significant power-control issues involved in multi-user spread spectrum (*e.g.* to address "near-far" problems). It also offers potential for reducing overall transmitter power requirements, which is highly desirable for battery-powered mobile units.

These advantages are illustrated in Figures 1 and 2. Figure 1 compares frequency spectra of typical Walsh and PONS sequences. Figure 2 compares their autocorrelation structures.

Since PONS sequences can be as long as desired (any power of two), it is possible to use them in place of long PN sequences as spreading codes. We are currently investigating the potential value of direct-sequence spreading

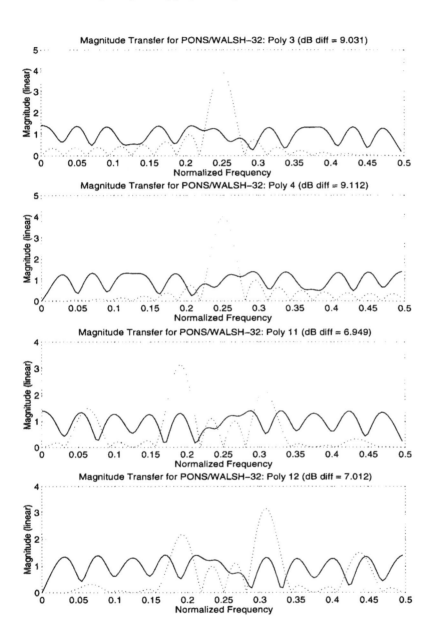

Figure 1. Magnitude spectra of representative 32-coefficient Walsh sequences (dotted curves) and PONS sequences (solid curves). Note the substantial reduction in peak dB for the PONS sequences.

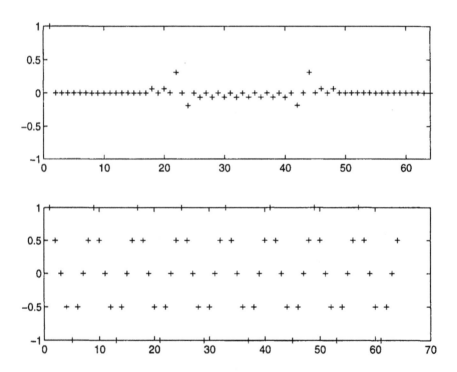

Figure 2. Autocorrelation functions of a representative 32-coefficient PONS sequence (top) and a representative Walsh sequence (bottom).

based entirely on PONS sequences. One feature appears to be an improvement in short-time characteristics of the sequences. PN sequences can have arbitrarily long constant subintervals (*i.e.* consecutive terms all of which are 0 or all of which are 1), whereas PONS sequences cannot have constant subintervals of length greater than 5.

§3   PONS advantages

The basic features of PONS, which have led to our initial success when applying it to one-dimensional signals and which are behind our optimism with regard to multi-dimensional applications, are:

- Very robust

- Easily extensible to any dimension

- Localizable (see Section 4.4 and Figures 5 and 6)

- Secure at multi-levels

- Computational

    Extremely efficient computations

    All basic operations are integer adds and subtracts only

    Fast systolic algorithm

    Amenable to massively parallel processing

- Mathematical

    Elementary construction

    Quadrature mirror filter

    Low crest-factor array

    Yields optimal uncertainty principle bounds.

There are many potential benefits which may result from these features, and which we have begun to exploit in our initial one-dimensional applications. The benefits of this one-dimensional coder, when no quantization is used, include:

- No aliasing errors

- No amplitude or phase distortion

- Transform accomplished via adds/subtracts using integer (2's complement) arithmetic

- Exact reconstruction (no round-off error due to coefficient round-off or truncation error due to bit allocation based on source statistics)

- Near optimum coding gain for sufficiently wideband input (independent of input statistics or spectrum).

When quantization is employed, the coder has the following advantages over conventional coders:

- Quantizes all transform domain coefficients to the same precision

- Aliasing errors are canceled to the same precision in all transform domain subspaces

- Quantization errors appearing in the reconstruction are approximately Gaussian (as a linear sum of independent uniform errors)

- No bit-allocation computation

- A real-time algorithm that has been implemented on general-purpose hardware.

Among the many application areas that can take advantage of these benefits, the most promising are those that have requirements for:

- Wideband coding where the statistics of the source are either not known or not constrained to be anything specific

- Simple coding/decoding hardware

- Good performance over unreliable media (good transmission-error properties)

- A simple progressive transmission scheme.

§4   Image processing

In order to exploit the benefits indicated in Section 3 when processing multidimensional signals, we have generalized the one-dimensional basis given in [4] as well as derived a new tensor decomposition theorem. For simplicity we present only the two-dimensional versions of these two results here. Further details and proofs will appear elsewhere.

4.1 A two-dimensional PONS basis

Let the inner product $\langle A, B \rangle$ of the two matrices A, B (of the same dimension, say $a \times b$) be the sum of component-wise multiplication of the two matrices,

$$\langle A, B \rangle = \sum_{i=1}^{a} \sum_{h=1}^{b} a_{i,h} \cdot b_{i,h}.$$

Let $A_1 = \begin{bmatrix} 1 & 1 \\ 1 & -1 \end{bmatrix}$, $A_2 = \begin{bmatrix} 1 & 1 \\ -1 & 1 \end{bmatrix}$, $A_3 = \begin{bmatrix} 1 & -1 \\ 1 & 1 \end{bmatrix}$, $A_4 = \begin{bmatrix} 1 & -1 \\ -1 & -1 \end{bmatrix}$.

Observe that $\langle A_j, A_k \rangle = 4\delta_{jk}$. Thus, A_j, $1 \leq j \leq 4$, span the characteristic functions of the four sub-quartersquares, such as

0	1
0	0

(ignoring boundaries).

Theorem 1. *For any $n > 0$, there exist 2^{2n} matrices E_i, $1 \leq i \leq 2^{2n}$, with dimension $2^n \times 2^n$, with entries ± 1, and satisfying $\langle E_j, E_k \rangle = 2^{2n}\delta_{jk}$. Furthermore, these matrices satisfy properties corresponding to the uncertainty principle optimality, quadrature mirror filter, and low-crest factor properties of their one-dimensional analog given in [4]. Finally, the sets of piecewise constant functions on the square, determined in the obvious way by these matrices, form a complete orthonormal set for $C\big([0,1] \times [0,1]\big)$.*

We present the first step of the proof since it gives a straightforward inductive definition of the E_i. A_i, $1 \leq i \leq 4$ above is the case $n = 1$. Suppose now that it is true for n. We must prove it for $n + 1$. So we have the E_i as stated in the theorem.

For each $j = 1, 2, \ldots, 2^{2n-2}$, form the $2^{n+1} \times 2^{n+1}$ matrices

$$F_1 = \begin{bmatrix} E_{4j-3} & E_{4j-2} \\ E_{4j-1} & E_{4j} \end{bmatrix}, \quad F_2 = \begin{bmatrix} E_{4j-3} & E_{4j-2} \\ -E_{4j-1} & -E_{4j} \end{bmatrix},$$

$$F_3 = \begin{bmatrix} E_{4j-2} & E_{4j-3} \\ E_{4j} & E_{4j-1} \end{bmatrix}, \quad F_4 = \begin{bmatrix} E_{4j-2} & E_{4j-3} \\ -E_{4j} & -E_{4j-1} \end{bmatrix},$$

$$F_9 = \begin{bmatrix} E_{4j-3} & -E_{4j-2} \\ E_{4j-2} & -E_{4j-3} \end{bmatrix}, \quad F_{10} = \begin{bmatrix} E_{4j-3} & -E_{4j-2} \\ -E_{4j-2} & E_{4j-3} \end{bmatrix},$$

$$F_{11} = \begin{bmatrix} -E_{4j-2} & E_{4j-3} \\ -E_{4j-3} & E_{4j-2} \end{bmatrix}, \quad F_{12} = \begin{bmatrix} -E_{4j-2} & E_{4j-3} \\ E_{4j-3} & -E_{4j-2} \end{bmatrix}.$$

F_5–F_8 are obtained from F_1–F_4 by exchanging E_{4j-3} with E_{4j-1} and E_{4j-2} with E_{4j}. F_{13}–F_{16} are the same as F_9–F_{12}, with E_{4j-3} replaced by E_{4j-1} and E_{4j-2} replaced by E_{4j}. Then there are $2^{2(n+1)}$ F_i's, and a straightforward calculation gives $\langle F_j, F_k \rangle = 2^{2(n+1)} \delta_{jk}$.

4.2 A two-dimensional tensor decomposition theorem

To take advantage of bases such as that constructed in Subsection 4.1, we have developed a method of decomposing a matrix into a sum of tensor products of smaller matrices. Once again we present only the result. The details and the generalization to higher dimensions, due mainly to B. Saffari and motivated by a conjecture of Izidor Gertner, will appear elsewhere.

Theorem 2. *Let $L(m,n)$ denote the set of all $m \times n$ real matrices, let a, b be divisors of m, n respectively, let $m = \mu \cdot a, n = \nu \cdot b$, let $A_j \in L(a,b)$, $(j = 1, 2, \ldots, a \cdot b)$ form an orthonormal basis of $L(a,b)$ under the inner product defined above, let $\otimes$ denote the standard tensor product, and let $M \in L(m,n)$. Then there exist $a \cdot b$ unique matrices $D_j \in L(\mu, \nu)$, $(j = 1, 2, \ldots, a \cdot b)$, such that*

$$M = \sum_{j=1}^{a \cdot b} D_j \otimes A_j.$$

4.3 Uniresolution and localization

Two of the unique and most important features of PONS that can be exploited for image processing are *uniresolution* and *localization*. Uniresolution means that when a decomposition of a high-resolution signal into a sum of low-resolution signals is performed, each of these low-resolution components contains essentially the same amount of energy. This is completely different from all classical signal processing techniques, and yields a possibly important advantage. Namely, since each low-resolution component contains significant energy from all portions of the signal, for applications such as automatic target detection/recognition and telebrowsing, just one or a few of these components might be sufficient. Thus, in certain instances, extremely high data compression is attainable even without employing any of the current sophisticated and often computationally intensive techniques. Moreover, PONS can be employed in conjunction with these other techniques if desired, thereby preserving their advantages while gaining the distinct properties of this new decomposition. A precise mathematical formulation of this uniresolution feature, developed mostly by H. S. Shapiro, will appear elsewhere.

Localization is the capability to reconstruct with high resolution any desired signal portion, while other parts of the signal remain at lower resolution. When applied to images, the result is a clear picture of the area that is wanted, while the background is recognizable but hazy. In tele-browsing, for example, it saves overall transmission and browsing time, and no redundant information need be stored or transmitted.

These properties combine to allow, with small computational effort involving only adds and subtracts, the decomposition of a high-resolution image into a collection of low-resolution images, which can then be processed individually or in parallel to determine where in the image a further search is necessary. Then the localization property can be exploited to exactly reconstruct the locations of interest. Due to the features of PONS discussed above and illustrated in Figures 3–10, this can be accomplished with minimal CPU time and computer memory. The figures require some explanation, which will be better understood if it is read in its entirety *before* they are examined.

Figure 3 is a low-resolution image of a particular scene. In fact, it is just one of the 256 low-resolution images that, when combined, yield the high-resolution image shown in Figure 4. Thus, if Figure 3 could be used to recognize the image, we would immediately have a compression ratio of 256 to 1. Even if recognition were not possible directly from Figure 3, if one could determine where in the overall image there are regions of interest, and then focus on these regions while ignoring the remainder of the image, significant data compression would still be achieved.

However, when the reader first examines Figure 3 it appears that nothing whatsoever can be recognized there. Certainly there are variations in grayscale occurring throughout, but just as certainly (or so it *appears*) there is no particular shape to be found. But, if one looks at Figure 4 and then re-examines Figure 3, the scene has appeared there as well! It might require slight changes in the distance or angle between the figure and the eye in order to see this, but it was indeed there all the time. A focus of our current research is to modify the PONS algorithm so that the computer can see what is clearly in the very low-resolution image, but which the eye first misses.

The localization feature is illustrated in Figures 5 and 6. In Figure 5 the lower right portion of the image has been isolated by using only one thirty-second of the area of Figure 4 for the reconstruction. The remainder of the image is compressed at 8 to 1, for the purpose of illustration. Thus, the compression ratio achieved in obtaining an exact reconstruction of a crucial portion of the image is almost 32 to 1. Similar remarks apply to Figure 6.

Finally, robustness is illustrated in Figures 7–14. In Figure 7, the 256-term expansion whose sum yields the exact reconstruction of Figure 4 was scrambled by randomly permuting the coefficients. In spite of the fact that every single coefficient was thereby matched to the wrong basis element, the resulting image is still recognizable!

Lest the reader feel that the natural redundancy of the original image has contributed to this robustness, we present an additional scrambled and exactly reconstructed image in Figures 8 and 9. There is certainly no question as to which scrambled image belongs to which original. In fact, the scrambled images are somewhat recognizable on their own, even without seeing the full reconstructions. Robustness is further illustrated in Figure 10, where the reproduced image is reasonably good despite the loss of half of the terms in the expansion (*i.e.* half of the data).

4.4   Applications

All of the benefits indicated in Section 3 apply to the two-dimensional PONS algorithms as well. Our current image processing research deals with utilizing the distinctive properties of computational simplicity, robustness, localization, speed, and amenability to parallel computation in areas such as ATD/R, image compression and transmission over narrow bandwidth channels, and image characterization for the purpose of efficiently searching large databases.

Automatic target detection and recognition

Fundamental ATD/R problems require effective feature extraction, often in the presence of heavy clutter and noise, for their solution in terms of detection and recognition. One potential application of PONS here is to use its data compression and real-time localization capabilities to preprocess the image in order to determine quickly which areas within the image require further investigation and to localize further processing to these areas.

Another possibility is to compare a highly compressed (say one, or at most a few, PONS coefficients) version of the unknown image with a library of images stored in this same highly compressed manner. Because of the energy-spreading feature, in which each individual coefficient contains essentially the same amount of hazy information about the entire image, in many situations a standard pattern-matching technique may be able to determine which, if any, of the stored images matches the unknown one.

Progressive image transmission

Progressive image transmission is a method of encoding, transmitting, and decoding digitized data representing an image in such a way that the main features of the image, for example outlines, may be displayed first at low resolution and subsequently refined to higher and higher resolution. In progressive transmission, an image is encoded by an electronic analog to multiple scans of the same image at differing resolutions.

At the receiving end, progressive image decoding results in an initial approximate reconstruction of the image, followed by successively better images whose fidelity is gradually built-up from succeeding scans. In one embodiment of this general scheme, the receiver or viewer can abort the transmission sequence at a less than perfect resolution or can decide on the basis of a partial image to proceed with further transmission and reconstruction. The objective is to show significant but broad features of an image at an early stage of transmission so that a viewer can interactively respond. There is a kind of coarse data compression resulting from this method of transmitting images that arises from the fact that in appropriate applications, only a small part of the total data needs to be sent if a low resolution image is desired.

The order in which the image data are selected, transmitted, and presented to the end user may be dynamically prioritized in response to image content and immediate user interest. Such transmission can result in a display which has a non-uniform resolution. Regions containing visually or operationally significant information may be rendered at a much higher resolution, with refinement deferred for areas of uniform intensity or lesser importance.

Of the many progressive image schemes in existence, the technique in most widespread use in transform domain progressive image transmission is based on the Discrete Cosine Transform (DCT). DCT progressive image coding is incorporated in the recently approved JPEG standard. These methods of implementing progressive image transmission, however, have the following disadvantages:

- The computational burden of the inverse progressive image transmission is high, since it requires the computation of the Inverse DCT (in addition to the forward DCT at the transmission end). To satisfy these computational requirements, the DCT is often implemented using specialized VLSI circuits.

- Using a given JPEG encoding of an image, it is not possible to reconstruct efficiently an arbitrary subimage of high interest with full resolution while other low interest areas remain at low resolution. Contrast this with Figures 5 and 6, produced using PONS.

- When progressive image transmission is applied to telebrowsing of archived images, highly compressed data are kept in on-line storage and used to browse the data efficiently to determine potentially useful data sets for further processing. Once the selection is made, the original data are obtained from off-line storage. Here the browse quality data and the corresponding original data contain redundant information, causing a fraction of this information to be transmitted twice, thereby reducing performance.

Progressive image transmission employing PONS is being designed to overcome these and other deficiencies. Localization and robustness are the two most important features in this regard. They are demonstrated in Figures 5–10. The reason behind the robustness of this coding technique, particularly clear in Figure 10, is that PONS provides a "balanced" representation of typical images, *i.e.* one in which the energy is spread evenly among all the transform coefficients. In contrast, Fourier and wavelet representations of typical images concentrate energy in low-frequency terms. If any of these terms are lost in transmission, the image will be essentially destroyed.

Medical image processing

PONS offers several important potential benefits for the processing of medical images. Such images will typically be large (as much as 1 gigabyte for a single pap smear slide at 40× magnification). On the other hand, much of the key diagnostic information will be contained in small portions of the image. The localization feature will facilitate quick telemedicine access to the significant portions while minimizing the load on the network. Similar savings can be imagined in medical image storage. Further, in the context of a preprocessor which highlights anomalies in the image, such a facility will aid diagnosis.

Image database search

As databases of images grow, the problem of searching them to locate desired images becomes more and more intractable. A basic constraint is the amount of data, typically on the order of a megabyte of 8-bit numbers for each black-and-white image. Thus, identifying and retrieving a specific image can require processing involving an enormous number of pixels. With current databases consisting of tens of thousands of images, an efficient means of image characterization for search purposes is essential. Employing properly designed image snippets, which enable significant data reduction, is one promising approach.

To achieve both uniqueness and data reduction, a procedure is required which employs an image snippet that is both a highly compressed version of the original image and also contains information from the entire image. These are exactly the properties that characterize the uniresolution feature discussed in Subsection 4.3.

Another desirable property, for example in telebrowsing, is to be able to examine in full detail a portion of an image without having to manipulate

Figure 3. Test image I, 256:1 compression.

the large amount of data required to view it in its entirety. The localization feature can be immediately employed to perform this required task.

§5 Conclusion

Any digital signal of any dimension can be simply and uniquely expressed in terms of PONS. This fact, together with its unique localization capability and its robustness, indicate considerable potential for this new digital coding and transmission tool. Further research, blending the mathematical ideas with the engineering implementation, will determine whether this promise can be brought to fruition.

Figure 4. Test image I, full reconstruction.

Figure 5. Test image I, localized.

Figure 6. Test image II, localized. Full resolution features in a low resolution (compressed) background.

Figure 7. Test image I, scrambled.

Figure 8. Test image III, scrambled.

Figure 9. Test image III, full reconstruction.

Figure 10. Test image III reconstructed using half of the terms (128 out of 256).

References

[1] Benedetto, J. and A. Teolis, Wavelet auditory models and data compression, *Applied and Computational Harmonic Analysis* **1** (1993), 3–28.

[2] Benedetto, J. and A. Teolis, Nonlinear method and apparatus for coding and decoding acoustic signals with data compression and noise suppression using cochlear filters, wavelet analysis, and irregular sampling reconstruction, U.S. Patent 5388182, 1995.

[3] Benedetto, J., J.S. Byrnes, D.J. Newman, H.S. Shapiro, and A. Teolis, Wavelet auditory models and irregular sampling, Tech. Report AR-92-2, Prometheus Inc., 1992.

[4] Byrnes, J.S., Quadrature mirror filters, low crest factor arrays, functions achieving optimal uncertainty principle bounds, and complete orthonormal sequences – a unified approach, *Applied and Computational Harmonic Analysis* **2** (1994), 261–266.

[5] Byrnes, J.S., M. A. Ramalho, G. Ostheimer, and I. Gertner, Discrete one dimensional signal processing method and apparatus using energy spreading coding, U.S. Patent pending.

[6] Le Floch, B., M. Alard, and C. Berrou, Coded orthogonal frequency division multiplex, *Proc. of the IEEE*, **83** no. 6, Special Issue on Digital Television (1995), 982–996.

[7] Shapiro, H.S., Extremal problems for polynomials and power series, Sc.M. Thesis, Massachusetts Institute of Technology, 1951.

[8] Veldhuis, R.N.J., Bit rates in audio source coding, *IEEE Journal on Selected Areas in Communications* **10(1)** (1992).

J. S. Byrnes
Prometheus Inc.
and the University of Massachusetts at Boston
jbyrnes@cs.umb.edu

Nonstationary Subdivision Schemes, Multiresolution Analysis, and Wavelet Packets

Albert Cohen and Nira Dyn

Abstract. Nonstationary subdivision schemes consist of recursive refinements of an initial sparse sequence with the use of masks that may vary from one scale to the next finer one. We show that such schemes can be used to construct C^∞ compactly supported orthonormal scaling functions, wavelets, and wavelet-packets with better control on the frequency localization.

§1   Introduction

Wavelets and wavelet packets constitute useful tools for the decomposition of complicated functions or signals into a small number of elementary waveforms that are localized both in time and frequency. Numerically, these decompositions are performed by iterative application of digital filter-banks followed by decimation.

In the other direction, the reconstruction of the signal is performed by iterative refinements with linear rules involving filter coefficients. Such refinement or subdivision algorithms have also been studied in the context of computer-aided geometric design.

In the standard construction of wavelets and wavelets packets, the filters are the same at every iteration, resulting in specific limitations in the regularity and frequency localization of the generating functions. In particular, there exists no C^∞ wavelet with compact support.

In contrast, it is known that subdivision schemes that are nonstationary in scale, *i.e.* for which the filter may vary from one step of the refinement process to the next one, can converge to C^∞ limit functions with compact support.

In this paper we give a survey of results (detailed in Cohen and Dyn [4] and Cohen and Séré [6]) that show the advantages of using different filters at different splitting or subdivision stages, in the context of wavelets and wavelet packets.

In Section 2, we review the existing connections between stationary subdivision algorithms, wavelets, and wavelet packets, and we describe the limitation that are inherent to the use of the same filters at every scale. We

Signal and Image Representation in Combined Spaces
Y. Y. Zeevi and R. R. Coifman (Eds.), pp. 189–200.

show in Section 3 that certain nonstationary subdivision schemes can be used to construct C^∞ compactly supported scaling functions and wavelets bases. In Section 4, we show that this approach can also be used to recover some properties of time-frequency localization of wavelet packets.

§2   From subdivision schemes to wavelet packets

Subdivision schemes constitute a useful tool for the fast generation of smooth curves and surfaces from a set of control points by means of iterative refinements. In the most-often considered binary univariate case, one starts from a sequence $s_0(k)$ and obtains at step j a sequence $s_j(2^{-j}k)$, generated from the previous one by linear rules:

$$s_j(2^{-j}k) = 2 \sum_{n \in k+2\mathbf{Z}} h_{j,k}(n)s_{j-1}(2^{-j}(k-n)). \qquad (2.1)$$

The masks $h_{j,k} = \{h_{j,k}(n)\}_{n\in\mathbf{Z}}$ are in general finite sequences, a property that is clearly useful for the practical implementation of (2.1).

A natural problem is then to study the convergence of such an algorithm to a limit function. In particular, the scheme is said to be strongly convergent if and only if there exists a continuous function $f(x)$ such that $\lim_{j\to+\infty}(\sup_k |s_j(2^{-j}k) - f(2^{-j}k)|) = 0$. One can study more general types of convergence with the use of a smooth function u that is well localized in space (for example compactly supported) and satisfies the interpolation property $u(k) = \delta_k$. One can then define $f_j(x) = \sum_k s_j(2^{-j}k)u(2^j x - k)$ and study the convergence in a functional sense of f_j to f.

A subdivision scheme is said to be stationary and uniform when the masks $h_{j,k}(n) = h_n$ are independent of the parameters j and k. In that case, one can rewrite (2.1) as

$$s_j(2^{-j}k) = 2 \sum_n h_{k-2n}s_{j-1}(2^{-j+1}n). \qquad (2.2)$$

Note that (2.2) is equivalent to filling in the sequence s_{j-1} with zeros at the intermediate points $2^{-j}(2k+1)$ and applying a discrete convolution with the sequence (h_k). Detailed reviews of stationary subdivision have been done by Cavaretta et al. [2] and Dyn [12].

These algorithms apply in a natural way to computer-aided geometric design. Moreover, the interest in stationary subdivision schemes has grown in the digital image processing and numerical analysis communities since they have been connected to multiresolution analysis, wavelet bases, and wavelet packets.

A multiresolution analysis consists of a nested sequence of approximation subspaces

$$\{0\} \to \cdots V_{-2} \subset V_{-1} \subsetneq V_0 \subset V_1 \subset V_2 \cdots \to L^2(\mathbb{R}), \qquad (2.3)$$

that are generated by a "scaling function" $\varphi \in V_0$ in the sense that the set $\{\varphi_{j,k}\}_{k\in\mathbb{Z}} = \{\varphi(2^j \cdot -k)\}_{k\in\mathbb{Z}}$ constitutes a Riesz basis for V_j. By $V_j \to L^2(\mathbb{R})$, we mean here that for any f in $L^2(\mathbb{R})$, $\lim_{j\to+\infty} \|P_j f - f\|_0 = 0$ where $P_j f$ is the L^2-projection of f onto V_j and $\|\cdot\|_0$ is the L^2 norm (we shall use the notation $\|\cdot\|_s$ for the Sobolev $H^s = W_2^s$ norm). Here again, many generalizations are possible (see Meyer [14]) or Daubechies [9] for a detailed review of this concept).

Since the spaces V_j are embedded, the scaling function satisfies an equation of the type

$$\varphi(x) = 2 \sum_n h_n \varphi(2x - n). \tag{2.4}$$

We shall assume here that φ is compactly supported so that the h_n's are finite in number. In that case, φ is also an L^1 function and by taking the Fourier transform of (2.4), we have

$$\hat{\varphi}(\omega) = m_0(\omega/2)\hat{\varphi}(\omega/2) \tag{2.5}$$

where $m_0(\omega) = \sum_{n\in\mathbb{Z}} h_n e^{-in\omega}$. Assuming that φ is normalized in the sense that $\int \varphi = \hat{\varphi}(0) = 1$, one obtains by iterating (2.5),

$$\hat{\varphi}(\omega) = \prod_{k=1}^{+\infty} m_0(2^{-k}\omega). \tag{2.6}$$

This last formula indicates that φ is the limit, in the weak (or distribution) sense, of a stationary subdivision scheme since it represents, in the Fourier domain, the refinement of an initial Dirac sequence by iterative convolutions with h_n. Note also that the support of φ is contained in the convex hull of the support of the mask (h_k). Conversely, any refinable function, *i.e.* weak limit of such a scheme, satisfies a "refinement equation" of the type described above and is a potential candidate to generate a multiresolution analysis (see also Derfel *et al.* [11]).

In this framework, one can introduce constraints that ensure the orthonormality of $\{\varphi(x-k)\}_{k\in\mathbb{Z}}$, namely

$$|m_0(\omega)|^2 + |m_0(\omega + \pi)|^2 = 1 \tag{2.7}$$

and the existence of a compact set K, congruent to $[-\pi, \pi]$ modulo 2π, such that $m_0(2^{-j}\omega) \neq 0$ for all $\omega \in K$ and $j > 0$ (see Cohen and Ryan [5]).

From this setting, one derives an orthonormal wavelet by

$$\psi(x) = 2 \sum_n g_n \varphi(2x - n), \tag{2.8}$$

with $g_n = (-1)^n h_{1-n}$. The family

$$\{\psi_{j,k}\}_{k\in\mathbb{Z}} = \{\psi(2^j \cdot -k)\}_{k\in\mathbb{Z}}$$

is then an orthonormal basis of the orthogonal complement W_j of V_j into V_{j+1}: it characterizes the missing details between two successive levels of approximation. It follows that for all $J \in \mathbb{Z}$, $\{\varphi_{J,k}\}_{k\in\mathbb{Z}} \cup \{\psi_{j,k}\}_{j\geq J,k\in\mathbb{Z}}$ are orthonormal bases of $L^2(\mathbb{R})$.

The decomposition of V_1 into V_0 and W_0, as expressed by (2.4) and (2.8), reveals a more general "splitting trick":

If $\{e_n\}_{n\in\mathbb{Z}}$ is an orthonormal basis of a Hilbert space H, and if h_n and g_n are coefficients that satisfy the previous constraints, then the sequences $u_n = \sqrt{2}\sum_k h_{k-2n}e_k$ and $v_n = \sqrt{2}\sum_k g_{k-2n}e_k$ are orthonormal bases of two orthogonal closed subspaces U and V such that $H = U \oplus V$.

Wavelet packets are obtained by using this trick to split the W_j spaces. More precisely, one can define a family $\{w_n\}_{n\in\mathbb{N}}$ by taking $w_0 = \varphi$, $w_1 = \psi$ and applying the following recursion:

$$w_{2n} = 2\sum_k h_k w_n(2x - k) \quad \text{and} \quad w_{2n+1} = 2\sum_k g_k w_n(2x - k). \qquad (2.9)$$

The families $\{w_n(x - m)\}_{m\in\mathbb{Z}} = \{w_{m,n}\}$ for $2^j \leq n < 2^{j+1}$ are the results of j splitting of the space W_j. Consequently, $\{w_{m,n}\}_{n\in\mathbb{N},m\in\mathbb{Z}}$ is an orthonormal basis of $L^2(\mathbb{R})$.

This is only a particular example of a wavelet packet basis: many other bases can be obtained by splitting W_j more or less than j times. However, all the elements of these bases have the general form

$$w_{j,m,n}(x) = 2^{j/2}w_n(2^j x - m). \qquad (2.10)$$

Such bases yield arbitrary decompositions of any discrete signal in the time-frequency plane, by means of fast algorithms: one assumes that the initial data $\{s_k\}_{k\in\mathbb{Z}}$ represents a function in V_j, i.e. $f = \sum_k s_k w_{j,k,0}$. The splitting trick indicates that the coordinates of f in the basis $\{w_{m,n}\}_{m\in\mathbb{Z},0\leq n<2^j}$ can be obtained by j iterative convolutions of s_k with the discrete filters h_n and g_n, followed by a decimation of one sample out of two. Again, one has the possibility of splitting more or less than j times and use entropy tests to select the sparsest decomposition (see Coifman *et al.* [7]). The computational cost is of order $CN \log N$, where N is the size of the initial data and C is the length of the filters h_n and g_n.

The construction that we have described suffers, however, from two limitations:

- The regularity of the functions φ and ψ are bounded by the number of coefficients in the filter. More precisely, one can prove (see Dyn

and Levin [13], Cavaretta *et al.* [2], Daubechies and Lagarias [10])
that the convergence of the subdivision scheme to a C^r function for
some $r \geq 0$ implies the factorization

$$m_0(\omega) = \left(\frac{1 + e^{-i\omega}}{2}\right)^{r+1} p(\omega). \tag{2.11}$$

In particular, Daubechies scaling functions φ_L correspond to $r = L-1$
and

$$|p_L(\omega)|^2 = \sum_{j=0}^{L-1} \binom{L-1+j}{j} \sin^{2j}(\omega/2), \tag{2.12}$$

where $p_L(\omega)$ is the second factor in (2.11). Thus, there exists no C^∞
scaling function with compact support.

- For finitely supported filters, the wavelet packets w_n suffer from a
  lack of frequency localization as n grows. More precisely, if we define
  the variance

$$\sigma_n = \min_{\omega_0 \geq 0} \int_0^{+\infty} |\hat{w}_n(\omega)|^2 |\omega - \omega_0|^2 d\omega \tag{2.13}$$

a result of Séré [16] states that for any $M > 0$, we have

$$\lim_{n \to \infty} \frac{1}{n} \operatorname{card}\{k \leq n \,:\, \sigma_k > M\,\} = 1. \tag{2.14}$$

Note that for $m_0(\omega) = \chi_{[-\pi/2, \pi/2]}(\omega)$, the wavelet packets are well
localized in frequency (but poorly in space): $\hat{w}_n(\omega) = \chi_{-J_n \cup J_n}(\omega)$
where the interval $J_n = [\pi g(n), \pi(g(n) + 1)]$ is determined by the
"Gray code" $g(n) = \sum_{k=1}^{j} \gamma_k 2^{k-1}$, $\gamma_j = \varepsilon_j$, and $\gamma_k = |\varepsilon_{k+1} - \varepsilon_k|$ for
$1 \leq k < j$.

Recently, attention has been given to subdivision schemes that are nonstationary in scale, *i.e.* for which the masks may vary from one step of
the refinement process to the next one. This is the case where in (2.1)
$h_{j,k}(n) = h_n^{(j)}$, for $j \in \mathbf{Z}_+$ and for $n \in \mathbf{Z}$. A model case is the scheme
that uses at step j the same mask $h_n^j = \binom{j}{n} 2^{-j+1}$ $(0 \leq n \leq j)$, that would
give rise, in the stationary subdivision case, to B-splines of degree $j - 1$.
It is proved by Derfer *et al.* [11] that such a scheme converges strongly to
the "up-function" introduced by Rvachev [15]. This function is a C^∞ and
supported on $[0, 2]$. We shall now apply a similar idea to the construction
of wavelets, cardinal interpolatory functions, and wavelet packets.

§3 Nonstationary scaling functions and wavelets

Let $\{m_0^k(\omega)\}_{k>0}$ be a family of trigonometric polynomials of degree $d(k)$ that satisfy the following properties:

$$|m_0^k(\omega)|^2 + |m_0^k(\omega + \pi)|^2 = 1, \tag{3.1}$$

$$m_0^k(0) = 1, \tag{3.2}$$

$$|\omega| \leq \pi/2 \Rightarrow m_0^k(\omega) \neq 0, \tag{3.3}$$

and

$$\sum_{k>0} 2^{-k} d(k) < +\infty. \tag{3.4}$$

(Condition (3.3) could be replaced by a weaker one: the existence of a compact set K congruous to $[-\pi, \pi]$ modulo 2π such that $m_0^k(2^{-l}\omega) \neq 0$ for all $k, l \geq 1$, and $\omega \in K$, see Cohen and Ryan [5] for more details.)

With these hypotheses the following results hold:

Theorem 1. *For all $j \in \mathbf{Z}_+$, the tempered distributions φ^j defined by*

$$\hat{\varphi}^j(\omega) = \prod_{k=1}^{+\infty} m_0^{k+j}(2^{-k}\omega) \tag{3.5}$$

are compactly supported L^2 functions and the associated subdivision schemes converge in L^2. The families $\{\varphi^j(x - k)\}_{k\in\mathbf{Z}}$ are orthonormal. Moreover, the spaces V_j generated by $\{2^{j/2}\varphi^j(2^j x - k)\}_{k\in\mathbf{Z}}$ constitute a "half multiresolution analysis" in the sense that $V_j \subset V_{j+1}$ and the projection $P_j(f)$ of any L^2 function f onto V_j goes to f in L^2 as j goes to $+\infty$. Finally, the functions $\{2^{j/2}\psi^j(2^j x - k)\}_{k\in\mathbf{Z}}$, where

$$\hat{\psi}^j(\omega) = e^{-i\omega/2}\overline{m_0^{j+1}(\omega/2 + \pi)}\hat{\varphi}^{j+1}(\omega/2) ,$$

are an orthonormal basis for the orthogonal complement W_j of V_j into V_{j+1}.

Theorem 2. *If $\{m_0^k(\omega)\} = \{m_{0,L(k)}(\omega)\}$ is a subset of the family given by (2.11) and (2.12), and if $\lim_{k\to+\infty} L(k) = +\infty$, then the functions φ^j are smooth, i.e. contained in C^∞.*

The proofs of these results are straightforward consequences of the results in Cohen and Dyn [4]. The case where $L(k) = k$ in Theorem 2 has also been considered by Berkolaiko and Novikov [1].

The proof of Theorem 1 goes in four steps:

- Using Bernstein inequality, one remarks that for a fixed ω, the sequence

$$s_k = |m_0^{k+j}(2^{-k}\omega) - 1| \le Cd(k)2^{-k}|\omega| \qquad (3.6)$$

is absolutely summable and thus the infinite product (3.5) converges pointwise and in the sense of tempered distributions. By (3.4), φ^j is a tempered distribution with compact support.

- As in the standard construction of wavelets (see Cohen [3]), one defines the approximants φ_n^j of φ^j by

$$\hat{\varphi}_n^j(\omega) = \prod_{k=1}^{n} m_0^{k+j}(2^{-k}\omega)\chi_{[-2^n\pi, 2^n\pi]}(\omega) \qquad (3.7)$$

and checks by recursion, using (3.1), that $\{\varphi_n^j(x - k)\}_{k\in\mathbb{Z}}$ is an orthonormal system. By Fatou's lemma, it follows that φ^j is in $L^2(\mathbb{R})$. We note that φ_n^j is an interpolation of the nonstationary subdivision with the choice $u(x) = \sin(\pi x)/(\pi x)$.

- By (3.3), we have

$$|\omega| \le \pi \Rightarrow |\hat{\varphi}^j(\omega)| > C > 0 \qquad (3.8)$$

and $|\hat{\varphi}_n^j(\omega)| \le C^{-1}|\hat{\varphi}^j(\omega)|$. By dominated convergence, φ_n^j converges to φ^j in L^2 and it follows that $\{\varphi^j(x - k)\}_{k\in\mathbb{Z}}$ is an orthonormal system.

- By the definition of φ^j, it is clear that $V_j \subset V_{j+1}$. To check that $P_j(f)$ goes to f in L^2 for any f, it suffices to prove that $\|P_j(\chi_I)\|$ goes to $|I|$ for any finite interval I. This follows from $\int \varphi^j = 1$ for all j and from the decay of the support length of $\varphi^j(2^j x)$ to zero as j increases.

The proof of Theorem 2 is based on an important estimate on the trigonometric polynomials given by (2.1) and (2.2):

Lemma 1. *The trigonometric polynomials constructed by Daubechies satisfy*

$$|m_{0,L}(\omega)| \le [m(\omega)]^{L-1} \qquad (3.9)$$

with $m(\omega) = 1$ if $|\omega| \le \pi/2$ and $m(\omega) = |\sin(\omega)|$ if $\pi/2 \le |\omega| \le \pi$.

Proof: This is obvious for $|\omega| \leq \pi/2$. For $\pi/2 \leq |\omega| \leq \pi$, we have $2\sin^2(\omega/2) \geq 1$ and consequently

$$
\begin{aligned}
|p_L(\omega)|^2 &= \sum_{j=0}^{L-1} \binom{L-1+j}{j} \sin^{2j}(\omega/2) \\
&= \sum_{j=0}^{L-1} \binom{L-1+j}{j} [2\sin^2(\omega/2)]^j 2^{-j} \\
&\leq [2\sin^2(\omega/2)]^{L-1} \sum_{j=0}^{L-1} \binom{L-1+j}{j} 2^{-j} \\
&= [2\sin^2(\omega/2)]^{L-1} |p_L(\pi/2)|^2 = [4\sin^2(\omega/2)]^{L-1}
\end{aligned}
$$

(we have $|p_L(\pi/2)|^2 = 2^{L-1}$ by (3.1)). It follows that

$$
\begin{aligned}
|m_{0,L}(\omega)| &= \left(\cos^{2L}(\omega/2)|p_L(\omega)|^2 \right)^{1/2} \\
&\leq \left([4\cos^2(\omega/2)\sin^2(\omega/2)]^{L-1} \right)^{1/2} = |\sin(\omega)|^{L-1}.
\end{aligned}
$$

The proof of Theorem 2 goes then in two steps:

- By using a technique introduced by Daubechies [8], one checks that the function $P(\omega) = \prod_{k=1}^{+\infty} m(2^{-k}\omega)$ satisfies

$$
|P(\omega)| \leq C(1 + |\omega|)^{-s}, \tag{3.10}
$$

with $s > 0$ (it can be shown that the largest value for s is $1 - \frac{\log 3}{2\log 2}$).

- For a fixed j and for any $A > 0$, it is possible to find $j_A > j$ such that, for all $k > j_A$, $s(L(k) - 1) \geq A$. It follows that φ^{j_A} satisfies

$$
\begin{aligned}
|\hat{\varphi}^{j_A}(\omega)| &= \prod_{k=1}^{+\infty} |m_{0,L(k+j_A)}(2^{-k}\omega)| \\
&\leq \prod_{k=1}^{+\infty} |m(2^{-k}\omega)|^{L(k+j_A)-1} \\
&\leq \prod_{k=1}^{+\infty} |m(2^{-k}\omega)|^{A/s} \\
&\leq |P(\omega)|^{A/s} \\
&\leq C_A (1 + |\omega|)^{-A}.
\end{aligned}
$$

Since $\hat{\varphi}^j(\omega) = \hat{\varphi}^{j_A}(2^{j-j_A}\omega) \prod_{k=1}^{j_A-j} m_{0,L(k+j)}(2^{-k}\omega)$, it follows that we have

$$|\hat{\varphi}^j(\omega)| \leq C_A'(1+|\omega|)^{-A}, \tag{3.11}$$

and since this holds for any $A > 0$, it follows that φ^j is a C^∞ function.

The same proofs of Theorems 1 and 2 can be used to obtain an analogous result on interpolatory subdivision schemes and the corresponding cardinal interpolatory functions:

Theorem 3. *Let $M_0^k(\omega) = |m_0^k(\omega)|^2$, where the trigonometric polynomials $\{m_0^k\}_{k>0}$ satisfy assumptions (3.1)–(3.4). Then for all $j \in \mathbf{Z}_+$, the tempered distributions φ^j defined by*

$$\hat{\varphi}^j(\omega) = \prod_{k=1}^{+\infty} M_0^{k+j}(2^{-k}\omega), \tag{3.12}$$

are compactly supported continuous functions, satisfying the cardinal interpolation conditions $\varphi^j(i) = \delta_{0,i}$ for $i \in \mathbf{Z}$ and $j \in \mathbf{Z}_+$.

Moreover, if the trigonometric polynomials $\{m_0^k(\omega)\} = \{m_{0,L(k)}(\omega)\}$ is a subset of the family given by (2.11) and (2.12), and if $\lim_{k\to+\infty} L(k) = +\infty$, then the functions φ^j in (3.12) are smooth, i.e. contained in C^∞, and the nonstationary subdivision scheme with masks $\{h_n^{(k)} : n \in \mathbf{Z}\}_{k>0}$, determined by the coefficients of the trigonometric polynomials $M_0^k(\omega) = \sum_{n\in\mathbf{Z}} h_n^{(k)} e^{-in\omega}$, is strongly convergent to the limit function

$$\sum_{j\in\mathbf{Z}} s_0(j)\varphi^0(\cdot - j),$$

for the initial data $\{s_0(j) : j \in \mathbf{Z}\}$.

Finally, let us mention that, under the conditions of Theorem 2, the half-multiresolution analysis $\{V_j\}_{j\geq 0}$, generated by either the functions in (3.5) or the functions in (3.12), has the property of "spectral approximation": for all $f \in H^s$ and $0 \leq t \leq s$ one has

$$\|P_j f - f\|_{H^t} \leq C(s,t) 2^{-j(s-t)} \|f\|_{H^s} \varepsilon(j,f) \tag{3.13}$$

where $\varepsilon(j,f) \in [0,1]$ and tends to 0 as j goes to $+\infty$ (see Cohen and Dyn [4]).

§4   Nonstationary wavelet packets

The previous results allow us to construct "nonstationary wavelet packets" with the following definition:

Definition 1. Let $\{m_0^k\}_{k>0}$ be a family of positive trigonometric polynomials that satisfy the set of conditions (3.1)–(3.4) and define $m_1^k(\omega) = e^{-i\omega}\overline{m_0^k(\omega + \pi)}$. The nonstationary wavelet packets associated to $\{m_0^k\}_{k>0}$ are a sequence $\{w_n\}_{n\geq0}$ of tempered distributions given by

$$\hat{w}_n(\omega) = \prod_{k=1}^{+\infty} m_{\varepsilon_k}^k(2^{-k}\omega), \tag{4.1}$$

with $n = \sum_{k>0} 2^{k-1}\varepsilon_k$, $\varepsilon_k \in \{0,1\}$.

From Theorems 1 and 2, we can easily derive:

Proposition 1. The distributions $\{w_n\}_{n\geq0}$ are compactly supported L^2 functions and the length of their support is independent of n.

The family $w_{n,m}(x) = w_n(x - m)$, $n \geq 0$, $m \in \mathbf{Z}$, constitutes an orthonormal basis of $L^2(\mathbb{R})$.

If $\{m_0^k(\omega)\} = \{m_{0,L(k)}(\omega)\}$ is a subset of the family given by (2.1) and (2.2) such that $\lim_{k\to+\infty} L(k) = +\infty$, then the functions w_n are infinitely smooth, i.e. contained in C^∞.

Proof: From the definition of w_n, we see that for $0 \leq n < 2^j$, the families $\{w_n(x - m)\}_{m\in\mathbf{Z}}$ are the result of j splitting of the space V_j. It follows that these functions are finite combinations of the functions $\varphi^j(2^j x - k)$. As a consequence, they are in L^2. They are also in C^∞ as soon as the hypotheses of Theorem 2 are satisfied.

From the denseness of the spaces V_j, it follows that the family $w_{n,m}(x) = w_n(x - m)$, $n \geq 0$, $m \in \mathbf{Z}$, constitutes an orthonormal basis of $L^2(\mathbb{R})$.

Finally, since m_1^k has the same degree as m_0^k, the support of w_n cannot be larger than $L = \sum_{k>0} 2^{-k}d(k) < +\infty$.

Remark. Since the generating functions of the V_j spaces are different at every scale, one can no longer use a recursion formula such as (2.9) to derive the nonstationary wavelet packets.

However, fast algorithms can still be used to compute the nonstationary wavelet packet coefficients. The principle is exactly the same as in the standard case except that one uses different filters at every stage of the decomposition. Note that in the case where the degree $d(k)$ of m_0^k is increasing as k goes to $+\infty$, the decomposition starts at the finest scale with a long filter and ends at a coarser scale with a smaller filter. The choice of using longer filters for finer scales offers three important advantages:

- Since the functions φ^j are in C^∞, the reconstruction from the low frequency components will have a smooth aspect.

- In the algorithm, one needs to deal properly with the boundaries of the signal. Specific methods have been developed by Cohen *et al.* [3] to adapt the filtering process near the edges. These methods require, in particular, that the size of the filter is smaller than the size of the signal: this can only be achieved by using smaller filters in the coarse scales.

- Finally, these techniques can be used to combine time and frequency localization properties as shown by Theorem 4 below.

A positive result on frequency localization can be obtained with nonstationary wavelet packets, provided that the filter length grows sufficiently fast. We assume once again that the hypotheses (3.1)–(3.4) are satisfied and that $\{m_0^k(\omega)\} = \{m_{0,L(k)}(\omega)\}$ is a subset of the family given by (2.11) and (2.12).

Theorem 4. *Assume that* $L(k) \geq Ck^{3+r}$ *for some* $r > 0$. *Then, for any* $\varepsilon > 0$ *there are* $M_\varepsilon > 0$ *and* $n_\varepsilon \in \mathbb{N}$ *such that, if* $n \geq n_\varepsilon$, *then*

$$\frac{1}{n} \operatorname{card} \{k \leq n \,:\, \sigma_k > M_\varepsilon \} \leq \varepsilon. \tag{4.2}$$

The proof of this result is detailed in Cohen and Séré [6]. It is rather technical, but essentially exploits the fact that, as $L(k)$ goes to $+\infty$, the function $m_{0,L(k)}(\omega)$ converges pointwise to 1 on $(-\pi/2, \pi/2)$, an immediate consequence of Lemma 1.

References

[1] Berkolaiko, S. and I. Novikov, On infinitely smooth almost-wavelets with compact support, *Dokl. Russ. Acal. Neuk.* **326** (1992), 935–938.

[2] Cavaretta, A., W. Dahmen, and C. A. Micchelli, Stationary subdivision, *Mem. Amer. Math. Soc.* **93** (1991), 1–186.

[3] Cohen, A., I. Daubechies, and P. Vial, Wavelets and fast wavelet transforms on an interval, *Appl. Comp. Harmonic Anal.* **1** (1993).

[4] Cohen, A. and N. Dyn, Nonstationary subdivision schemes and multiresolution analysis, *SIAM J. Math. Anal.* (1995), to appear.

[5] Cohen, A. and R. Ryan, *Wavelets and Multiscale Signal Processing*, Chapman & Hall, London, 1995.

[6] Cohen A. and E. Séré, Time-frequency localization with nonstationary wavelet packets, in *Wavelet, Filter Banks and Applications*, A. Akansu (ed.), Academic Press, 1995, to appear.

[7] Coifman, R. R., Y. Meyer, and V. M. Wickerhauser, Entropy-based algorithms for best basis selection, *IEEE Trans. Inform. Theory* **38**(2) (1992), 713–718.

[8] Daubechies, I., Orthonormal bases of compactly supported wavelets, *Comm. Pure Appl. Math.* **41** (1988), 909–996.

[9] Daubechies, I., *Ten Lectures on Wavelets*, SIAM, Philadelphia, 1992.

[10] Daubechies, I. and J. Lagarias, Two scale difference equations I. Existence and global regularity of solutions, *SIAM J. Math. Anal.* **22**(5) (1991), 1388–1410.

[11] Derfel, G., N. Dyn, and D. Levin, Generalized functional equations and subdivision processes, *J. Approx. Theory* **80** (1995), 272–297.

[12] Dyn, N., Subdivision schemes in computer-aided geometric design, in *Advances in Numerical Analysis II, Wavelets, Subdivision Algorithms and Radial Functions*, W. A. Light (ed.), Clarendon Press, Oxford, 1992, pp. 36–104.

[13] Dyn N. and D. Levin, Interpolatory subdivision schemes for the generation of curves and surfaces, in *Multivariate Approximation and Interpolation*, W. Haussmann and K. Jetter (eds.), Birkauser Verlag, Basel, 1990, pp. 91–106.

[14] Meyer, Y., *Ondelettes et Opérateurs*, Hermann, Paris, 1990.

[15] Rvachev, V. L. and V. A. Rvachev, On a function with compact support, *Dopov. Akad. Nauk. URSR* **8** (1971), 705–707 (in Ukrainian).

[16] Séré, E., Localisation fréquentielle des paquets d'ondelettes, *Revista Matemática Iberoamericana* (1993).

Albert Cohen
Laboratoire d'Analyse Numérique
Université Pierre et Marie Curie
4 Place Jussieu
75005 Paris, France
cohen@ann.jussieu.fr

Nira Dyn
School of Mathematical Sciences
Sackler Faculty of Exact Sciences
Tel Aviv University
Tel Aviv 69978, Israel
niradyn@math.tau.ac.il

Interpolatory and Orthonormal
Trigonometric Wavelets

Jürgen Prestin and Kathi Selig

Abstract. The aim of this paper is the detailed investigation of trigonometric polynomial spaces as a tool for approximation and signal analysis. Sample spaces are generated by equidistant translates of certain de la Vallée Poussin means. The different de la Vallée Poussin means enable us to choose between better time or frequency localization. For nested sample spaces and corresponding wavelet spaces, we discuss different bases and their transformations.

§1 Introduction

Trigonometric polynomials and the approximation of periodic functions by polynomials play an important role in harmonic analysis. Here we are interested in constructing time-localized bases for certain spaces of trigonometric polynomials. We use de la Vallée Poussin means of the usual Dirichlet kernel, which allow the investigation of simple projections onto these spaces. Interpolation and orthogonal projection are discussed. With the different de la Vallée Poussin means, the operator norms of these projections as well as the time-frequency localization of the basis functions can be controlled. In order to improve the time-localization, the side oscillations are reduced by including more frequencies and averaging the highest ones. Adapting basic ideas of wavelet theory, interpolatory and orthonormal bases are employed, both of which are constructed from equidistant shifts of a single polynomial.

One of our main goals is the investigation of basis transformations in a form that facilitates fast algorithms. We observe that the corresponding transformation matrices have a circulant structure and can be diagonalized by Fourier matrices. The resulting diagonal matrices contain the eigenvalues, which are computed explicitly. So, the algorithms can be easily realized using the Fast Fourier Transform (FFT).

Further, we consider the nesting of the sample spaces to obtain multiresolution analyses (MRA's). For the resulting orthogonal wavelet spaces, we proceed as above and find wavelet bases consisting of translates of a single polynomial. Again, interpolatory and orthonormal bases are constructed,

Signal and Image Representation in Combined Spaces
Y. Y. Zeevi and R. R. Coifman (Eds.), pp. 201–255.

which show the same time-frequency behavior as the sample bases do. The basis transformations can be described analogously by circulant matrices.

Most important for practical reasons are the decomposition of signals in frequency bands, which correspond to the wavelet spaces, and their reconstruction. Focusing on basis transformations, the two-scale relations and decomposition formulas are also given in matrix notation suitable for the use of FFT methods. Then, the transformations for signal data follow easily (see [19]).

This direct approach to trigonometric wavelets is an alternative to the construction of periodic wavelets by periodization of cardinal wavelets (see [1, 3, 6, 8, 10]). In particular, the de la Vallée Poussin means can be obtained by periodization of cardinal functions which are products of sinus cardinalis functions.

A first constructive approach to trigonometric wavelets was introduced by C. K. Chui and H. N. Mhaskar in [2] taking very simple nested multiresolution spaces and corresponding wavelet spaces. They constructed the scaling functions by means of the partial Fourier sum operator applied to a Haar function with small support, in order to get time-localized basis functions. The two disadvantages of this procedure are, first, that the resulting approximation in the sample space is only a quasi-interpolation and, second, that the basis functions are just as localized as the simplest de la Vallée Poussin mean, the modified Dirichlet kernel.

The idea that we followed was initiated by the paper [18] of A. A. Privalov, where he considered the de la Vallée Poussin means in a different context. Namely, he was interested in an orthonormal polynomial Schauder basis for $C_{2\pi}$ of optimal degree. This problem has also been considered by P. Wojtaszczyk and K. Woźniakowski in [23], by D. Offin and K. Oskolkov in [9], and was finally solved by R. Lorentz and A. A. Sahakian in [7] (see also K. Woźniakowski [24]).

Privalov's approach was also the starting point for E. Quak and the first named author to investigate in [14], [15], and [16] particular cases of trigonometric wavelet spaces. A more general concept of periodic wavelets was given by G. Plonka and M. Tasche in [12], where both constructive and periodized versions are included.

The present paper is organized as follows. In Section 2 we introduce sample spaces and discuss different bases and their transformations. Besides the interpolatory basis, its dual basis and the orthonormalized basis (which all consist of translates of a single function) we also consider a frequency basis. We conclude this section by general $L_{2\pi}^p$-stability arguments and asymptotic estimates for the basis functions, which are connected to approximation estimates for the corresponding projections.

In Section 3 we restrict ourselves to certain parameters which allow nested sample spaces and the construction of wavelet spaces. Then the ideas of Section 2 concerning scaling function bases are adapted to the corresponding wavelet bases, their transformations and asymptotics.

Finally, the reconstruction and decomposition formulas are developed in Section 4 for the interpolatory and orthonormal bases, respectively, in order to provide the relevant algorithms for signal analysis.

§2 Sample spaces

We denote by $C_{2\pi}$ the space of all continuous 2π-periodic functions with the maximum norm, by $L_{2\pi}^2$ the space of square-integrable 2π-periodic functions and by T_n the set of all trigonometric polynomials of degree at most n. The inner product in $L_{2\pi}^2$ is given by

$$\langle f, g \rangle := \frac{1}{2\pi} \int_0^{2\pi} f(x)\, \overline{g(x)}\, dx, \qquad \text{for all } f, g \in L_{2\pi}^2.$$

At first we consider interpolation at equidistant nodal points in the interval $[0, 2\pi)$ in its simplest form. This can be achieved by a discretization of the Fourier sum operator S_n with kernel D_n, $n \in \mathbb{N}$,

$$S_n f(x) := \frac{a_0}{2} + \sum_{k=1}^{n} (a_k \cos kx + b_k \sin kx) = \langle f, D_n(\circ - x) \rangle, \qquad (2.1)$$

with the coefficients

$$a_0 := \frac{1}{\pi} \int_0^{2\pi} f(t)\, dt, \quad a_k := \frac{1}{\pi} \int_0^{2\pi} f(t) \cos kt\, dt, \quad b_k := \frac{1}{\pi} \int_0^{2\pi} f(t) \sin kt\, dt.$$

The Dirichlet kernel

$$D_n(x) := 1 + 2 \sum_{s=1}^{n} \cos sx = \begin{cases} \dfrac{\sin(2n+1)\frac{x}{2}}{\sin \frac{x}{2}}, & \text{for } x \notin 2\pi\mathbf{Z}, \\[2mm] 2n+1, & \text{for } x \in 2\pi\mathbf{Z} \end{cases}$$

is vanishing at all nodal points $\frac{2k\pi}{2n+1}$, for $k = 1, \ldots, 2n$. Taking corresponding translates of D_n as fundamental polynomials, we can define the Lagrange interpolation operator, for any $f \in C_{2\pi}$, at $2n+1$ equidistant points.

The disadvantages of this approach are, firstly, that periodic multiresolution needs an even (dyadic) number of nodal points, and secondly, that the kernel has no bounded $L^1_{2\pi}$-norm, for $n \to \infty$, or, in other words, is not sufficiently local. Both of these problems can be resolved by using certain de la Vallée Poussin means of D_n instead of the Dirichlet kernel itself.

2.1   Trigonometric interpolation

We define the de la Vallée Poussin means φ_N^M, for $N, M \in \mathbb{N}$ and $N \geq M$, by

$$
\begin{aligned}
\varphi_N^M(x) &:= \frac{1}{2M\sqrt{2N}} \sum_{m=N-M}^{N+M-1} D_m(x) \\
&= \frac{1}{\sqrt{2N}} \left(1 + 2 \sum_{\ell=1}^{N-M} \cos \ell x + 2 \sum_{\ell=N-M+1}^{N+M-1} \frac{N+M-\ell}{2M} \cos \ell x \right) \quad (2.1.1) \\
&= \frac{1}{\sqrt{2N}} \left(D_{N-M}(x) + \sum_{k=-M+1}^{M-1} \frac{M-k}{M} \cos(N+k)x \right) . \quad (2.1.2)
\end{aligned}
$$

By induction, one easily proves that they can be rewritten in the form

$$
\varphi_N^M(x) =
\begin{cases}
\dfrac{\sin Nx \sin Mx}{2M\sqrt{2N}\sin^2 \frac{x}{2}}, & \text{for } x \notin 2\pi\mathbb{Z}, \\[2ex]
\sqrt{2N}, & \text{for } x \in 2\pi\mathbb{Z},
\end{cases}
\quad (2.1.3)
$$

which guarantees zero values at the nodes $\frac{k\pi}{N}$, $k = 1, \ldots, 2N-1$. Figure 1 shows the interpolating sampling functions.

The corresponding trigonometric interpolation operator $L_N^M : C_{2\pi} \to C_{2\pi}$ is defined by

$$
L_N^M f(x) := \frac{1}{\sqrt{2N}} \sum_{s=0}^{2N-1} f\left(\frac{s\pi}{N}\right) \varphi_N^M\left(x - \frac{s\pi}{N}\right) \quad (2.1.4)
$$

satisfying, for all $k \in \mathbb{Z}$,

$$
L_N^M f\left(\frac{k\pi}{N}\right) = f\left(\frac{k\pi}{N}\right) .
$$

For $s = 0, \ldots, 2N-1$, the translates

$$
\varphi_{N,s}^M(x) := \varphi_N^M\left(x - \frac{s\pi}{N}\right)
$$

of the de la Vallée Poussin means are linearly independent since $\varphi_{N,s}^M\left(\frac{k\pi}{N}\right) = \sqrt{2N}\,\delta_{k,s}$, for $k, s = 0, \ldots, 2N-1$, but they are not orthogonal (see Proposition 1). Let

$$
V_N^M := \text{span}\left\{ \varphi_{N,s}^M : s = 0, \ldots, 2N-1 \right\} .
$$

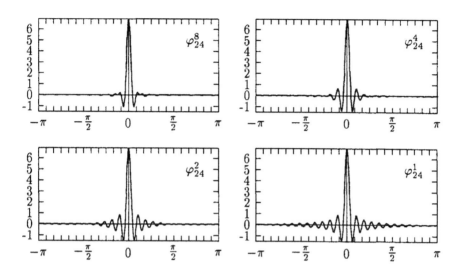

Figure 1. Interpolating sampling functions.

Then the dimension of V_N^M is $2N$ independently of M. Being an interpolation operator, $L_N^M : C_{2\pi} \to V_N^M$ is the identity on V_N^M.

It is instructive to know how single frequencies are mapped by L_N^M.

Lemma 1. *For all* $k \in \mathbb{N}_0$, *let* $k \equiv \ell \bmod 2N$, *with* $0 \le \ell \le 2N - 1$. *Then,*

$$L_N^M \cos(k \circ)(x)$$
$$= \begin{cases} \cos \ell x, & \text{if } 0 \le \ell \le N - M, \\ \frac{M + (N - \ell)}{2M} \cos \ell x + \frac{M - (N - \ell)}{2M} \cos(2N - \ell)x, & \text{if } N - M < \ell < N + M, \\ \cos(2N - \ell)x, & \text{if } N + M \le \ell \le 2N - 1, \end{cases}$$

$$L_N^M \sin(k \circ)(x)$$
$$= \begin{cases} \sin \ell x, & \text{if } 0 < \ell \le NM, \\ \frac{M + (N - \ell)}{2M} \sin \ell x - \frac{M - (N - \ell)}{2M} \sin(2N - \ell)x, & \text{if } N - M < \ell < N + M, \\ -\sin(2N - \ell)x, & \text{if } N + M \le \ell \le 2N - 1. \end{cases}$$

Proof: We use the well-known fact that, for all $N \in \mathbb{N}$, $k \in \mathbf{Z}$,

$$\sum_{s=0}^{2N-1} \cos\left(x + \frac{ks\pi}{N}\right) = \begin{cases} 2N \cos x, & \text{if } k \equiv 0 \bmod 2N, \\ 0, & \text{if } k \not\equiv 0 \bmod 2N, \end{cases} \qquad (2.1.5)$$

$$\sum_{s=0}^{2N-1} \sin\left(x + \frac{ks\pi}{N}\right) = \begin{cases} 2N\sin x, & \text{if } k \equiv 0 \bmod 2N, \\ 0, & \text{if } k \not\equiv 0 \bmod 2N. \end{cases} \quad (2.1.6)$$

By definition of L_N^M, we have, for any $k \in \mathbb{N}_0$,

$$L_N^M \cos(k \circ)(x) = \frac{1}{\sqrt{2N}} \sum_{s=0}^{2N-1} \cos k\left(\frac{s\pi}{N}\right) \varphi_N^M\left(x - \frac{s\pi}{N}\right)$$

$$= \frac{1}{4NM} \sum_{s=0}^{2N-1} \cos\frac{ks\pi}{N} \sum_{m=N-M}^{N+M-1} \left(1 + 2\sum_{r=1}^{m} \cos r\left(x - \frac{s\pi}{N}\right)\right)$$

$$= \delta_{0,k \bmod 2N}$$

$$+ \frac{1}{4NM} \sum_{m=N-M}^{N+M-1} \sum_{r=1}^{m} \sum_{s=0}^{2N-1} \left(\cos\left(rx + \frac{(k-r)s\pi}{N}\right) + \cos\left(rx - \frac{(k+r)s\pi}{N}\right)\right)$$

$$= \delta_{0,k \bmod 2N} + \frac{1}{2M} \sum_{m=N-M}^{N+M-1} \sum_{r=1}^{m} (\delta_{0,(k-r) \bmod 2N} + \delta_{0,(k+r) \bmod 2N}) \cos rx.$$

Now we have to consider three different cases for ℓ, with $k \equiv \ell \bmod 2N$. If $0 \le \ell \le N - M$, then

$$L_N^M \cos(k \circ)(x) = \delta_{0,\ell} + \frac{1}{2M} \sum_{m=N-M}^{N+M-1} \sum_{r=1}^{m} \delta_{r,\ell} \cos rx = \cos \ell x.$$

If $N - M < \ell < N + M$, then

$$L_N^M \cos(k \circ)(x) = \frac{1}{2M} \sum_{m=N-M}^{N+M-1} \sum_{r=1}^{m} (\delta_{r,\ell} + \delta_{r,2N-\ell}) \cos rx$$

$$= \frac{(N+M)-\ell}{2M} \cos \ell x + \frac{N+M-(2N-\ell)}{2M} \cos(2N-\ell)x$$

$$= \left(\frac{1}{2} + \frac{N-\ell}{2M}\right) \cos \ell x + \left(\frac{1}{2} - \frac{N-\ell}{2M}\right) \cos(2N-\ell)x.$$

If $N + M \le \ell \le 2N - 1$, then

$$L_N^M \cos(k \circ)(x) = \frac{1}{2M} \sum_{m=N-M}^{N+M-1} \sum_{r=1}^{m} \delta_{r,2N-\ell} \cos rx = \cos(2N-\ell)x.$$

The formula for the sine frequencies is proved analogously. ■

From Lemma 1, it follows immediately that L_N^M is also the identity on T_{N-M}, and

$$T_{N-M} \subset V_N^M \subset T_{N+M-1}, \quad (2.1.7)$$

which is basic for nesting subspaces to form a trigonometric multiresolution analysis.

A more detailed description of V_N^M in terms of frequencies is given by the following result:

Theorem 1. *For* $M < N$, *the set*

$$\{\varrho_{N,k}^M : \; k = 0,\ldots, 2N - 1\},$$

with

$$\varrho_{N,0}^M(x) := \sqrt{2}/2, \;\; \varrho_{N,k}^M(x) := \sqrt{2}\cos kx, \;\; \varrho_{N,2N-k}^M(x) := \sqrt{2}\sin kx, \tag{2.1.8}$$

where $k = 1,\ldots, N - M$, *and*

$$\varrho_{N,N}^M(x) \;\; := \tfrac{\sqrt{2}}{2}\cos Nx,$$

$$\varrho_{N,N-k}^M(x) \;\; := \sqrt{2}\Big(\tfrac{M+k}{2M}\cos(N - k)x + \tfrac{M-k}{2M}\cos(N + k)x\Big), \tag{2.1.9}$$

$$\varrho_{N,N+k}^M(x) \;\; := \sqrt{2}\Big(\tfrac{M+k}{2M}\sin(N - k)x - \tfrac{M-k}{2M}\sin(N + k)x\Big),$$

where $k = 1,\ldots, M - 1$, *constitutes an orthogonal basis of* V_N^M.

Proof: By (2.1.7), the orthogonal basis (2.1.8) of T_{N-M} is a basis of $V_N^M \cap T_{N-M}$. In order to find the remaining orthogonal basis of $V_N^M \ominus T_{N-M}$, where $\ominus$ denotes the orthogonal difference, we rewrite the high-frequency part of (2.1.2) for the translates $\varphi_{N,\ell}^M$ as follows:

$$\sum_{k=-M+1}^{M-1} \tfrac{M-k}{M}\cos(N + k)\left(x - \tfrac{\ell\pi}{N}\right)$$

$$= \cos\tfrac{N\ell\pi}{N}\cos Nx$$

$$+ \sum_{k=1}^{M-1} \tfrac{M-k}{M}\left(\cos\tfrac{(N+k)\ell\pi}{N}\cos(N + k)x + \sin\tfrac{(N+k)\ell\pi}{N}\sin(N + k)x\right)$$

$$+ \sum_{k=1}^{M-1} \tfrac{M+k}{M}\left(\cos\tfrac{(N-k)\ell\pi}{N}\cos(N - k)x - \sin\tfrac{(N-k)\ell\pi}{N}\sin(N - k)x\right)$$

$$= \cos\tfrac{N\ell\pi}{N}\cos Nx$$

$$+ \sum_{k=1}^{M-1} \cos\tfrac{(N-k)\ell\pi}{N}\left(\tfrac{M+k}{M}\cos(N - k)x + \tfrac{M-k}{M}\cos(N + k)x\right)$$

$$+ \sum_{k=1}^{M-1} \sin\tfrac{(N-k)\ell\pi}{N}\left(\tfrac{M+k}{M}\sin(N - k)x - \tfrac{M-k}{M}\sin(N + k)x\right).$$

Thus, we see that the functions in (2.1.9) are elements of V_N^M and clearly orthogonal to all functions of T_{N-M}. The mutual orthogonality of these functions is obvious, and their number is $2M - 1 = \dim V_N^M - \dim T_{N-M}$. ∎

Let us introduce vector notations for the sequence of basis functions,

$$\underline{\varphi}_N^M := \left(\varphi_{N,\ell}^M\right)_{\ell=0}^{2N-1} \quad \text{and} \quad \underline{\varrho}_N^M := \left(\varrho_{N,k}^M\right)_{k=0}^{2N-1}.$$

Then the corresponding basis transformations can be described by their transformation matrices.

Theorem 2. *The translates of the de la Vallée Poussin means satisfy the relations*

$$\underline{\varphi}_N^M = U_N \, \underline{\varrho}_N^M \quad \text{and} \quad \underline{\varrho}_N^M = U_N^{-1} \, \underline{\varphi}_N^M,$$

where the matrices $U_N = \left(u_{\ell,k}^N\right)_{\ell,k=0}^{2N-1}$ *and* $U_N^{-1} = \left(\breve{u}_{k,\ell}^N\right)_{k,\ell=0}^{2N-1}$ *have the entries, for all* $\ell = 0, \ldots, 2N - 1$,

$$u_{\ell,0}^N = \frac{1}{\sqrt{N}}, \qquad \breve{u}_{0,\ell}^N = \frac{1}{2\sqrt{N}},$$

$$u_{\ell,N}^N = \frac{(-1)^\ell}{\sqrt{N}}, \qquad \breve{u}_{N,\ell}^N = \frac{(-1)^\ell}{2\sqrt{N}},$$

$$u_{\ell,k}^N = \breve{u}_{k,\ell}^N = \begin{cases} \dfrac{1}{\sqrt{N}} \cos \dfrac{\ell k \pi}{N}, & \text{if} \quad 0 < k < N, \\[2ex] \dfrac{-1}{\sqrt{N}} \sin \dfrac{\ell k \pi}{N}, & \text{if} \quad N < k < 2N. \end{cases}$$

Proof: From (2.1.2) and the definition of the basis elements $\varrho_{N,k}^M$, it follows immediately that

$$\varphi_{N,\ell}^M(x) = \frac{\sqrt{2}}{\sqrt{2N}} \left(\sum_{k=0}^{N} \cos \frac{\ell k \pi}{N} \, \varrho_{N,k}^M(x) + \sum_{k=1}^{N-1} \sin \frac{\ell k \pi}{N} \, \varrho_{N,2N-k}^M(x) \right),$$

which confirms the entries of U_N. The second relation is determined by the inverse transformation. From the interpolation

$$\varrho_{N,k}^M(x) = L_N^M \, \varrho_{N,k}^M(x)$$

$$= \frac{1}{\sqrt{2N}} \sum_{\ell=0}^{2N-1} \varrho_{N,k}^M\left(\frac{\ell \pi}{N}\right) \varphi_{N,\ell}^M(x)$$

and from Lemma 1, we obtain the entries of U_N^{-1}. ∎

2.2 Gram matrix and dual basis

For a more detailed investigation of the sample spaces, we introduce the
Gram matrix

$$G_N^M \quad := \quad \left(\langle \varphi_{N,r}^M , \varphi_{N,s}^M \rangle \right)_{r,s=0}^{2N-1} .$$

To simplify the further computations, we will employ the particular structure of this Gram matrix. It is easily seen that G_N^M is symmetric and
circulant. The latter fact allows us, according to [4], Chap. 3, to diagonalize G_N^M by means of the 2N-th Fourier matrix; *i.e.*

$$G_N^M \quad = \quad \overline{F}_{2N} \, D_N^M \, F_{2N} ,$$

where

$$F_{2N} \quad := \quad \frac{1}{\sqrt{2N}} \left(e^{-\frac{2\pi i r s}{2N}} \right)_{r,s=0}^{2N-1} , \qquad \overline{F}_{2N} \ = \ (F_{2N})^{-1} . \qquad (2.2.1)$$

Clearly, the diagonal matrix $D_N^M = \mathrm{diag} \left(d_{N,r}^M \right)_{r=0}^{2N-1}$ contains the eigenvalues of G_N^M .

Lemma 2. *The eigenvalues of the Gram matrix* G_N^M *are*

$$d_{N,r}^M \quad = \quad \begin{cases} \dfrac{M^2 + (N-r)^2}{2M^2} , & \text{if} \quad N-M < r < N+M , \\[2mm] 1 , & \text{otherwise.} \end{cases}$$

Proof: We compute the elements of the diagonal matrix $D_N^M = F_{2N} \, G_N^M \, \overline{F}_{2N}$ in the form

$$d_{N,r}^M \quad = \quad \sum_{k=0}^{2N-1} \langle \varphi_{N,0}^M , \varphi_{N,k}^M \rangle \, e^{\frac{ikr\pi}{N}} .$$

Since G_N^M is symmetric, the eigenvalues are real-valued, and we obtain by
the interpolation formula (2.1.4)

$$\begin{aligned}
d_{N,r}^M \quad &= \quad \sum_{k=0}^{2N-1} \langle \varphi_{N,0}^M , \varphi_{N,k}^M \rangle \ \cos \frac{kr\pi}{N} \\
&= \quad \langle \varphi_{N,0}^M , \sum_{k=0}^{2N-1} \cos \frac{kr\pi}{N} \ \varphi_{N,k}^M \rangle \\
&= \quad \sqrt{2N} \langle \varphi_{N,0}^M , L_N^M \cos(r \circ) \rangle .
\end{aligned}$$

Using Lemma 1 and the representation (2.1.2) of φ_N^M, we consider again three cases. If $0 \leq r \leq N - M$, then

$$d_{N,r}^M = \langle D_{N-M}, \cos(r \circ) \rangle = 1,$$

if $N + M \leq r \leq 2N - 1$, then

$$d_{N,r}^M = \langle D_{N-M}, \cos((2N - r) \circ) \rangle = 1,$$

and if $N - M < r < N + M$, then

$$d_{N,r}^M = 2 \left\langle \sum_{\ell=N-M+1}^{N+M-1} \frac{M+N-\ell}{2M} \cos(\ell \circ) \right.$$

$$\left. \frac{M+(N-r)}{2M} \cos(r \circ) + \frac{M-(N-r)}{2M} \cos((2N - r) \circ) \right\rangle$$

$$= \frac{(M+N-r)^2}{4M^2} + \frac{(M-(N-r))^2}{4M^2} = \frac{M^2+(N-r)^2}{2M^2}. \quad \blacksquare$$

The functions $\varphi_{N,k}^M$ being non-orthogonal, we are now interested in the dual functions. They are needed in order to find the orthogonal projection of a function $f \in L_{2\pi}^2$ with respect to the interpolating basis. The dual functions $\tilde{\varphi}_{N,r}^M \in V_N^M$ of the basis functions $\varphi_{N,\ell}^M$ $(r, \ell = 0, \ldots, 2N - 1)$ are uniquely determined by the orthonormality conditions

$$\langle \tilde{\varphi}_{N,r}^M, \varphi_{N,\ell}^M \rangle = \delta_{r,\ell}.$$

As the duals are also in V_N^M, they possess an expansion in V_N^M. Hence, let

$$\tilde{\varphi}_{N,r}^M(x) = \sum_{s=0}^{2N-1} \alpha_{N,r,s}^M \varphi_{N,s}^M(x). \tag{2.2.2}$$

To simplify the main result on the dual functions, we denote

$$\underline{\tilde{\varphi}}_N^M := \left(\tilde{\varphi}_{N,r}^M \right)_{r=0}^{2N-1}.$$

Theorem 3. *The dual functions satisfy the transformation equations*

$$\underline{\tilde{\varphi}}_N^M = \left(G_N^M \right)^{-1} \underline{\varphi}_N^M \quad \text{and} \quad \underline{\varphi}_N^M = G_N^M \, \underline{\tilde{\varphi}}_N^M, \tag{2.2.3}$$

where

$$\left(G_N^M \right)^{-1} = \overline{F}_{2N} \left(D_N^M \right)^{-1} F_{2N}$$

$$= \left(\delta_{r,s} + \frac{(-1)^{s-r}}{2N} \sum_{k=-M+1}^{M-1} \left(\frac{2M^2}{M^2+k^2} - 1 \right) \cos \frac{k(s-r)\pi}{N} \right)_{r,s=0}^{2N-1}$$

(2.2.4)

$$G_N^M = \overline{F}_{2N} \, D_N^M \, F_{2N}$$

$$= \left(\delta_{r,s} + \frac{(-1)^{s-r}}{2N} \sum_{k=-M+1}^{M-1} \left(\frac{M^2+k^2}{2M^2} - 1 \right) \cos \frac{k(s-r)\pi}{N} \right)_{r,s=0}^{2N-1},$$

with

$$D_N^M = \text{diag} \left(d_{N,r}^M \right)_{r=0}^{2N-1}, \qquad \left(D_N^M \right)^{-1} = \text{diag} \left(1/d_{N,r}^M \right)_{r=0}^{2N-1},$$

the entries of which are known from Lemma 2.

Proof: From the definition in (2.2.2), it follows that

$$\langle \tilde{\varphi}_{N,r}^M , \varphi_{N,\ell}^M \rangle = \sum_{s=0}^{2N-1} \alpha_{N,r,s}^M \langle \varphi_{N,s}^M , \varphi_{N,\ell}^M \rangle = \delta_{r,\ell},$$

which establishes the coefficient matrix $\left(\alpha_{N,r,s}^M \right)_{r,s=0}^{2N-1} = \left(G_N^M \right)^{-1}$. Furthermore, the inverse of a circulant is again a circulant, and hence,

$$\left(G_N^M \right)^{-1} = \overline{F}_{2N} \left(D_N^M \right)^{-1} F_{2N},$$

with $\left(D_N^M \right)^{-1} = \text{diag} \left(1/d_{N,r}^M \right)_{r=0}^{2N-1}$. From there we calculate

$$\alpha_{N,r,s}^M = \frac{1}{2N} \sum_{\ell=0}^{2N-1} \frac{1}{d_{N,\ell}^M} e^{\frac{i\ell(s-r)\pi}{N}}$$

$$= \frac{1}{2N} \left(\sum_{\ell=0}^{2N-1} e^{\frac{i\ell(s-r)\pi}{N}} + \sum_{\ell=N-M+1}^{N+M-1} \left(\frac{2M^2}{M^2+(N-\ell)^2} - 1 \right) e^{\frac{i\ell(s-r)\pi}{N}} \right)$$

$$= \frac{1}{2N} \left(2N \delta_{s,r} + \sum_{k=-M+1}^{M-1} \left(\frac{2M^2}{M^2+k^2} - 1 \right) e^{\frac{i(N+k)(s-r)\pi}{N}} \right)$$

$$= \delta_{s,r} + \frac{(-1)^{s-r}}{2N} \sum_{k=-M+1}^{M-1} \left(\frac{2M^2}{M^2+k^2} - 1 \right) \cos \frac{k(s-r)\pi}{N}.$$

Inverting this relation yields the second formula, and we can compute

$$\langle \varphi_{N,r}^M, \varphi_{N,s}^M \rangle$$

$$= \frac{1}{2N} \sum_{\ell=0}^{2N-1} d_{N,\ell}^M \, e^{\frac{i\ell(s-r)\pi}{N}}$$

$$= \frac{1}{2N} \left(\sum_{\ell=0}^{2N-1} e^{\frac{i\ell(s-r)\pi}{N}} + \sum_{\ell=N-M+1}^{N+M-1} \left(\frac{M^2+(N-\ell)^2}{2M^2} - 1 \right) e^{\frac{i\ell(s-r)\pi}{N}} \right)$$

$$= \frac{1}{2N} \left(2N \, \delta_{s,r} + \sum_{k=-M+1}^{M-1} \left(\frac{M^2+k^2}{2M^2} - 1 \right) e^{\frac{i(N+k)(s-r)\pi}{N}} \right)$$

$$= \delta_{s,r} + \frac{(-1)^{s-r}}{2N} \sum_{k=-M+1}^{M-1} \left(\frac{M^2+k^2}{2M^2} - 1 \right) \cos \frac{k(s-r)\pi}{N}. \quad \blacksquare$$

For the entries of G_N^M, we also know another form from the paper [18] of A. A. Privalov.

Proposition 1. ([18]) *Suppose* $M < N \in \mathbb{N}$. *Then*

$$\langle \varphi_{N,r}^M, \varphi_{N,r}^M \rangle = 1 - \frac{M}{3N} + \frac{1}{12NM}$$

and, for $r \neq s$,

$$\langle \varphi_{N,r}^M, \varphi_{N,s}^M \rangle = (-1)^{r-s} \frac{2M \cos \frac{M(r-s)\pi}{N} \sin \frac{(r-s)\pi}{2N} - \sin \frac{M(r-s)\pi}{N} \cos \frac{(r-s)\pi}{2N}}{8NM^2 \sin^3 \frac{(r-s)\pi}{2N}}.$$

In Section 2.4 we will use the fact that G_N^M is positive definite. In particular, we proved the following inequality.

Proposition 2. ([17]) *For* $N, M \in \mathbb{N}$, *with* $M|N$, *and for* $r = 0, \ldots, 2N-1$,

$$\langle \varphi_{N,r}^M, \varphi_{N,r}^M \rangle - \sum_{\substack{s=0 \\ s \neq r}}^{2N-1} |\langle \varphi_{N,s}^M, \varphi_{N,r}^M \rangle| > \frac{1}{4}.$$

Let us end with the representation of the dual functions in the orthogonal basis of frequencies.

Theorem 4. *For the dual functions, we have*

$$\tilde{\vec{\varphi}}_N^M = U_N \left(D_N^M \right)^{-1} \vec{\varrho}_N^M \quad \text{and} \quad \vec{\varrho}_N^M = D_N^M \, U_N^{-1} \, \tilde{\vec{\varphi}}_N^M,$$

with U_N, U_N^{-1} *from Theorem 2 and* $D_N^M, (D_N^M)^{-1}$ *as given in Theorem 3.*

Proof: We insert the coefficients of (2.2.4) in (2.2.2) and obtain

$$
\tilde\varphi_{N,r}^M(x) = \varphi_{N,r}^M(x) + \frac{1}{2N} \sum_{k=-M+1}^{M-1} \left(\frac{2M^2}{M^2+k^2} - 1 \right)
$$

$$
\sum_{s=0}^{2N-1} \cos \frac{(N+k)(s-r)\pi}{N} \varphi_{N,s}^M(x)
$$

$$
= \frac{1}{\sqrt{2N}} \left(D_{N-M}\left(x - \tfrac{r\pi}{N}\right) + \sum_{k=-M+1}^{M-1} \frac{M-k}{M} \cos(N+k)\left(x - \tfrac{r\pi}{N}\right) \right.
$$

$$
+ \sum_{k=-M+1}^{M-1} \left(\frac{2M^2}{M^2+k^2} - 1 \right)
$$

$$
\left. \times \left(\tfrac{M-k}{2M} \cos(N+k)\left(x - \tfrac{r\pi}{N}\right) + \tfrac{M+k}{2M} \cos(N-k)\left(x - \tfrac{r\pi}{N}\right) \right) \right).
$$

Thus, the dual functions are

$$
\tilde\varphi_{N,r}^M(x) = \frac{1}{\sqrt{2N}} \left(D_{N-M}\left(x - \tfrac{r\pi}{N}\right) \right.
$$

$$
\left. + 2M \sum_{k=-M+1}^{M-1} \tfrac{M-k}{M^2+k^2} \cos(N+k)\left(x - \tfrac{r\pi}{N}\right) \right). \tag{2.2.5}
$$

By the definition of $\varrho_{N,k}^M$, we can conclude that

$$
\tilde\varphi_{N,r}^M = \frac{\sqrt{2}}{\sqrt{2N}} \left(\sum_{k=0}^{N} \frac{1}{d_{N,k}^M} \cos \frac{rk\pi}{N} \varrho_{N,k}^M + \sum_{k=1}^{N-1} \frac{1}{d_{N,2N-k}^M} \sin \frac{rk\pi}{N} \varrho_{N,2N-k}^M \right).
$$

Writing this in matrix notation yields the first equation. Then the inverse transformation works by means of the identity $(U_N(D_N^M)^{-1})^{-1} = D_N^M U_N^{-1}$. ∎

From (2.2.5), it follows that the dual functions are shifts of $\tilde\varphi_{N,0}^M$ again, which is an even function like $\varphi_{N,0}^M$. Note also that the coefficients $\alpha_{N,r,s}^M$ are equal to the function values

$$
\alpha_{N,r,s}^M = (2N)^{-1/2} \, \tilde\varphi_{N,0}^M \left(\tfrac{(s-r)\pi}{N} \right), \tag{2.2.6}
$$

which follows from the interpolation property of the basis in the expansion (2.2.2).

2.3 Orthonormal basis of translates

We also seek a basis of V_N^M consisting of orthonormal translates $\mathcal{O}\varphi_{N,r}^M(x) :=$ $\mathcal{O}\varphi_N^M \left(x - \frac{r\pi}{N}\right)$ of a function $\mathcal{O}\varphi_N^M \in V_N^M$. Writing

$$\mathcal{O}\varphi_{N,r}^M(x) = \sum_{s=0}^{2N-1} \gamma_{N,r,s}^M \, \varphi_{N,s}^M(x), \qquad (2.3.1)$$

for all $r = 0, \dots, 2N-1$, we have to determine the coefficients $\gamma_{N,r,s}^M$. The coefficient matrix

$$\boldsymbol{\Gamma}_N^M := \left(\gamma_{N,r,s}^M\right)_{r,s=0}^{2N-1} = (2N)^{-1/2} \left(\mathcal{O}\varphi_N^M \left(\frac{(s-r)\pi}{N}\right)\right)_{r,s=0}^{2N-1}$$

is obviously circulant. Hence, we can write $\boldsymbol{\Gamma}_N^M = \overline{\boldsymbol{F}}_{2N} \, \boldsymbol{\Delta}_N^M \, \boldsymbol{F}_{2N}$, with a diagonal matrix $\boldsymbol{\Delta}_N^M$. If we require the orthonormality

$$\langle \mathcal{O}\varphi_{N,k}^M, \mathcal{O}\varphi_{N,r}^M \rangle = \sum_{\ell=0}^{2N-1}\sum_{s=0}^{2N-1} \gamma_{N,k,\ell}^M \gamma_{N,r,s}^M \langle \varphi_{N,\ell}^M, \varphi_{N,s}^M \rangle = \delta_{k,r}, \quad (2.3.2)$$

and denote

$$\underline{\mathcal{O}\varphi}_N^M := \left(\mathcal{O}\varphi_{N,r}^M\right)_{r=0}^{2N-1},$$

then we can summarize the representation as follows.

Theorem 5. *In V_N^M, there exists a basis of orthonormal translates of an even function $\mathcal{O}\varphi_N^M$ satisfying the relations*

$$\underline{\mathcal{O}\varphi}_N^M = \boldsymbol{\Gamma}_N^M \, \underline{\varphi}_N^M, \qquad \underline{\varphi}_N^M = \left(\boldsymbol{\Gamma}_N^M\right)^{-1} \underline{\mathcal{O}\varphi}_N^M, \quad (2.3.3)$$

$$\underline{\mathcal{O}\varphi}_N^M = \left(\boldsymbol{\Gamma}_N^M\right)^{-1} \underline{\tilde{\varphi}}_N^M \qquad \text{and} \qquad \underline{\tilde{\varphi}}_N^M = \boldsymbol{\Gamma}_N^M \, \underline{\mathcal{O}\varphi}_N^M, \quad (2.3.4)$$

where

$$\boldsymbol{\Gamma}_N^M = \overline{\boldsymbol{F}}_{2N} \, \boldsymbol{\Delta}_N^M \, \boldsymbol{F}_{2N}$$

$$= \left(\delta_{r,s} + \frac{(-1)^{s-r}}{2N} \sum_{k=-M+1}^{M-1} \left(\frac{2M}{\sqrt{2M^2+2k^2}} - 1\right) \cos\frac{k(s-r)\pi}{N}\right)_{r,s=0}^{2N-1} \quad (2.3.5)$$

and

$$\left(\boldsymbol{\Gamma}_N^M\right)^{-1} = \overline{\boldsymbol{F}}_{2N} \left(\boldsymbol{\Delta}_N^M\right)^{-1} \boldsymbol{F}_{2N}$$

$$= \left(\delta_{r,s} + \frac{(-1)^{s-r}}{2N} \sum_{k=-M+1}^{M-1} \left(\frac{\sqrt{2M^2+2k^2}}{2M} - 1\right) \cos\frac{k(s-r)\pi}{N}\right)_{r,s=0}^{2N-1},$$

with

$$\Delta_N^M = \text{diag}\left(1/\sqrt{d_{N,\ell}^M}\right)_{\ell=0}^{2N-1}, \qquad \left(\Delta_N^M\right)^{-1} = \text{diag}\left(\sqrt{d_{N,\ell}^M}\right)_{\ell=0}^{2N-1}$$

and $d_{N,\ell}^M$ from Lemma 2.

Proof: With $\Delta_N^M = \text{diag}\left(1/\sqrt{d_{N,\ell}^M}\right)_{\ell=0}^{2N-1}$, we are able to compute the coefficients $\gamma_{N,r,s}^M$, for $r, s = 0, \ldots, 2N-1$, from the equation $\Gamma_N^M = \overline{F}_{2N}\,\Delta_N^M\,F_{2N}$ and obtain

$$
\begin{aligned}
\gamma_{N,r,s}^M &= \frac{1}{2N}\sum_{\ell=0}^{2N-1}\frac{1}{\sqrt{d_{N,\ell}^M}}\,e^{\frac{i\ell(s-r)\pi}{N}} \\
&= \frac{1}{2N}\left(\sum_{\ell=0}^{2N-1}e^{\frac{i\ell(s-r)\pi}{N}} + \sum_{\ell=N-M+1}^{N+M-1}\left(\frac{1}{\sqrt{d_{N,\ell}^M}}-1\right)e^{\frac{i\ell(s-r)\pi}{N}}\right) \\
&= \frac{1}{2N}\left(2N\,\delta_{s,r} + \sum_{k=-M+1}^{M-1}\left(\frac{\sqrt{2M^2}}{\sqrt{M^2+k^2}}-1\right)e^{\frac{i(N+k)(s-r)\pi}{N}}\right) \\
&= \delta_{s,r} + \frac{(-1)^{s-r}}{2N}\sum_{k=-M+1}^{M-1}\left(\frac{2M}{\sqrt{2M^2+2k^2}}-1\right)\cos\frac{k(s-r)\pi}{N}.
\end{aligned}
$$

The entries of $\left(\Gamma_N^M\right)^{-1} = \left(\breve{\gamma}_{N,s,r}^M\right)_{s,r=0}^{2N-1} = \overline{F}_{2N}\left(\Delta_N^M\right)^{-1}F_{2N}$ follow from

$$
\begin{aligned}
\breve{\gamma}_{N,s,r}^M &= \frac{1}{2N}\sum_{\ell=0}^{2N-1}\sqrt{d_{N,\ell}^M}\,e^{\frac{i\ell(s-r)\pi}{N}} \\
&= \frac{1}{2N}\left(\sum_{\ell=0}^{2N-1}e^{\frac{i\ell(s-r)\pi}{N}} + \sum_{k=-M+1}^{M-1}\left(\frac{\sqrt{M^2+k^2}}{\sqrt{2M^2}}-1\right)e^{\frac{i(N+k)(s-r)\pi}{N}}\right) \\
&= \delta_{s,r} + \frac{(-1)^{s-r}}{2N}\sum_{k=-M+1}^{M-1}\left(\frac{\sqrt{2M^2+2k^2}}{2M}-1\right)\cos\frac{k(s-r)\pi}{N}.
\end{aligned}
$$

Now we prove the orthonormality of the translates for the representations in (2.3.3). From (2.3.5), we derive that $\gamma_{N,r,s}^M = \gamma_{N,s,r}^M$. Hence, $O\varphi_N^M$ is even, and we have

$$\Gamma_N^M\,G_N^M\left(\Gamma_N^M\right)^T = \Gamma_N^M\,G_N^M\,\Gamma_N^M = \overline{F}_{2N}\,\Delta_N^M\,D_N^M\,\Delta_N^M\,F_{2N} = I_{2N}$$

being the $2N$-th identity matrix, which proves (2.3.2). The relations in (2.3.4) are easy to derive from (2.3.3) and (2.2.3). ∎

Finally, we want to represent the orthonormal translates in the basis of frequencies.

Theorem 6. *For the orthonormal translates, we have the relations*

$$\underline{O\varphi}_N^M = U_N \, \Delta_N^M \, \underline{\varrho}_N^M \quad \text{and} \quad \underline{\varrho}_N^M = \left(\Delta_N^M\right)^{-1} U_N^{-1} \, \underline{O\varphi}_N^M , \quad (2.3.6)$$

with U_N, U_N^{-1} from Theorem 2 and $\Delta_N^M, (\Delta_N^M)^{-1}$ as given in Theorem 5.

Proof: We obtain an explicit form for the orthogonal translates of $O\varphi_N^M$ by evaluating

$$O\varphi_{N,r}^M(x) = \sum_{s=0}^{2N-1} \gamma_{N,r,s}^M \varphi_{N,s}^M(x)$$

$$= \varphi_{N,r}^M + \sum_{s=0}^{2N-1} \frac{1}{2N} \sum_{k=-M+1}^{M-1} \left(\frac{2M}{\sqrt{2M^2+2k^2}} - 1\right) \cos \frac{(N+k)(s-r)\pi}{N} \, \varphi_{N,s}^M(x)$$

$$= \varphi_{N,r}^M + \frac{1}{2N} \sum_{k=-M+1}^{M-1} \left(\frac{2M}{\sqrt{2M^2+2k^2}} - 1\right) \sum_{s=0}^{2N-1} \cos \frac{(N+k)(s-r)\pi}{N} \, \varphi_{N,s}^M(x)$$

with

$$\sum_{s=0}^{2N-1} \cos \frac{(N+k)(s-r)\pi}{N} \, \varphi_{N,s}^M(x) = L_N^M \cos((N+k)\circ) \left(x - \tfrac{r\pi}{N}\right)$$

$$= \tfrac{M-k}{2M} \cos(N+k) \left(x - \tfrac{r\pi}{N}\right) + \tfrac{M+k}{2M} \cos(N-k) \left(x - \tfrac{r\pi}{N}\right).$$

Thus,

$$O\varphi_{N,r}^M(x) = \frac{1}{\sqrt{2N}} \left(D_{N-M}(x - \tfrac{r\pi}{N}) + 2 \sum_{k=-M+1}^{M-1} \frac{M-k}{\sqrt{2M^2+2k^2}} \cos(N+k)(x - \tfrac{r\pi}{N})\right).$$

This can be rewritten:

$$O\varphi_{N,r}^M = \frac{1}{\sqrt{2N}} \left(\sum_{k=0}^{N} \frac{1}{\sqrt{d_{N,k}^M}} \cos \frac{kr\pi}{N} \varrho_{N,k}^M + \sum_{k=1}^{N-1} \frac{1}{\sqrt{d_{N,2N-k}^M}} \sin \frac{kr\pi}{N} \varrho_{N,2N-k}^M\right),$$

which yields the matrices in (2.3.6). ∎

2.4 Time-frequency localization and projection estimates

$L_{2\pi}^p$-Norms and stability

Our first aim is to describe the time localization in terms of the $L_{2\pi}^p$-norms of the scaling functions. Therefore, let us define for $1 \le p < \infty$,

$$\|f\|_p := \left(\frac{1}{2\pi}\int_0^{2\pi}|f(x)|^p dx\right)^{\frac{1}{p}} \quad \text{and} \quad \|f\|_\infty := \operatorname*{ess\,sup}_{x\in[0,2\pi]}|f(x)|,$$

and analogously for vectors $\{\alpha_k\} \in \mathbb{C}^{2N}$

$$\|\{\alpha_k\}\|_{\ell^p} := \left(\sum_{k=0}^{2N-1}|\alpha_k|^p\right)^{\frac{1}{p}} \quad \text{and} \quad \|\{\alpha_k\}\|_{\ell^\infty} := \max_k|\alpha_k|.$$

From the well-known inequalities of Hölder and Nikolskii (see *e.g.* A. F. Timan [22], Chap. 4.9), one knows that

$$1 \le \frac{\|t_n\|_q}{\|t_n\|_p} \le C n^{\frac{1}{p}-\frac{1}{q}},$$

for all polynomials, $t_n \in T_n$ and $1 \le p \le q \le \infty$. We want to have basis functions with the order $n^{\frac{1}{p}-\frac{1}{q}}$ on the right-hand side, which ensures the best possible time localization with respect to a polynomial MRA. In particular, we are looking for polynomials satisfying

$$\|t_n\|_\infty \sim n^{\frac{1}{p}}\|t_n\|_p \sim n\|t_n\|_1.$$

The notation $A_n \sim B_n$ means $|A_n^{-1}B_n| \le C$ and $|A_n B_n^{-1}| \le C$. Here and in the following, C denotes positive constants depending on fixed parameters involved, but their values may be different at each occurrence.

Let us recall the basic result of an $L_{2\pi}^1$-estimate for the interpolating functions established by S.B. Stečkin [20]

$$\|\varphi_{N,0}^M\|_1 = \frac{2}{\pi^2\sqrt{2N}}\log\frac{N+M}{2M} + C + \mathcal{O}\left(\frac{M}{N+M}\right).$$

In order to obtain explicit constants, we proceed with the following simple derivation. For $2M \le N$,

$$\|\varphi_{N,0}^M\|_1 = \frac{2}{\pi\,2M\sqrt{2N}}\int_0^{\frac{\pi}{2}}\left|\frac{\sin(2Mt)\sin(2Nt)}{\sin^2 t}\right|dt$$

$$< \frac{2}{\pi\sqrt{2N}}\left(\frac{2N\pi}{2N} + \int_{\frac{\pi}{2N}}^{\frac{\pi}{4M}}\left|\frac{\sin(2Nt)}{\sin t}\right|dt + \frac{1}{2M}\int_{\frac{\pi}{4M}}^{\frac{\pi}{2}}\frac{1}{\sin^2 t}dt\right)$$

$$< \frac{1}{\pi\sqrt{2N}}\left(2\pi + \pi\ln\frac{N}{2M} + \pi\right) = \frac{1}{\sqrt{2N}}\left(3 + \ln\frac{N}{2M}\right). \qquad (2.4.1)$$

This computation contains some information about the time localization of $\varphi_{N,0}^M$ in the three different intervals of integration. Moreover, we know the maximum value of $\varphi_{N,0}^M$, and thus

$$\left\|\varphi_{N,0}^M\right\|_\infty = \sqrt{2N}. \tag{2.4.2}$$

In Section 3.1 we will define N_j and M_j in such a way that $3M_j \leq N_j$ and that N_j/M_j is bounded from above by a constant. Assuming $N/2M$ to be constant, we conclude from (2.4.1) and (2.4.2) by means of Nikolskii's inequality, for arbitrary $1 \leq p \leq \infty$, that

$$\left\|\varphi_{N,0}^M\right\|_p \sim N^{\frac{1}{2}-\frac{1}{p}}. \tag{2.4.3}$$

As a further step, we formulate the Riesz stability in the more general $L_{2\pi}^p$-setting.

Theorem 7. *For arbitrary sequences* $\{\alpha_k\} \in \mathbb{C}^{2N}$ *and* $1 \leq p \leq \infty$, *the inequalities*

$$\frac{3}{4\pi + 3}\left\|\{\alpha_k\}\right\|_{\ell^p} \leq (2N)^{\frac{1}{p}-\frac{1}{2}}\left\|\sum_{k=0}^{2N-1}\alpha_k\,\varphi_{N,k}^M\right\|_p \leq A\left\|\{\alpha_k\}\right\|_{\ell^p} \tag{2.4.4}$$

are satisfied, with

$$A = \frac{4\pi + 3}{3}\sqrt{2N}\left\|\varphi_{N,0}^M\right\|_1,$$

independent of p, *or*

$$A = \begin{cases} \dfrac{C}{p-1}, & \text{if } 1 < p < 2, \\[2mm] Cp, & \text{if } 2 \leq p < \infty. \end{cases} \tag{2.4.5}$$

As noted above, for simplicity, we assume

$$3 \leq \frac{N}{M} \leq K, \tag{2.4.6}$$

where K is a constant independent of N. This implies the uniform boundedness of A in (2.4.4), for $1 \leq p \leq \infty$, because of

$$A \leq \frac{4\pi + 3}{3}\left(3 + \ln\frac{N}{2M}\right) < 12 + 5.2\,\ln K.$$

Proof: a) We start with the inequality on the left-hand side, which can be proved by using a well-known inequality for trigonometric polynomials $t_n \in T_n$ (see *e.g.* [22], Chap. 4.9), namely:

$$\sup_x \left(\frac{1}{2N} \sum_{\ell=0}^{2N-1} \left| t_n \left(x - \frac{\ell\pi}{N} \right) \right|^p \right)^{\frac{1}{p}} \leq \left(1 + \frac{n\pi}{N} \right) \| t_n \|_p , \quad (2.4.7)$$

for $1 \leq p \leq \infty$ and $N \in \mathbb{N}$. Then, by applying the interpolatory properties of $\varphi_{N,k}^M$, we conclude that

$$\left(\frac{1}{2N} \sum_{\ell=0}^{2N-1} |\alpha_\ell|^p \right)^{\frac{1}{p}} = \left(\frac{1}{2N} \sum_{\ell=0}^{2N-1} \left| \frac{1}{\sqrt{2N}} \sum_{k=0}^{2N-1} \alpha_k \, \varphi_{N,k}^M \left(\frac{\ell\pi}{N} \right) \right|^p \right)^{\frac{1}{p}}$$

$$\leq \left(1 + \frac{(N+M-1)\pi}{N} \right) \left\| \frac{1}{\sqrt{2N}} \sum_{k=0}^{2N-1} \alpha_k \, \varphi_{N,k}^M \right\|_p$$

$$\leq \left(1 + \frac{4\pi}{3} \right) \left\| \frac{1}{\sqrt{2N}} \sum_{k=0}^{2N-1} \alpha_k \, \varphi_{N,k}^M \right\|_p .$$

b) In order to prove the inequality on the right-hand side, we use, for $1 \leq p < \infty$, the function g from the space $L_{2\pi}^q$, where $1/p + 1/q = 1$, with $\|g\|_q = 1$ such that

$$\left\| \sum_{k=0}^{2N-1} \alpha_k \, \varphi_{N,k}^M \right\|_p = \frac{1}{2\pi} \int_0^{2\pi} g(x) \sum_{k=0}^{2N-1} \alpha_k \, \varphi_{N,k}^M(x) \, dx .$$

Then Hölder's inequality gives

$$\sum_{k=0}^{2N-1} \alpha_k \frac{1}{2\pi} \int_0^{2\pi} g(x) \, \varphi_{N,k}^M(x) \, dx$$

$$\leq 2N \left(\frac{1}{2N} \sum_{k=0}^{2N-1} |\alpha_k|^p \right)^{\frac{1}{p}} \left(\frac{1}{2N} \sum_{k=0}^{2N-1} \left| (g * \varphi_{N,0}^M) \left(\frac{k\pi}{N} \right) \right|^q \right)^{\frac{1}{q}} ,$$

where $*$ denotes the usual 2π-periodic convolution,

$$\left(g * \varphi_{N,0}^M \right)(t) := \frac{1}{2\pi} \int_0^{2\pi} g(x) \, \varphi_{N,0}^M(t - x) \, dx .$$

Now we apply (2.4.7) and obtain two different estimates,

$$\left(\frac{1}{2N}\sum_{k=0}^{2N-1}\left|(g*\varphi_{N,0}^M)\left(\tfrac{k\pi}{N}\right)\right|^q\right)^{\frac{1}{q}} \leq \left(1+\tfrac{4\pi}{3}\right)\left\|g*\varphi_{N,0}^M\right\|_q$$

$$\leq \left(1+\frac{4\pi}{3}\right)\times\begin{cases} \|g\|_q\,\|\varphi_{N,0}^M\|_1\,, & \text{if } 1\leq q<\infty, \\[2mm] \dfrac{1}{2M\sqrt{2N}}\displaystyle\sum_{m=N-M}^{N+M-1}\|S_m g\|_q\,, & \text{if } 1<q<\infty. \end{cases}$$

The second estimate in (2.4.5) follows from the boundedness of the Fourier sum operator S_m in $L_{2\pi}^q$, for $1<q<\infty$ (see *e.g.* [25]). The proof is completed by a similar derivation for $p=\infty$. Applying (2.4.7) again, we find that

$$\left\|\frac{1}{\sqrt{2N}}\sum_{k=0}^{2N-1}\alpha_k\,\varphi_{N,k}^M\right\|_\infty \leq \sqrt{2N}\max_k|\alpha_k|\sup_x\left(\frac{1}{2N}\sum_{k=0}^{2N-1}\left|\varphi_{N,0}^M\left(x-\tfrac{k\pi}{N}\right)\right|\right)$$

$$\leq \|\{\alpha_k\}\|_{\ell^\infty}\,\tfrac{4\pi+3}{3}\,\sqrt{2N}\,\|\varphi_{N,0}^M\|_1\,. \quad\blacksquare$$

Asymptotics for the dual and orthogonal functions

Our next goal is to achieve the same time localization results for the dual functions $\tilde{\varphi}_N^M$ and for the orthogonal functions $O\varphi_N^M$ as we have in (2.4.3) for the interpolatory functions. At first we determine estimates for the coefficients $\alpha_{N,r,s}^M$ and $\gamma_{N,r,s}^M$.

Lemma 3. *The coefficients in the basis representations (2.2.2) and (2.3.1) satisfy the decay conditions, for $r,s=0,\ldots,2N-1$,*

$$|\alpha_{N,r,s}^M|,\,|\gamma_{N,r,s}^M| \leq C\max\left\{(|r-s|+1)^{-2},\,(2N-|r+s|)^{-2}\right\}. \quad (2.4.8)$$

Proof: For $r=s$, both of the coefficients given in (2.2.4) and (2.3.5) are less than 2, since $M<N$. For $r\neq s$, let us focus on

$$|\gamma_{N,r,s}^M| = \left|\frac{1}{2N}\sum_{k=-M+1}^{M-1}\left(\frac{2M}{\sqrt{2M^2+2k^2}}-1\right)\cos\frac{k(r-s)\pi}{N}\right|.$$

We define the 2π-periodic even function f_γ by

$$f_\gamma(x) := \begin{cases} \dfrac{M\pi\sqrt{2}}{\sqrt{(\pi M)^2+N^2x^2}}-1\,, & \text{for } |x|\leq\tfrac{\pi M}{N}\,, \\[3mm] 0\,, & \text{for } \tfrac{\pi M}{N}<|x|\leq\pi, \end{cases}$$

The derivative of f_γ is of bounded variation and possesses two jumps in $[-\pi, \pi]$. Hence, for the Fourier coefficients a_{r-s}, with $r \geq s$, we have (see [25], Chap. 10)

$$|a_{|r-s|}| = \left| \frac{1}{\pi} \int_{-\pi}^{\pi} f_\gamma(x) \cos(r-s)x\, dx \right| \leq C(|r-s|+1)^{-2},$$

and by the aliasing formula for discrete Fourier coefficients,

$$
\begin{aligned}
|\gamma_{N,r,s}^M| &= \left| \frac{1}{2N} \sum_{k=-N}^{N-1} f_\gamma(\tfrac{k\pi}{N}) \cos(r-s)\tfrac{k\pi}{N} \right| \\
&= \frac{1}{4} \left| \sum_{\ell=-\infty}^{\infty} a_{|2N\ell+r-s|} + a_{|2N\ell-r+s|} \right| \\
&\leq C \max\left\{ (|r-s|)^{-2}, \; |2N-r+s|^{-2} \right\}.
\end{aligned}
$$

Thus, (2.4.8) is proved for $\gamma_{N,r,s}^M$. The same idea works for $\alpha_{N,r,s}^M$ (see Theorem 3), if one replaces f_γ by

$$
f_\alpha(x) := \begin{cases} \dfrac{2M^2\pi^2}{M^2\pi^2+N^2x^2} - 1, & \text{for} \quad |x| \leq \frac{\pi M}{N}, \\[2mm] 0, & \text{for} \quad \frac{\pi M}{N} < |x| \leq \pi. \end{cases} \quad \blacksquare
$$

Now it is easy to obtain the same time-localization result for the dual and orthonormal functions as we have in (2.4.3) for φ_N^M.

Theorem 8. *For $1 \leq p \leq \infty$, we have the asymptotic behavior*

$$\|\tilde{\varphi}_{N,0}^M\|_p \sim \|\mathcal{O}\varphi_{N,0}^M\|_p \sim N^{\frac{1}{2}-\frac{1}{p}}.$$

Proof: The proof simply consists of the application of Lemma 3 to (2.4.4). For example, we obtain

$$\|\mathcal{O}\varphi_{N,0}^M\|_p = \left\| \sum_{k=0}^{2N-1} \gamma_{N,0,k}^M \varphi_{N,k}^M \right\|_p \sim \sqrt{N} \left(\frac{1}{2N} \sum_{k=0}^{2N-1} |\gamma_{N,0,k}^M|^p \right)^{\frac{1}{p}} \sim N^{\frac{1}{2}-\frac{1}{p}}. \quad \blacksquare$$

Let us end this subsection with a quite different description of time localization (see also [16]). In $[0, 2\pi]$, we define, for $k = 1, \ldots, 2N$, the intervals

$$I_{N,k} := \left[\frac{(k-1)\pi}{N}, \frac{k\pi}{N} \right]$$

and describe the decay in terms of the distance from the peak. Again, we start with the interpolating function $\varphi_{N,0}^M(x)$. From (2.1.3), one obtains by standard estimates and (2.4.6) that

$$\max_{x \in I_{N,k}} |\varphi_{N,0}^M(x)| \sim \begin{cases} \dfrac{\sqrt{N}}{k^2}, & \text{if} \quad 0 < k \leq N, \\[3mm] \dfrac{\sqrt{N}}{(2N-k)^2}, & \text{if} \quad N < k < 2N. \end{cases} \tag{2.4.9}$$

Now, for simplicity, we restrict ourselves to $0 < k \leq N$. The same time localization results can be shown for the dual and orthonormal functions.

Theorem 9. *Let* $0 < k \leq N$. *Then*

$$\|\tilde{\varphi}_{N,0}^M\|_\infty \sim \|\mathcal{O}\varphi_N^M\|_\infty \sim \sqrt{N} \tag{2.4.10}$$

and

$$\max_{x \in I_{N,k}} \{|\tilde{\varphi}_{N,0}^M(x)|, |\mathcal{O}\varphi_{N,0}^M(x)|\} \leq C \frac{\sqrt{N}}{k^2}. \tag{2.4.11}$$

Proof: Formula (2.4.10) is already included in Theorem 8. Here we only show the estimate in (2.4.11) for $\mathcal{O}\varphi_N^M$. With Lemma 3, (2.3.1), and (2.4.9) we obtain

$$\max_{x \in I_{N,k}} |\mathcal{O}\varphi_{N,0}^M(x)|$$

$$\leq C \max_{x \in I_{N,k}} \left(\sum_{r=0}^{N} \frac{1}{(r+1)^2} |\varphi_{N,r}^M(x)| + \sum_{r=N+1}^{2N-1} \frac{1}{(2N-r)^2} |\varphi_{N,r}^M(x)| \right)$$

$$\leq C\sqrt{N} \Big(\sum_{r=0}^{N} \frac{1}{((r+1)(|r-k|+1))^2}$$

$$+ \sum_{r=N+1}^{2N-1} \frac{1}{(2N-r)^2 (\min\{(r-k),(2N-r+k)\})^2} \Big).$$

Then (2.4.11) follows directly by splitting the sums on the right-hand side into

$$\sum_{r=0}^{2N-1} = \sum_{r=0}^{k/2} + \sum_{r=1+k/2}^{3k/2} + \sum_{r=1+3k/2}^{2N-1} . \qquad \blacksquare$$

Error estimates

Here we want to summarize approximation properties of the interpolatory and orthogonal projections onto the spaces V_N^M. For simplicity, we only deal with the most interesting case of continuous functions and estimates in the sup-norm. Roughly speaking, the same results are available for $L_{2\pi}^1$ and especially for $L_{2\pi}^p, 1 < p < \infty$, avoiding the logarithmic term in the estimates.

In particular, we want to compare the approximation order with the best approximation by trigonometric polynomials of corresponding degree, as A. A. Privalov mentioned in [18]. Further results connecting the convergence order of a given function f with smoothness conditions on f can be easily deduced. Denote, for $f \in C_{2\pi}$,

$$E_n(f) := \inf \{\|f - t_n\|_\infty : t_n \in T_n\} \, .$$

In this subsection we do not necessarily assume the uniform boundedness of N/M as done in (2.4.6). But, for further use, we restrict ourselves to $N/M = 3, 4, 5, \ldots$. However, in the case of (2.4.6), we obtain error estimates in the order of the best approximation.

Theorem 10. *Let $f \in C_{2\pi}$. For the interpolatory projection L_N^M onto V_N^M, we have the estimates*

$$E_{N+M-1}(f) \leq \|f - L_N^M f\|_\infty \leq (A+1) E_{N-M}(f), \qquad (2.4.12)$$

with A as defined in (2.4.5).

Proof: The inequality on the left-hand side follows immediately from $L_N^M f \in T_{N+M-1}$. For the inequality on the right-hand side of (2.4.12), we write

$$E_{N-M}(f) = \|f - t_{N-M}\|_\infty$$

to conclude with the help of (2.4.4) that

$$\begin{aligned}
\|f - L_N^M f\|_\infty &\leq E_{N-M}(f) + \|t_{N-M} - L_N^M f\|_\infty \\
&\leq E_{N-M}(f) + A \|t_{N-M} - f\|_\infty \, . \quad \blacksquare
\end{aligned}$$

Now we are looking for error estimates for the orthogonal projection. The proof of Theorem 10 shows that we only have to deal with the operator norm of the orthogonal projection operator $P_N^M : L_{2\pi}^2 \to V_N^M$,

$$P_N^M f := \sum_{k=0}^{2N-1} \langle f, \tilde{\varphi}_{N,k}^M \rangle \, \varphi_{N,k}^M = \sum_{k=0}^{2N-1} \langle f, \mathcal{O}\varphi_{N,k}^M \rangle \, \mathcal{O}\varphi_{N,k}^M \, .$$

Lemma 4. *The orthogonal projection P_N^M onto V_N^M satisfies*

$$\|P_N^M\|_{C_{2\pi} \to C_{2\pi}} \leq C \log^2 \frac{N}{M}.$$

Proof: Writing

$$P_N^M f = \sum_{k=0}^{2N-1} \epsilon_k \varphi_{N,k}^M, \qquad (2.4.13)$$

we obtain by (2.4.4)

$$\|P_N^M f\|_\infty \leq C N \|\varphi_{N,0}^M\|_1 \max_{0 \leq k < 2N} |\epsilon_k| =: C N \|\varphi_{N,0}^M\|_1 |\epsilon_\ell|.$$

Taking the inner product with $\varphi_{N,\ell}^M$ in (2.4.13), we estimate

$$|\langle P_N^M f, \varphi_{N,\ell}^M \rangle| = \left| \sum_{k=0}^{2N-1} \epsilon_k \langle \varphi_{N,k}^M, \varphi_{N,\ell}^M \rangle \right|$$

$$\geq |\epsilon_\ell| \left(\langle \varphi_{N,\ell}^M, \varphi_{N,\ell}^M \rangle - \sum_{\substack{k=0 \\ k \neq \ell}}^{2N-1} |\langle \varphi_{N,k}^M, \varphi_{N,\ell}^M \rangle| \right).$$

Applying Proposition 2 yields

$$|\epsilon_\ell| \leq 4 |\langle P_N^M f, \varphi_{N,\ell}^M \rangle| = 4 |\langle f, \varphi_{N,\ell}^M \rangle| \leq 4 \|f\|_\infty \|\varphi_{N,\ell}^M\|_1.$$

Hence,

$$\|P_N^M f\|_\infty \leq C N \|\varphi_{N,0}^M\|_1^2 \|f\|_\infty,$$

which, together with (2.4.1), proves the lemma. ∎

Using Lemma 4 and standard projection estimates, we arrive at the following final result.

Theorem 11. *Let $f \in L_{2\pi}^2$. For the orthogonal projection P_N^M onto V_N^M, we have the estimates*

$$E_{N+M-1}(f) \leq \|f - P_N^M f\|_\infty \leq C \log^2 \frac{N}{M} E_{N-M}(f). \qquad (2.4.14)$$

It remains an open question whether one can replace the factor $\log^2 \frac{N}{M}$ in (2.4.14) by $\log \frac{N}{M}$, the best possible constant in (2.4.12).

§3   Wavelet spaces

3.1   Trigonometric MRA

For periodic polynomial multiresolution analyses (MRA), it is natural to consider a dyadic number of nodal points (see [10]). To allow more generality for practical applications, we additionally allow any (odd) factor $c \in \mathbb{N}$ and define

$$N_j \; := \; c\, 2^j, \qquad M_j \; := \; \begin{cases} 2^{j-\lambda}, & \text{if } j \geq \lambda, \\[2mm] 1, & \text{if } j < \lambda, \end{cases} \qquad (3.1.1)$$

for all $j \in \mathbb{N}_0$ and any constant $\lambda \in \mathbb{N}_0$. For fixed c and λ, we define a trigonometric MRA $\{V_j\}_{j=0}^{\infty}$ of $L_{2\pi}^2$ by

$$V_j \; := \; V_{N_j}^{M_j}.$$

Theorem 12. *If* $c\, 2^{\lambda} \geq 3$, *then all properties of a periodic MRA are satisfied, namely:*

(i) $V_j \subset V_{j+1}$,   for all $j \in \mathbb{N}_0$, $\dim V_j = 2N_j$,

(ii) $\operatorname{clos}_{L_{2\pi}^2} \left(\displaystyle\bigcup_{j=0}^{\infty} V_j \right) = L_{2\pi}^2$,

(iii) $f \in V_j \;\Rightarrow\; f\!\left(\circ - \dfrac{k\pi}{N_j} \right) \in V_j$,   for all $k \in \mathbb{Z}$, $j \in \mathbb{N}_0$,

(iv) $\exists\, \phi_j \in V_j : \quad V_j = \operatorname{span}\{\phi_{j,k} := \phi_j\!\left(\circ - \dfrac{k\pi}{N_j} \right) : \; k = 0, \ldots, 2N_j - 1\}$.

Taking $\phi_j := \varphi_{N_j}^{M_j}$ as the generating function, for any $j \in \mathbb{N}_0$ and $\{\alpha_k\} \in \mathbb{C}^{2N_j}$, we have the stability condition

$$A\, \|\{\alpha_k\}\|_{\ell^2} \; \leq \; \left\| \sum_{k=0}^{2N_j-1} \alpha_k\, \phi_{j,k} \right\|_2 \; \leq \; B\, \|\{\alpha_k\}\|_{\ell^2}, \qquad (3.1.2)$$

with best possible constants $A = 1/\sqrt{2}$ and $B = 1$.

Note that, if we replace $\phi_{j,k}$ by the orthonormal translates $\mathcal{O}\phi_{j,k} := \mathcal{O}\varphi_{N_j,k}^{M_j}$, then we obtain best stability; *i.e.* $A = B = 1$.

Proof: The inclusion relation $V_j \subset V_{j+1}$, for all $j \in \mathbb{N}_0$, follows from (2.1.7), if

$$N_j + M_j - 1 \leq N_{j+1} - M_{j+1}.$$

For $j < \lambda$, this is obviously true, and for $j \geq \lambda$, it means that

$$M_j + M_{j+1} \leq 3 \cdot 2^{j-\lambda} \leq c\, 2^j + 1.$$

This yields the condition

$$3 \leq c\, 2^\lambda; \quad i.e. \quad N_j \geq 3M_j, \quad \text{for all } j \geq \lambda.$$

Then we have $N_j - M_j \geq 2^{j+1-\lambda}$, which, by (2.1.7), implies property (ii). The third and fourth properties follow directly from the construction in Subsection 2.1. It remains to establish the stability constants in (3.1.2).

Let us denote by $\underline{v}_{j,r}$, for $r = 0, \ldots, 2N_j - 1$, the eigenvectors of the Gram matrix $\boldsymbol{G}_j = (\langle \phi_{j,k}, \phi_{j,\ell} \rangle)_{k,\ell=0}^{2N_j-1}$ such that $\boldsymbol{G}_j \underline{v}_{j,r} = d_{j,r} \underline{v}_{j,r}$, where $d_{j,r} = d_{N_j,r}^{M_j}$ are the eigenvalues known from Lemma 2. The eigenvectors form an orthogonal basis of $\mathbb{C}^{2N_j}$. Therefore, we can write any $2N_j$-dimensional vector $\underline{\alpha}$ as a linear combination $\underline{\alpha} = \sum_{r=0}^{2N_j-1} a_r \underline{v}_{j,r}$, with complex coefficients a_r. Let $\langle \underline{\alpha}, \underline{\beta} \rangle_E = \sum_{k=0}^{2N_j-1} \alpha_k \overline{\beta}_k$ denote the Euclidian inner product of $\underline{\alpha}, \underline{\beta} \in \mathbb{C}^{2N_j}$. Then we have

$$\left\| \sum_{k=0}^{2N_j-1} \alpha_k\, \phi_{j,k} \right\|_2^2 = \left\langle \sum_{k=0}^{2N_j-1} \alpha_k\, \phi_{j,k}, \sum_{l=0}^{2N_j-1} \alpha_l\, \phi_{j,l} \right\rangle$$

$$= \langle \boldsymbol{G}_j \underline{\alpha}, \underline{\alpha} \rangle_E = \sum_{r=0}^{2N_j-1} d_{j,r}\, |a_r|^2\, \langle \underline{v}_{j,r}, \underline{v}_{j,r} \rangle_E,$$

and since

$$\sum_{k=0}^{2N_j-1} |\alpha_k|^2 = \langle \underline{\alpha}, \underline{\alpha} \rangle_E = \sum_{r=0}^{2N_j-1} |a_r|^2\, \langle \underline{v}_{j,r}, \underline{v}_{j,r} \rangle_E,$$

the quotient $\left\| \sum_{k=0}^{2N_j-1} \alpha_k\, \phi_{j,k} \right\|_2^2 \big/ \left(\sum_{k=0}^{2N_j-1} |\alpha_k|^2 \right)$ is bounded by the minimal and maximal eigenvalue of $\boldsymbol{G}_j$, respectively. ∎

Let $L_j := L_{N_j}^{M_j}$ denote the interpolation operator mapping $C_{2\pi}$ onto V_j. We can prove the following two-scale relations for the shifted scaling functions.

Theorem 13. *For* $r = 0, \ldots, 2N_j - 1$, *we have the refinement equations*

$$\phi_{j,r}(x) = \tfrac{1}{\sqrt{2}} \phi_{j+1,2r}(x) + \frac{1}{2\sqrt{N_j}} \sum_{s=0}^{2N_j - 1} \phi_{j,r}\left(\frac{(2s+1)\pi}{2N_j}\right) \phi_{j+1,2s+1}(x) \quad (3.1.3)$$

and, for $j \geq \lambda$, *the dilation relation*

$$\phi_{j+1}(x) = \sqrt{2}\, \phi_j(2x)\, \frac{1 + \cos x}{2}. \quad (3.1.4)$$

Note that in (3.1.4) the function $(1 + \cos x)/2$ maintains the values of $\phi_j(2\circ)$ around $2k\pi$ and suppresses the interposed oscillation of $\phi_j(2\circ)$ around $(2k+1)\pi$, where $k \in \mathbf{Z}$.

Proof: By the interpolation property of $\phi_{j,r}$, we have

$$
\begin{aligned}
\phi_{j,r}(x) &= L_{j+1}\, \phi_{j,r}(x) \\
&= \frac{1}{\sqrt{2N_{j+1}}} \left(\sum_{k=0}^{4N_j - 1} \phi_{j,r}\left(\frac{k\pi}{2N_j}\right) \phi_{j+1,k}(x) \right) \\
&= \frac{1}{\sqrt{2}} \left(\phi_{j+1,2r}(x) + \frac{1}{\sqrt{2N_j}} \sum_{s=0}^{2N_j - 1} \phi_{j,r}\left(\frac{(2s+1)\pi}{2N_j}\right) \phi_{j+1,2s+1}(x) \right).
\end{aligned}
$$

For all levels $j \geq \lambda$, we have $N_{j+1} = 2N_j$ and $M_{j+1} = 2M_j$, which yields

$$
\begin{aligned}
\phi_{j+1}(x) &= \frac{\sin N_j(2x) \sin M_j(2x)}{M_j \sqrt{4N_j} \sin^2 x} \cos^2 \frac{x}{2} \\
&= \sqrt{2}\, \phi_j(2x)\, \frac{1 + \cos x}{2}. \quad \blacksquare
\end{aligned}
$$

Furthermore, we need the following two-scale inner product.

Lemma 5. *For* $\ell = 0, \ldots, 2N_j - 1$, *and* $m = 0, \ldots, 2N_{j+1} - 1$, *we have*

$$\langle \phi_{j,\ell}, \phi_{j+1,m} \rangle = \frac{1}{\sqrt{2N_{j+1}}} \phi_{j,\ell}\left(\frac{m\pi}{2N_j}\right). \quad (3.1.5)$$

Proof: By (2.1), (2.1.2) and $\langle \cos(m\circ), \cos(n\circ) \rangle = \tfrac{1}{2}\delta_{m,n}$, we can write

$$
\begin{aligned}
\langle \phi_{j,\ell}, \phi_{j+1,m} \rangle &= \frac{1}{\sqrt{2N_{j+1}}} \left\langle \phi_{j,\ell}, D_{N_j + M_j - 1}\left(\circ - \frac{m\pi}{N_{j+1}}\right) \right\rangle \\
&= \frac{1}{\sqrt{2N_{j+1}}} \phi_{j,\ell}\left(\frac{m\pi}{2N_j}\right). \quad \blacksquare
\end{aligned}
$$

3.2   Interpolating wavelets

The wavelet spaces are defined as the relevant orthogonal complements of V_j in V_{j+1}; *i.e.* $V_{j+1} = V_j \oplus W_j$, for all $j \in \mathbb{N}_0$, where $\oplus$ denotes the orthogonal sum. From Theorem 1, we derive a simple orthogonal basis of the wavelet spaces.

Theorem 14. *An orthogonal basis of W_j is given by the set*

$$\{\sigma_{j,\ell} : \ell = 0, \ldots, 2N_j - 1\},$$

with the functions

$$\sigma_{j,0}(x) := \tfrac{\sqrt{2}}{2} \cos 2N_j x, \qquad \sigma_{j,N_j}(x) := \tfrac{\sqrt{2}}{2} \sin N_j x,$$

$$\sigma_{j,2N_j-k}(x) := \sqrt{2} \cos kx, \qquad \sigma_{j,k}(x) := \sqrt{2} \sin kx, \tag{3.2.1}$$

where $k = N_j + M_j, \ldots, N_{j+1} - M_{j+1}$;

$$\sigma_{j,k}(x) := \sqrt{2}\Big(\tfrac{M_{j+1}+k}{2M_{j+1}} \cos(2N_j - k)x + \tfrac{M_{j+1}-k}{2M_{j+1}} \cos(2N_j + k)x\Big),$$

$$\sigma_{j,2N_j-k}(x) := \sqrt{2}\Big(\tfrac{M_{j+1}+k}{2M_{j+1}} \sin(2N_j - k)x - \tfrac{M_{j+1}-k}{2M_{j+1}} \sin(2N_j + k)x\Big), \tag{3.2.2}$$

where $k = 1, \ldots, M_{j+1} - 1$; and

$$\sigma_{j,N_j-k}(x) := \sqrt{2}\Big(\tfrac{M_j+k}{2M_j} \cos(N_j + k)x - \tfrac{M_j-k}{2M_j} \cos(N_j - k)x\Big),$$

$$\sigma_{j,N_j+k}(x) := \sqrt{2}\Big(\tfrac{M_j+k}{2M_j} \sin(N_j + k)x + \tfrac{M_j-k}{2M_j} \sin(N_j - k)x\Big), \tag{3.2.3}$$

where $k = 1, \ldots, M_j - 1$.

Proof: First, we check that all functions $\sigma_{j,\ell}$, for $\ell = 0, \ldots, 2N_j - 1$, belong to V_{j+1}. By Theorem 1, this is obvious for the polynomials in (3.2.1) and (3.2.3). Let $\rho_{j,\ell} := \varrho_{N_j,\ell}^{M_j}$. In (3.2.2), we have, for $k = 1, \ldots, M_{j+1} - 1$,

$$\sigma_{j,k} = \rho_{j+1,2N_j-k}, \qquad \sigma_{j,2N_j-k} = \rho_{j+1,2N_j+k}.$$

Second, we verify that $\langle \sigma_{j,\ell}, \rho_{j,k} \rangle = 0$, for all $k, \ell = 0, \ldots, 2N_j - 1$. This is evident for all $\sigma_{j,\ell}$ in (3.2.1) and (3.2.2). For (3.2.3), it follows by simple calculations. ■

Regarding shift-invariant spaces, we are now looking for a localized wavelet function $\psi_j \in V_{j+1}$ that generates the wavelet space W_j; *i.e.*

$$W_j = \text{span}\Big\{\psi_j\Big(\circ - \tfrac{n\pi}{N_j}\Big) : \quad n = 0, \ldots, 2N_j - 1\Big\}.$$

To determine the wavelet ψ_j uniquely, we may additionally require inter-polation properties for the translates $\psi_{j,n}$, in this case at the $2N_j$ interpolation nodes of V_{j+1} which are not interpolation nodes of the translates of ϕ_j. Hence, we demand that $\psi_j \in V_{j+1}$,

$$\langle \phi_{j,k}, \psi_j \rangle = 0, \qquad \text{for all } k = 0, \ldots, 2N_j - 1, \qquad (3.2.4)$$

and, in analogy to the scaling function, that

$$\psi_j \left(\frac{(2m+1)\pi}{2N_j} \right) = \sqrt{2N_j}\, \delta_{m,0}, \qquad \text{for all } m = 0, \ldots, 2N_j - 1. \quad (3.2.5)$$

Theorem 15. *There exists a uniquely determined function $\psi_j \in V_{j+1}$ satisfying (3.2.4) and (3.2.5), namely:*

$$\psi_j(x) := \sqrt{2}\, \phi_{j+1}\left(x - \frac{\pi}{2N_j}\right) - \phi_j\left(x - \frac{\pi}{2N_j}\right). \qquad (3.2.6)$$

For its translates $\psi_{j,n}(x) := \psi_j(x - \frac{n\pi}{N_j})$, $n = 0, \ldots, 2N_j - 1$, the refinement equations are given by

$$\psi_{j,n}(x) = \frac{1}{\sqrt{2}} \phi_{j+1,2n+1}(x) - \frac{1}{2\sqrt{N_j}} \sum_{s=0}^{2N_j-1} \phi_{j,s}\left(\frac{(2n+1)\pi}{2N_j} \right) \phi_{j+1,2s}(x). \; (3.2.7)$$

For $j \geq \lambda$, we have the special dilation equation

$$\psi_{j+1}(x) = \sqrt{2}\, \psi_j(2x)\, \frac{1 + \cos(x - \frac{\pi}{2N_{j+1}})}{2}. \qquad (3.2.8)$$

Proof: For an arbitrary $f \in V_{j+1}$, there is an expansion

$$f = L_{j+1} f = \frac{1}{\sqrt{2N_{j+1}}} \sum_{s=0}^{4N_j-1} \mu_s\, \phi_{j+1,s},$$

with $4N_j$ coefficients $\mu_s = f(\frac{s\pi}{N_{j+1}})$. By means of (3.1.5), the orthogonality condition in (3.2.4) is equivalent to

$$\begin{aligned}
0 &= \langle \phi_{j,k}, f \rangle \\
&= \frac{1}{\sqrt{2N_{j+1}}} \sum_{s=0}^{4N_j-1} \mu_s\, \langle \phi_{j,k}, \phi_{j+1,s} \rangle \\
&= \frac{1}{2N_{j+1}} \sum_{s=0}^{4N_j-1} \mu_s\, \phi_{j,k}\left(\frac{s\pi}{2N_j} \right),
\end{aligned}$$

for all $k = 0, \ldots, 2N_j - 1$. The interpolation property in (3.2.5) holds if and only if

$$f\left(\frac{(2m+1)\pi}{2N_j}\right) = \mu_{2m+1} = \sqrt{2N_j}\,\delta_{m,0}\,,$$

for all $m = 0, \ldots, 2N_j - 1$. Hence,

$$f = \frac{1}{\sqrt{2}}\left(\phi_{j+1,1} - \frac{1}{\sqrt{2N_j}}\sum_{k=0}^{2N_j-1}\phi_{j,k}\left(\frac{\pi}{2N_j}\right)\phi_{j+1,2k}\right)$$

is the only function in V_{j+1} satisfying (3.2.4) and (3.2.5). By (3.1.3), it follows $f = \psi_j$, where ψ_j is defined in (3.2.6). For its translates, (3.2.7) is obtained using that ϕ_j is an even function. Finally, by means of (3.1.4), we can write for $j \geq \lambda$,

$$\begin{aligned}
\psi_{j+1}(x) &= \sqrt{2}\,\phi_{j+2}\left(x - \frac{\pi}{2N_{j+1}}\right) - \phi_{j+1}\left(x - \frac{\pi}{2N_{j+1}}\right) \\
&= \left(2\,\phi_{j+1}\left(2x - \frac{\pi}{2N_j}\right) - \sqrt{2}\,\phi_j\left(2x - \frac{\pi}{2N_j}\right)\right) \\
&\quad \times \left(\frac{1 + \cos(x - \frac{\pi}{2N_{j+1}})}{2}\right),
\end{aligned}$$

which proves (3.2.8). ∎

Figure 2 shows the interpolating wavelet functions.

Now we can define the interpolation operator $R_j : C_{2\pi} \to W_j$,

$$R_j f := \frac{1}{\sqrt{2N_j}}\sum_{k=0}^{2N_j-1}f\left(\frac{(2k+1)\pi}{2N_j}\right)\psi_{j,k}\,, \tag{3.2.9}$$

which is the identity in W_j and fulfills, for $k \in \mathbf{Z}$,

$$R_j f\left(\frac{(2k+1)\pi}{2N_j}\right) = f\left(\frac{(2k+1)\pi}{2N_j}\right)\,.$$

Further, we are interested in the expansion of the localized basis in the orthogonal basis from Theorem 14. Let us introduce the notations for the vectors of the functions

$$\underline{\psi}_j := (\psi_{j,r})_{r=0}^{2N_j-1} \quad \text{and} \quad \underline{\sigma}_j := (\sigma_{j,k})_{k=0}^{2N_j-1}\,.$$

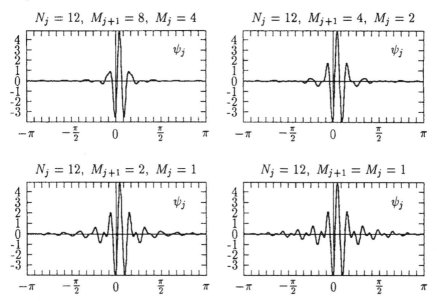

Figure 2. Interpolating wavelet functions.

Theorem 16. *The translates of the wavelet ψ_j defined in (3.2.6) satisfy the relations*

$$\underline{\psi}_j \; = \; O_j \, \underline{\sigma}_j \quad \text{and} \quad \underline{\sigma}_j \; = \; O_j^{-1} \, \underline{\psi}_j,$$

where the matrices $O_j = (o_{j,r,s})_{r,s=0}^{2N_j-1}$ and $O_j^{-1} = (\breve{o}_{j,s,r})_{s,r=0}^{2N_j-1}$ have, for $r = 0, \ldots, 2N_j - 1$, the entries

$$o_{j,r,0} \; = \; \frac{-1}{\sqrt{N_j}}, \qquad \breve{o}_{j,0,r} \; = \; \frac{-1}{2\sqrt{N_j}},$$

$$o_{j,r,N_j} \; = \; \frac{(-1)^r}{\sqrt{N_j}}, \qquad \breve{o}_{j,N_j,r} \; = \; \frac{(-1)^r}{2\sqrt{N_j}},$$

$$o_{j,r,s} \; = \; \breve{o}_{j,s,r} \; = \; \begin{cases} \dfrac{-1}{\sqrt{N_j}} \cos \dfrac{(2r+1)s\pi}{2N_j}, & \text{if} \quad 0 < s < N_j, \\[3mm] \dfrac{1}{\sqrt{N_j}} \sin \dfrac{(2r+1)s\pi}{2N_j}, & \text{if} \quad N_j < s < 2N_j. \end{cases}$$

Proof: Similar to (2.1.1), we can write

$$\psi_{j,r}(x) \; = \; \frac{1}{\sqrt{2N_j}} \left(D_{N_{j+1}-M_{j+1}} \left(x - \frac{(2r+1)\pi}{2N_j} \right) - D_{N_j-M_j} \left(x - \frac{(2r+1)\pi}{2N_j} \right) \right)$$

$$+ \sum_{k=-M_{j+1}+1}^{M_{j+1}-1} \frac{M_{j+1}-k}{M_{j+1}} \cos(N_{j+1}+k)\left(x - \frac{(2r+1)\pi}{2N_j}\right)$$

$$- \sum_{k=-M_j+1}^{M_j-1} \frac{M_j-k}{M_j} \cos(N_j+k)\left(x - \frac{(2r+1)\pi}{2N_j}\right)\Bigg)$$

$$= \frac{1}{\sqrt{2N_j}} \Bigg(\sum_{k=-M_j+1}^{M_j-1} \frac{M_j+k}{M_j} \cos(N_j+k)\left(x - \frac{(2r+1)\pi}{2N_j}\right)$$

$$+ 2 \sum_{\ell=N_j+M_j}^{N_{j+1}-M_{j+1}} \cos\ell\left(x - \frac{(2r+1)\pi}{2N_j}\right)$$

$$+ \sum_{k=-M_{j+1}+1}^{M_{j+1}-1} \frac{M_{j+1}-k}{M_{j+1}} \cos(N_{j+1}+k)\left(x - \frac{(2r+1)\pi}{2N_j}\right)\Bigg).$$

Rewriting this in the basis defined in Theorem 14 gives

$$\psi_{j,r}(x) = \frac{\sqrt{2}}{\sqrt{2N_j}} \Bigg(\sum_{k=0}^{N_j-1} \cos\frac{(2N_j-k)(2r+1)\pi}{2N_j} \, \sigma_{j,k}(x)$$

$$+ \sum_{k=1}^{N_j} \sin\frac{(2N_j-k)(2r+1)\pi}{2N_j} \, \sigma_{j,2N_j-k}(x) \Bigg),$$

which yields the entries of $\boldsymbol{O}_j$. The inverse transformation matrix $\boldsymbol{O}_j^{-1}$ is easily obtained by the interpolation

$$\sigma_{j,s}(x) = R_j \, \sigma_{j,s}(x)$$

$$= \frac{1}{\sqrt{2N_j}} \sum_{r=0}^{2N_j-1} \sigma_{j,s}\left(\frac{(2r+1)\pi}{2N_j}\right) \psi_{j,r}(x). \quad \blacksquare$$

Figure 3 shows a frequency spectrum.

3.3   Gram matrix and dual wavelets

In the sequel, the index j of a matrix indicates that its dimension is $2N_j \times 2N_j$. In particular, let $\boldsymbol{I}_j$ be the $2N_j$-th identity matrix and $\boldsymbol{F}_j$ be the $2N_j$-th Fourier matrix defined in (2.2.1). We denote the Gram matrix of the translates of the wavelet function $\psi_j \in W_j$ by

$$\boldsymbol{H}_j := \left(\langle \psi_{j,r}, \psi_{j,s} \rangle \right)_{r,s=0}^{2N_j-1}.$$

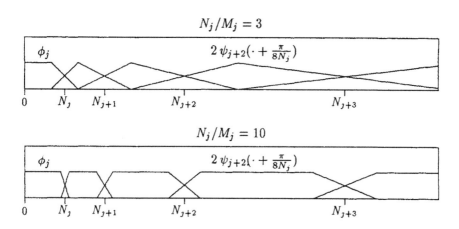

Figure 3. Frequency spectrum of ϕ_j and $2^{(k-j)/2}\psi_k(\circ + \frac{\pi}{2N_k})$, $k = 0, 1, 2, \ldots$.

The Gram matrix is symmetric and circulant, and so we can diagonalize it as well by means of the $2N_j$-th Fourier matrix F_j such that

$$H_j =: \overline{F}_j\, E_j\, F_j, \qquad (3.3.1)$$

with a diagonal matrix $E_j =: \mathrm{diag}\,(e_{j,k})_{k=0}^{2N_j-1}$.

Lemma 6. *The eigenvalues of the Gram matrix H_j are*

$$e_{j,k} = \begin{cases} \frac{1}{2} + \frac{k^2}{2M_{j+1}^2}, & \text{if } \ 0 \le k < M_{j+1}, \\[2mm] 1, & \text{if } \ M_{j+1} \le k \le N_j - M_{j+1}, \\[2mm] \frac{1}{2} + \frac{(N_j-k)^2}{2M_j^2}, & \text{if } \ N_j - M_j < k < N_j + M_j, \\[2mm] 1, & \text{if } \ N_j + M_j \le k \le 2N_j - M_{j+1}, \\[2mm] \frac{1}{2} + \frac{(2N_j-k)^2}{2M_{j+1}^2}, & \text{if } \ 2N_j - M_{j+1} < k \le 2N_j - 1. \end{cases}$$

Proof: By means of (3.2.6) and (3.3.1), it is possible to compute the eigenvalues as elements of the diagonal matrix $E_j = F_j\, H_j\, \overline{F}_j$, using that

$$\langle \psi_{j,r}, \psi_{j,s} \rangle$$
$$= \left\langle \sqrt{2}\phi_{j+1,2r} - \phi_{j,r}, \sqrt{2}\phi_{j+1,2s} - \phi_{j,s} \right\rangle$$

$$= 2 \langle \phi_{j+1,2r}, \phi_{j+1,2s} \rangle + \langle \phi_{j,r}, \phi_{j,s} \rangle - 2\sqrt{2} \left\langle \phi_{j+1}, \phi_j \left(\circ - \frac{(s-r)\pi}{N_j} \right) \right\rangle$$

$$= 2 \langle \phi_{j+1,2r}, \phi_{j+1,2s} \rangle + \langle \phi_{j,r}, \phi_{j,s} \rangle - 2\delta_{r,s}.$$

Hence,

$$\boldsymbol{H}_j = 2\boldsymbol{\mathcal{G}}_j + \boldsymbol{G}_j - 2\boldsymbol{I}_j,$$

where $\boldsymbol{G}_j = (\langle \phi_{j,r}, \phi_{j,s} \rangle)_{r,s=0}^{2N_j-1} = \boldsymbol{G}_{N_j}^{M_j}$ and $\boldsymbol{\mathcal{G}}_j := (\langle \phi_{j+1,2r}, \phi_{j+1,2s} \rangle)_{r,s=0}^{2N_j-1}$ are known to be circulant. By the distributive law for matrices, we have

$$\boldsymbol{H}_j = \overline{\boldsymbol{F}}_j \, \boldsymbol{E}_j \, \boldsymbol{F}_j = \overline{\boldsymbol{F}}_j \left(2\boldsymbol{\mathcal{D}}_j + \boldsymbol{D}_j - 2\boldsymbol{I}_j \right) \boldsymbol{F}_j,$$

with $\boldsymbol{D}_j = \mathrm{diag} \left(d_{N_j,r}^{M_j} \right)_{r=0}^{2N_j-1}$ and $\boldsymbol{\mathcal{D}}_j := \mathrm{diag} \left(\eta_{j,\ell} \right)_{\ell=0}^{2N_j-1}$. To compute the elements of $\boldsymbol{\mathcal{D}}_j$, we rewrite the elements of the circulant $\boldsymbol{\mathcal{G}}_j$,

$$\frac{1}{2N_j} \sum_{\ell=0}^{2N_j-1} \eta_{j,\ell} \, e^{\frac{i\ell(s-r)\pi}{N_j}} = \langle \phi_{j+1,2r}, \phi_{j+1,2s} \rangle$$

$$= \frac{1}{2N_{j+1}} \sum_{\ell=0}^{2N_{j+1}-1} d_{j+1,\ell} \, e^{\frac{i\ell(2s-2r)\pi}{N_{j+1}}}$$

$$= \frac{1}{4N_j} \sum_{\ell=0}^{2N_{j+1}-1} d_{j+1,\ell} \, e^{\frac{i\ell(s-r)\pi}{N_j}}$$

$$= \frac{1}{2N_j} \sum_{\ell=0}^{2N_j-1} \frac{d_{j+1,\ell} + d_{j+1,2N_j+\ell}}{2} \, e^{\frac{i\ell(s-r)\pi}{N_j}}.$$

Then, by Lemma 2 we obtain

$$\eta_{j,\ell} = \frac{d_{j+1,\ell} + d_{j+1,2N_j+\ell}}{2}$$

$$= \begin{cases} \dfrac{3M_{j+1}^2+\ell^2}{4M_{j+1}^2}, & \text{if} \quad 0 \le \ell < M_{j+1}, \\[2ex] 1, & \text{if} \quad M_{j+1} \le \ell \le 2N_j - M_{j+1}, \\[2ex] \dfrac{3M_{j+1}^2+(2N_j-\ell)^2}{4M_{j+1}^2}, & \text{if} \quad 2N_j - M_{j+1} < \ell \le 2N_j - 1. \end{cases}$$

The eigenvalues of $\boldsymbol{H}_j$ follow immediately, for $k = 0, \ldots, 2N_j - 1$, from

$$e_{j,k} = 2\eta_{j,k} + d_{j,k} - 2. \quad \blacksquare$$

Now, we define the dual functions $\tilde{\psi}_{j,r} \in W_j$ of the interpolating translates $\psi_{j,\ell}$, for $r, \ell = 0, \ldots, 2N_j - 1$, by the orthogonality conditions

$$\left\langle \tilde{\psi}_{j,r}, \psi_{j,\ell} \right\rangle = \delta_{r,\ell}. \tag{3.3.2}$$

As elements of W_j, they have a representation

$$\tilde{\psi}_{j,r}(x) = \sum_{s=0}^{2N_j-1} \beta_{j,r,s} \, \psi_{j,s}(x). \tag{3.3.3}$$

We denote the function vector

$$\vec{\underline{\psi}}_j := \left(\tilde{\psi}_{j,r} \right)_{r=0}^{2N_j-1}.$$

Theorem 17. *For the dual functions, we have the relations*

$$\vec{\underline{\psi}}_j = \boldsymbol{H}_j^{-1} \, \underline{\psi}_j \quad \text{and} \quad \underline{\psi}_j = \boldsymbol{H}_j \, \vec{\underline{\psi}}_j,$$

with

$$
\begin{aligned}
\boldsymbol{H}_j &= \overline{\boldsymbol{F}}_j \, \boldsymbol{E}_j \, \boldsymbol{F}_j \\
&= \left(\delta_{r,s} - \frac{1}{2N_j} \left(\sum_{k=-M_j+1}^{M_j-1} \frac{M_j^2-k^2}{2M_j^2} \cos \frac{(N_j+k)(s-r)\pi}{N_j} \right. \right. \\
&\qquad \left. \left. + \sum_{k=-M_{j+1}+1}^{M_{j+1}-1} \frac{M_{j+1}^2-k^2}{2M_{j+1}^2} \cos \frac{k(s-r)\pi}{N_j} \right) \right)_{r,s=0}^{2N_j-1}
\end{aligned}
$$

and

$$
\begin{aligned}
\boldsymbol{H}_j^{-1} &= \overline{\boldsymbol{F}}_j \, \boldsymbol{E}_j^{-1} \, \boldsymbol{F}_j \\
&= \left(\delta_{r,s} + \frac{1}{2N_j} \left(\sum_{k=-M_j+1}^{M_j-1} \frac{M_j^2-k^2}{M_j^2+k^2} \cos \frac{(N_j+k)(s-r)\pi}{N_j} \right. \right. \\
&\qquad \left. \left. + \sum_{k=-M_{j+1}+1}^{M_{j+1}-1} \frac{M_{j+1}^2-k^2}{M_{j+1}^2+k^2} \cos \frac{k(s-r)\pi}{N_j} \right) \right)_{r,s=0}^{2N_j-1} \tag{3.3.4}
\end{aligned}
$$

where

$$\boldsymbol{E}_j = \operatorname{diag}(e_{j,r})_{r=0}^{2N_j-1}, \quad \boldsymbol{E}_j^{-1} = \operatorname{diag}(1/e_{j,r})_{r=0}^{2N_j-1},$$

the entries of which are known from Lemma 6.

Proof: By (3.3.2) and (3.3.3), we have

$$\left\langle \tilde{\psi}_{j,r}, \psi_{j,\ell} \right\rangle = \sum_{s=0}^{2N_j-1} \beta_{j,r,s} \left\langle \psi_{j,s}, \psi_{j,\ell} \right\rangle = \delta_{r,\ell},$$

which yields

$$(\beta_{j,r,s})_{r,s=0}^{2N_j-1} = H_j^{-1} = \overline{F}_j E_j^{-1} F_j.$$

Due to the circulant structure of the inverse $H_j^{-1} = \overline{F}_j E_j^{-1} F_j$, we can compute its elements as

$$
\begin{aligned}
\beta_{j,r,s} &= \frac{1}{2N_j} \sum_{k=0}^{2N_j-1} \frac{1}{e_{j,k}} e^{\frac{ik(s-r)\pi}{N_j}} \\
&= \frac{1}{2N_j} \left(\sum_{k=0}^{2N_j-1} e^{\frac{ik(s-r)\pi}{N_j}} + \sum_{k=-M_j+1}^{M_j-1} \left(\frac{2M_j^2}{M_j^2+k^2} - 1 \right) e^{\frac{i(N_j+k)(s-r)\pi}{N_j}} \right. \\
&\qquad \left. + \sum_{k=-M_{j+1}+1}^{M_{j+1}-1} \left(\frac{2M_{j+1}^2}{M_{j+1}^2+k^2} - 1 \right) e^{\frac{ik(s-r)\pi}{N_j}} \right) \\
&= \delta_{s,r} + \frac{1}{2N_j} \left(\sum_{k=-M_j+1}^{M_j-1} \frac{M_j^2-k^2}{M_j^2+k^2} \cos \frac{(N_j+k)(s-r)\pi}{N_j} \right. \\
&\qquad \left. + \sum_{k=-M_{j+1}+1}^{M_{j+1}-1} \frac{M_{j+1}^2-k^2}{M_{j+1}^2+k^2} \cos \frac{k(s-r)\pi}{N_j} \right).
\end{aligned}
$$

Finally, we need the entries of $H_j = \overline{F}_j E_j F_j$, which are

$$
\begin{aligned}
\left\langle \psi_{j,r}, \psi_{j,s} \right\rangle &= \frac{1}{2N_j} \sum_{k=0}^{2N_j-1} e_{j,k} e^{\frac{ik(s-r)\pi}{N_j}} \\
&= \frac{1}{2N_j} \left(\sum_{k=0}^{2N_j-1} e^{\frac{ik(s-r)\pi}{N_j}} + \sum_{k=-M_{j+1}+1}^{M_{j+1}-1} \frac{k^2-M_{j+1}^2}{2M_{j+1}^2} e^{\frac{ik(s-r)\pi}{N_j}} \right. \\
&\qquad \left. + \sum_{k=-M_j+1}^{M_j-1} \frac{k^2-M_j^2}{2M_j^2} e^{\frac{i(N_j+k)(s-r)\pi}{N_j}} \right) \\
&= \delta_{s,r} + \frac{1}{2N_j} \left(\sum_{k=-M_{j+1}+1}^{M_{j+1}-1} \frac{k^2-M_{j+1}^2}{2M_{j+1}^2} \cos \frac{k(s-r)\pi}{N_j} \right)
\end{aligned}
$$

$$+ \sum_{k=-M_j+1}^{M_j-1} \frac{k^2-M_j^2}{2M_j^2} \cos \frac{(N_j+k)(s-r)\pi}{N_j} \Bigg) . \quad \blacksquare$$

A. A. Privalov computed the entries of H_j as follows.

Proposition 3. (see [18]) *If* $M_{j+1} = 2M_j$; *i.e.* $j \geq \lambda$, *and if* $3M_j \leq N_j$,
then

$$\langle \psi_{j,r}, \psi_{j,r} \rangle = 1 - \frac{M_j}{N_j} + \frac{1}{8N_j M_j}$$

and, for $r \neq s$,

$$\langle \psi_{j,r}, \psi_{j,s} \rangle = \frac{1}{32 N_j M_j^2 \sin^3 \frac{(r-s)\pi}{2N_j}}$$

$$\times \left(4M_j \sin \frac{(r-s)\pi}{2N_j} \left(\cos \frac{2M_j(r-s)\pi}{N_j} + 2(-1)^{r-s} \cos \frac{M_j(r-s)\pi}{N_j} \right) \right.$$

$$\left. - \cos \frac{(r-s)\pi}{2N_j} \left(\sin \frac{2M_j(r-s)\pi}{N_j} + 4(-1)^{r-s} \sin \frac{M_j(r-s)\pi}{N_j} \right) \right) .$$

For completeness, we deduce the representation of the dual translates
in terms of the frequencies.

Theorem 18. *We can transform*

$$\vec{\tilde{\psi}}_j = O_j E_j^{-1} \underline{\sigma}_j \quad \text{and} \quad \underline{\sigma}_j = E_j O_j^{-1} \vec{\tilde{\psi}}_j,$$

with O_j, O_j^{-1} *from Theorem 16 and* E_j, E_j^{-1} *as given in Theorem 17.*

Proof: First, we compute the dual wavelets from (3.3.3),

$$\tilde{\psi}_{j,r}(x)$$

$$= \psi_{j,r}(x) + \frac{1}{2N_j} \left(\sum_{k=-M_j+1}^{M_j-1} \left(\frac{2M_j^2}{M_j^2+k^2} - 1 \right) \sum_{s=0}^{2N_j-1} \cos \frac{(N_j+k)(s-r)\pi}{N_j} \psi_{j,s}(x) \right.$$

$$\left. + \sum_{k=-M_{j+1}+1}^{M_{j+1}-1} \left(\frac{2M_{j+1}^2}{M_{j+1}^2+k^2} - 1 \right) \sum_{s=0}^{2N_j-1} \cos \frac{k(s-r)\pi}{N_j} \psi_{j,s}(x) \right)$$

$$
= \frac{1}{\sqrt{2N_j}} \left(\sum_{k=-M_j+1}^{M_j-1} \frac{M_j+k}{M_j} \cos(N_j+k)\left(x - \frac{(2r+1)\pi}{2N_j}\right) \right.
$$

$$
+ 2 \sum_{\ell=N_j+M_j}^{N_{j+1}-M_{j+1}} \cos \ell \left(x - \frac{(2r+1)\pi}{2N_j}\right)
$$

$$
\left. + \sum_{k=-M_{j+1}+1}^{M_{j+1}-1} \frac{M_{j+1}-k}{M_{j+1}} \cos(N_{j+1}+k)\left(x - \frac{(2r+1)\pi}{2N_j}\right) \right)
$$

$$
+ \frac{1}{\sqrt{2N_j}} \left(\sum_{k=-M_j+1}^{M_j-1} \left(\frac{2M_j^2}{M_j^2+k^2}-1\right)\left(\frac{M_j+k}{2M_j}\cos(N_j+k)\left(x - \frac{(2r+1)\pi}{2N_j}\right)\right.\right.
$$

$$
\left.+\frac{M_j-k}{2M_j}\cos(N_j-k)\left(x - \frac{(2r+1)\pi}{2N_j}\right)\right)
$$

$$
+ \sum_{k=-M_{j+1}+1}^{M_{j+1}-1} \left(\frac{2M_{j+1}^2}{M_{j+1}^2+k^2}-1\right)\left(\frac{M_{j+1}-k}{2M_{j+1}}\cos(N_{j+1}+k)\right.
$$

$$
\left.\left.\times\left(x - \frac{(2r+1)\pi}{2N_j}\right)+\frac{M_{j+1}+k}{2M_{j+1}}\cos(N_{j+1}-k)\left(x - \frac{(2r+1)\pi}{2N_j}\right)\right)\right),
$$

and therefore,

$$
\tilde{\psi}_{j,r}(x) = \frac{1}{\sqrt{2N_j}} \left(2M_j \sum_{k=-M_j+1}^{M_j-1} \frac{M_j+k}{M_j^2+k^2} \cos(N_j+k)\left(x - \frac{(2r+1)\pi}{2N_j}\right) \right.
$$

$$
+ 2 \sum_{\ell=N_j+M_j}^{N_{j+1}-M_{j+1}} \cos \ell \left(x - \frac{(2r+1)\pi}{2N_j}\right)
$$

$$
\left. + 2M_{j+1} \sum_{k=-M_{j+1}+1}^{M_{j+1}-1} \frac{M_{j+1}-k}{M_{j+1}^2+k^2} \cos(N_{j+1}+k)\left(x - \frac{(2r+1)\pi}{2N_j}\right) \right).
$$

From there the transformation matrices follow easily. ∎

Note that $\tilde{\psi}_{j,r}(x) = \tilde{\psi}_{j,0}(x - \frac{r\pi}{N_j})$, for $r = 0,\ldots,2N_j-1$, are translates of one function, again. Moreover, since $V_j \perp W_j$, we have

$$
\left\langle \tilde{\phi}_{j,r}, \psi_{j,s} \right\rangle = \left\langle \tilde{\psi}_{j,r}, \phi_{j,s} \right\rangle = 0,
$$

for all $r, s = 0,\ldots,2N_j-1$ and $j \in \mathbb{N}_0$.

3.4 Orthonormal translates

In order to determine wavelets with orthonormal translates, we proceed in the same way as for the scaling functions; *i.e.* via the Gram matrix. We need orthonormal translates $\mathcal{O}\psi_{j,r}(x) := \mathcal{O}\psi_j \left(x - \frac{r\pi}{N_j}\right)$ of a function $\mathcal{O}\psi_j \in W_j$; *i.e.* coefficients $\nu_{j,r,s}$, such that

$$\mathcal{O}\psi_{j,r}(x) = \sum_{s=0}^{2N_j-1} \nu_{j,r,s}\, \psi_{j,s}(x)$$

and

$$\langle \mathcal{O}\psi_{j,r}, \mathcal{O}\psi_{j,k} \rangle = \delta_{r,k}, \tag{3.4.1}$$

for all $r, k = 0, \dots, 2N_j - 1$. As in Section 2.2, we use the matrix notation and introduce the coefficient matrix $\mathcal{V}_j := (\nu_{j,r,s})_{r,s=0}^{2N_j-1}$ and the function vector

$$\underline{\mathcal{O}\psi}_j := (\mathcal{O}\psi_{j,r})_{r=0}^{2N_j-1}.$$

Theorem 19. *There exists an orthonormal wavelet basis of W_j with the relations*

$$\underline{\mathcal{O}\psi}_j = \mathcal{V}_j\, \underline{\psi}_j, \qquad\qquad \underline{\psi}_j = \mathcal{V}_j^{-1}\, \underline{\mathcal{O}\psi}_j, \tag{3.4.2}$$

$$\underline{\mathcal{O}\vec{\psi}}_j = \mathcal{V}_j^{-1}\, \underline{\vec{\psi}}_j \qquad \text{and} \qquad \underline{\vec{\psi}}_j = \mathcal{V}_j\, \underline{\mathcal{O}\vec{\psi}}_j, \tag{3.4.3}$$

where

$$\mathcal{V}_j = \overline{F}_j\, \Lambda_j\, F_j$$

$$= \left(\delta_{r,s} + \frac{1}{2N_j} \sum_{k=-M_{j+1}+1}^{M_{j+1}-1} \left(\frac{2M_{j+1}}{\sqrt{2M_{j+1}^2 + 2k^2}} - 1 \right) \cos \frac{k(s-r)\pi}{N_j} \right.$$

$$\left. + \frac{1}{2N_j} \sum_{k=-M_j+1}^{M_j-1} \left(\frac{2M_j}{\sqrt{2M_j^2 + 2k^2}} - 1 \right) \cos \frac{(N_j+k)(s-r)\pi}{N_j} \right)_{r,s=0}^{2N_j-1} \tag{3.4.4}$$

$$\mathcal{V}_j^{-1} = \overline{F}_j\, \Lambda_j^{-1}\, F_j$$

$$= \left(\delta_{r,s} + \frac{1}{2N_j} \sum_{k=-M_{j+1}+1}^{M_{j+1}-1} \left(\frac{\sqrt{2M_{j+1}^2 + 2k^2}}{2M_{j+1}} - 1 \right) \cos \frac{k(s-r)\pi}{N_j} \right.$$

$$\left. + \frac{(-1)^{s-r}}{2N_j} \sum_{k=-M_j+1}^{M_j-1} \left(\frac{\sqrt{2M_j^2 + 2k^2}}{2M_j} - 1 \right) \cos \frac{k(s-r)\pi}{N_j} \right)_{r,s=0}^{2N_j-1} \tag{3.4.5}$$

with

$$\Lambda_j := \operatorname{diag}\left(1/\sqrt{e_{j,\ell}}\right)_{\ell=0}^{2N_j-1} \quad \text{and} \quad \Lambda_j^{-1} = \operatorname{diag}\left(\sqrt{e_{j,\ell}}\right)_{\ell=0}^{2N_j-1},$$

the entries of which are given in Lemma 6.

Proof: Here we follow the same line as in the proof of Theorem 5. At first, we show (3.4.4) and (3.4.5). With $\mathcal{V}_j = \overline{F}_j \Lambda_j F_j = (\nu_{j,r,s})_{r,s=0}^{2N_j-1}$, we compute

$$
\begin{aligned}
\nu_{j,r,s} &= \frac{1}{2N_j} \sum_{k=0}^{2N_j-1} \frac{1}{\sqrt{e_{j,k}}} e^{\frac{ik(s-r)\pi}{N_j}} \\
&= \frac{1}{2N_j} \left(\sum_{k=0}^{2N_j-1} e^{\frac{ik(s-r)\pi}{N_j}} + \sum_{k=-M_{j+1}+1}^{M_{j+1}-1} \left(\frac{\sqrt{2M_{j+1}^2}}{\sqrt{M_{j+1}^2+k^2}} - 1 \right) e^{\frac{ik(s-r)\pi}{N_j}} \right. \\
&\qquad\qquad \left. + \sum_{k=-M_j+1}^{M_j-1} \left(\frac{\sqrt{2M_j^2}}{\sqrt{M_j^2+k^2}} - 1 \right) e^{\frac{i(N_j+k)(s-r)\pi}{N_j}} \right) \\
&= \delta_{s,r} + \frac{1}{2N_j} \left(\sum_{k=-M_{j+1}+1}^{M_{j+1}-1} \left(\frac{2M_{j+1}}{\sqrt{2M_{j+1}^2+2k^2}} - 1 \right) \cos \frac{k(s-r)\pi}{N_j} \right. \\
&\qquad\qquad \left. + \sum_{k=-M_j+1}^{M_j-1} \left(\frac{2M_j}{\sqrt{2M_j^2+2k^2}} - 1 \right) \cos \frac{(N_j+k)(s-r)\pi}{N_j} \right).
\end{aligned}
$$

Then, for the entries of

$$\mathcal{V}_j^{-1} = \overline{F}_j \Lambda_j^{-1} F_j =: (\breve{\nu}_{j,s,r})_{s,r=0}^{2N_j-1},$$

we have

$$
\begin{aligned}
\breve{\nu}_{j,s,r} &= \frac{1}{2N_j} \sum_{k=0}^{2N_j-1} \sqrt{e_{j,k}}\, e^{\frac{ik(s-r)\pi}{N_j}} \\
&= \frac{1}{2N_j} \left(\sum_{k=0}^{2N_j-1} e^{\frac{ik(s-r)\pi}{N_j}} + \sum_{k=-M_{j+1}+1}^{M_{j+1}-1} \left(\frac{\sqrt{M_{j+1}^2+k^2}}{\sqrt{2M_{j+1}^2}} - 1 \right) e^{\frac{ik(s-r)\pi}{N_j}} \right. \\
&\qquad\qquad \left. + \sum_{k=-M_j+1}^{M_j-1} \left(\frac{\sqrt{M_j^2+k^2}}{\sqrt{2M_j^2}} - 1 \right) e^{\frac{i(N_j+k)(s-r)\pi}{N_j}} \right) \\
&= \delta_{s,r} + \frac{1}{2N_j} \sum_{k=-M_{j+1}+1}^{M_{j+1}-1} \left(\frac{\sqrt{2M_{j+1}^2+2k^2}}{2M_{j+1}} - 1 \right) \cos \frac{k(s-r)\pi}{N_j}
\end{aligned}
$$

$$+ \frac{(-1)^{s-r}}{2N_j} \sum_{k=-M_j+1}^{M_j-1} \left(\frac{\sqrt{2M_j^2+2k^2}}{2M_j} - 1 \right) \cos \frac{k(s-r)\pi}{N_j} .$$

So, $\mathcal{V}_j$ and $\mathcal{V}_j^{-1}$ are symmetric. Hence, the orthonormality of the translates $\mathcal{O}\psi_{j,r}$, required in (3.4.1), follows easily from

$$\mathcal{V}_j H_j \mathcal{V}_j^T = \overline{F}_j \Lambda_j E_j \Lambda_j F_j = I_j .$$

This, together with (3.4.2) and Theorem 17, proves (3.4.3), too. ∎

Last but not least, we give the representation of the orthonormal translates in the basis of frequencies.

Theorem 20. *We have*

$$\underline{\mathcal{O}\psi}_j = O_j \Lambda_j \underline{\sigma}_j \quad and \quad \underline{\sigma}_j = \Lambda_j^{-1} O_j^{-1} \underline{\mathcal{O}\psi}_j , \qquad (3.4.6)$$

with O_j, O_j^{-1} *from Theorem 16 and* Λ_j, Λ_j^{-1} *as given in Theorem 19.*

Proof: First, we need an explicit formula for

$$\mathcal{O}\psi_j(x) = \sum_{s=0}^{2N_j-1} \nu_{j,0,s} \, \psi_{j,s}(x)$$

$$= \psi_{j,0}(x) + \frac{1}{2N_j} \left(\sum_{k=-M_{j+1}+1}^{M_{j+1}-1} \left(\frac{2M_{j+1}}{\sqrt{2M_{j+1}^2+2k^2}} - 1 \right) \sum_{s=0}^{2N_j-1} \cos \frac{ks\pi}{N_j} \, \psi_{j,s}(x) \right.$$

$$\left. + \sum_{k=-M_j+1}^{M_j-1} \left(\frac{2M_j}{\sqrt{2M_j^2+2k^2}} - 1 \right) \sum_{s=0}^{2N_j-1} \cos \frac{(N_j+k)s\pi}{N_j} \, \psi_{j,s}(x) \right).$$

Therefore, we consider for $-M_{j+1} < k < M_{j+1}$ and $N_j - M_j < k < N_j + M_j$, the sum

$$\sum_{s=0}^{2N_j-1} \cos \frac{ks\pi}{N_j} \, \psi_{j,s}(x) = \sqrt{2} \sum_{s=0}^{2N_j-1} \cos \frac{ks\pi}{N_j} \, \phi_{j+1,2s} \left(x - \frac{\pi}{2N_j} \right)$$

$$- \sum_{s=0}^{2N_j-1} \cos \frac{ks\pi}{N_j} \, \phi_{j,s} \left(x - \frac{\pi}{2N_j} \right) .$$

By means of (2.1.5) and (2.1.6), we calculate the first part,

$$\sqrt{2} \sum_{s=0}^{2N_j-1} \cos \frac{ks\pi}{N_j} \, \phi_{j+1,2s}(x)$$

$$= \frac{\sqrt{2}}{\sqrt{2N_{j+1}}} \sum_{s=0}^{2N_j-1} \cos \frac{ks\pi}{N_j} \left(1 + 2 \sum_{\ell=1}^{N_{j+1}-M_{j+1}} \cos \ell \left(x - \frac{2s\pi}{N_{j+1}} \right) \right.$$

$$+ 2 \sum_{\ell=-M_{j+1}+1}^{M_{j+1}-1} \frac{M_{j+1}-\ell}{2M_{j+1}} \cos(N_{j+1} + \ell) \left(x - \frac{2s\pi}{N_{j+1}} \right) \bigg)$$

$$= \sqrt{2N_j}\, \delta_{k,0} + \frac{1}{\sqrt{2N_j}} \left(\sum_{\ell=1}^{N_{j+1}-M_{j+1}} \sum_{s=0}^{2N_j-1} \left(\cos \left(\ell x + \frac{(k-\ell)s\pi}{N_j} \right) \right. \right.$$

$$+ \cos \left(\ell x - \frac{(k+\ell)s\pi}{N_j} \right) \bigg) + \sum_{\ell=-M_{j+1}+1}^{M_{j+1}-1} \frac{M_{j+1}-\ell}{2M_{j+1}}$$

$$\times \sum_{s=0}^{2N_j-1} \left(\cos \left((N_{j+1} + \ell)x + \frac{(k-\ell)s\pi}{N_j} \right) + \cos \left((N_{j+1} + \ell)x - \frac{(k+\ell)s\pi}{N_j} \right) \right) \bigg)$$

$$= \sqrt{2N_j} \left(\delta_{k,0} + \sum_{\ell=1}^{N_{j+1}-M_{j+1}} \left(\delta_{0,(k-\ell)\bmod 2N_j} + \delta_{0,(k+\ell)\bmod 2N_j} \right) \cos \ell x \right.$$

$$+ \sum_{\ell=-M_{j+1}+1}^{M_{j+1}-1} \frac{M_{j+1}-\ell}{2M_{j+1}} \left(\delta_{0,(k-\ell)\bmod 2N_j} + \delta_{0,(k+\ell)\bmod 2N_j} \right) \cos(2N_j + \ell)x \bigg)$$

$$= \begin{cases} \sqrt{2N_j} \left(\cos kx + \frac{M_{j+1}-k}{2M_{j+1}} \cos(2N_j + k)x \right), & \text{if } |k| < M_{j+1}, \\ \sqrt{2N_j} \left(\cos kx + \cos(2N_j - k)x \right), & \text{if } |k - N_j| < M_j. \end{cases}$$

The second part is known from Lemma 1. Hence,

$$\frac{1}{\sqrt{2N_j}} \sum_{s=0}^{2N_j-1} \cos \frac{ks\pi}{N_j}\, \psi_{j,s}(x)$$

$$= \begin{cases} \dfrac{M_{j+1}-k}{2M_{j+1}} \cos(2N_j + k)\left(x - \dfrac{\pi}{N_{j+1}} \right) + \dfrac{M_{j+1}+k}{2M_{j+1}} \cos(2N_j - k)\left(x - \dfrac{\pi}{N_{j+1}} \right), \\ \hfill \text{if } -M_{j+1} < k < M_{j+1}, \\[2mm] \dfrac{M_j-(N_j-k)}{2M_j} \cos k\left(x - \dfrac{\pi}{2N_j} \right) + \dfrac{M_j+(N_j-k)}{2M_j} \cos(2N_j - k)\left(x - \dfrac{\pi}{2N_j} \right), \\ \hfill \text{if } N_j - M_j < k < N_j + M_j. \end{cases}$$

Consequently,

$$O\psi_j(x) = \frac{1}{\sqrt{2N_j}} \left(\sum_{k=-M_j+1}^{M_j-1} \frac{M_j+k}{M_j} \cos(N_j + k)\left(x - \frac{\pi}{2N_j} \right) \right.$$

$$
+\, 2 \sum_{\ell=N_j+M_j}^{N_{j+1}-M_{j+1}} \cos \ell \left(x - \tfrac{\pi}{2N_j} \right)
$$

$$
+\, \sum_{k=-M_{j+1}+1}^{M_{j+1}-1} \tfrac{M_{j+1}-k}{M_{j+1}} \cos(N_{j+1}+k) \left(x - \tfrac{\pi}{2N_j} \right)
$$

$$
+\, \sum_{k=-M_{j+1}+1}^{M_{j+1}-1} \left(\tfrac{2M_{j+1}}{\sqrt{2M_{j+1}^2+2k^2}} - 1 \right) \left(\tfrac{M_{j+1}-k}{2M_{j+1}} \cos(N_{j+1}+k) \left(x - \tfrac{\pi}{2N_j} \right) \right.
$$

$$
\left. +\, \tfrac{M_{j+1}+k}{2M_{j+1}} \cos(N_{j+1}-k) \left(x - \tfrac{\pi}{2N_j} \right) \right)
$$

$$
+\, \sum_{k=-M_j+1}^{M_j-1} \left(\tfrac{2M_{j+1}}{\sqrt{2M_j^2+2k^2}} - 1 \right) \left(\tfrac{M_{j+1}-(N_j-k)}{2M_{j+1}} \cos k \left(x - \tfrac{\pi}{2N_j} \right) \right.
$$

$$
\left. +\, \tfrac{M_j+(N_j-k)}{2M_j} \cos(2N_j-k) \left(x - \tfrac{\pi}{2N_j} \right) \right) \bigg),
$$

and finally,

$$
\mathcal{O}\psi_j(x) \;=\; \frac{1}{\sqrt{2N_j}} \left(\sum_{k=-M_j+1}^{M_j-1} \tfrac{2M_j+2k}{\sqrt{2M_j^2+2k^2}} \cos(N_j+k) \left(x - \tfrac{\pi}{2N_j} \right) \right.
$$

$$
+\, 2 \sum_{\ell=N_j+M_j}^{N_{j+1}-M_{j+1}} \cos \ell \left(x - \tfrac{\pi}{2N_j} \right)
$$

$$
\left. +\, \sum_{k=-M_{j+1}+1}^{M_{j+1}-1} \tfrac{2M_{j+1}-2k}{\sqrt{2M_{j+1}^2+2k^2}} \cos(N_{j+1}+k) \left(x - \tfrac{\pi}{2N_j} \right) \right).
$$

From this, the matrices in (3.4.6) can be easily derived. ∎

3.5   Localization and projection estimates

We present results for the wavelet spaces, which are analogous to the ones of Section 2.4. Here, we can be brief because most of the proofs of that section can be adapted easily.

$L_{2\pi}^p$-Norms and stability

As in (2.4.6) in Section 2.4, we assume N_j/M_j to be bounded from above by a constant. Then the $L_{2\pi}^p$-stability of the wavelets $\psi_j \in W_j$, for $1 \le p \le \infty$, is given by

$$\left\|\sum_{k=0}^{2N_j-1} \alpha_k\, \psi_{j,k}\right\|_p \quad \sim \quad (2N_j)^{\frac{1}{2}-\frac{1}{p}}\, \|\{\alpha_k\}\|_{\ell^p}\,.$$

This can be proved by applying (3.2.6) and (2.4.4).

For the special case $\alpha_k = \delta_{k,0}$, we obtain immediately the same asymptotics as we have in (2.4.4), namely:

$$\|\psi_{j,0}\|_p \ \sim\ (2N_j)^{\frac{1}{2}-\frac{1}{p}}\,, \qquad \text{for any } 1 \le p \le \infty\,.$$

In order to determine the asymptotics of the dual and orthonormal wavelets, we follow the same line as in Subsection 2.4. Thus, we have to estimate the coefficients $\beta_{j,r,s}$ and $\nu_{j,r,s}$.

Lemma 7. *For the entries of the circulant matrices $\boldsymbol{H}_j^{-1}$ and $\boldsymbol{\mathcal{V}}_j$, the inequalities*

$$|\beta_{j,r,s}|\,,\ |\nu_{j,r,s}| \ \le\ C \max\left\{(|r-s|+1)^{-2}\,,\ (2N_j-|r-s|)^{-2}\right\}$$

hold, for $r,s = 0,\ldots,2N_j-1$.

Proof: Again, we have to estimate discrete Fourier coefficients. But this time, with respect to (3.3.4) and (3.4.4), we consider the 2π-periodic functions

$$f_\beta(x) := \begin{cases} \dfrac{M_{j+1}^2\pi^2-N_j^2x^2}{M_{j+1}^2\pi^2+N_j^2x^2} + (-1)^{s-r}\dfrac{M_j^2\pi^2-N_j^2x^2}{M_j^2\pi^2+N_j^2x^2}, & \text{for } |x| \le \frac{\pi M_j}{N_j}, \\[3ex] \dfrac{M_{j+1}^2\pi^2-N_j^2x^2}{M_{j+1}^2\pi^2+N_j^2x^2}, & \text{for } \frac{\pi M_j}{N_j} < |x| \le \frac{\pi M_{j+1}}{N_j}, \\[3ex] 0, & \text{for } \frac{\pi M_{j+1}}{N_j} < |x| \le \pi, \end{cases}$$

and

$$f_\nu(x) := \begin{cases} \dfrac{\sqrt{2}\pi M_{j+1}}{\sqrt{\pi^2 M_{j+1}^2+N_j^2x^2}} - 1 + \dfrac{\sqrt{2}(-1)^{s-r}\pi M_j}{\sqrt{\pi^2 M_j^2+N_j^2x^2}} - (-1)^{s-r}, & \text{for } |x| \le \frac{\pi M_j}{N_j}, \\[3ex] \dfrac{\sqrt{2}\pi M_{j+1}}{\sqrt{\pi^2 M_{j+1}^2+N_j^2x^2}} - 1, & \text{for } \frac{\pi M_j}{N_j} < |x| \le \frac{\pi M_{j+1}}{N_j}, \\[3ex] 0, & \text{for } \frac{\pi M_{j+1}}{N_j} < |x| \le \pi, \end{cases}$$

which have a piecewise continuous first derivative. ∎

The application of Lemma 7 gives the time-localization result for the dual and orthonormal wavelets.

Theorem 21. *For* $1 \leq p \leq \infty$,

$$\|\tilde{\psi}_{j,r}\|_p \sim \|\mathcal{O}\psi_{j,r}\|_p \sim (2N_j)^{\frac{1}{2}-\frac{1}{p}} .$$

The pointwise estimates, which correspond to (2.4.9) and (2.4.11), can be deduced using (3.2.6).

Theorem 22. *For simplicity, we restrict ourselves to* $0 < k \leq N_j$. *Then,*

$$\max_{x \in I_{N_j,k}} \left\{ |\psi_{j,0}(x)|, |\mathcal{O}\psi_{j,0}(x)|, |\tilde{\psi}_{j,0}(x)| \right\} \leq \frac{C\sqrt{N_j}}{k^2} .$$

Error estimates

In Subsection 2.4 we investigated the approximation of continuous functions by elements from the sample space V_j. Now, we consider the corresponding results for approximation processes in W_j. Of particular interest are the interpolation operator R_j defined in (3.2.9) and the orthogonal projection Q_j that can be easily handled by

$$Q_j = P^{M_{j+1}}_{N_{j+1}} - P^{M_j}_{N_j} .$$

Here, we only assume (3.1.1) for N_j and M_j, with arbitrary $\lambda \in \mathbb{N}$.

Theorem 23. *For the interpolatory and orthogonal projections, we have*

$$\|R_j\|_{C_{2\pi} \to C_{2\pi}} \leq C \log \frac{N_j}{M_j} \quad \text{and} \quad \|Q_j\|_{C_{2\pi} \to C_{2\pi}} \leq C \log^2 \frac{N_j}{M_j} .$$

Note that this Theorem is a key to finding polynomial bases of $C_{2\pi}$. We discuss this in the paper [17].

§4   Decomposition and reconstruction

The wavelet analysis of functions is based on the transformations between a sufficiently large level sample space and the wavelet spaces of lower levels, *i.e.* the iterative decomposition of V_{j+1} into the orthogonal sum $V_j \oplus W_j$. Starting from an approximation of a given function in a specific sample space, either by interpolation or by orthogonal projection, we only have to know the corresponding basis transformations to calculate the wavelet coefficients of the function. Conversely, we can reconstruct the projection of the function from the wavelet coefficients.

4.1   Interpolating bases

For the interpolatory scaling and wavelet functions, we have already determined the refinement relations in Theorems 13 and 15. Now, we present the reverse basis transformation.

Theorem 24. *The decomposition formulas, for $\ell = 0, \ldots, 2N_j - 1$, are*

$$\phi_{j+1,2\ell}(x) = \frac{1}{\sqrt{2N_{j+1}}} \left(\sum_{s=0}^{2N_j-1} \tilde{\phi}_{j,s}\left(\frac{\ell\pi}{N_j}\right) \phi_{j,s}(x) \right.$$

$$(4.1.1)$$

$$\left. - \sum_{s=0}^{2N_j-1} \tilde{\phi}_{j,\ell}\left(\frac{(2s+1)\pi}{2N_j}\right) \psi_{j,s}(x) \right)$$

and

$$\phi_{j+1,2\ell+1}(x) = \frac{1}{\sqrt{2N_{j+1}}} \left(\sum_{s=0}^{2N_j-1} \tilde{\phi}_{j,s}\left(\frac{(2\ell+1)\pi}{2N_j}\right) \phi_{j,s}(x) \right.$$

$$(4.1.2)$$

$$\left. + \sum_{s=0}^{2N_j-1} \tilde{\phi}_{j,\ell}\left(\frac{s\pi}{N_j}\right) \psi_{j,s}(x) \right).$$

Proof: We only have to calculate the coefficients $a_{j,\ell,s}$ and $b_{j,\ell,s}$, such that

$$\phi_{j+1,\ell}(x) = \sum_{s=0}^{2N_j-1} a_{j,\ell,s}\, \phi_{j,s}(x) + \sum_{s=0}^{2N_j-1} b_{j,\ell,s}\, \psi_{j,s}(x)$$

for $\ell = 0, \ldots, 4N_j - 1$. These coefficients are the inner products

$$a_{j,\ell,s} = \left\langle \phi_{j+1,\ell}, \tilde{\phi}_{j,s} \right\rangle$$

$$= \frac{1}{\sqrt{2N_{j+1}}} \left\langle D_{N_{j+1}-M_{j+1}}\left(\circ - \frac{\ell\pi}{N_{j+1}}\right), \tilde{\phi}_{j,s} \right\rangle$$

$$= \frac{1}{\sqrt{2N_{j+1}}} \tilde{\phi}_{j,s}\left(\frac{\ell\pi}{2N_j}\right)$$

and

$$b_{j,\ell,s} = \left\langle \phi_{j+1,\ell}, \tilde{\psi}_{j,s} \right\rangle$$

$$= \frac{1}{\sqrt{2}\, N_{j+1}} \left\langle D_{N_{j+1}-M_{j+1}}\left(\circ - \frac{\ell\pi}{N_{j+1}}\right) \right.$$

$$\left. + \sum_{k=-M_{j+1}+1}^{M_{j+1}-1} \frac{M_{j+1}-k}{M_{j+1}} \cos(N_{j+1}+k)\left(\circ - \frac{\ell\pi}{N_{j+1}}\right), \right.$$

$$2M_j \sum_{k=-M_j+1}^{M_j-1} \frac{M_j+k}{M_j^2+k^2} \cos(N_j + k)\left(\circ - \frac{(2s+1)\pi}{2N_j}\right)$$

$$+ 2 \sum_{k=N_j+M_j}^{N_{j+1}-M_{j+1}} \cos k\left(\circ - \frac{(2s+1)\pi}{2N_j}\right)$$

$$\left. + 2M_{j+1} \sum_{k=-M_{j+1}+1}^{M_{j+1}-1} \frac{M_{j+1}-k}{M_{j+1}^2+k^2} \cos(N_{j+1} + k)\left(\circ - \frac{(2s+1)\pi}{2N_j}\right)\right)$$

$$= \frac{1}{\sqrt{2}\,N_{j+1}}\left(2M_j \sum_{k=-M_j+1}^{M_j-1} \frac{M_j+k}{M_j^2+k^2} \cos\frac{(N_j+k)(\ell-1-2s)\pi}{2N_j}\right.$$

$$+ 2 \sum_{k=N_j+M_j}^{2N_j-M_{j+1}} \cos\frac{k(\ell-1-2s)\pi}{2N_j}$$

$$\left. + \sum_{k=-M_{j+1}+1}^{M_{j+1}-1} \frac{(M_{j+1}-k)^2}{M_{j+1}^2+k^2} \cos\frac{(2N_j+k)(\ell-1-2s)\pi}{2N_j}\right)$$

$$= \frac{(-1)^{\ell-1}}{\sqrt{2}\,N_{j+1}}\left(1 + 2\sum_{k=1}^{M_{j+1}} \cos\frac{k(\ell-1-2s)\pi}{2N_j} + 2\sum_{k=M_{j+1}}^{N_j-M_j} \cos\frac{k(\ell-1-2s)\pi}{2N_j}\right.$$

$$\left. + 2M_j \sum_{k=-M_j+1}^{M_j-1} \frac{M_j-k}{M_j^2+k^2} \cos\frac{(N_j+k)(\ell-1-2s)\pi}{2N_j}\right)$$

$$= \frac{(-1)^{\ell-1}}{\sqrt{2N_{j+1}}}\, \tilde{\phi}_{j,s}\left(\frac{(\ell-1)\pi}{2N_j}\right) = \frac{(-1)^{\ell-1}}{\sqrt{2N_{j+1}}}\, \tilde{\phi}_{j,0}\left(\frac{(2s-\ell-1)\pi}{2N_j}\right)$$

$$= \frac{1}{\sqrt{2N_{j+1}}} \times \begin{cases} \tilde{\phi}_{j,\frac{\ell-1}{2}}\left(\frac{s\pi}{N_j}\right), & \text{for odd } \ell, \\[2mm] -\tilde{\phi}_{j,\frac{\ell}{2}}\left(\frac{(2s+1)\pi}{2N_j}\right), & \text{for even } \ell. \quad\blacksquare \end{cases}$$

Motivated by the different structures for translates $\phi_{j+1,\ell}$ with even and odd index ℓ, respectively, we introduce a permutation matrix

$$
P_{j+1} := \left(\begin{array}{cccccccc}
1&0&0&0 & \cdots & 0&0&0&0 \\
0&0&1&0 & \cdots & 0&0&0&0 \\
\vdots&\vdots&\vdots&\vdots & \ddots & \vdots&\vdots&\vdots&\vdots \\
0&0&0&0 & \cdots & 1&0&0&0 \\
0&0&0&0 & \cdots & 0&0&1&0 \\
\hline
0&1&0&0 & \cdots & 0&0&0&0 \\
0&0&0&1 & \cdots & 0&0&0&0 \\
\vdots&\vdots&\vdots&\vdots & \ddots & \vdots&\vdots&\vdots&\vdots \\
0&0&0&0 & \cdots & 0&1&0&0 \\
0&0&0&0 & \cdots & 0&0&0&1
\end{array}\right) = \left(\delta_{2r,s} + \delta_{2r,2N_{j+1}+s-1}\right)_{r,s=0}^{2N_{j+1}-1},
$$

that yields

$$
P_{j+1}\,\underline{\phi}_{j+1} \tag{4.1.3}
$$
$$
= \left(\phi_{j+1,0}, \phi_{j+1,2}, \ldots, \phi_{j+1,4N_j-2}, \phi_{j+1,1}, \phi_{j+1,3}, \ldots, \phi_{j+1,4N_j-1}\right)^T.
$$

In order to write the basis transformations in matrix-vector-notations analogous to Sections 2 and 3, we have to investigate the following matrix of function values:

Lemma 8. *The singular matrix*

$$
K_j := \left(\frac{1}{\sqrt{2N_j}}\,\phi_{j,r}\left(\frac{(2s+1)\pi}{2N_j}\right)\right)_{r,s=0}^{2N_j-1} = \overline{F}_j\,Q_j\,F_j,
$$

with the diagonal matrix $Q_j := \mathrm{diag}\,(q_{j,r})_{r=0}^{2N_j-1}$, *has the eigenvalues*

$$
q_{j,r} = e^{-\frac{ir\pi}{2N_j}} \times \begin{cases}
1, & \text{if } 0 \le r \le N_j - M_j, \\[2mm]
\frac{N_j-r}{M_j}, & \text{if } N_j - M_j < r < N_j + M_j, \\[2mm]
-1, & \text{if } N_j + M_j \le r \le 2N_j - 1,
\end{cases}
$$

and the transpose $K_j^T = \overline{F}_j\,\overline{Q}_j\,F_j$.

Proof: From $K_j = \left(\frac{1}{\sqrt{2N_j}}\,\phi_{j,0}\left(\frac{(2(s-r)+1)\pi}{2N_j}\right)\right)_{r,s=0}^{2N_j-1}$, we conclude that the matrix is circulant and can be written as $K_j = \overline{F}_j\,Q_j\,F_j$, where Q_j

is a diagonal matrix. The entries of Q_j follow from

$$
\begin{aligned}
q_{j,r} &= \frac{1}{\sqrt{2N_j}} \sum_{s=0}^{2N_j-1} \phi_{j,0}\left(\frac{(2s+1)\pi}{2N_j}\right) e^{\frac{irs\pi}{N_j}} \\
&= \frac{1}{\sqrt{2N_j}} \sum_{s=0}^{2N_j-1} \cos r\frac{s\pi}{N_j}\, \phi_{j,s}\left(-\frac{\pi}{2N_j}\right) + i \sum_{s=0}^{2N_j-1} \sin r\frac{s\pi}{N_j}\, \phi_{j,s}\left(-\frac{\pi}{2N_j}\right) \\
&= L_j \cos(r\circ)\left(-\frac{\pi}{2N_j}\right) + i\, L_j \sin(r\circ)\left(-\frac{\pi}{2N_j}\right).
\end{aligned}
$$

and the application of Lemma 1. Since K_j is real-valued,

$$
K_j^T = \overline{K_j^T} = \overline{F}_j \overline{Q}_j F_j. \quad \blacksquare
$$

Now we can formulate the decomposition and reconstruction in matrix notation to be used in numerical algorithms.

Theorem 25. *Let A_{j+1} and B_{j+1} be the transformation matrices in the decomposition and the reconstruction equations such that*

$$
\underline{\phi}_{j+1} = P_{j+1}^T B_{j+1} \begin{pmatrix} \underline{\phi}_j \\ \underline{\psi}_j \end{pmatrix} \quad \text{and} \quad \begin{pmatrix} \underline{\phi}_j \\ \underline{\psi}_j \end{pmatrix} = A_{j+1} P_{j+1} \underline{\phi}_{j+1}.
$$

Then A_{j+1} and B_{j+1} have circulant blocks, namely:

$$
A_{j+1} = \frac{1}{\sqrt{2}} \left(\begin{array}{c|c} I_j & \overline{F}_j\, Q_j\, F_j \\ \hline -\,\overline{F}_j\, \overline{Q}_j\, F_j & I_j \end{array} \right),
$$

$$
B_{j+1} = \frac{1}{\sqrt{2}} \left(\begin{array}{c|c} \overline{F}_j\, D_j^{-1}\, F_j & -\,\overline{F}_j\, D_j^{-1}\, Q_j\, F_j \\ \hline \overline{F}_j\, D_j^{-1}\, \overline{Q}_j\, F_j & \overline{F}_j\, D_j^{-1}\, F_j \end{array} \right),
$$

with Q_j known from Lemma 8 and $D_j = D_{N_j}^{M_j}$ as given in Lemma 2.

Proof: From the relations (3.1.3) and (3.2.7), together with (4.1.3), we derive the matrix

$$
\begin{aligned}
& A_{j+1} \\
&= \frac{1}{\sqrt{2}} \left(\begin{array}{c|c} \left(\frac{1}{\sqrt{2N_j}} \phi_{j,r}\left(\frac{s\pi}{N_j}\right)\right)_{r,s=0}^{2N_j-1} & \left(\frac{1}{\sqrt{2N_j}} \phi_{j,r}\left(\frac{(2s+1)\pi}{2N_j}\right)\right)_{r,s=0}^{2N_j-1} \\ \hline -\left(\frac{1}{\sqrt{2N_j}} \phi_{j,s}\left(\frac{(2r+1)\pi}{2N_j}\right)\right)_{r,s=0}^{2N_j-1} & \left(\frac{1}{\sqrt{2N_j}} \phi_{j,r}\left(\frac{s\pi}{2N_j}\right)\right)_{r,s=0}^{2N_j-1} \end{array} \right) \\
&= \frac{1}{\sqrt{2}} \left(\begin{array}{c|c} I_j & K_j \\ \hline -\,K_j^T & I_j \end{array} \right),
\end{aligned}
$$

where K_j is the matrix described in Lemma 8. Analogously, we deduce from (4.1.1) and (4.1.2) that

$$
B_{j+1}
$$

$$
= \frac{1}{\sqrt{2}} \left(\begin{array}{c|c} \left(\frac{1}{\sqrt{2N_j}} \, \tilde{\phi}_{j,r} \left(\frac{s\pi}{N_j} \right) \right)_{r,s=0}^{2N_j-1} & - \left(\frac{1}{\sqrt{2N_j}} \, \tilde{\phi}_{j,r} \left(\frac{(2s+1)\pi}{2N_j} \right) \right)_{r,s=0}^{2N_j-1} \\ \hline \left(\frac{1}{\sqrt{2N_j}} \, \tilde{\phi}_{j,s} \left(\frac{(2r+1)\pi}{2N_j} \right) \right)_{r,s=0}^{2N_j-1} & \left(\frac{1}{\sqrt{2N_j}} \, \tilde{\phi}_{j,r} \left(\frac{s\pi}{2N_j} \right) \right)_{r,s=0}^{2N_j-1} \end{array} \right)
$$

$$
= \frac{1}{\sqrt{2}} \left(\begin{array}{c|c} G_j^{-1} & - G_j^{-1} \, K_j \\ \hline \left(G_j^{-1} \, K_j \right)^T & G_j^{-1} \end{array} \right) \,,
$$

using (2.2.6) and Theorem 3. ∎

4.2   Orthonormal bases

For orthonormal bases, the transformation matrices are unitary. The decomposition and reconstruction relations for the orthonormal translates are given, for $\ell = 0, \ldots, 4N_j - 1$, by

$$
\mathcal{O}\phi_{j+1,\ell} = \sum_{r=0}^{2N_j-1} \langle \mathcal{O}\phi_{j+1,\ell}, \mathcal{O}\phi_{j,r} \rangle \, \mathcal{O}\phi_{j,r} + \sum_{r=0}^{2N_j-1} \langle \mathcal{O}\phi_{j+1,\ell}, \mathcal{O}\psi_{j,r} \rangle \, \mathcal{O}\psi_{j,r}, \quad (4.2.1)
$$

from which we deduce the transformation matrices.

Theorem 26. *For the matrices* $\mathcal{O\!A}_{j+1}$ *and* $\mathcal{O\!B}_{j+1}$ *in the relations*

$$
\underline{\mathcal{O}\phi}_{j+1} = P_{j+1}^T \, \mathcal{O\!B}_{j+1} \left(\begin{array}{c} \underline{\mathcal{O}\phi}_j \\ \underline{\mathcal{O}\psi}_j \end{array} \right) \qquad (4.2.2)
$$

and

$$
\left(\begin{array}{c} \underline{\mathcal{O}\phi}_j \\ \underline{\mathcal{O}\psi}_j \end{array} \right) = \mathcal{O\!A}_{j+1} \, P_{j+1} \, \underline{\mathcal{O}\phi}_{j+1} \,, \qquad (4.2.3)
$$

we have

$$
\mathcal{O\!A}_{j+1} = \mathcal{O\!B}_{j+1}^T = \frac{1}{\sqrt{2}} \left(\begin{array}{c|c} \overline{F}_j \, \Delta_j \, F_j & \overline{F}_j \, \Delta_j \, Q_j \, F_j \\ \hline -F_j \, \Delta_j \, Q_j \, \overline{F}_j & \overline{F}_j \, \Delta_j \, F_j \end{array} \right) \,,
$$

with $\Delta_j = \Delta_{N_j}^{M_j}$ *as given in Theorem 5 and* Q_j *from Lemma 8.*

Proof: Analogous to the proof of Theorem 25, we distinguish between the translates $\mathcal{O}\phi_{j+1,\ell}$ with even and odd index ℓ. Then, formula (4.2.2) is relation (4.2.1) in matrix notation, with

$$
\mathcal{CB}_{j+1} = \left(\begin{array}{c|c} \left(\langle \mathcal{O}\phi_{j+1,2s}, \mathcal{O}\phi_{j,r} \rangle \right)_{s,r=0}^{2N_j-1} & -\left(\langle \mathcal{O}\phi_{j+1,2s}, \mathcal{O}\psi_{j,r} \rangle \right)_{s,r=0}^{2N_j-1} \\ \hline \left(\langle \mathcal{O}\phi_{j+1,2s+1}, \mathcal{O}\phi_{j,r} \rangle \right)_{s,r=0}^{2N_j-1} & \left(\langle \mathcal{O}\phi_{j+1,2s+1}, \mathcal{O}\psi_{j,r} \rangle \right)_{s,r=0}^{2N_j-1} \end{array} \right).
$$

Similarly, we obtain (4.2.3) with $\mathcal{CA}_{j+1} = \mathcal{CB}_{j+1}^T$.

We compute, for $\ell = 0, \ldots, 4N_j - 1$ and $r = 0, \ldots, 2N_j - 1$,

$$
\begin{aligned}
&\langle \mathcal{O}\phi_{j+1,\ell}, \mathcal{O}\phi_{j,r} \rangle \\
&= \frac{1}{2\sqrt{N_j N_{j+1}}} \Big\langle D_{N_{j+1}-M_{j+1}}\left(\circ - \frac{\ell\pi}{N_{j+1}}\right), \\
&\qquad D_{N_j-M_j}\left(\circ - \frac{r\pi}{N_j}\right) + \sum_{k=-M_j+1}^{M_j-1} \frac{2M_j-2k}{\sqrt{2M_j^2+2k^2}} \cos(N_j+k)\left(\circ - \frac{r\pi}{N_j}\right) \Big\rangle \\
&= \frac{1}{2\sqrt{N_j N_{j+1}}} D_{N_j-M_j}\left(\frac{(\ell-2r)\pi}{2N_j}\right) + 2 \sum_{k=-M_j+1}^{M_j-1} \frac{M_j-k}{\sqrt{2M_j^2+2k^2}} \cos \frac{(N_j+k)(\ell-2r)\pi}{2N_j} \\
&= \frac{1}{\sqrt{2N_{j+1}}} \mathcal{O}\phi_{j,r}\left(\frac{\ell\pi}{2N_j}\right)
\end{aligned}
$$

and

$$
\begin{aligned}
\langle \mathcal{O}\phi_{j+1,\ell}, \mathcal{O}\psi_{j,r} \rangle &= \frac{1}{2\sqrt{N_j N_{j+1}}} \Big\langle D_{N_{j+1}-M_{j+1}}\left(\circ - \frac{\ell\pi}{N_{j+1}}\right) \\
&\quad + \sum_{k=-M_{j+1}+1}^{M_{j+1}-1} \frac{2M_{j+1}-2k}{\sqrt{2M_{j+1}^2+2k^2}} \cos(N_{j+1}+k)\left(\circ - \frac{\ell\pi}{N_{j+1}}\right), \\
&\quad \sum_{k=-M_j+1}^{M_j-1} \frac{2M_j+2k}{\sqrt{2M_j^2+2k^2}} \cos(N_j+k)\left(\circ - \frac{(2r+1)\pi}{2N_j}\right) \\
&\quad + 2 \sum_{k=N_j+M_j}^{N_{j+1}-M_{j+1}} \cos k \left(\circ - \frac{(2r+1)\pi}{2N_j}\right) \\
&\quad + \sum_{k=-M_{j+1}+1}^{M_{j+1}-1} \frac{2M_{j+1}-2k}{\sqrt{2M_{j+1}^2+2k^2}} \cos(N_{j+1}+k)\left(\circ - \frac{(2r+1)\pi}{2N_j}\right) \Big\rangle
\end{aligned}
$$

$$= \frac{1}{2\sqrt{N_j N_{j+1}}} \left(\sum_{k=-M_j+1}^{M_j-1} \frac{2M_j+2k}{\sqrt{2M_j^2+2k^2}} \cos \frac{(N_j+k)(\ell-2r-1)\pi}{2N_j} \right.$$

$$+ 2 \sum_{k=N_j+M_j}^{N_{j+1}-M_{j+1}} \cos \frac{k(\ell-2r-1)\pi}{2N_j}$$

$$\left. + \sum_{k=-M_{j+1}+1}^{M_{j+1}-1} \frac{(M_{j+1}-k)^2}{M_{j+1}^2+k^2} \cos \frac{(N_{j+1}+k)(\ell-2r-1)\pi}{2N_j} \right)$$

$$= \frac{1}{\sqrt{2N_{j+1}}} \times \begin{cases} \mathcal{O}\phi_{j,r}\left(\frac{\ell\pi}{2N_j}\right), & \text{for odd } \ell, \\ -\mathcal{O}\phi_{j,\frac{\ell}{2}}\left(\frac{(2r+1)\pi}{2N_j}\right), & \text{for even } \ell. \end{cases}$$

Hence, from Theorem 5 and Theorem 25 we conclude that

$$\mathcal{O}\mathcal{A}_{j+1} = \frac{1}{\sqrt{2}} \left(\begin{array}{c|c} \boldsymbol{\Gamma}_j & \boldsymbol{\Gamma}_j \, \boldsymbol{K}_j \\ \hline -(\boldsymbol{\Gamma}_j \, \boldsymbol{K}_j)^T & \boldsymbol{\Gamma}_j \end{array} \right),$$

which finally proves the desired form of the transformation matrices. ∎

§5   Conclusion

In this paper we have studied nested subspaces $V_j = V_{N_j}^{M_j}$ of trigonometric polynomials constructed from translates of de la Vallée Poussin kernels $\varphi_{N_j}^{M_j}$ based on averaging over $2M_j$ Dirichlet kernels. Moreover, we considered orthogonal polynomial wavelet spaces $W_j = V_{j+1} \ominus V_j$ and their biorthogonal wavelet bases. The greater the number M_j for fixed N_j, the better time localized those basis functions behave, as it was investigated in Sections 2.4 and 3.5. On the other hand, the best frequency-localized translates as well as the best frequency splittings between the wavelet spaces W_j are obtained for $M_j = 1$; i.e. $\lambda = \infty$, which is the original Fourier case at an even number of nodes.

Our considerations have been focused not only on the interpolatory bases and their dual bases, but also on two orthogonal bases in each space. Among all orthogonal bases functions in V_j and W_j, respectively, $\mathcal{O}\phi_{j,r}$ and $\mathcal{O}\psi_{j,s}$ are the most time-localized (and translation invariant) bases functions, and $\rho_{j,k}$ and $\sigma_{j,\ell}$ are the most frequency-localized ones. The main part of this paper has been concerned with the algorithms for the basis transformations being described by their transformation matrices with respect to fast numerical implementations. For the circulant matrices in i) and ii) below, the knowledge of the eigenvalues reduces the computations to the application of the FFT. In *ii)*, we also need to permute the vector entries.

Let us summarize the relationship between the bases by the relevant matrices in the following schemes

i) for the sample spaces V_N^M or V_j and the wavelet spaces W_j on one level:

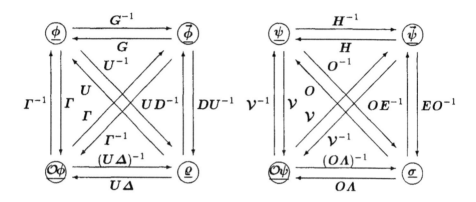

ii) for the decomposition and reconstruction $V_{j+1} = V_j \oplus W_j$:

Error estimates for the initial approximation of a given function by the interpolatory or the orthogonal projection onto a sample space of a sufficiently large level are included in Subsection 2.4.

Let us end with the remark that $\phi_{j,r}(t) + \phi_{j,-r}(t)$ and $\psi_{j,r}(t) + \psi_{j,-r-1}(t)$ are even functions. Therefore, this approach can be transformed by $x = \cos t$, for $t \in [0, \pi]$, to the algebraic case which yields polynomial wavelet spaces on $[-1, 1]$ orthogonal with respect to the Chebyshev weight $(1 - x^2)^{-1/2}$. For this, see [5], [11], and [21].

References

[1] Chui, C. K., *An introduction to wavelets*, Academic Press, Boston, 1992.

[2] Chui, C. K. and H. N. Mhaskar, On trigonometric wavelets, *Constr. Approx.* **9** (1993), 167–190.

[3] Daubechies, I., *Ten lectures on wavelets*, CBMS-NSF Series in Appl. Math. #61, SIAM, Philadelphia, 1992.

[4] Davis, P. J., *Circulant Matrices*, Wiley Interscience, New York, 1979.

[5] Kilgore, T. and J. Prestin, Polynomial wavelets on the interval, *Constr. Approx.*, 1995, to appear.

[6] Lemarié P. G. and Y. Meyer, Ondelettes et bases hilbertiennes, *Rev. Mat. Iberoamericana* **2** (1986), 1–18.

[7] Lorentz, R. A. and A. A. Sahakian, Orthogonal trigonometric Schauder bases of optimal degree for $C(0, 2\pi)$, *J. Fourier Anal. Appl.* **1** (1994), 103–112.

[8] Meyer, Y., *Ondelettes*, Herman, Paris, 1990.

[9] Offin, D. and K. Oskolkov, A note on orthonormal polynomial bases and wavelets, *Constr. Approx.* **9** (1993), 319–325.

[10] Perrier, V. and C. Basdevant, Periodic wavelet analysis, a tool for inhomogeneous field investigation, theory and algorithms, *Rech. Aérospat.* **3** (1989), 53–67.

[11] Plonka, G., K. Selig, and M. Tasche, On the construction of wavelets on the interval, *Adv. Comp. Math.*, 1995, to appear.

[12] Plonka, G. and M. Tasche, A unified approach to periodic wavelets, in *Wavelets: Theory, Algorithms, and Applications*, C. K. Chui, L. Montefusco, and L. Puccio (eds.), Academic Press, Boston, 1994, 137–151.

[13] Prestin, J., On the approximation by de la Vallée Poussin sums and interpolatory polynomials in Lipschitz norms, *Anal. Math.* **13** (1987), 251–259.

[14] Prestin, J. and E. Quak, Trigonometric interpolation and wavelet decomposition, *Numer. Algorithms* **9** (1995), 293– 317.

[15] Prestin, J. and E. Quak, A duality principle for trigonometric wavelets, in *Wavelets, Images, and Surface Fitting*, P. J. Laurent, A. Le Méhauté, and L. L. Schumaker (eds.), A K Peters, Wellesley, 1994, pp. 407–418.

[16] Prestin, J. and E. Quak, Decay Properties of trigonometric Wavelets, in *Proceedings of the Cornelius Lanczos International Centenary Conference*, J. D. Brown, M. T. Chu, D. C. Ellison, and R. J. Plemmons (eds.), SIAM, 1994, pp. 413–415.

[17] Prestin, J. and K. Selig, On a constructive representation of an orthogonal trigonometric Schauder basis for $C_{2\pi}$, in preparation.

[18] Privalov, A. A., On an orthogonal trigonometric basis, *Math. USSR Sbornik* **72** (1992), 363–372.

[19] Selig, K., Trigonometric wavelets for time–frequency– analysis, in *Approximation Theory, Wavelets and Applications*, S. P. Singh (ed.), Kluwer Academic Publ., Dordrecht, 1995, pp. 453–464.

[20] Stečkin, S. B., On de la Vallée Poussin sums, *Doklady Akad. Nauk SSSR* **80** (1951), 545–548 (Russ.).

[21] Tasche, M., Polynomial wavelets on $[-1, 1]$, in *Approximation Theory, Wavelets and Applications*, S. P. Singh (ed.), Kluwer Academic Publ., Dordrecht, 1995, pp. 497–512.

[22] Timan, A. F., *Theory of Approximation of Functions of a Real Variable*, English translation Pergamon Press, 1963.

[23] Wojtaszczyk, P. and K. Woźniakowski, Orthonormal polynomial bases in function spaces, *Isr. J. Math.* **75** (1991), 167–191.

[24] Woźniakowski, K., Orthonormal polynomial basis in $C(\Pi)$ with optimal growth of degrees, preprint.

[25] Zygmund, A., *Trigonometric Series*, Cambridge University Press, Second Edition, 1959.

Jürgen Prestin
FB Mathematik
Universität Rostock
D–18051 Rostock
Germany
prestin@mathematik.uni-rostock.d400.de

Kathi Selig
FB Mathematik
Universität Rostock
D–18051 Rostock
Germany
selig@mathematik.uni-rostock.d400.de

III.

Redundant Waveform Representation for Signal Processing and Image Analysis

Noise Reduction in Terms of the Theory of Frames

John J. Benedetto

Abstract. Frames were introduced by Duffin and Schaeffer in 1952 to deal with problems in nonharmonic Fourier series, and have been used more recently in signal analysis. Frames provide a useful starting point to obtain signal reconstruction for signals embedded in certain noises. We develop two methods to solve a class of noise reduction and signal reconstruction problems in a computationally efficient way. These methods go beyond the elementary properties of frames; and they deal with compression nonlinearities, a quantitative characterization of frames of translates, irregular sampling theorems, and generalized spatio-temporal Laplacians. Applications of these methods are made on EEG, ECoG, TIMIT, and MRI data.

§1 Introduction

The theory of frames was introduced by Duffin and Schaeffer [16] in the early 1950s to deal with problems in nonharmonic Fourier series. There has been renewed interest in the subject related to its role in wavelet theory [14, 13]. Frames provide a useful *starting point* to obtain signal decompositions and reconstructions in cases where redundancy, robustness, oversampling, irregular sampling, and certain types of noise play a role [6, 2, 3, 13, 7, 8, 9, 20].

Our goal is to solve a class of noise reduction and signal reconstruction problems in a computationally efficient way.

We have used the theory of frames (Section 2) to develop two noise reduction methods (Sections 3 and 4), as well as a signal reconstruction formula for signals whose values are known at an irregularly spaced set of points (Section 6). These methods and formula are implemented in the FrameTool, a software package developed by Melissa Harrison [17, 5].

The theory of frames is quantified to make it an effective mathematical device. In the process, we analyze various parameters to show how they can be adjusted to deal with particular applications in a computationally effective way.

Signal and Image Representation in Combined Spaces
Y. Y. Zeevi and R. R. Coifman (Eds.), pp. 259–284.

The noise reduction methods that we develop provide better suppression of broadband noise than quadrature mirror filter (wavelet multiresolution) multirate systems; and the iterative algorithms converge exponentially. The irregular sampling formula provides perfect reconstruction theoretically, whereas its implementation in FrameTool is aimed at achieving realtime performance.

There are many unresolved higher-dimensional irregular sampling problems; and our goal is to achieve simultaneous signal reconstruction and noise reduction for such problems. The problems we have in mind are in the areas of image processing, MRI, multi-channel control theory, error analysis of gridding algorithms, and seizure prediction and localization based on bioelectric data. As such, in Section 5 we outline a new multidimensional method which generalizes the classical spatio-temporal Laplacian for time-varying systems.

§2   Theory of frames

2.1   Frames

A sequence $\{\theta_n\}$ in a space H of finite energy signals is a *frame* if there exist $A, B > 0$ such that for all $f \in H$,

$$A\|f\|^2 \le \sum_n |\langle f, \theta_n \rangle|^2 \le B\|f\|^2, \tag{2.1.1}$$

where $\langle \cdot, \cdot \rangle$ is an inner product and $\|f\|^2 = \langle f, f \rangle$. If $\{\theta_n\}$ is a frame then the *frame operator* $S : H \to H$ is defined by $Sf = \sum_n \langle f, \theta_n \rangle \theta_n$. The main elementary fact about frames is that the inverse S^{-1} is well defined and

$$f = \sum_n \langle f, S^{-1}\theta_n \rangle \theta_n \tag{2.1.2}$$

for all $f \in H$ [16, 13, 14, 3]. Frames generalize orthonormal bases.

2.2   Frame effectiveness

It is natural to ask how the inequalities (2.1.1), which are only a simple extension of the Plancherel-Parseval relation, could be of any use. Without further processing or a deeper analysis of S^{-1}, the answer is that frames and the corresponding reconstruction formula (2.1.2) do *not* provide substantial progress in applications. In fact, the unadulterated frame reconstruction formula (2.1.2) is sometimes used in comparisons to enhance features of other methods!

However, with proper processing, we establish the effectiveness of frames in noise reduction and/or irregular sampling problems. This is done in Sections 3 and 4, where two algorithms are given.

The first algorithm is our WAM Noise Suppression algorithm [11, 12]. It achieves simultaneous compression, noise reduction, and signal reconstruction. (WAM is an acronym for *Wavelet Auditory Model*.) It depends on a filter design (for filter-dependent noises), irregular sampling, and nonlinear compression operations.

The second algorithm is our Frame Multiresolution Analysis algorithm [7], [8, 9, 10]. It depends on a useful but surprisingly complicated characterization of frames which are not bases. This characterization leads to new multirate systems, which allow us to obtain simultaneous noise reduction and signal reconstruction for broadband noises, *e.g.* quantization noise.

The simple formula (2.1.2) can also be used to prove sampling formulas which are more useful in some applications than the Classical (Shannon) Sampling Theorem, *e.g.* the following equation (2.3.1).

2.3   Frame reconstruction, sampling, and "infinite frequencies"

William Heller and I proved the validity of the following equation as well as an irregular sampling version:

$$f = T \sum_{m,n} \left\langle \hat{f}, e_{nT} \tau_{mb} \hat{g} \right\rangle \tau_{-nT}(e_{mb}s). \tag{2.3.1}$$

Notationally, $e_a(\gamma) = e^{2\pi i a \gamma}$, $\tau_b h(t) = h(t - b)$, and $\hat{g}$ is the Fourier transform of g. Equation (2.3.1) is valid for *all* finite energy signals when we assume $0 < 2T\Omega \le 1$, $\hat{g} = 1$ on $[-\Omega, \Omega]$, and some natural conditions on b and the sampling function s, *e.g.* [3, pages 269–270]. If f is Ω-bandlimited then (2.3.1) reduces to the Classical Sampling Theorem with sampling function s:

$$f = T \sum_{n} f(nT)\tau_{nT}s.$$

2.4   Aliasing

In high frequency information, we must sample closely to capture all of the fluctuations. Thus, in the case of very high frequencies ("infinite frequencies"), *i.e.* nonbandlimitedness, we cannot reconstruct f with discrete sampling.

The frame reconstruction formula (2.3.1) gives the ordinary sampling formula for bandlimited functions *as well as* giving a signal representation for nonbandlimited functions. In this latter case, there is added complexity, as illustrated by (2.3.1). This complexity is required to deal with "infinite frequencies," and has the advantage of quantifying aliasing.

2.5   Remark on S^{-1}

The main reason the implementation of (2.1.2) requires processing to achieve effectiveness is because of the subtlety in quantifying the operator S^{-1}. Recently, there has been an important new analysis of S^{-1} by A.J.E.M. Janssen [19] of Philips Laboratories. In fact, Volume 1, Issue 4 of the *Journal of Fourier Analysis and Applications*, in which Janssen's article appears, contains up-to-date and important information on Gabor expansions such as (2.3.1).

§3   WAM noise suppression algorithm

This algorithm, constructed with Anthony Teolis, consists of the following components:

1. A continuous wavelet transform W_g with adaptive analyzing impulse response g;

2. Signal dependent irregular sampling of W_g;

3. Modification of sampled coefficients by nonlinear and thresholding operations to achieve compression and noise reduction, respectively;

4. Iterative irregular sampling reconstruction method.

3.1   Continuous wavelet transform

The *continuous wavelet transform* W_g of a signal f for a given analyzing impulse response g is the function of two variables

$$W_g f(t, s) = (f * D_s g)(t),$$

where t is time, s is scale, $D_s g(t) = s^{\frac{1}{2}} g(ts)$, and $*$ is convolution. The design of the filter $\hat{g}$ (Fourier transform of g) is important, and the continuous wavelet transform is a reasonable mathematical model in the case of overlapping filter banks such as arise in auditory and visual systems. In auditory systems, the impulse response is causal, and the magnitude of the corresponding filter is shark-fin shaped. Consequently, by a calculation involving the Hilbert transform and the Paley-Wiener logarithmic integral theorem, the impulse response has the form of Figure 1.

3.2   System output

Suppose we are dealing with a system of overlapping dilation related filters, with a mechanical to electrical transference, and a sigmoidal nonlinearity

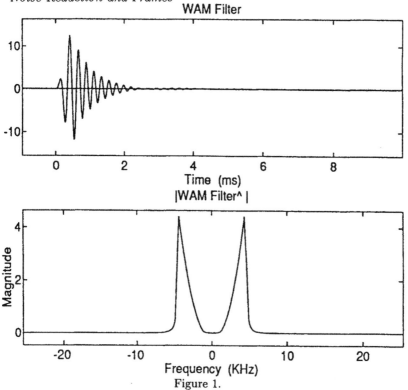

Figure 1.

to achieve compression. An example is a mammalian auditory system. An analysis of the derivative of the nonlinearity, and a lateral scale operation, produce sampling sets $X = \{t_{m,n} : n = 1, ...\}$ for each scale s_m. Our irregular sampling theorem allows us to reconstruct the original signal from the discrete set Lf of sampled values on X and to obtain compression in the process.

It turns out that L can be thought of as an operator with adjoint operator L^*, and that $S = L^*L$ is a frame operator of the type described in Section 2. This factorization is the basis of one of the ideas behind our irregular sampling formulas.

3.3 Thresholding

It is natural to introduce thresholding methods, in conjunction with wavelet theory, to achieve noise reduction, *e.g.* [11, 15]. As indicated in Section 2, we have generalized the original idea of WAM beyond the original auditory applications. It can now be used as a noise suppressant in the following way.

Suppose we are given a class of signals f which are to be transmitted in a noisy environment caused by additive noise w. The main theoret-

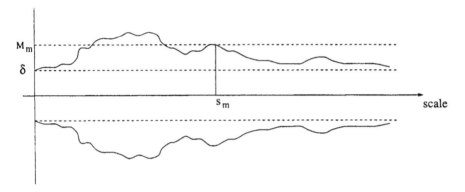

Figure 2.

ical problem is to design a filter $\hat{g}$ so that w is incoherent with regard to the "wavelets" $\{\psi_{m,n}\}$ determined by the impulse response g. Using the notation L from the section on *System Output*, this means that $Lw = \{\langle w, \psi_{m,n} \rangle\}$ and each $|\langle w, \psi_{m,n} \rangle| \leq \delta$.

In our experiments with TIMIT data, we obtained symmetric envelopes as in Figure 2. For each scale s_m, M_m is defined as $\max_n\{\langle f + w, \psi_{m,n} \rangle\}$, and the maximum is taken over a specified portion of the aforementioned irregular sampling pattern. The noise threshold δ is then subtracted from each nonnegative element of the data set $\{\langle f + w, \psi_{m,n} \rangle\}$. The resulting data set is used in conjunction with our irregular sampling theory to form a signal $\tilde{f}$ which, mathematically, is a reasonable representation of f, *e.g.* [3]. These representations are effective, both graphically and in audio-taped versions, in the case of speech signals f and white noise w.

3.4 Discretization, noise reduction, and iterative reconstruction

In terms of the L operator, the description in the section on *Thresholding* amounts to the following. In the case of a signal space H and of frames which are not bases, L is not subjective onto the space of finite energy sequences. Thus, the filter $\hat{g}$ can be designed so that $Lw \notin L(H)$. In this setting we construct the frame correlation $R = LL^* : L(H) \to L(H)$ so that the discrete set R^{-1} can be stored off-line.

Consequently,

$$f = L^* R^{-1} L f, \qquad (3.4.1)$$

e.g. f can be reconstructed in terms of the sequence Lf. Therefore, since

$Lw \notin L(H)$ we expect $L^*R^{-1}L(f + w)$ to give a reasonable representation of f.

This approach makes thresholding viable for noise reduction, and ensures that such reduction is not due to low-pass filtering.

Equation (3.4.1) can be implemented by an iterative procedure with exponentially decreasing error. The examples in Figures 3–5 were generated in this way.

§4 Frame multiresolution analysis algorithm

This algorithm was constructed with Shidong Li.

4.1 Frame multiresolution analysis

A *frame multiresolution analysis* of the space L^2 of finite energy signals on $\mathbb{R}$ is an increasing sequence of closed linear subspaces $V_j \subseteq L^2$ and an element $\varphi \in V_0$, for which the following hold:

i. $\overline{\cup V_j} = L^2$ and $\cap V_j = \{0\}$,

ii. $f(t) \in V_j$ if and only if $f(2t) \in V_{j+1}$,

iii. $f \in V_0$ implies $\tau_k f \in V_0$ for all integers k, where $\tau_k f(t) = f(t - k)$,

iv. $\{\tau_k \varphi\}$ is a frame for V_0.

Using this notion of an FMRA, we can construct a corresponding subband coding system, which suppresses channel/quantization noise, and yields signal reconstruction.

Because of the importance of narrow-band signals in topics as varied as EEG analysis and spread-spectrum communications, we are using FMRAs to achieve *simultaneous* narrow-band signal reconstruction and channel/quantization noise reduction, in a subband coding scheme.

Our main theoretical tool is our characterization of frames of translates in terms of periodizations of the spectral content of scaling functions associated with FMRAs [7, 9]. For the frames of interest, these periodizations have spectral gaps which allow for the suppression of broadband channel noise *and* give perfect signal reconstruction. For many other signals, FMRA systems should also yield perfect reconstruction in the case where channel noise is negligible. This is achieved by a preprocessing method associated with the scaling properties of FMRAs. It should be pointed out that wavelet subband coding, or even more general PRFBs, can not be expected to achieve the simultaneity mentioned above; and this simultaneity plays a role in effecting computational efficiency. The reason for this deficiency in an MRA (say) is that the whole frequency band is covered by the

Numerical Example: Packet - Low Noise ─────────────■

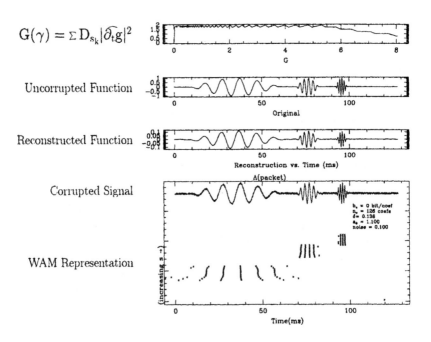

$$G(\gamma) = \Sigma\,D_{s_k}|\widehat{\partial_t g}|^2$$

Uncorrupted Function

Reconstructed Function

Corrupted Signal

WAM Representation

Figure 3.

Numerical Example: Packet - High Noise ────────────────────■

$$G(\gamma) = \Sigma \, D_{s_k} |\widehat{\partial_t g}|^2$$

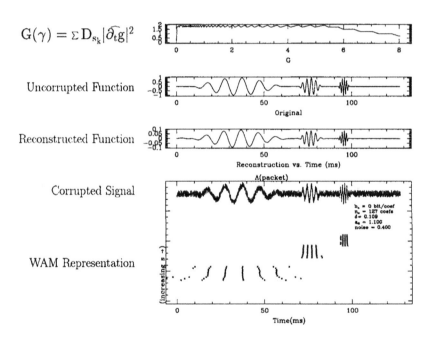

Uncorrupted Function

Reconstructed Function

Corrupted Signal

WAM Representation

Figure 4.

Numerical Example: "Water" in Noise ────────────────────────●

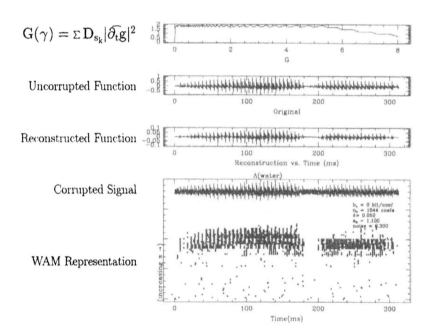

$$G(\gamma) = \Sigma \, \mathbf{D}_{s_k} |\widehat{\partial_t g}|^2$$

Uncorrupted Function

Reconstructed Function

Corrupted Signal

WAM Representation

Figure 5.

analysis and synthesis filters; and hence the PR in such systems includes PR of the accompanying noise.

4.2 Processing for continuous signals

Let $f \in L^2(\mathbb{R})$, and choose $\epsilon > 0$. There exist $J = J_\epsilon$ and a function $g \in V_J$ such that

$$\|f - g\| < \epsilon$$

and

$$g = \sum_k \langle g, \phi_{Jk} \rangle \, \phi^0_{Jk},$$

where $\{\phi^0_{Jk} \equiv 2^{1/2} \phi^0(2^J t - k)\}$ is the dual frame of $\{\phi_{Jk} \equiv 2^{1/2} \phi(2^J t - k)\}$ in V_J, and $\| \cdot \|$ is the finite energy norm. The *pre-processing (sampling)* step is the computation of

$$\forall k, \quad c_J(k) \equiv \langle f, \phi_{Jk} \rangle.$$

This has the effect of squeezing the bandwidth of the signal f into the bandwidth of the associated filter bank.

The *subband processing* step consists of decomposition and reconstruction computations, viz.,

(a) Decomposition

$$c_{J-1}(k) \equiv \langle f, \phi_{J-1,k} \rangle = \sum_n \overline{h_0(n - 2k)} c_J(n),$$

$$d_{J-1}(k) = \sum_n \overline{h_1(n - 2k)} c_J(n),$$

and

(b) Reconstruction

$$c_J(n) = \sum_k \overline{g_0(2k - n)} c_{J-1}(k) + \sum_k \overline{g_1(2k - n)} d_{J-1}(k).$$

Finally, the *post-processing (interpolation)* step is the computation of the projection

$$\tilde{f} = \sum_k c_J(k) \phi^0_{Jk} \in V_J.$$

It is easy to prove that $\|f - \tilde{f}\| < \epsilon$. Note that this approximation also arises in an ordinary MRA (either an orthonormal basis or exact frame).

The system can be summarized in Figure 6.

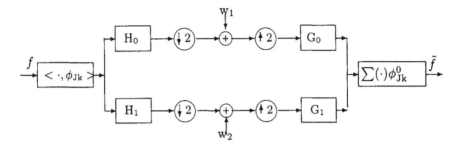

Figure 6. A subband processing system associated with an FMRA for analog signals f, where w_1 and w_2 model the quantization noise in subbands.

4.3 Simulation

The FMRA used in the following simulation is generated by $\hat{\phi} = \mathbf{1}_{[-\frac{1}{4}, \frac{1}{4})}$. We can choose $H_0 = \mathbf{1}_{[-\frac{1}{8}, \frac{1}{8})}$ and $H_1 = \mathbf{1}_\Omega$ where $\Omega = [-\frac{1}{4}, -\frac{1}{8}) \cup [\frac{1}{8}, \frac{1}{4})$. The G_0 and G_1 can be computed from our general theory, *e.g.* [7, 9].

This example illustrates an FMRA whose associated filter bank alone is not a PRFB for all discrete finite energy signals. In fact, it gives rise to a frame which is not exact. In particular, the filters H_0 and H_1 are not essentially half-band and are not associated with an MRA.

Figures 7 and 8 are a chirp signal and its power spectrum, respectively. The signal is chosen so that the spectrum goes beyond the bandwidth of the associated filter bank. Figure 9 is the reconstruction of the signal through an FMRA subband processing system (with pre- and post-processing) as in Figure 6. The reconstruction is excellent. For comparison, Figure 10 is the output of the same signal through the filter banks only, without pre- and post-processing.

4.4 Quantization noise reduction with FMRA subband systems

Since one of the fundamental purposes of subband processing is to achieve greater data compression, quantization at subbands is a key element in subband coding systems. Therefore, the effect of noise introduced by quantization, and its consequences at the output of the system, is an extremely important issue. There is a large literature dealing with the subject, where the quantization effect is integrated into the subband filtering system to optimize the design of the filter bank.

We have a new point-of-view. While almost all existing subband filter banks correspond to some type of orthogonal or biorthonormal basis in terms of wavelet theory, we shall demonstrate that significant noise reduction is achieved by simply using wavelet frames and their associated filter

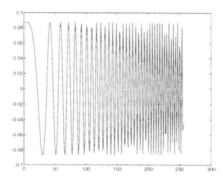

Figure 7. A chirp signal chosen for numerical experiment.

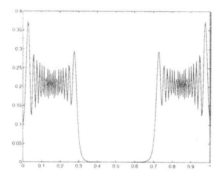

Figure 8. The power spectrum of the chirp signal in Figure 7. The spectrum is wider than the bandwidth of the filter bank associated with the FMRA.

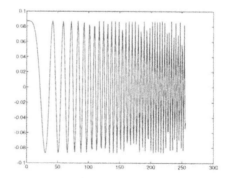

Figure 9. The reconstruction of the signal in Figure 7 through an FMRA subband processing system as in Figure 6. The pre- and post-processing is taken in the subspace V_1 ($J = 1$). The reconstruction error (in absolute value) is on the order of 10^{-4}. Notice that a certain error is expected since the signal is not band limited.

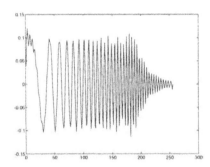

Figure 10. The output of the chirp signal through the corresponding filter bank, without pre- and post-processing. The loss of high frequency is expected.

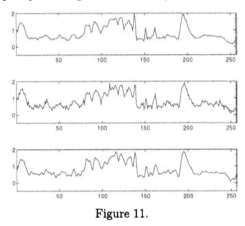

Figure 11.

banks. The structure of FMRAs and the redundancy they model provide a natural tool for noise suppression (see [8]), where a probabilistic study of this phenomenon is made.

In Figures 11, 12, and 13, the top picture is the original signal. The middle picture corresponds to Daubechies QMFs having 4, 8, and 16 taps, respectively. The bottom picture is the result when FMRA filters are used. The noise introduced at the quantization stage is a white Gaussian noise having mean 0 and variance 0.01, 0.04, and 0.09, respectively.

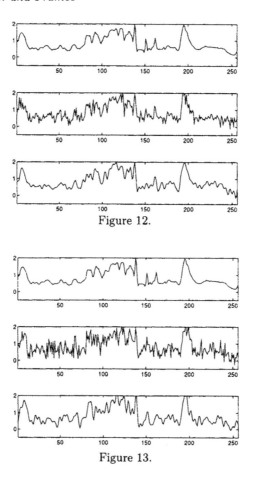

Figure 12.

Figure 13.

4.5 Remark

FMRAs include narrow-band filter banks which, in concert with our pre-
and post-processing technique, allow for simultaneous reconstruction of
narrow-band signals and quantization noise suppression. As in the case of
MRAs, the signal reconstruction is perfect if the signal, narrow-band or not,
is an element of some V_J and if channel/quantization noise is discounted.
Further, because of the possibility of narrow bandwidth, FMRAs are essentially different from MRAs as well as more general half-band PRFBs.
Our current examples of filter banks arising from FMRAs are mostly of
ideal type. Ongoing research includes the construction of FIR filter banks
for FMRAs and studies on a more general structure, due to Li [20].

§5   Wavelet-frame processing and generalized spatio-temporal Laplacians

We made the following spectral and wavelet analysis on electrocorticogram (ECoG) data with David Colella [4]. Our wavelet analysis involves the design of a *wavelet-frame* and a generalization of classical spatio-temporal Laplacians.

5.1   Spectral analysis of ECoG data

Spectral content and the evolution of waveform morphologies are important clues that are used by ECoG readers to identify seizure activity in ECoG waveforms. Figure 14 shows a typical example of a single-channel ECoG waveform along with individual segments highlighting a nonseizure epoch and a seizure epoch. Figure 15 provides a spectrogram of the data shown in Figure 14. The chirp-like structures seen in that figure are associated with the seizure activity.

5.2   Wavelet analysis of ECoG data

The seizure chirp patterns illustrated in the spectogram in Figure 15 can be corroborated by a wavelet analysis. This identification is in terms of horizontal bands of striations seen in Figure 16. These striations reflect a relationship between particular scales and fundamental frequencies. Seizure activity can be distinguished from nonseizure activity through the existence (seizure) or nonexistence (nonseizure) of these striations. This is quantified by the following analysis.

Let h be a function defined on the real line $\mathbb{R}$, and let the dilation h_s be defined by $h_s(t) = sh(st)$, where $s > 0$ and $t \in \mathbb{R}$. The continuous wavelet transform $W_h f$ of a function f defined on $\mathbb{R}$ is the function $W_h f(t, s) = f * h_s(t)$ defined on $\mathbb{R} \times (0, \infty)$, where $*$ denotes the convolution operator. A discretized version of $W_h f$ is obtained in the case that h is the Haar function on the integers, namely, $h(-1) = 2^{1/2}$, $h(0) = -2^{1/2}$, and h equals 0 otherwise. The scalograms show the (discrete) wavelet transform $W_h f(x, 1/s)$, where the abscissa x is time and the ordinate $1/s$ is scale. For our purposes, since our interest is primarily in feature identification, we employ not only dyadic scales in the wavelet transform scalogram but all integer scales as well.

We examine the case of a pure tone, *i.e.* where the signal $f(t) = \sin(2\pi\gamma t + \theta)$, with γ a fixed frequency and θ the random phase. The (continuous) wavelet transform is then

$$W_h f(x, 1/s) = \frac{2}{\pi\gamma\sqrt{s}}\,\sin^2(\pi\gamma s/2)\cos(2\pi x\gamma + \theta). \qquad (5.2.1)$$

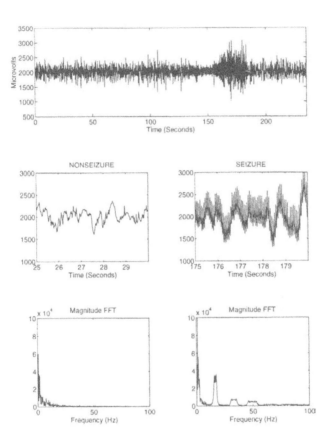

Figure 14. Top: Time series trace of electrocorticogram data. Middle: Time series for nonseizure epoch (left) and seizure epoch (right). Bottom: Magnitude FFT for the time series shown in the middle.

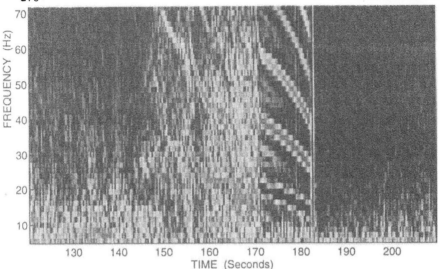

Figure 15. Spectrogram for the ECoG time series shown in Figure 14.

This expression can be examined from two perspectives. For a fixed frequency, one can ask in which scales this frequency is most dominant; one can also consider which frequencies significantly contribute to any given scale.

For example, whenever γs is an even integer, the expression (5.2.1) is zero for all x, and if γs is an odd integer, $\gamma s = (1 + 2k)$, the wavelet transform becomes:

$$W_h f(x, 1/s) = \frac{\sqrt{s}}{(1 + 2k)\pi} \cos(2\pi x\gamma + \theta). \qquad (5.2.2)$$

Hence, the amplitude tends to zero as k increases. This phenomenon can be observed in Figure 16, where, for a fixed frequency γ, the scales $s = \gamma/(1 + 2k)$ should provide evidence of the frequency γ for all $k = 1, 2, 3, \ldots$, but with decreasing amplitude. This matches precisely the pattern of diminishing striations seen in the scalogram. Of course, there should be no evidence of the frequency γ at scales $s = \gamma/sk$; this accounts for the gaps seen in the scales between the horizontal bands associated with $s = \gamma/(1 + 2k)$.

Conversely, for a given fixed scale, s, one can identify fundamental frequencies that are associated with that scale. These are $\gamma = (1 + 2k)/s$, $k = 1, 2, 3, \ldots$; however, once again the amplitude associated with a given frequency is seen to be a decreasing function of k. This property can be considered a form of aliasing of frequencies within a given scale. In general, the striation patterns in the scalogram span across several scales. Estimates of the relevant frequencies can be made by averaging across scale (*i.e.* vertically) and then using the above characterization of amplitude variation.

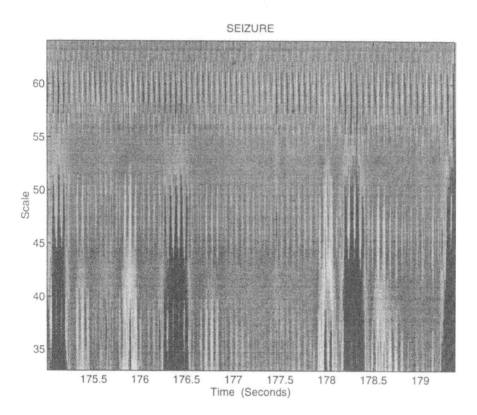

Figure 16.

These ideas can be extended to multiple frequencies, such as when the signal f is of the form

$$f(t) = A_1 \sin(2\pi\gamma_1 t) + A_2 \sin(2\pi\gamma_2 t),$$

and γ_1 and γ_2 are unknown frequencies. Thus, if $s(\gamma_1 + \gamma_2)$ is an even integer, then nonstriations in the scalogram give estimates of $\gamma_1 + \gamma_2$. Similar analysis in striational regions give estimates of γ_1 and γ_2, as well as spectral information compatible with the seizure chirp patterns.

5.3 Edge detection and Gaussian wavelet frames

In computer vision, a standard method of feature extraction is to "pass a Laplacian Δ" over a planar image f in $\mathbb{R}^2$ to detect edges, homogeneous areas, and orientation, *e.g.* [18]. Digitally, we compute

$$\Delta_d f \equiv \Delta_d * f,$$

where Δ_d is the matrix

0	1	0
1	−4	1
0	1	0

Consistent with the above 3×3 matrix "mask," we note that if $g(x) = \frac{1}{\pi^{d/2}} e^{-|x|^2}$ on $\mathbb{R}^d$ then

$$\int \Delta g(x)\, dx = 0.$$

It is easy to see that $\{g''(x-n) : n \in \mathbf{Z}\}$ is not orthonormal on $\mathbb{R}$; thus, Δg will not lead to orthonormal wavelet bases, but will, in fact, yield wavelet frames.

Next, consider the *heat equation* initial value problem:

$$\Delta u(x,t) = \partial_t u(x,t), \qquad (x,t) \in \mathbb{R}^d \times (0,\infty),$$

$$u(x,0) = f(x),$$

where f is the given initial data. We know that

$$\Delta g_{\sqrt{1/4s}}(x) = \partial_s g_{\sqrt{1/4s}}(x). \tag{5.3.1}$$

Because of Equation (5.3.1), it is natural to approximate $\Delta g_{\sqrt{1/4s}}$ by an approximation of $\partial_s g_{\sqrt{1/4s}}$. Using the heat equation and assuming natural hypotheses on the signal f, we obtain the following result that is a wavelet-based generalization of the Laplacian operator: if f is defined on $\mathbb{R}^d$, then

$$\lim_{s \to 0} \partial_s W_{\Delta g} f(x, \sqrt{1/4s}) = 4\Delta f. \tag{5.3.2}$$

5.4 Spatio-temporal Laplacian

The spatio-temporal Laplacian is defined as follows. Let $f(x_1, x_2, t)$ be a time-varying planar image. The *spatio-temporal Laplacian (STL)* of f is

$$STL(f)(x_1, x_2, t) = \Delta f(x_1, x_2, t)\partial_t^2 f(x_1, x_2, t),$$

where Δ is the 2-D Laplacian on the (x_1, x_2)-plane. The *STL* has found applications in the analysis of EEG and ECoG data, *e.g.* [1].

Because of Equation (5.3.2), we define the *generalized spatio-temporal Laplacian (GSTL)* of f as

$$GSTL(f)(x_1, x_2, t) = \frac{1}{4}\partial_s W_{\Delta g}f(x_1, x_2, t, \sqrt{1/4s})\partial_t^2 f(x_1, x_2, t), \quad (5.4.1)$$

where g and Δ are defined on the (x_1, x_2)-plane, and $W_{\Delta g}f$ is calculated for each fixed t, *i.e.* for a fixed t we write $f(x_1, x_2, t)$ as $f(x_1, x_2)$ so that $W_{\Delta g}f(x_1, x_2, t, \sqrt{1/4s})$ is really $W_{\Delta g}f(x_1, x_2, \sqrt{1/4s})$.

Note that Equation (5.4.1) is s-dependent, and, because of (5.3.2), *GSTL* is *STL* in the limiting case $s = 0$. Thus, with notation as above, for large m where $\lim_{m\to\infty} s_m = 0$, we have

$$\Delta f(x_1, x_2, t) \approx \frac{1}{4}\left\{\frac{W_{\Delta g}f(x_1, x_2, t, \sqrt{1/4s_m}) - W_{\Delta g}f(x_1, x_2, t, \sqrt{1/4s_{m+1}})}{s_m - s_{m+1}}\right\}$$

for each fixed t. For actual implementations, $GSTL(f)$ has the form

$$GSTL(f)(x_1, x_2, t, n, m) =$$
$$\frac{1}{4}\left\{\frac{W_{\Delta g}f(x_1, x_2, t_n, \sqrt{1/4s_m}) - W_{\Delta g}f(x_1, x_2, t_n, \sqrt{1/4s_{m+1}})}{s_m - s_{m+1}}\right\}$$
$$\times\frac{1}{(t_{n+1} - t_n)^2}\left\{f(x_1, x_2, t_{n+1}, \sqrt{1/4s_m}) - 2f(x_1, x_2, t_n, \sqrt{1/4s_m})\right.$$
$$\left. + f(x_1, x_2, t_{n-1}, \sqrt{1/4s_m})\right\}.$$

§6 Mathematical background for the FrameTool

6.1 Double window theorem

Our formula (2.3.1) and the irregular sampling version of (2.3.1) was extended by William Heller in the following way. PW_Ω denotes the Paley-Wiener class of bandwidth $[-\Omega, \Omega]$, supp denotes support, and $L^2(X)$ is the space of finite energy functions on X.

Let $0 < \Omega_1 < \Omega_2 < \Omega_3$ and $\Omega_2 + \Omega_3 \leq b < 2\Omega_3$, and suppose that the signal f is in PW_{Ω_1}. Let $\{-t_n\}_{n\in\mathbb{Z}}$ be a sampling sequence which generates an exponential frame, $\{e_{-t_n}\}_{n\in\mathbf{Z}}$ for $L^2[-\Omega_3, \Omega_3]$. We choose bounded windows $\hat{g}, \hat{h}$ with the following properties:

$$\hat{h} = 1 \text{ on } [-\Omega_1, \Omega_1], \qquad \hat{h} > 0 \text{ on } (-\Omega_2, \Omega_2), \qquad \text{supp } \hat{h} \subseteq [-\Omega_2, \Omega_2]$$

$$\hat{g} = 1 \text{ on } [-\Omega_2, \Omega_2], \qquad \hat{g} > 0 \text{ on } (-\Omega_3, \Omega_3), \qquad \text{supp } \hat{g} \subseteq [-\Omega_3, \Omega_3].$$

Then $\{\tau_{mb} e_{-t_n} \hat{g}\}_{n\in\mathbf{Z}}$ is a frame for $L^2(\mathbb{R})$ with frame operator S, and for each $f \in PW_{\Omega_1}$,

$$f = \sum_{n=-\infty}^{\infty} \left\langle \hat{f}, S^{-1}(e_{-t_n}\hat{g}) \right\rangle \tau_{t_n} h \qquad (6.1.1)$$

converges in $L^2(\mathbb{R})$.

Smooth filters $\hat{g}$ allow for fast coefficient computation, and small support of $\hat{h}$ allows for economy in reconstruction bandwidth.

6.2 Regular sampling

If $\{t_n\} = \{nT\}$ for some $0 < T \le \frac{1}{2\Omega_3}$, then the Double Window Theorem reduces to the Shannon Sampling Theorem with sampling function h. In this case, truncation error is elementary to estimate.

6.3 Computable matrices for irregular sampling

We can generate matrices for the computation of irregular sampling coefficients, and store them off-line.

Beginning with the Double Window Theorem,

$$\hat{f}(\gamma) = \sum_n c_n e_{-t_n} \hat{h}(\gamma)$$

where

$$c_n = \left\langle \hat{f}, S^{-1}(e_{-t_n}\hat{g}) \right\rangle,$$

we compute

$$c_n = \left\langle \sum_m c_m e_{t_m} \hat{h}, S^{-1}(e_{-t_n}\hat{g}) \right\rangle$$

$$= \sum_m c_m \left\langle e_{t_m} \hat{h}, S^{-1}(e_{-t_n}\hat{g}) \right\rangle = \sum_m c_m A_{m,n}.$$

In other words, $c = cA$ for all coefficient vectors $c = \{c_n\}$; and, hence, A is *idempotent on the space of coefficient vectors*. The entry $A_{m,n}$ may also be viewed as the nth coefficient for the representation of $e_{t_m}\hat{h}$ in terms of the frame $\{\tau_{jb} e_{t_k} \hat{g}\}_{j,k}$.

When the sampling sequence is regular, so is the matrix A. Furthermore, it has a band-like structure with most entries of significant magnitude being in one of three bands: one along the main diagonal, and smaller parallel bands in the opposing corners.

6.4 Error analysis

The effectiveness of the Double Window Theorem, and its implementation in the FrameTool, depends on a procedure involving parameter choice and algorithm design.

For a given error margin ϵ, parameters N_1, N_2, and M_m must be chosen so that

$$\left\| f - \sum_{N_1 \leq n \leq N_2} \left\langle \hat{f}, S_{M_n}^{-1}(e_{-t_n}\hat{g}) \right\rangle \tau_{t_n} h \right\| < \epsilon, \tag{6.4.1}$$

where $\| \cdots \|$ is the finite energy L^2-norm on $\mathbb{R}$ and $S_{M_n}^{-1}$ is the Neumann partial sum of S^{-1} for a given t_n. By adding and subtracting the approximation of the double window frame decomposition over the range $[N_1, N_2]$, the goal is to construct the parameters so that

$$I_1 = \left\| \sum_{n < N_1 < N_2 < n} \left\langle \hat{f}, S^{-1}(e_{-t_n}\hat{g}) \right\rangle \tau_{t_n} h \right\| < \frac{\epsilon}{2}$$

and

$$I_2 = \left\| \sum_{N_1 \leq n \leq N_2} \left(\left\langle \hat{f}, S^{-1}(e_{-t_n}\hat{g}) \right\rangle - \left\langle \hat{f}, S_{M_n}^{-1}(e_{-t_n}\hat{g}) \right\rangle \right) \tau_{t_n} h \right\| < \frac{\epsilon}{2}.$$

Clearly,

$$I_1 \leq \left(\int_{-\Omega_2}^{\Omega_2} | \sum_{n < N_1 < N_2 < n} \left\langle \hat{f}, S^{-1}(e_{-t_n}\hat{g}) \right\rangle e^{-2\pi i t_n \gamma} |^2 \, d\gamma \right)^{\frac{1}{2}}, \tag{6.4.2}$$

and invoking Hölder's Inequality at this point would lose too much information. Note that $F = (S^{-1}\hat{f})\bar{\hat{g}} \mathbf{1}_{[-\Omega_2, \Omega_2]} \in L^2[-\Omega_3, \Omega_3]$; and if S_3 is the frame operator for $\{e_{-t_n}\}$, then $S_3(F)(\gamma) = \sum_n F(t_n) e^{-2\pi i t_n \gamma} \in L^2[-\Omega_3, \Omega_3]$. Combining this with (6.4.2), we see that N_1, N_2 can be chosen so that $I_1 < \frac{\epsilon}{3}$. The FrameTool involves a constructive mechanism for this choice.

Let A, B be frame bounds for the frame $\{\tau_{mb} e_{-t_n}\hat{g}\}$ and frame operator S of the Double Window Theorem. Best possible bounds can be estimated, e.g. [3]. If $\rho = \frac{2}{(A+B)}$ then the operator norm $\|I - \rho S\| < 1$, where I is the identity operator. Setting $R_M = \sum_{m=0}^{M}(I - \rho S)^m$, we see that

$$\lim_{M \to \infty} \|R_M(e_{-t_n}\hat{g}) - R(e_{-t_n}\hat{g})\| = \lim_{M \to \infty} r(M, n) = 0 \tag{6.4.3}$$

uniformly in n. Also for $M = M_n$,

$$I_2 \le \|h\| \|f\| (N_2 - N_1) \sup_{N_1 \le n \le N_2} r(M, n). \qquad (6.4.4)$$

Combining (6.4.3) and (6.4.4) we can choose M independent of n so that $I_2 < \frac{\epsilon}{2}$. M can be constructed off-line and used in real-time reconstructions. Finally, these calculations give (6.4.1).

6.5 Support expansion

We shall close by making an observation about *support expansion*, which we are in the process of quantifying because of its potential usefulness in coefficient computation for irregular sampling reconstructions.

Using the notation of the Double Window Theorem and the frame operator S_3, we ask under what circumstances supp $S(\hat{f}) \subseteq [-\Omega_1, \Omega_1]$ for $f \in PW_{\Omega_1}$; and, in the case inclusion does not occur, how fast do compositions $S^n \hat{f}$ expand? Precise estimates in answering this latter question are closely related to the optimal algorithm efficiency of irregular sampling reconstruction for a given sampling sequence $\{t_n\}$. Further, coefficient computation in terms of S^{-1} can be effected by sums of compositions of S.

The first question concerning the inclusion supp $S(\hat{f}) \subseteq [-\Omega_1, \Omega_1]$ can be rephrased as follows. When does $\left\langle S_3 \hat{f}, \phi \right\rangle = 0$ for all L^2 functions ϕ for which supp $\phi \subseteq [-\Omega_3, -\Omega_1] \cup [\Omega_1, \Omega_3]$? The following calculation shows that the support of $S(\hat{f})$ does not necessarily remain contained in $[-\Omega_1, \Omega_1]$:

$$
\begin{aligned}
\left\langle S_3 \hat{f}, \varphi \right\rangle &= \int \left(\sum_n \left\langle \hat{f}, e_{t_n} \right\rangle e_{t_n}(\gamma) \right) \overline{\varphi(\gamma)} \, d\gamma \\
&= \int \left(\sum_n \left(\int \hat{f}(\lambda) \overline{e_{t_n}(\lambda)} \, d\lambda \right) e_{t_n}(\gamma) \right) \overline{\varphi(\gamma)} \, d\gamma \\
&= \int \overline{\varphi(\gamma)} \int \hat{f}(\lambda) \left(\sum_n e^{2\pi i t_n (\gamma - \lambda)} \right) d\lambda \, d\gamma.
\end{aligned}
$$

We note that $\sum_n e^{2\pi i t_n \gamma} = \delta$ distributionally when $t_n = n$. In this case, $\left\langle S_3 \hat{f}, \varphi \right\rangle = 0$. Since we are also dealing with $K(\gamma - \lambda) = \sum_n e^{2\pi i t_n (\gamma - \lambda)}$ where $\{t_n\}$ is not regular, we may expect to obtain kernels K for which $\hat{f} * K$ spreads beyond $[-\Omega_1, \Omega_1]$. This represents ongoing work with Melissa Harrison.

References

[1] Barreto, A. J. Principe, and S. Reid, STL: A spatio-temporal characterization of focal interictal events, *Brain Topography* **5** (1993), 215–228.

[2] Benedetto, J. J., Irregular sampling and frames, in *Wavelets: a Tutorial in Theory and Applications*, C. K. Chui (ed.), Academic Press, New York, 1992, volume 2, 445–507.

[3] Benedetto, J. J., Frame decompositions, sampling, and uncertainty principle inequalities, in *Wavelets: Mathematics and Applications*, J. J. Benedetto and M. Frazier (eds.), CRC Press, Boca Raton, FL, 1994, 247–304.

[4] Benedetto, J. J. and D. Colella, Wavelet analysis of spectrogram seizure chirps, in *Proceedings of the SPIE*, San Diego, 1995.

[5] Benedetto, J. J., M. L. Harrison, and W. Heller, A double window frame decomposition and irregular sampling, 1996, preprint.

[6] Benedetto, J. J. and W. Heller, Irregular sampling and the theory of frames, *Note di Matematica, X, Suppl. n.* **1** (1990), 103–125.

[7] Benedetto, J. J. and S. Li, Multiresolution analysis frames with applications, in *Proc. ICASSP-93*, 1993.

[8] Benedetto, J. J. and S. Li, Narrow band frame multiresolution analysis with perfect reconstruction, in *Proceedings of IEEE-SP International Symposium on Time Frequency and Time Scale Analysis*, Philadelphia, 1994.

[9] Benedetto, J. J. and S. Li, Subband coding and noise reduction in multiresolution analysis frames, in *Proceedings of the SPIE Wavelet Applications in Signal and Image Processing*, San Diego, 199, volume 2303, pp. 154–1654.

[10] Benedetto, J. J. and S. Li, The theory of multiresolution analysis frames and applications to filter banks, 1996, preprint.

[11] Benedetto, J. J. and A. Teolis, A wavelet auditory model and data compression, *Appl. Comp. Harmonic Analysis* **1** (1993), 3–28.

[12] Benedetto, J. J. and A. Teolis, U.S. Patent Number 5, pages 388 and 182, 1995.

[13] Daubechies, I., *Ten Lectures on Wavelets*, CBMS-NSF Regional Conference Series in Appl. Math. #61, SIAM, Philadelphia, 1992.

[14] Daubechies, I., A. Grossmann, and Y. Meyer, Painless nonorthogonal expansions, *J. Math. Physics* **27** (1986), 1271–1283.

[15] Donoho, D.L., De-noising by soft-thresholding, *IEEE Trans. Information Theory* **41** (1995), 613– 627.

[16] Duffin, R.J. and A. C. Schaeffer, A class of nonharmonic Fourier series, *Trans. Amer. Math. Soc.* **72** (1952), 341–366.

[17] Harrison, M. L., Ph.D. Thesis, University of Maryland, College Park, MD, 1996.

[18] Hummel, R. and R. Moniot, Reconstructions from zero crossings in scale space, *IEEE Trans. Acoustics, Speech, Signal Processing* **37** (1989), 2111–2130.

[19] Janssen, A. J. E. M., Duality and biorthogonality for Weyl-Heisenberg frames, *J. Fourier Anal. Appl.* **1** (1995), 403–436.

[20] Li, S., General frame decompositions, pseudo-duals and its application to Weyl-Heisenberg frames, *Numer. Functional Anal. Optimization*, 1995 to appear.

John J. Benedetto
Department of Mathematics
University of Maryland
College Park, Maryland 29742
jjb@math.umd.edu

Matching Pursuit of Images

François Bergeaud and Stéphane Mallat

Abstract. A crucial problem in image analysis is to construct efficient low-level representations of an image, providing precise characterization of its features, such as edges and texture components.

An image usually contains very different types of features, which have been successfully modeled by the very redundant family of two-dimensional Gabor-oriented wavelets, describing the local properties of the image: localization, scale, preferred orientation, amplitude, and phase of the discontinuity.

However, this model generates representations of very large size. Instead of decomposing a given image over this whole set of Gabor functions, we use an adaptive algorithm (called Matching Pursuit) to select the Gabor elements which approximate at best the image, corresponding to the main features of the image.

This produces compact representation in terms of few features that reveal the local image properties. Results proved that the elements are precisely localized on the edges of the images, and give a local decomposition as linear combinations of "textons" in the textured regions.

We introduce a fast algorithm to compute the Matching Pursuit decomposition for images with a complexity of $\mathcal{O}(N \log^2 N)$ per iteration for an image of N^2 pixels.

§1 Introduction

The complexity of image structures including different types of textures and edges requires flexible image representations. Although an image is entirely characterized by its decomposition in a basis, any such basis is not rich enough to represent efficiently all potentially interesting low-level structures. Some image components are diffused across many bases elements and are then difficult to analyze from the basis representation. This is like trying to express oneself in a language including a small dictionary. Unavailable words must be replaced by long awkward sentences. To provide explicit information on important local properties, the image is represented as a sum of waveforms selected from an extremely redundant dictionary of oriented Gabor functions. As opposed to previous approaches, we do not

Signal and Image Representation in Combined Spaces
Y. Y. Zeevi and R. R. Coifman (Eds.), pp. 285–300.
ISBN 0-12-777830-6

decompose the image over the whole dictionary, but like in a sentence formation, we select the most appropriate Gabor waveforms to represent the image. Instead of increasing the representation by a large factor as in typical multiscale Gabor representations [14], the adaptive choice of dictionary vectors defines a compact representation that takes advantage of the flexibility offered by the dictionary redundancy.

There is an infinite number of ways to decompose an image over a redundant dictionary of waveforms. The selection of appropriate waveforms to construct the image representation is obtained by constructing efficient image approximations from few dictionary vectors. The optimization of the approximation is not intended for data compression but as a criteria for feature selection. If most of the image is recovered as a sum of few dictionary vectors, these vectors must closely match the local image properties. One can, however, prove that finding optimal approximations in redundant dictionaries is an NP complete problem. The redundancy opens a combinatorial explosion. This explosion is avoided by the matching pursuit algorithm that uses a non-optimal greedy strategy to select each dictionary element. For a dictionary of Gabor functions, the greedy optimization of the image approximation leads to an efficient image representation where each Gabor waveform reflects the orientation, scale, and phase of local image variations. For textures, the selected Gabor elements can be interpreted as textons, where along the edges, the multiscale properties of these Gabor elements reflect the edge properties. When the image is translated or rotated, the selected Gabor elements are translated and rotated. A fast implementation of this algorithm and numerical examples are presented.

§2 The two-dimensional Gabor wavelet dictionary

Image decompositions in families of Gabor functions characterize the local scale, orientation, and phase of the image variations. Gabor functions are constructed from a window $b(x, y)$, modulated by sinusoidal waves of fixed frequency ω_0 that propagate along different direction θ with different phases ϕ

$$b_{\theta,\phi}(x, y) = b(x, y) \cos(\omega_0(x \cos \theta + y \sin \theta) + \phi). \tag{2.1}$$

Each of these modulated windows can be interpreted as wavelets having different orientation selectivities. Figure 1 shows the surface of such a two-dimensional Gabor wavelet for $\phi = \theta = 0$. The window $b(x, y)$ is not chosen to be a Gaussian but is a compactly supported box spline that is adjusted so that the average of $b_{\theta,\phi}(x, y)$ is zero for all orientations and phases. The phase modifies the profile of the window oscillations. When $\phi = 0$, the oscillations are symmetrical and the window can be interpreted as the second-order derivative of a two-dimensional window, along the direction θ. When $\phi = \pi/2$, the oscillations are antisymmetrical and the corresponding wavelet rather behaves as a first-order derivative along the direction θ. Figure 2 shows the cross-section of an element of the dictionary for different values of the phase ($\phi = -\frac{\pi}{4}$, $\phi = 0$, $\phi = \frac{\pi}{4}$, $\phi = \frac{\pi}{2}$).

These oriented wavelets are then scaled by s and translated to define a whole family of Gabor wavelets $\{g_\gamma\}_{\gamma \in \Gamma}$ with:

$$g_\gamma(x, y) = \frac{1}{s} b_{\theta, \phi} \left(\frac{x - u}{s}, \frac{y - v}{s} \right) \tag{2.2}$$

where the multi-index parameter $\gamma = (\theta, \phi, s, u, v)$ carries the orientation, phase, scale, and position of the corresponding Gabor function.

The Gabor transform of an image $f(x, y)$ is defined by the inner product

$$Gf(\gamma) = \; < f, g_\gamma > = \iint_{\mathbf{R}^2} f(x, y) g_\gamma(x, y) \, dx \, dy. \tag{2.3}$$

The orientation parameter θ allows us to match the different orientation of image structures. The scaling factor s allows us to zoom into singularities at fine scales but also to recover large-scale image variations. The phase is a convenient parameter to modify the number of vanishing moments of the wavelet. When $\phi = 0$, the Gabor transform provides a second-order partial derivative along θ of the image that is smoothed by a dilated window. When $\phi = \pi/2$, the Gabor transform can be interpreted as a first-order partial derivative, and thus responds particularly to discontinuities and edges.

The Fourier transform of a Gabor function is a waveform whose energy is well concentrated in the Fourier plane. In numerical computations, the scale is restricted to powers of two $\{2^j\}_{j \in \mathbf{Z}}$ and the angles are discretized. The Gabor dictionary used in this paper includes eight orientations (see Figure 3). To define a complete representation, we guarantee that the whole Fourier plane is covered by dilations of the eight elementary Gabor wavelets.

Besides engineering and mathematical motivations, families of Gabor waveforms have been used as models for the receptive fields of simple cells. The representation of the visual information is described as moving from the retina to the primary visual cortex: at the retina level, the receptive fields are homogeneous and provide a broad bandwidth for the visual information received from the photoreceptors. In the cortex, the receptive fields are narrow-band and direction-oriented. The cortex has multiple representations, generated by distinct neural populations.

The properties of simple cells of the V1 area of the cortex suggests some requirements for the image representation: they perform a linear spatial summation of their inputs. They are local in space and frequency

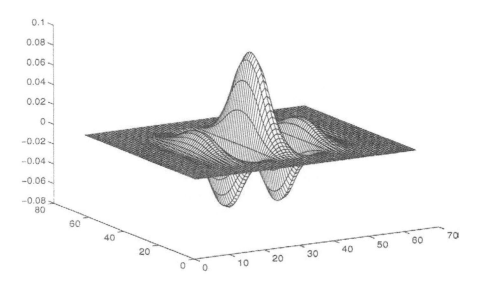

Figure 1. The surface of the two-dimensional Gabor function.

and cover the entire spectral visual field. They are divided among several types, each of which being selective to the orientation (with a maximum range of 45°, which suggests a minimum of four orientations), and having a limited frequency bandpass (1.5 octave), which requires self similarity among the different V1 cells class. The receptive fields profile of simple cells, which appears to be the descriptors of the features, have been shown to be strikingly similar to the two-dimensional Gabor functions [13].

The decomposition of images in a Gabor dictionary defines a very redundant representation. An image of 512 by 512 pixels, 8 orientations, 6 octaves, and a two-phase ($\phi = 0$ and $\phi = \frac{\pi}{2}$) representation, would correspond to 96 images of 512 by 512 pixels. These images could be subsampled but we then lose the translation invariance of the representation. Instead of decomposing the image over the whole dictionary, we select specific Gabor waveforms that provide an efficient image approximation.

§3   Matching pursuit

We consider the general problem of decomposing a signal f over a dictionary of unit vectors $\{g_\gamma\}_{\gamma \in \Gamma}$ whose linear combinations are dense in the signal space $\mathcal{H}$. The smallest possible dictionary is a basis of $\mathcal{H}$; general dictionaries are redundant families of vectors. When the dictionary

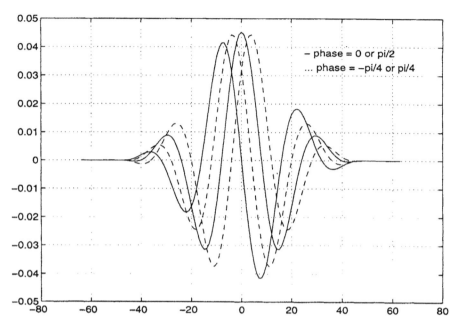

Figure 2. Cross-section of a two-dimensional Gabor function for $\phi = -\frac{\pi}{4}$, 0, $\frac{\pi}{4}$, $\frac{\pi}{2}$.

is redundant, unlike the case of a basis, we have some degree of freedom in choosing a signal's particular representation. This freedom allows us to choose few dictionary vectors, whose linear combinations approximate efficiently the signal. The chosen vectors highlight the predominant signal features. For any fixed approximation error ϵ, when the dictionary is redundant, we can show that finding the minimum number of dictionary elements that approximates the image with an error smaller than ϵ is an NP hard problem.

Because of the difficulty of finding optimal solutions, we use a greedy matching pursuit algorithm that has previously been tested on a one-dimensional signal. The matching pursuit [11] uses a greedy strategy that computes a good suboptimal approximation. It successively approximates a signal f with orthogonal projections onto dictionary elements. The first step is to approximate f by projecting it on a vector $g_{\gamma_0} \in \mathcal{D}$:

$$f = <f, g_{\gamma_0}> g_{\gamma_0} + Rf, \tag{3.1}$$

Since the residue Rf is orthogonal to g_{γ_0},

$$\|f\|^2 = |<f, g_{\gamma_0}>|^2 + \|Rf\|^2. \tag{3.2}$$

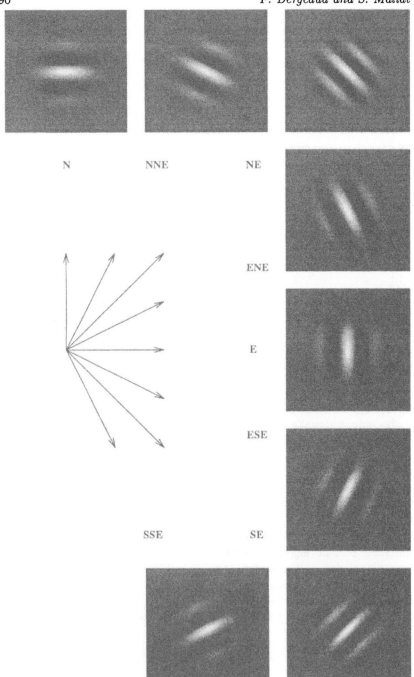

Figure 3. Discretization of the angle θ.

We minimize $\|Rf\|$ by choosing g_{γ_0}, which maximizes $|<f, g_\gamma>|$. We choose g_{γ_0} such that

$$|<f, g_{\gamma_0}>| = \sup_{\gamma \in \Gamma} |<f, g_\gamma>|. \tag{3.3}$$

The pursuit iterates this procedure by sub-decomposing the residue. Let $R^0 f = f$. Suppose that we have already computed the residue $R^k f$.

We choose $g_{\gamma_k} \in \mathcal{D}$ such that:

$$|<R^k f, g_{\gamma_k}>| = \sup_{\gamma \in \Gamma} |<R^k f, g_\gamma>|, \tag{3.4}$$

and project $R^k f$ on g_{γ_k}

$$R^{k+1} f = R^k f - <R^n k f, g_{\gamma_k}> g_{\gamma_k}, \tag{3.5}$$

which defines the residue at the order $k + 1$. The orthogonality of $R^{k+1} f$ and g_{γ_k} implies

$$\|R^{k+1} f\|^2 = \|R^k f\|^2 - |<R^k f, g_{\gamma_k}>|^2. \tag{3.6}$$

By summing (3.5) for k between 0 and $n - 1$, we obtain

$$f = \sum_{k=0}^{n-1} <R^k f, g_{\gamma_k}> g_{\gamma_k} + R^n f. \tag{3.7}$$

Similarly, summing (3.6) for k between 0 and $n - 1$ yields

$$\|f\|^2 = \sum_{n=0}^{n-1} |<R^k f, g_{\gamma_k}>|^2 + \|R^n f\|^2. \tag{3.8}$$

The residue $R^n f$ is the approximation error of f after choosing n vectors in the dictionary, and the energy of this error is given by (3.8).

In infinite dimensional spaces, the convergence of the error to zero is shown [11] to be a consequence of a theorem proved by Jones [8]:

$$\lim_{m \to +\infty} \|R^m f\| = 0. \tag{3.9}$$

Hence,

$$f = \sum_{n=0}^{+\infty} <R^n f, g_{\gamma_n}> g_{\gamma_n}, \tag{3.10}$$

and we obtain an energy conservation

$$\|f\|^2 = \sum_{n=0}^{+\infty} |<R^n f, g_{\gamma_n}>|^2. \tag{3.11}$$

In finite-dimensional signal spaces, the convergence is proved to be exponential [11].

§4 Fast calculation

Despite the apparent brute-force strategy of a matching pursuit, this algorithm can be implemented in an efficient way that requires on average $\mathcal{O}(N \log^2 N)$ operations per iteration in a Gabor dictionary, for an image of N^2 pixels.

The fast algorithm is based on three major points: as the search for the maximal coefficient (scalar product of the residue with a dictionary vector) is very time-consuming, we first make all the computations on a much smaller dictionary $\mathcal{D}' \subset \mathcal{D}$. It is constructed such that the maximal coefficient in $\mathcal{D}'$ gives a good estimation of the maximal coefficient in the whole dictionary. We then use a local Newton search of complexity $\mathcal{O}(1)$ to find the best local coefficient of $\mathcal{D}$ in the neighborhood of the element selected in $\mathcal{D}'$.

As we want to find the maximal coefficient in $\mathcal{D}'$ in a quick and efficient way, we sort and store the scalar products of the current residue with the elements of $\mathcal{D}'$ in a Hash table. The maximal coefficient is always at the top of the table, so that we don't need to search for it.

The coarse dictionary $\mathcal{D}' = (g_{\gamma'})_{\gamma' \in \Gamma'}$ is obtained by subsampling the position of the elements of the dictionary by a factor of 4 at each scale. It is nearly a tight frame of $\mathcal{H}$ and thus complete, because we only remove the excess of redundancy from the initial dictionary. The size of such a coarse dictionary goes down to $\frac{64}{3} N^2$ (instead of $16 N^2 \log N$ for the whole dictionary $\mathcal{D}$).

Once we have an estimation of the best vector g_{γ_n} such that:

$$| < R^n f, g_{\gamma_n} > | = \sup_{\gamma \in \Gamma} | < R^n f, g_\gamma > | \qquad (4.1)$$

we compute the inner product of the new residue $R^{n+1}f$ with any $g_{\gamma'} \in \mathcal{D}'$, with a linear updating formula derived from (3.5)

$$< R^{n+1}f, g_{\gamma'} > = < R^n f, g_{\gamma'} > - < R^n f, g_{\gamma_n} > < g_{\gamma_n}, g_{\gamma'} > . \qquad (4.2)$$

This updating equation can be interpreted as an inhibition of $< R^n f, g_\gamma >$ by the cross correlation of g_{γ_n} and g_γ. The only coefficients $< R^n f, g_{\gamma'} >$ of the Hash table which are modified are such that $< g_{\gamma_n}, g_{\gamma'} >$ is non-zero. As we use compact-supported Gabor wavelets, this set is indeed very sparse, and the updating formula only applies to few coefficients ($\mathcal{O}(N \log^2 N)$ in the average), compared to the size of the table ($\mathcal{O}(N^2)$). We store the set of all non-null inner products $< g_\gamma, g_{\gamma'} >$. Since the Gabor wavelets used in these numerical experiments are separable, we only need to compute and store the set of inner products of one-dimensional Gabor functions, thus requiring an order of magnitude less memory.

§5   Results

The matching pursuit algorithm applied to a Gabor dictionary selects iteratively the Gabor waveforms, also called atoms, whose scales, phases, orientations, and positions best match the local image variations.

The distribution of Gabor functions at the scales 2^j for $1 \leq j \leq 4$ is shown in Figures 4 and 5. For the lady image (256×256), light atoms correspond to high amplitude inner products. As clearly shown by these images, the fine-scale atoms are distributed along the edges, and the texture regions and their orientation indicate the local orientation of the image transitions.

In order to display the edge information (localization, orientation, scale, amplitude), we adopt the following convention: each selected Gabor vector g_γ for $\gamma_n = (\theta, \phi, 2^j, u, v)$ is symbolized by an elongated Gaussian function of width proportional to the scale 2^j, centered at (u, v), of orientation θ. The mean gray level of each symbol is proportional to $|< R^n f, g_{\gamma_n} >|$.

Figure 6 displays the reconstruction of the Lena image with the 2500 and 5000 first Gabor waveforms selected by the algorithm. The image reconstruction is simply obtained with the truncated sum of the infinite series decomposition (3.10). With 2500 atoms, we already recover a good image quality and with 5000 atoms, the reconstructed image has no visual degradation. The latter image is constructed without using any compression method, and achieves a 1.5 bit/pixel ratio with a perfect visual quality. This shows clearly that, while specifying precisely the important information such as the local scale and orientation, this representation is very compact and can be used as the input to a high-level processing. Moreover, the degradation of the image quality is very small when using 2500 atoms instead of 5000 atoms, which suggests that the most relevant features of the image lie in a very small set of atoms.

Figure 7 illustrates the texture discrimination properties of the representation: a straw texture image is inserted into a paper texture image. The paper texture has no orientation specificity. On the other hand, the straw texture has horizontal and vertical structures. At fine scales, most structures are vertical because the horizontal variations are relatively smooth. At the intermediate scale 2^2 we clearly see the horizontal and vertical image structures. At the larger scale 2^3, the vertical structures dominate again because the vertical straws are of larger sizes.

The symbolic representations at scale 2^1, 2^2, 2^3, 2^4 show the distinct behavior of the two textures relative to the scale: although the paper texture shows a scale-invariant uniform distribution, the straw texture representation points out the vertical features of the straw at finer and larger scale, and the horizontal and vertical structures of the straw at medium scale. Moreover, at scale 2^1 we clearly distinguish the texture edge as the boundary between the two textures.

As noticed by Turner in [14], the time-frequency analysis performed using Gabor functions corresponds to a tiling of the four-dimensional space (two dimensions for spatial position, two dimensions for the frequency position) called "information hyperspace." Each element of the dictionary occupies a particular "volume" or cell in this space, and combines information in space and frequency related to the structural description of the texture: the response to the spatially large Gabor filters and small spatial frequency extent give precise information on the periodicity of the texture. On the other hand, the response to the spatially small filters and large spatial frequency extent distinguish texture elements and provide a mean to characterize the density of elements. The intermediate Gabor filters allow simultaneous measurements of textons and their distribution.

The fundamental idea, introduced in previous work by Julesz [9] and Beck [1], and stated by Malik and Perona in [10] is that "in preattentive vision, precise positional relationships between textons are not important; only densities matter." The discrimination between textures will consist in averaging over the response of the dictionary elements.

In the example of the straw texture inserted into a paper texture, the discrimination is achieved using the information concerning the textons, and mainly on the energy density relative to the orientation and the scale. Indeed, the energy distribution of the paper is nearly constant at a given scale for different orientations, and is concentrated in the vertical "cells" at low and high scales and in the vertical and horizontal "cells" at intermediate scales.

§6   Vision applications

We introduced here a method to construct a decomposition of images into its main features. We showed that this transform provides a precise and complete characterization of the edges and texture components in terms of localization, orientation, scale, and amplitude. By reconstructing high-visual quality images with very few atoms, we also showed that this representation is compact.

Another advantage of Matching Pursuit is the flexibility of the dictionary choice allowing us to explicitly introduce *a priori* knowledge on the features of object classes into the dictionary to solve specific vision problems.

First experiments on image denoising suggest the use of a higher level information on edges obtained by atom linking to discriminate image features from noise.

A promising aspect of this method is its aptitude to model some early vision mechanisms in the visual cortex, and especially those involved in

Figure 4. Symbolic representation of the Gabor vectors: each vector is symbolized by an elongated Gaussian of width proportional to the scale 2^j, oriented along the modulation direction, of gray level proportional to the amplitude $|< R^n f, g_{\gamma_n} >|$. Light atoms correspond to high amplitude. The left figure corresponds to the scale 2^1 and the right figure to the scale 2^2.

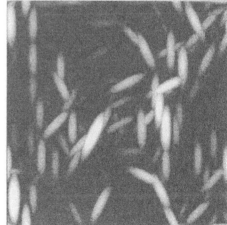

Figure 5. Symbolic representation of the Gabor vectors: the left and right figures corresponds respectively to the scales 2^3 and 2^4.

Figure 6. Reconstruction of the Lena image with 2500 atoms (left figure) and 5000 atoms (right figure). The reconstruction is obtained as a truncated sum of the decomposition.

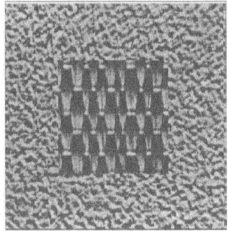

Figure 7. Original paper and straw texture mosaic (256 × 256, left image) and reconstructions with 5000 atoms (right image).

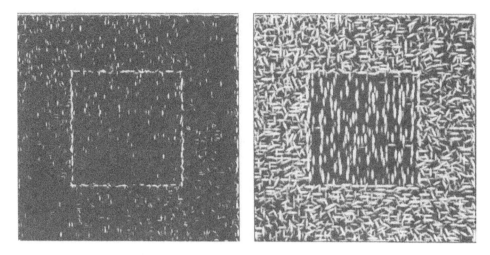

Figure 8. Symbolic representation of the Gabor vectors for the texture mosaic: scale 2^1 (left image) and 2^2 (right image).

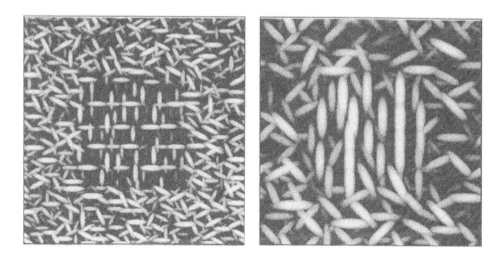

Figure 9. Symbolic representation of the Gabor vectors for the texture mosaic: scale 2^3 (left image) and 2^4 (right image).

texture discrimination. The Matching Pursuit with a Gabor dictionary acts as a linear filter followed by a nonlinearity (selection of the maximal coefficient) and some local inhibitions in the representation (update), which is very similar to the recent model of early vision mechanisms involved in texture discrimination as introduced by Malik and Perona [10].

As intrinsic steps of the Matching Pursuit, these two fundamental operations appear here in a very natural way as a consequence of the redundancy of the dictionary. First, the selection of the best matching element of the dictionary yields the amplitude of the vector, including its sign through the phase information. This corresponds to a full-wave rectification, but contrary to some other work in the same field, is combined with the phase extraction and thus keeps the essential information (we still can discriminate the two symmetrical textures, one composed of the repetition of a micro-pattern M, and the other of the repetition of the opposite micro-pattern -M). It is also biologically plausible since Pollen and Ronner [13] suggested the existence of cell pairs in the quadrature phase in the cortex. Second, the update of the representation at each step locally removes the coefficients of the Gabor functions that do not correspond to an optimal match with the selected image feature. This mechanism appears to be very similar to the interactions among the neurons in the primary visual cortex, as described in [10]. This model introduce the inhibition among the neurons as a simple consequence of the selection of the best-responding neuron (the dictionary element in our model): once a predominant neuron is selected, all the spurious responses in its neighborhood are removed, preventing the choice of neurons that are not optimally tuned to the visual stimulus. This local enhancement of the best-responding neuron is a variant of the winner-take-all mechanism previously introduced in the neural network literature and hypothesized by Barlow: "Our perceptions are caused by the activity of a rather small number of neurons selected from a very large population of predominantly silent cells." Moreover, the intrinsic inhibition law of the matching pursuit gives an explicit formula for the interactions among neurons as the inner product of the dictionary elements, which does not depend on any parameter.

The decomposition of textured images proved the efficient discrimination properties of the method and suggests the computation of a texture gradient by an appropriate pooling of the gradient responses for each scale and orientation. However, instead of using a gradient computed with a smoothed version of the matching pursuit representation (see [10]), which assumes an arbitrarily fixed smoothness parameter, we are studying some clustering techniques based on the uniformity of the atom density, the predominant orientation, and the amplitude. We then compute a gradient between these regions, without making any *a priori* assumption on their size.

Acknowledgments. This work was supported by the AFOSR grant F49620- 93-1-0102, ONR grant N00014-91-J-1967 and the Alfred Sloan Foundation.

References

[1] Beck, J., Textural segmentation, in *Representation in Perception*, J. Beck (ed.), Lawrence Erlbaum Associates, Hillsdale, NJ, pp. 285–317.

[2] Daugman, J., Uncertainty relation for resolution in space, spatial-frequency and oriented optimized by two-dimensional visual cortex filters, *J. Opt. Soc. Amer.* **2** (1985), 1160–1169.

[3] Daugman, J., Complete discrete 2D Gabor transform by neural networks for image analysis and compression, *IEEE Trans. Acoustic, Speech, and Signal Processing* **ASSP-36** (1988), 1169–1179.

[4] Davis, G., S. Mallat, and M. Avenaleda, Chaos in adaptive approximations, Technical Report, Computer Science, No. 325, April 1994.

[5] Gabor, D., Theory of communication, *J. Inst. Elect. Ing.* **93** (1946), 429–457.

[6] Huber, P. J., Projection pursuit, *The Annals of Statistics* **134**(2) (1985), 435–475.

[7] Jones, J. and L. Palmer, An evaluation of two-dimensional Gabor filters model of simple receptive fields in cat striate cortex, *J. Neurophisiol.* **58** (1987), 538–539.

[8] Jones, L. K., On a conjecture of Huber concerning the convergence of projection pursuit regression, *The Annals of Statistics* **15**(2) (1987), 880–882.

[9] Julesz, B. and J. R. Bergen, Textons, the fundamental elements in preattentive vision and perception of textures, *The Bell System Tech. J.* **62** (1983), 1619–1645.

[10] Malik, J. and P. Perona, Preattentive texture discrimination with early vision mechanisms, *J. Opt. Soc. Amer.* **7**(5) (1990), 923–932.

[11] Mallat, S. and Z. Zhang, Matching pursuit with time-frequency dictionaries, *IEEE Trans. Signal Processing* (1993).

[12] Marcelja, S., Mathematical description of the response of simple cortical cells, *J. Opt. Soc. Amer.* **70**(11) (1980), 1297–1300.

[13] Pollen, D. A. and S. F. Ronner, Visual cortical neurons as localized spatial filters, *IEEE Trans. Systems, Man, and Cybernetics* **SMC-13**(5) (1983), 907–936.

[14] Turner, M. R., Texture discrimination by Gabor functions, *Biol. Cybern.* **55** (1986), 71–82.

[15] Watson, A. B and A. J. Ahumada, Jr., A hexagonal orthogonal-oriented pyramid as a model of image representation in visual cortex, *IEEE Trans. Biomed. Eng.* **36**(1) (1989), 97–106.

François Bergeaud
Ecole Centrale Paris
Applied Mathematics Laboratory
Grande Voie des Vignes
F-92290 Châtenay-Malabry
France
francois@mas.ecp.fr

Stéphane Mallat
Ecole Polytechnique
CMAP
91128 Palaiseau Cedex
France
mallat@cmapx.polytechnique.fr

Overcomplete Expansions and Robustness

Zoran Cvetković and Martin Vetterli

Abstract. Manifestations of robustness of overcomplete expansions, acquired as a result of the redundancy, have been observed for two primary sources of degradation: white additive noise and quantization. These two cases require different treatments since quantization error has a certain structure which is obscured by statistical analysis. The central issue of this chapter is a deterministic analysis of quantization error of overcomplete expansions with a focus on Weyl-Heisenberg frames in $L^2(\mathbb{R})$. This analysis demonstrates that overcomplete expansions exhibit a higher degree of robustness to quantization error than to a white additive noise, and indicates how this additional robustness can be exploited for improvement of representation accuracy.

§1 Introduction

The development of the theory of signal expansions with respect to time-frequency or time-scale localized atoms revealed early on limitations of orthonormal expansions. Requiring orthogonality or linear independence of expansion vectors imposes considerable constraints which are often in conflict with application-related specifications. Perhaps the most striking example stems from Gabor's pioneering work in this field. The result known as the Balian-Low theorem asserts that Gabor analysis with orthonormal bases having good time-frequency resolution is not possible [1, 12, 17]. Design flexibility was one of the main reasons for the development of the theory beyond the scope of orthonormal or biorthonormal bases, with a focus on wavelet and Weyl-Heisenberg frames. Along with these developments it was observed that redundant expansions exhibit a higher degree of robustness.

The fact that overcomplete expansions are less sensitive to degradations than nonredundant expansions comes as no surprise considering that redundancy in engineering systems usually provides robustness. This principle is exploited in modern techniques for analog-to-digital (A/D) conversion, which use oversampling in order to compensate for coarse quantization, resulting in a highly accurate conversion overall. Another example, from communications, is channel coding, where signal dependent patterns of

Signal and Image Representation in Combined Spaces
Y. Y. Zeevi and R. R. Coifman (Eds.), pp. 301–338.

control bits are added to data bits in order to combat channel noise. Error correction capability, *i.e.* the robustness of the code, is proportional to redundancy, which is reflected in the number of control bits. Although the principle of redundancy-robustness trade-off seems close to one's intuition, it is not always simple to unravel underlying mechanisms and give a quantitative characterization.

Two relevant cases of degradation of expansion coefficients are additive noise and quantization. The goal of this chapter is to provide some insights into the effect of increased robustness of overcomplete expansions to these kinds of degradations and give a quantitative characterization in the case of quantization of Weyl-Heisenberg expansions in $L^2(\mathbb{R})$. Quantization error is commonly modeled as an additive white noise, and this approach gives satisfactory results when applied to orthogonal expansions. However, the two cases require different treatments if overcomplete expansions are considered, since then the deterministic nature of the quantization error is more pronounced. The treatment of the quantization error reduction property of frames in this chapter is based on the concept of *consistent reconstruction* and deterministic analysis, which have been pointed out first by Thao and one of us in the context of oversampled A/D conversion [14, 16, 15]. The main point of this approach is to demonstrate that traditional linear reconstruction from quantized coefficients is suboptimal in the case of overcomplete expansions, giving an error which is inversely proportional to the redundancy factor r, $\|e(t)\|^2 = O(1/r)$, and that information contained in the quantized coefficients allows for more accurate reconstruction, with an error $\|e(t)\|^2 = O(1/r^2)$. The higher accuracy can be attained using a consistent reconstruction strategy. This has been demonstrated either analytically or experimentally in several particular cases of overcomplete expansions [15, 9, 8, 2], and we conjecture that these results have a more general scope.

We first review relevant results on frames in Section 2. General considerations about the error reduction property of overcomplete expansions are the content of Section 3. Quantization error in Weyl-Heisenberg expansions is studied in Section 5 as a generalization of a deterministic analysis of oversampled A/D conversion, which is the subject of Section 4.

Notations

The convolution of two signals $f(t)$ and $g(t)$ will be denoted as $f(t) * g(t)$,

$$f(t) * g(t) = \int_{\infty}^{\infty} f(s)g(t-s)ds.$$

The Fourier transform of a signal $f(t)$, $\mathcal{F}\{f(t)\}$, will be written as $\hat{f}(\omega)$.

We say that a signal $f(t)$ is σ-bandlimited if

$$\hat{f}(\omega) = 0 \quad \text{for} \quad |\omega| > \sigma, \quad \text{and} \quad \|f\|^2 = \int_{-\infty}^{\infty} |f(t)|^2 dt < \infty.$$

Similarly, a signal $f(t)$ is said to be T-time limited if

$$f(t) = 0 \quad \text{for} \quad |t| > T, \quad \text{and} \quad \|f\|^2 = \int_{-\infty}^{\infty} |f(t)|^2 dt < \infty.$$

§2   Frames in a Hilbert space

A common signal processing framework is to consider signals as vectors in a Hilbert space $\mathcal{H}$, usually $L^2(\mathbb{R})$, $\ell^2(\mathbb{Z})$, or $\mathbb{R}^n$. Linear expansion refers to a representation of a signal as a linear combination of vectors of a family $\{\varphi_j\}_{j \in J}$ which is complete in $\mathcal{H}$. In addition to completeness, for applications in signal processing, it is essential that $\{\varphi_j\}_{j \in J}$ also has some stability properties. First, coefficients of the expansion should constitute a sequence of finite energy for any signal in the space. Moreover, signals in any given bounded set cannot have expansion coefficients of arbitrarily high energy. The second requirement is that the expansions should provide a unique description of signals in $\mathcal{H}$, and enable discrimination with a certain reasonable resolution. In other words, expansion coefficients corresponding to two different signals x and y should not be close in $\ell^2(J)$ if x and y are not close themselves. A family $\{\varphi_j\}_{j \in J}$ which satisfies these requirements is said to be a *frame* in $\mathcal{H}$.

The review of results on frames in this section is based on the material presented in [4, 3, 5, 18]. First, we give a rigorous definition of frames.

Definition 1. *A family of vectors $\{\varphi_j\}_{j \in J}$ in a Hilbert space $\mathcal{H}$ is called a frame if there exist $A > 0$ and $B < \infty$ such that for all f in $\mathcal{H}$,*

$$A\|f\|^2 \leq \sum_{j \in J} |\langle f, \varphi_j \rangle|^2 \leq B\|f\|^2. \tag{2.1}$$

A and B are called frame bounds.

For a frame $\{\varphi_j\}_{j \in J}$ there exists a dual frame, $\{\psi_j\}_{j \in J}$, such that any f in the space can be reconstructed from the inner products $\{\langle f, \varphi_j \rangle\}_{j \in J}$ as

$$f = \sum_{j \in J} \langle f, \varphi_j \rangle \psi_j. \tag{2.2}$$

The frames $\{\varphi_j\}_{j \in J}$ and $\{\psi_j\}_{j \in J}$ have interchangeable roles, so that any f can also be written as a superposition of φ_j vectors, as

$$f = \sum_{j \in J} \langle f, \psi_j \rangle \varphi_j. \tag{2.3}$$

Implications of this expansion formula and the frame condition in (2.1) are the stability requirements on frames.

The existence of a dual frame can be established using the frame operator. With the frame $\{\varphi_j\}_{j \in J}$ we can associate a linear operator F which maps f in $\mathcal{H}$ to the set of inner products with the frame vectors, $Ff = c$, where $c = \{c_j : c_j = \langle f, \varphi_j \rangle\}_{j \in J}$. This operator is called the *frame operator*. The frame condition (2.1) implies that F is a bounded invertible operator from $\mathcal{H}$ to $\ell^2(J)$. The Hilbert adjoint F^* of F is a linear operator from $\ell^2(J)$ to $\mathcal{H}$ and is given by

$$F^* c = \sum_{j \in J} c_j \varphi_j. \qquad (2.4)$$

The image of $\{\varphi_j\}_{j \in J}$ under $(F^*F)^{-1}$, $\{\tilde{\varphi}_j : \tilde{\varphi}_j = (F^*F)^{-1}\varphi_j\}_{j \in J}$, turns out to be a frame dual to $\{\varphi_j\}_{j \in J}$. So, we can write $\tilde{F}^*F = F^*\tilde{F} = I$, where $\tilde{F}$ is the frame operator associated to $\{\tilde{\varphi}_j\}_{j \in J}$, and I stands for the identity operator on $\mathcal{H}$. If the frame $\{\varphi_j\}_{j \in J}$ is overcomplete, that is φ_j are linearly dependent, then its dual is not unique. The dual $\{\tilde{\varphi}_j\}_{j \in J}$ has the distinctive property that for any $c = \{c_j\}_{j \in J}$ in the orthogonal complement of $\text{Ran}(F)$, $\sum_{j \in J} c_j \tilde{\varphi}_j = 0$. This is particularly important for reconstruction of signals from coefficients degraded by an additive noise. If a signal f is synthesized from noisy coefficients $\{\langle f, \varphi_j \rangle + n_j\}$ using $\{\tilde{\varphi}_j\}_{j \in J}$ as

$$f' = \sum_{j \in J} (\langle f, \varphi_j \rangle + n_j) \tilde{\varphi}_j, \qquad (2.5)$$

the noise component which is in the orthogonal complement of $\text{Ran}(F)$ is automatically reduced to zero. No other dual of $\{\varphi_j\}_{j \in J}$ implicitly performs this projection while effecting the reconstruction. Being so special, $\{\tilde{\varphi}_j\}_{j \in J}$ deserves a special name, and is said to be the *minimal dual* of $\{\varphi_j\}_{j \in J}$.[1]

A frame with equal frame bounds is called a *tight frame*. For a tight frame and any f in the space

$$\sum_{j \in J} |\langle f, \varphi_j \rangle|^2 = A\|f\|^2. \qquad (2.6)$$

The minimal dual of a tight frame is up to a multiplicative factor equal to the frame itself, $\tilde{\varphi}_j = A^{-1}\varphi_j$, so that any signal in $\mathcal{H}$ can be represented as

$$f = \frac{1}{A} \sum_{j \in J} \langle f, \varphi_j \rangle \varphi_j. \qquad (2.7)$$

---

[1] Duffin and Schaeffer used the term *conjugate frames* for a frame and its minimal dual [5].

This formula is reminiscent of orthogonal expansions. However, tight frames are in general not orthonormal bases, but overcomplete families in $\mathcal{H}$. If vectors of the tight frame are normalized to unit norm, $\|\varphi_j\| = 1$, the frame bound A gives a measure of redundancy of the frame. Tight frames that are not redundant are orthonormal bases.

A similar relationship exists between frames in general and Riesz bases. The removal of a vector from a frame leaves either a frame or an incomplete set. A frame which ceases to be a frame when any of its vectors is removed is said to be an *exact frame*. It turns out that the class exact frames is equivalent to the class of Riesz bases [18].

Frames which are designed for a particular application or arise from engineering practice usually exhibit certain structure. Two classes which are of concern here are Weyl-Heisenberg frames and frames of complex exponentials. They had a paramount impact on the development of the theory of time-frequency localized expansions and the theory of frames.

Weyl-Heisenberg frames are the tool for Short-Time Fourier analysis and have been also used in many areas of theoretical physics under the name of coherent states. They consist of vectors which are obtained by translating and modulating a single prototype window function,

$$\left\{ \varphi_{mn}(t) : \varphi_{mn}(t) = e^{jm\omega_0 t} \varphi(t - nt_0) \right\}_{m,n \in Z}. \tag{2.8}$$

Vectors $\varphi_{mn}(t)$ are distributed on a rectangular lattice in the time-frequency plane. The lattice cell size is given by the time and frequency steps, t_0 and ω_0, of the frame. All vectors have the same effective support, which is determined by the shape of the window. The window initially proposed by Gabor [6] was Gaussian, since it achieves the lower bound of the uncertainty in the joint time-frequency plane, facilitating signal description with the best possible joint resolution. Other window functions are admissible, too. In fact, the only restriction which the window function of a Weyl-Heisenberg frame with bounds A and B has to obey is

$$A \le \frac{2\pi}{\omega_0 t_0} \|\varphi\|^2 \le B. \tag{2.9}$$

The restriction on the time and frequency steps is

$$\omega_0 t_0 \le 2\pi. \tag{2.10}$$

The meaning of this condition is that if the Weyl-Heisenberg vectors are to cover the whole time-frequency plane, so that any signal in $L^2(\mathbf{R})$ can be represented in a numerically stable way as their superposition, then they have to be "sufficiently dense" in the space. The minimal density corresponds to $\omega_0 t_0 = 2\pi$, so that $2\pi/\omega_0 t_0$ represent redundancy of the frame. Note that in order to attain a good time-frequency resolution of Weyl-Heisenberg expansions, certain redundancy has to be allowed [4, 3].

The general theory of frames originated from studying the characterizability of bandlimited signals from a sequence of samples [5]. Given a space of σ-bandlimited signals, it is of interest to characterize sequences of sampling instants which uniquely describe signals in the space and moreover allow for a numerically stable reconstruction. Closely related to this problem are frames of complex exponentials in $L^2[-\sigma, \sigma]$.

A classical result of Fourier analysis, known as the sampling theorem in communications, asserts that regular sampling of a σ-bandlimited signal $f(t)$ at points $\{n\tau\}_{n\in\mathbf{Z}}$ introduces no loss of information as long as $\tau \leq \pi/\sigma$. Recall that $f(t)$ can be recovered from the samples as

$$f(t) = \frac{1}{\tau} \sum_n f(n\tau)\mathrm{sinc}_\sigma(t - n\tau), \qquad (2.11)$$

where

$$\mathrm{sinc}_\sigma(x) = \frac{\sigma}{\pi}\frac{\sin(\sigma x)}{\sigma x}.$$

The value which $f(t)$ attains at the sampling instant $n\tau$ represents the inner product

$$f(n\tau) = \langle f, \mathrm{sinc}_\sigma(t - n\tau)\rangle,$$

so that $f(t)$ can be written as

$$f(t) = \frac{1}{\tau} \sum_n \langle f, \varphi_{n,\tau}\rangle\varphi_{n,\tau}(t), \qquad (2.12)$$

where

$$\varphi_{n,\tau}(t) = \mathrm{sinc}_\sigma(t - n\tau).$$

Therefore, sampling above the Nyquist rate, $\tau < \pi/\sigma$, is equivalent to expanding a signal with respect to

$$\{\varphi_{n,\tau}(t)\}_{n\in\mathbf{Z}},$$

which is a tight frame for the space of σ-bandlimited signals. Alternatively, this sampling can be viewed as a linear expansion of the Fourier transform of $f(t)$ in terms of complex exponentials $\{e^{j\omega n\tau}\}$, which constitute a tight frame for $L^2[-\sigma, \sigma]$. This interpretation comes from the fact that the space of σ-bandlimited signals is isometrically isomorphic to $L^2[-\sigma, \sigma]$. In general, σ-bandlimited signals can be reconstructed in a numerically stable way from the samples at points λ_n, $n \in \mathbf{Z}$, if and only if $\{e^{j\lambda_n\omega}\}$ is a frame in $L^2[-\sigma, \sigma]$. A sequence of complex exponentials, $\{e^{j\lambda_n \omega}\}$, is said to be a frame in $L^2[-\sigma, \sigma]$ if there exist constants $A > 0$ and $B < \infty$ such that

$$A \int_{-\sigma}^{\sigma} |\hat{f}(\omega)|^2 d\omega \leq \frac{1}{2\pi} \sum_n \left| \int_{-\sigma}^{\sigma} \hat{f}(\omega) e^{j\lambda_n \omega} d\omega \right|^2 \leq B \int_{-\sigma}^{\sigma} |\hat{f}(\omega)|^2 d\omega, \quad (2.13)$$

for any $\hat{f}(\omega) \in L^2[-\sigma, \sigma]$. This can be rephrased as the condition that

$$A \|f(t)\|^2 \leq \sum_n |f(\lambda_n)|^2 \leq B \|f(t)\|^2 \qquad (2.14)$$

for every σ-bandlimited function $f(t)$. If this is satisfied we shall also say that $\{\lambda_n\}$ is a frame sequence for the space of σ-bandlimited signals. Conditions under which a sequence $\{e^{j\lambda_n \omega}\}$ is complete and moreover a frame in an $L^2[-\sigma, \sigma]$ space are the core issues of nonharmonic Fourier analysis. An excellent introductory treatment of this subject can be found in [18]. Some relevant results will be reviewed in the course of this chapter.

§3 General considerations on robustness of frames

3.1 Reduction of additive noise

The first explanation of the noise reduction property of overcomplete expansions in general was given by Daubechies [4] and is reviewed here. Consider a frame $\{\varphi_j\}_{j \in J}$ in a Hilbert space $\mathcal{H}$. Let $\{\tilde{\varphi}_j\}_{j \in J}$ be its minimal dual frame. Expansion coefficients of a signal f in $\mathcal{H}$ with respect to $\{\varphi_j\}_{j \in J}$ are given as inner products with corresponding vectors of the dual frame,

$$f = \sum_{j \in J} \langle f, \tilde{\varphi}_j \rangle \varphi_j. \qquad (3.1.1)$$

In other words, the expansion coefficients are the image of f under the frame operator $\tilde{F}$ which is associated with the dual frame. Recall that $\tilde{F}$ is an operator from $\mathcal{H}$ to $\ell^2(J)$. If $\{\varphi_j\}$ is a linearly independent set, then the range of $\tilde{F}$ is all of $\ell^2(J)$; that is, any point in $\ell^2(J)$ represents a set of the expansion coefficients of some g in $\mathcal{H}$. However, if $\{\varphi_j\}$ is overcomplete, then $\tilde{\varphi}_j$ are linearly dependent. This linear dependency translates to expansion coefficients $\langle f, \tilde{\varphi}_j \rangle$. Consequently, the range of $\tilde{F}$ is no longer all of $\ell^2(J)$, but a proper subspace. A family of expansion coefficients $\langle f, \tilde{\varphi}_j \rangle$ is usually not in this subspace after degradation with some additive noise. Starting with the adulterated coefficients, $\{\langle f, \tilde{\varphi}_j \rangle + n_j\}$, another set closer to the originals can be obtained as the orthogonal projection of $\{\langle f, \tilde{\varphi}_j \rangle + n_j\}$ onto the range of $\tilde{F}$ (see Figure 1). Recall that the reconstruction formula

$$f_{\text{rec}} = \sum_{j \in J} (\langle f, \tilde{\varphi}_j \rangle + n_j) \varphi_j \qquad (3.1.2)$$

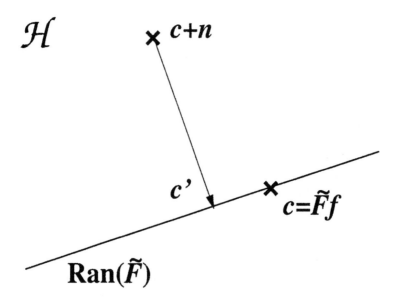

Figure 1. Noise reduction in frames using linear reconstruction. Linear reconstruction of f from noisy coefficients, $\tilde{F}f + \mathbf{n}$ gives a signal f_{rec}. The sequence of expansion coefficients of f_{rec}, $\mathbf{c}' = \tilde{F}f_{\text{rec}}$ is the orthogonal projection of degraded coefficients onto the range of $\tilde{F}$, and is therefore closer to the original.

implicitly involves this projection, thus reducing to zero the component of $\{n_j\}$ which is orthogonal to the range of $\tilde{F}$. As the frame redundancy increases, the range of $\tilde{F}$ becomes more and more constrained, "smaller" in some sense, so the noise reduction becomes more effective. These heuristics can be put in a quantitative framework in some particular but nevertheless significant cases. If we consider tight frames in a finite dimensional space, $\mathbb{R}^n$, and a white, zero-mean, additive noise, it can be shown that in this way the expected squared-error norm is reduced proportionally to the frame redundancy factor, r [4, 8]. Precisely

$$E(\|f - f_{\text{rec}}\|^2) = n\sigma^2/r, \qquad (3.1.3)$$

where σ^2 is the noise variance. Noise reduction in Weyl-Heisenberg and wavelet frames in $L^2(\mathbb{R})$ was also studied by Daubechies [4], under the same assumption of a white, additive, zero-mean noise. Suppose that a signal $f \in L^2(\mathbb{R})$ is well localized in a bounded region of the time-frequency plane, so that it can be well approximated using a finite number of elements of a wavelet or Weyl-Heisenberg expansion (those localized in the same region). If the expansion coefficients $\{\langle f, \tilde{\varphi}_j \rangle\}_{j \in J}$ are subject to the noise $\{n_j\}_{j \in J}$, then f can be reconstructed as

$$f_{\text{rec}} = \sum_{j \in B \subset J} (\langle f, \tilde{\varphi}_j \rangle + n_j)\varphi_j. \tag{3.1.4}$$

The reconstruction error produced in this way can be bounded as

$$E(\|f - f_{\text{rec}}\|^2) = \epsilon\|f\|^2 + O(\sigma^2/r), \tag{3.1.5}$$

where the $\epsilon\|f\|^2$ component is the result of the approximation of f using the finite set of the expansion terms, $\{\varphi_j\}_{j \in B \subset J}$. A rigorous proof of this result was given by Munch [13] for the case of tight Weyl-Heisenberg frames and integer frame redundancy factors.

Based on these results it may be conjectured that the $O(1/r)$ noise reduction property has a more general scope than discussed here. For instance, in $\mathbb{R}^n$ it might hold for frames in general, and that is indicated by results reported in [8]. Or in $L^2(\mathbb{R})$ (3.1.5) may be valid for frames other than Weyl-Heisenberg frames, and redundancy factors other than the integer ones.

3.2 Reduction of quantization error and suboptimality of the linear reconstruction

The work on the noise reduction effect, reviewed in the previous subsection, was actually aimed at estimating quantization error, which is modeled as an additive white noise. However, results on reconstruction from quantized frame expansions, reported by Morlet (see [4]), indicate that the error decays faster than could be expected from the $O(1/r)$ results. The purpose of this subsection is to elucidate why a higher reconstruction precision can be expected and how it can be attained.

Quantization of expansions in a Hilbert space $\mathcal{H}$, with respect to a given frame $\{\varphi_j\}_{j \in J}$, is a many-to-one mapping from $\mathcal{H}$ to $\mathcal{H}$,

$$Q : \mathcal{H} \rightarrow \mathcal{H}.$$

It defines a partition of $\mathcal{H}$ into disjoint cells $\{\mathcal{C}_i\}_{i \in \mathcal{I}}$. In the case of uniform scalar quantization, each of the cells is defined by a set of convex constraints of the type

$$\mathcal{C}_i = \{f : (n_{ij} - 1/2)q \leq \langle f, \tilde{\varphi}_j \rangle < (n_{ij} + 1/2)q, \quad j \in J\}, \tag{3.2.1}$$

where q is the quantization step. For each of the cells, the quantization maps all the signals in the cell to a single signal in its interior, usually its centroid. Roughly speaking, the expected value of quantization error reflects the fineness of the partition. One way to refine the partitioning is to tighten the constraints which define cells (3.2.1) by decreasing the quantization step. Alternatively, for a given quantization step, more constraints can be added, which corresponds to an increase in redundancy of the frame. This gives another explanation of the error reduction property in frames, this time for the quantization error. Effectiveness of the two approaches to partition refinement, that is error reduction, can be assessed based on the results reviewed in the previous subsection. If the white noise model for the quantization error is accepted, the error variance σ^2 is proportional to the square of the error maximum value, which is half of the quantization step. According to formulae (3.1.3) and (3.1.5), it seems that the quantization step refinement is more effective than the increase in frame redundancy for the error reduction, since the error decreases proportionally to q^2 in the former case, and proportionally to $1/r$ in the latter case. Is this always true, or are we underestimating the potential of frame redundancy for error reduction? Let us look at a simple example for the sake of providing some intuition about this effect.

Quantization of expansions in $\mathbb{R}^2$, with respect to the orthonormal basis

$$\{(0,1),(1,0)\}$$

induces the partition of the plane into square cells

$$C_i = \{(x,y) : (n_{ix} - 1/2)q \leq x < (n_{ix} + 1/2)q,$$
$$(n_{iy} - 1/2)q \leq y < (n_{iy} + 1/2)q\}. \tag{3.2.2}$$

Figure 2 illustrates partitions obtained after refining the initial partition, with $q = 1$, by reducing q by a factor k or introducing redundant expansions with redundancy factor k, for $k = 2, 3, 4$. In the first case, each cell of the initial partition is uniformly divided into k^2 square subcells. On the other hand, in the case of increased redundancy the number of subcells varies around k^2. Some of the cells induced by overcomplete expansions are larger while some of them are smaller than the cells of the orthonormal partition for the same k. Based on a simple inspection of the partitions in Figure 2, it is hard to tell which of the two approaches gives a finer partition on average. However, if we take a look at the partitions in Figure 3, generated by reducing q by the factor k or increasing the redundancy by the factor k^2, we can observe that in the latter case partition refinement is faster, contrary to what would be expected from relations (3.1.3) and (3.1.5).

The reason for this discrepancy is that the error estimates in the previous subsection were derived under the assumption of linear reconstruction

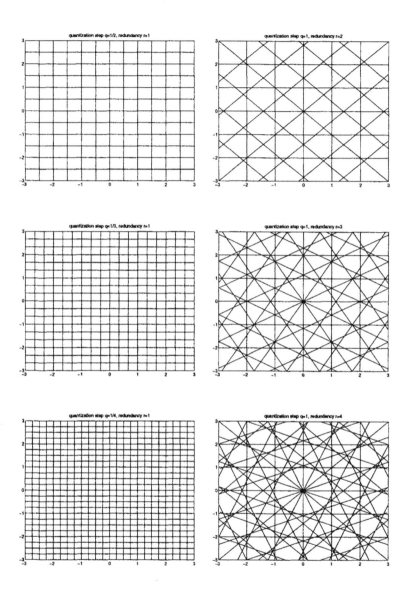

Figure 2.   Partitions of $\mathbb{R}^2$ induced by quantization of frame expansions. Partitions in the left column are obtained for the orthonormal basis $\{(1,0),(0,1)\}$ and quantization steps $q = 1/2,\ 1/3,\ and\ 1/4$. Partitions in the right column correspond to the quantization step $q = 1$ and tight frames $\{(\cos(i\pi/2r), \sin(i\pi/2r))\}_{i=0,\ldots,2r-1}$, for $r = 2,\ 3,\ and\ 4$.

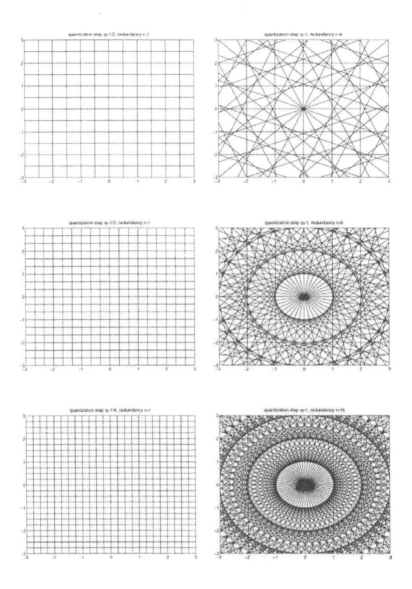

Figure 3. Partitions of $\mathbb{R}^2$ induced by quantization of frame expansions. Partitions in the left column are obtained for the orthonormal basis $\{(1,0),(0,1)\}$ and quantization steps $q = 1/2$, $1/3$, and $1/4$. Partitions in the right column correspond to the quantization step $q = 1$ and tight frames $\{(\cos(i\pi/2r), \sin(i\pi/2r))\}_{i=0,\dots,2r-1}$, for $r = 4$, 9, and 16.

given by (3.1.2). However, when dealing with quantization of overcomplete expansions, linear reconstruction is not optimal in the sense that it does not necessarily yield a signal which lies in the same quantization cell with the original. Hence, linear reconstruction does not fully utilize information which is contained in the quantized set of coefficients, so there is room for improvement. Figure 4 depicts such a scenario. Let c denote the set of quantized coefficients of a signal f, $\mathbf{c} = Q(\tilde{F}f)$, and let $\mathbf{c}$ be in the partition cell C_i. As the result of linear reconstruction we obtain a signal f_{rec}. Expansion coefficients of f_{rec}, $\tilde{F}f_{\text{rec}}$, are obviously in the range of frame expansions, and are closer to the original ones than $\mathbf{c}$ is, $\|\tilde{F}f_{\text{rec}} - \tilde{F}f\| < \|\mathbf{c} - \tilde{F}f\|$. However, $\tilde{F}f$ and $\tilde{F}f_{\text{rec}}$ do not lie in the same quantization cell. Reconstruction can be further improved by alternating

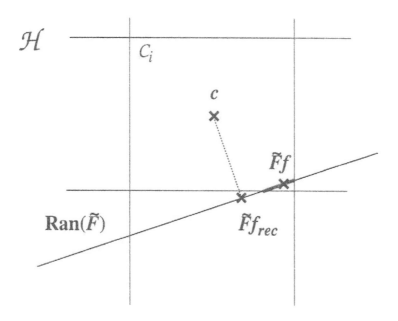

Figure 4. Inconsistency of linear reconstruction from quantized coefficients of an overcomplete expansion. Expansion coefficients of a signal f, $\tilde{F}f$, are located in the quantization cell C_i. Expansion coefficients $\tilde{F}f_{\text{rec}}$, which are obtained from the quantized coefficients $\mathbf{c}$ using the linear reconstruction, need not lie in the same quantization cell.

projections of $\tilde{F}f_{\mathrm{rec}}$ onto $\mathcal{C}_i$ and $\mathrm{Ran}(\tilde{F})$ until a signal in their intersection is reached. Note that all signals which have expansion coefficients in the intersection $\mathrm{Ran}(\tilde{F}) \bigcap \mathcal{C}_i$ produce the same result after the quantization, and therefore cannot be distinguished from f based on the quantized expansion coefficients. These signals constitute the so-called *reconstruction set* of f,

$$\Omega_r(f) = \{g : Q(\tilde{F}g) = Q(\tilde{F}f)\}, \tag{3.2.3}$$

and are denoted as its *consistent estimates*.

This has been pointed out first by Thao and one of us in the context of oversampled A/D conversion [14, 16, 15]. It was also shown there that in the case of the conversion of periodic bandlimited signals, for any consistent estimate f_c of f, $\|f - f_c\|^2 = O(1/r^2)$, where r is the oversampling ratio, provided that f has a sufficient number of quantization threshold crossings. An experimental evidence of the $O(1/r^2)$ behavior was also reported in [9, 10]. The question which naturally arises is whether this result has a wider scope. So far it has been numerically verified for frames in $\mathbb{R}^n$ [7, 8]. In the remainder of this chapter we prove that in a particular infinite-dimensional case, which is Weyl-Heisenberg expansions in $L^2(\mathbb{R})$, under certain reasonable assumptions, consistent reconstruction gives an error e which can be bounded as $\|e\|^2 = O(1/r^2)$. This result comes as an implication of the deterministic analysis of oversampled A/D conversion of bandlimited signals in $L^2(\mathbb{R})$, which is studied in more detail for the sake of clarity of presentation. Considerations in this section already have the flavor of deterministic analysis, and this is the approach which is taken here in the treatment of the error reduction effect. Statistical analysis and the white noise model are not appropriate since they ignore the deterministic nature of the error and consequently miss some details on its structure, which are essential for establishing more accurate bounds.

Before the details of this deterministic analysis are presented, a comment on rate-distortion tradeoffs in frame expansions is in order. Besides the suboptimality of linear reconstruction, traditional encoding of quantized coefficients using PCM is also not suitable in conjunction with overcomplete expansions if efficient signal representations are an issue. The reason lies in the fact that in the case of quantized orthonormal expansions, reduction of the quantization step k times, increases the bit rate by the factor $\log_2 k$, whereas the effect of a k-times higher redundancy of a frame expansion is a k-times higher bit rate. A sophisticated encoding algorithm is needed to overcome this disadvantage. However, as we pointed out earlier, frames which appear in engineering practice are not arbitrary frames but rather well-structured ones which may facilitate the encoding task. In the following section we discuss this problem in the case of oversampled A/D conversion and suggest an efficient encoding scheme.

§4 Oversampled A/D conversion

4.1 Traditional view and its limitations

Oversampled analog-to-digital conversion is the fundamental instance of signal representation based on quantization of overcomplete expansions. The aspects discussed in a general framework in the previous section are considered in more detail here while the tools for establishment of the $\|e\|^2 = O(1/r^2)$ result in the case of Weyl-Heisenberg frames are developed.

Oversampled A/D conversion as a process of digital encoding of analog signals involves discretization of a σ-bandlimited analog signal in time, implemented as sampling with a time interval τ, followed by discretization in amplitude, that is quantization of the sequence of samples with a quantization step q. The classical way to reconstruct the discretized analog signal is by low-pass filtering, with $\sigma/2\pi$ cutoff frequency, a pulse train which is modulated by the quantized sequence of samples. A block diagram of the converter together with a classical reconstruction scheme is illustrated in Figure 5. The discretization in time allows for perfect reconstruction of the analog signal as long as it is sampled at or above the Nyquist rate, $\tau < \tau_N$, where $\tau_N = \pi/\sigma$. However, the discretization in amplitude introduces an irreversible loss of information. The difference between the original analog signal, f, and its version, f_{rec}, recovered from the digital representation is therefore denoted as the quantization error, $e = f - f_{\text{rec}}$. Quantization error is commonly modeled as a uniformly distributed white noise independent of the input. Under certain conditions, such as a large number of quantization levels and small a quantization step q compared to the input amplitude range, this model provides a satisfactory error approximation giving the error variance $\sigma_e^2 = E(e(t)^2) = q^2/12$. Therefore,

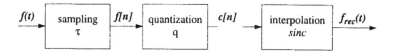

Figure 5. Block diagram of simple A/D conversion followed by classical reconstruction. Input σ-bandlimited signal $f(t)$ is first sampled at a frequency $f_s = 1/\tau$, which is above the Nyquist frequency $f_N = \sigma/\pi$. The sequence of samples $f[n]$ is then discretized in amplitude with a quantization step q. Classical reconstruction gives a signal $f_{\text{rec}}(t)$, which is obtained as a low-pass filtered version of some signal having the same digital version as the original $f(t)$, which amounts to a sinc interpolation between quantized samples.

a natural way to improve the accuracy of the conversion would be to reduce the quantization step q. However, complexity of quantizer circuitry and the precision of analog components impose limits on the quantization step refinement. Modern techniques for high resolution A/D conversion are based on oversampling. Sampling a signal at a rate higher than the Nyquist rate introduces redundancy which can be exploited for error reduction. The white noise model explains this effect in the following way. As a result of oversampling, the quantization error spectrum remains uniform and spreads over a wider frequency range $[-1/\tau, 1/\tau]$, while the error variance remains $q^2/12$. Only the τ/τ_N portion of the error spectrum is in the frequency range of the input signal. Therefore, after the low-pass filtering with σ/π cutoff frequency, the error variance is reduced to $\sigma_e^2 = q^2\tau/12\tau_N$, or

$$E(e(t)^2) = \frac{q^2}{12r},\qquad(4.1.1)$$

where $r = \tau/\tau_N$ is the oversampling ratio (see Figure 6). The error esti-

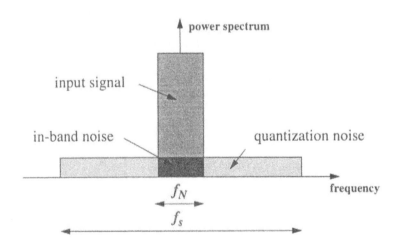

Figure 6. Statistical analysis of the MSE of oversampled A/D conversion with classical reconstruction. Under certain assumptions, the quantization error can be modeled as a white noise. As the sampling interval decreases, the noise spectrum remains flat and spreads out over the whole sampling frequency range, while keeping a constant total power equal to $q^2/12$. For a sampling frequency f_s, which is above the Nyquist rate, only the portion $q^2 f_N/12f_s$ is still present in the signal spectrum after the low-pass filtering with $f_N/2$ cut-off frequency.

mate in (4.1.1) indicates that although the oversampling can be exploited for improvement of conversion accuracy, it is inferior to quantization refinement. Furthermore, if the error is considered as a function of the bit rate, quantization refinement drastically outperforms oversampling. As an illustration, consider the case when the quantization step is halved. This reduces the error by the factor of four, at the expense of an additional bit per sample. In order to achieve the same error reduction, the oversampling ratio should be increased four times, and consequently the total number of bits increases by the same factor. This reasoning gives the following results. If the quantization step varies, for a fixed-sampling interval τ, the quantization error as a function of the bit rate, B, is given by

$$E\left(e(t)^2\right) = \frac{1}{12} A^2 \frac{\tau}{\tau_N} 2^{-2\tau B}, \qquad (4.1.2)$$

where A denotes signal amplitude range. Hence, as the quantization step decreases, the error decays as

$$E\left(e(t)^2\right) = O\left(2^{-2\tau B}\right). \qquad (4.1.3)$$

On the other hand, if the sampling is successively refined, $\tau \to 0$, for a fixed quantization step,

$$E\left(e(t)^2\right) = \frac{1}{12} \frac{q^2}{\tau_N} \left(\log_2 \frac{A}{q}\right) \frac{1}{B}, \qquad (4.1.4)$$

or in other words, the quantization error behaves as

$$E\left(e(t)^2\right) = O\left(\frac{1}{B}\right). \qquad (4.1.5)$$

In light of the discussion in the previous section we can infer that such inferior performance of oversampling with respect to quantization refinement is not a consequence of some fundamental phenomena, but is rather a consequence of inadequate reconstruction and coding. Sampling a signal at a rate higher than the Nyquist rate amounts to expanding it in terms of a family of sinc functions

$$\{\text{sinc}_\sigma (t - n\tau)\}_{n \in \mathbf{Z}}, \qquad (4.1.6)$$

which is a tight frame for the space of σ-bandlimited signals with a redundancy factor equal to the oversampling ratio. The classical reconstruction based on low-pass filtering is essentially the linear reconstruction (see (3.1.2)) and is therefore inconsistent and suboptimal. The main point of this section is a deterministic analysis of the quantization error which demonstrates that the error of consistent reconstruction can be, under certain conditions, bounded as $\|e(t)\|^2 = O(1/r^2)$. We also suggest an efficient scheme for lossless encoding of quantized samples, which when used together with consistent reconstruction gives error-rate characteristics of the form

$$\|e(t)\|^2 = O\left(2^{-2\beta B}\right). \tag{4.1.7}$$

The constant β in (4.1.7) is of the same order of magnitude as the Nyquist sampling interval, so that we have a performance comparable to that where a quantization step tends to zero (4.1.3), and significantly improved compared with linear reconstruction and traditional PCM encoding (4.1.5).

4.2   Reconstruction sets and asymptotic behavior

The deterministic analysis of oversampled A/D conversion is based on the study of the structure of the partition which it induces in $L^2[-\sigma, \sigma]$, when considered as a quantization of an overcomplete expansion. The knowledge on a signal, contained in its digitally encoded representation, defines a set of signals which cannot be distinguished from it after the A/D conversion. Recall that this set is referred to as the reconstruction set of the considered signal. In this context, the reconstruction set of a signal f can be represented as the intersection of two sets

$$\Omega_r(f) = \mathcal{V}_\sigma \cap \mathcal{C}_q(f), \tag{4.2.1}$$

where $\mathcal{V}_\sigma$ is the space of σ-bandlimited signals, and $\mathcal{C}_q(f)$ consists of signals in $L^2(\mathbb{R})$ which have amplitudes within the same quantization intervals as f at all sampling instants.

   The size of the reconstruction set can be considered as a measure of the uncertainty about the original analog signal. Let us assume that the quantization step is decreased by an integer factor m. As a result, the reconstruction set becomes smaller, since $\mathcal{C}_{q/m}(f)$ is a proper subset of $\mathcal{C}_q(f)$ (see Figure 7). In the limit, when $q \to 0$, there is no uncertainty about the signal amplitude at the sampling points ..., $-\tau$, 0, τ, 2τ, ..., and the signal can be perfectly reconstructed if sampled above the Nyquist rate. This means that the reconstruction set asymptotically reduces to a single point which is the considered analog signal itself. This view of the reconstruction set is based on the traditional interpretation of the digital version of an analog signal as a representation which bears information on its amplitude, with the uncertainty determined by the quantization step q, at the sequence of time instants $t = ..., -\tau, 0, \tau, 2\tau, ...$. An alternative interpretation comes into play if the quantization threshold crossings are separated and sampling is sufficiently fine so that at most one quantization threshold crossing can occur in each of the sampling intervals. If this is satisfied, quantized samples of the signal f are uniquely determined by

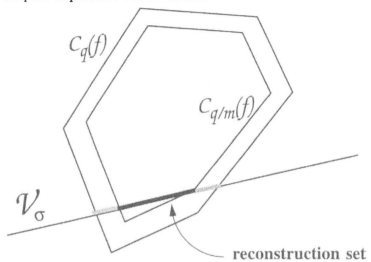

Figure 7. The effect of the quantization refinement on the reconstruction set. The set of signals which share the same digital version with the original, $C_q(f)$, becomes smaller as the result of quantization refinement, $C_{q/m}(f) \subset C_q(f)$. Consequently, the reconstruction set is also reduced giving a higher accuracy of the representation

the sampling intervals in which its quantization threshold crossings occur, and vice versa. So, we can think of the digital representation as carrying information on the instants, with uncertainty τ in time, when the signal assumes values $\dots, -q, 0, q, 2q, \dots$. Hence, the reconstruction set of f can also be viewed as lying in the intersection

$$\Omega_r(f) \subseteq \mathcal{V}_\sigma \cap \mathcal{D}_\tau(f), \qquad (4.2.2)$$

where $\mathcal{D}_\tau(f)$ is the set of all signals in $L^2(\mathbb{R})$ which have the same quantization threshold crossing as the original, in all of the sampling intervals where the original goes through a quantization threshold. If in addition we require that signals in $\mathcal{D}_\tau(f)$ cannot have more than one quantization threshold crossing per sampling interval, then in (4.2.2) equality holds.

Consider now the effect of reduction of the sampling interval by an integer factor m. The set of signals sharing the sampling intervals of quantization threshold crossings with the original, for this new sampling interval, is a proper subset of $\mathcal{D}_\tau(f)$, $\mathcal{D}_{\tau/m}(f) \subset \mathcal{D}_\tau(f)$. As a result, the size of the reconstruction set is also reduced implying a higher accuracy of the conversion. This reasoning gives a deterministic explanation for the error reduction as a consequence of oversampling in the A/D conversion. Quantization step refinement, which also improves the accuracy of the conversion, asymptotically gives the reconstruction set reduced to the original signal only. Does this also happen in this case when the sampling interval goes to zero? In other words, is it possible to reconstruct perfectly an analog σ-bandlimited signal after simple A/D conversion in the case of infinitely high oversampling?

In the limit, when the sampling interval assumes an infinitely small value, the time instants in which the signal goes through the quantization thresholds are known with infinite precision. The information on the input analog signal in this limiting case is its values at a sequence of irregularly spaced points $\{\lambda_n\}$, which are its quantization threshold crossings. If $\{e^{j\lambda_n\omega}\}$ is complete in $L^2[-\sigma,\sigma]$, then the reconstruction set asymptotically does reduce to a single point, giving the perfectly restored input analog signal. Completeness of $\{e^{j\lambda_n\omega}\}$ in $L^2[-\sigma,\sigma]$ means that any σ-bandlimited signal $f(t)$ is determined by the sequence of samples $\{f(\lambda_n)\}$. However, this does not necessarily mean that $f(t)$ can be reconstructed in a numerically stable way from the samples $\{f(\lambda_n)\}$, unless another constraint on $\{\lambda_n\}$ is introduced which would ensure that any two σ-bandlimited signals which are close at points $\{\lambda_n\}$ are also close in L^2 norms. In other words, we want to be sure that the reconstruction error does converge to zero as the oversampling tends to infinity, that is

$$\lim_{\tau \to 0} \|f - f_{\text{rec}}\| = 0, \qquad (4.2.3)$$

rather than only asymptotically becoming zero in some odd fashion. A precise formulation of the stability requirement is that $\{e^{j\lambda_n\omega}\}$ constitute a frame in $L^2[-\sigma,\sigma]$. The error bound derived in the following subsection is based on this assumption. In order to illustrate how restrictive this assumption is, at this point we briefly review some of the frame conditions. A more rigorous treatment can be found in [18].

Recall that exact frames in a separable Hilbert space, those which cease to be frames when any of the elements is removed, are Riesz bases. Are there any bases of complex exponentials other than Riesz bases? This is still an open problem. Every example of a basis of complex exponentials for $L^2[-\sigma,\sigma]$ so far has been proven to be a Riesz basis ([18], pp. 190–197). So, if we want a reasonably complete characterization of σ-bandlimited signals from samples at points $\{\lambda_n\}$, then $\{e^{j\lambda_n\omega}\}$ should be a frame in $L^2[-\sigma,\sigma]$.

If a sequence of real numbers $\{\lambda_n\}$ satisfies

$$\left|\lambda_n - n\frac{\pi}{\sigma}\right| \le L < \frac{1}{4}\frac{\pi}{\sigma}, \quad n = 0, \pm 1, \pm 2, \ldots, \qquad (4.2.4)$$

then $\{e^{j\lambda_n\omega}\}$ is a Riesz basis for $L^2[-\sigma,\sigma]$. This result is known as Kadec's 1/4 Theorem [11]. How realistic is this condition in the quantization context? For a quantization step small compared to a signal amplitude, a sufficiently dense sequence of quantization threshold crossings could be expected, thus satisfying (4.2.4), at least on a time interval before the signal magnitude eventually falls below the lowest nonzero quantization threshold. Suppose that the analog signal $f(t)$ at the input of an A/D converter satisfies: $\hat{f}(\omega)$ is continuous on $[-\sigma, \sigma]$ and $\hat{f}(\sigma) \neq 0$. Such a signal $f(t)$, in terms of its zero-crossings, asymptotically behaves as $\mathrm{sinc}(\sigma t)$. So, if one of the quantization thresholds is set at zero, then for large n, the quantization threshold crossings should be close to points $n\pi/\sigma$.

Another criterion is given by Duffin and Schaeffer in their pioneering work on frames [5]. It states that if $\{\lambda_n\}$ has a uniform density $d > \sigma/\pi$, then $\{e^{j\lambda_n \omega}\}$ is a frame in $L^2[-\sigma, \sigma]$. A sequence of real numbers $\{\lambda_n\}$ is said to have a uniform density $d > 0$ if there exist constants $L < \infty$ and $\delta > 0$ such that

$$|\lambda_n - \tfrac{n}{d}| \leq L, \quad n = 0, \pm 1, \pm 2, \ldots$$

$$|\lambda_n - \lambda_m| \geq \delta \qquad n \neq m.$$

(4.2.5)

For a quantization scheme with a fixed quantization step, a σ-bandlimited signal having quantization threshold crossings with a uniform density greater than σ/π is unlikely. The amplitude of a bandlimited signal $f(t)$ decays at least as fast as $1/t$, and outside of some finite time interval, $[-T, T]$, it is confined in the range between the two lowest nonzero quantization thresholds. Zero-crossings are the only possible quantization threshold crossings of $f(t)$ on $(-\infty, -T] \cup [T, \infty)$. However, it doesn't seem to be plausible that zero-crossings of a σ-bandlimited signal can have a uniform density greater than the density of zero-crossings of $\mathrm{sinc}(\sigma t)$, which is σ/π. Therefore, in order to meet this criterion, the quantization step has to change in time following the decay of the signal. One way to achieve this is a scheme with quantization steps which are fixed on given time segments but eventually decrease with segment order.

4.3 $O(1/r^2)$ error bound

Consider a σ-bandlimited signal, $f(t)$, at the input of an oversampled A/D converter with sampling interval τ. Suppose that its sequence of quantization threshold crossings, $\{x_n\}$, is a frame sequence for the space of σ-bandlimited signals. If $g(t)$ is a consistent estimate of $f(t)$, there exists a sequence $\{y_n\}$ of quantization threshold crossings of $g(t)$, such that every x_n has a corresponding y_n in the same sampling interval. Hence, for each pair (x_n, y_n), $f(x_n) = g(y_n)$ and $|x_n - y_n| \leq \tau$. At the points $\{x_n\}$, the error amplitude is bounded by the variation of $g(t)$ on the interval $[y_n, x_n]$ (without loss of generality we assume that $y_n < x_n$), as shown in Figure 8. Since $g(t)$ is bandlimited, which also means that it has finite energy, it can have only a limited variation on $[y_n, x_n]$. This variation is bounded by

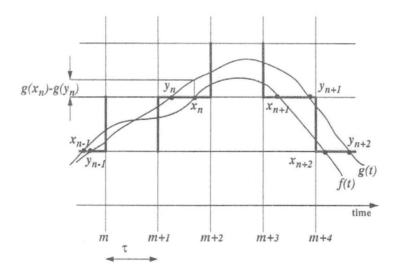

Figure 8. Quantization threshold crossings of an analog signal $f(t)$ and its consistent estimate $g(t)$. The sequence of quantization threshold crossings, $\{x_n\}$, of $f(t)$, uniquely determines its digital version and vice versa, provided that all the crossings occur in distinct sampling intervals. If $f(t)$ goes through a certain quantization threshold at the point x_n, then $g(t)$ has to cross the same threshold at a point y_n, which is in the same sampling interval with x_n. At a point x_n, the error amplitude is equal to $|g(x_n) - f(x_n)| = |g(y_n) - g(x_n)|$.

some value which is proportional to the sampling interval, so

$$|g(x_n) - g(y_n)| \leq c_n \cdot \tau. \tag{4.3.1}$$

The constant c_n in this relation can be the maximum slope of $g(t)$ on the interval $[y_n, x_n]$, $c_n = \sup_{t \in (y_n, x_n)} g'(t)$. The error signal, $e(t) = g(t) - f(t)$, is itself a σ-bandlimited signal. At the points $\{x_n\}$ its amplitude is bounded by values which are proportional to τ (4.3.1). If the sequence $\{c_n\}$ is square summable, it can be expected that the energy of the error signal is bounded as $\|e\|^2 \leq \text{const} \cdot \tau^2$, or in terms of the oversampling ratio r,

$$\|e\|^2 \leq \frac{\text{const}}{r^2}. \tag{4.3.2}$$

This result is the content of the following theorem.

Theorem 1. *Let $f(t)$ be a real σ-bandlimited signal at the input of an A/D converter with a sampling interval $\tau < \pi/\sigma$. If the sequence of quantization threshold crossings of $f(t)$, $\{x_n\}$, forms a frame sequence for the space of σ-bandlimited signals, then there exists a positive constant δ such that if $\tau < \delta$, for every consistent estimate of $f(t)$, $g(t) \in C^1$,*

$$\|f(t) - g(t)\|^2 \leq k\|f(t)\|^2 \tau^2, \tag{4.3.3}$$

where k is a constant which does not depend on τ.

Proof: Let A and B be bounds of the frame $\{e^{jx_n \omega}\}$ in $L^2[-\sigma, \sigma]$, so that for any σ-bandlimited $s(t)$

$$A\|s(t)\|^2 \leq \sum_n |s(x_n)|^2 \leq B\|s(t)\|^2. \tag{4.3.4}$$

At the points x_n, the error amplitude is bounded by the variation of the reconstructed signal on the corresponding sampling intervals, as described by the following relations:

$$
\begin{aligned}
|e(x_n)| &= |f(x_n) - g(x_n)| \\
&= |f(x_n) - g(y_n) + g(y_n) - g(x_n)| \\
&= |g(y_n) - g(x_n)| \\
&\leq \tau \cdot g'(\epsilon_n), \quad \min(x_n, y_n) \leq \epsilon_n \leq \max(x_n, y_n).
\end{aligned}
\tag{4.3.5}
$$

Here, ϵ_n denotes a point on the interval $[y_n, x_n]$. The error norm then satisfies

$$
\begin{aligned}
\|e(t)\|^2 &\leq \frac{1}{A} \sum_n |e(x_n)|^2 \\
&\leq \frac{\tau^2}{A} \sum_n |g'(\epsilon_n)|^2.
\end{aligned}
\tag{4.3.6}
$$

If τ is smaller than $\delta = \delta_{1/4}(\{x_n\}, \sigma)$ (see Lemma 2 in Appendix), then $|\epsilon_n - x_n| < \delta$ and consequently,

$$\sum_n |g'(\epsilon_n)|^2 \leq \frac{9B}{4} \|g'(t)\|^2. \tag{4.3.7}$$

This gives

$$\|e(t)\|^2 \leq \tau^2 \frac{9B}{4A} \|g'(t)\|^2$$
$$\leq \tau^2 \sigma^2 \frac{9B}{4A} \|g(t)\|^2 \tag{4.3.8}$$

so that the energy of the error can be bounded as

$$\|e(t)\|^2 \leq \frac{\tau^2 \sigma^2 9B}{4A} \|g(t)\|^2. \tag{4.3.9}$$

It remains to find a bound for the norm of $g(t)$.

According to Lemma 2 (see Appendix), since $|x_n - y_n| < \tau < \delta_{1/4}(\{x_n\}, \sigma)$, the following holds:

$$\|g(t)\|^2 \leq \frac{4}{A} \sum_n g(y_n)^2$$
$$= \frac{4}{A} \sum_n f(x_n)^2$$
$$\leq \frac{4B}{A} \|f(t)\|^2. \tag{4.3.10}$$

As a consequence of the last inequality and the error bound in (4.3.9), we obtain

$$\|e(t)\|^2 \leq 9\sigma^2 \frac{B^2}{A^2} \|f(t)\|^2 \tau^2. \quad \blacksquare \tag{4.3.11}$$

In the Appendix we give another more intuitive error bound for the case when the quantization threshold crossings form a sequence of uniform density greater than σ/π.

An extension of the result of Theorem 1 to complex bandlimited signals is straightforward, provided that the real and imaginary parts are quantized separately.

Corollary 1. *Let the real and imaginary parts of a complex σ-bandlimited signal $f(t)$, at the input of an A/D converter with the sampling interval $\tau < \pi/\sigma$, be quantized separately. If sequences of quantization threshold* crossings of both $\operatorname{Re}f(t)$ and $\operatorname{Im}f(t)$ form frame sequences for the space of σ-bandlimited signals, then there exists a positive constant δ such that if $\tau < \delta$ then for every consistent estimate $g(t) \in C^1$

$$\|f(t) - g(t)\|^2 \le k\|f(t)\|^2\tau^2, \tag{4.3.12}$$

for some constant k which does not depend on τ.

Proof: Let $\{x_n^r\}$ and $\{x_n^i\}$ be the sequences of quantization threshold crossings of the real and imaginary parts of $f(t)$, respectively, and $0 < A_r \le B_r < \infty$, $0 < A_i \le B_i < \infty$, the corresponding frame bounds. Then the relation (4.3.12) holds for

$$\tau < \min\left(\delta_{1/4}\left(\{x_n^r\}, \sigma\right), \delta_{1/4}\left(\{x_n^i\}, \sigma\right)\right)$$

and

$$k = 9\sigma^2\mu, \quad \mu = \max\left(\frac{B_i^2}{A_i^2}, \frac{B_r^2}{A_r^2}\right). \quad \blacksquare \tag{4.3.13}$$

4.4 Error-rate characteristics with optimal reconstruction and efficient coding

An efficient scheme for lossless encoding of the digital representation follows from the observation that, for sufficiently fine sampling, the quantized values of signal samples can be determined from the corresponding sequence of quantization threshold crossings.

Consider a bandlimited signal $f(t)$ and suppose that its quantization threshold crossings are separated, that is there is an $\epsilon > 0$ such that no two threshold crossings are closer than ϵ. Note that as a bandlimited signal of a finite energy, $f(t)$ has a bounded slope so that there is always an $\epsilon_1 > 0$ such that $f(t)$ can not go through more than one quantization threshold on any interval shorter than ϵ_1. The condition for separated quantization threshold crossings requires in addition that intervals between consecutive crossings through a same threshold are limited from below away from zero. For a sampling frequency higher than $1/\epsilon$, all quantization threshold crossings of $f(t)$ occur in distinct sampling intervals. Under this condition, quantized samples of $f(t)$ are completely determined by its sequence of quantization threshold crossings (see Figure 8). Another effect of high oversampling is that quantized values of consecutive samples differ with a small probability. In this case, an economical digital representation would encode incidences of data changes, that is, sampling intervals where quantization threshold crossings occur rather than quantized samples themselves. Quantization threshold crossings can be grouped on consecutive time intervals of a given length, for instance T. For each of the crossings at most $1 + \log_2(T/\tau)$ bits

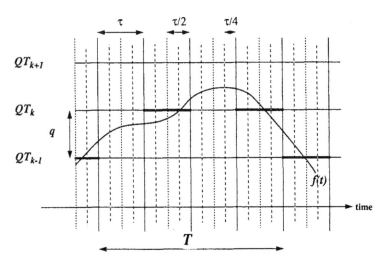

Figure 9. Quantization threshold crossings encoding. Quantization threshold crossings are grouped on intervals of a length T. Refining the sampling interval by a factor 2^k requires additional k bits per quantization threshold crossing to encode its position inside the interval T.

are then needed to record its position inside the interval T. The height of the threshold crossing can be given with respect to the previous one, so that for this information only one additional bit is needed, to denote direction of the crossing (upwards or downwards). Hence, for recording of the information on quantization threshold crossings on an interval where Q of them occur, less than $Q\left(2 + \log_2(T/\tau)\right)$ bits are needed (see Figure 9). The bit rate is then bounded as

$$B \le \frac{Q_m}{T}\left(2 + \log_2\left(\frac{T}{\tau}\right)\right),\qquad(4.4.1)$$

where Q_m denotes the maximal number of the crossings on an interval of length T. Note that if the samples themselves are recorded, the bit rate increases linearly with the oversampling ratio, $B = O(T/\tau)$, so quantization threshold crossings encoding is substantially more efficient. If this efficient coding is used together with consistent reconstruction, then the error of the oversampled A/D conversion becomes bounded as

$$\|e(t)\|^2 \le KT^2 2^{-2\frac{T}{Q_m}B},\qquad(4.4.2)$$

where K is a constant which depends on the input signal, as follows from (4.3.3) and (4.4.1).

In order to estimate the factor T/Q_m in this expression, we consider again the two types of quantization threshold crossings, denoting them as

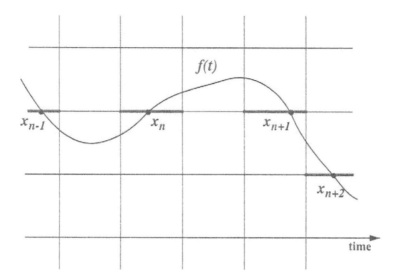

Figure 10. Quantization threshold crossing types. A quantization threshold crossing can be immediately preceded by a crossing of the same quantization threshold, as illustrated by the crossings at points x_n and x_{n+1}. Such threshold crossings are denoted as s-crossings, and each of them is preceded by a point where the signal assumes an extremum. The other type of quantization threshold crossings, d-crossings, are those which occur after a crossing of a different quantization threshold. The threshold crossing at x_{n+2} is of this type.

d-crossings and s-crossings. A quantization threshold crossing is said to be a d-crossing if it is preceded by a crossing of a different quantization threshold, or an s-crossing if it is preceded by a crossing of the same threshold (see Figure 10). The total number of quantization threshold crossings of a σ-bandlimited signal $f(t)$ on an interval T is sum of these two types of crossings. The count of d-crossings, Q_d, depends on the slope of $f(t)$ as well as the quantization step size, q. The slope of $f(t)$ can be bounded as

$$|f'(t)| \leq \|f\| \sigma^{\frac{3}{2}}, \tag{4.4.3}$$

which gives

$$\frac{Q_d}{T} \le \frac{\sigma^{\frac{3}{2}}}{q} \|f\|. \tag{4.4.4}$$

For the count of s-crossings, Q_s, we can investigate average behavior. Each of the s-crossings of $f(t)$ is preceded by a point where $f(t)$ assumes a local extremum. If s-crossings constitute a sequence of a uniform density d, then there is a subset of zeros of $f'(t)$ which also constitutes a sequence of uniform density d. Since a sequence of real numbers with a uniform density $d > \sigma/\pi$ is a frame sequence for the space of σ-bandlimited signals, it follows that d has to be less or equal to σ/π, except in a degenerate case when $f'(t)$ is identically equal to zero. If we assume that the sequence of quantization threshold crossings of $f(t)$ is a realization of an ergodic process, then as the interval T grows

$$\frac{Q_s}{T} \to c \le \frac{\sigma}{\pi}. \tag{4.4.5}$$

Hence, the error rate characteristics are given by $\|e(t)\|^2 = O(2^{-\frac{2}{\alpha_1 + \alpha_2}B})$ where $\alpha_1 \le \|f\|\sigma^{\frac{3}{2}}/q$ and α_2 is "close" to σ/π. Recall that for an A/D converter with a fixed sampling frequency $f_s \ge \sigma/\pi$ and quantization step refinement, the mean squared error is given by $E(e(t)^2) = O(2^{-\frac{2}{\alpha}B})$, where $\alpha = f_s$.

It is interesting to find the error-rate characteristics of oversampled A/D conversion for the four combinations of reconstruction and encoding, *i.e.* linear versus consistent reconstruction, and PCM versus quantization threshold crossings encoding. These are given in Table 1. These results demonstrate the importance of an appropriate lossless coding of the dig-

	Linear Reconstruction	Consistent Reconstruction
PCM	$E(e(t)^2) = O\left(\frac{1}{B}\right)$	$\|e(t)\|^2 = O\left(\frac{1}{B^2}\right)$
Efficient Encoding	$E(e(t)^2) = O\left(2^{-\frac{T}{q_m}B}\right)$	$\|e(t)\|^2 = O\left(2^{-2\frac{T}{q_m}B}\right)$

Table 1. Error-rate characteristics of oversampled A/D conversion, as the sampling interval tends to zero, for the four different combinations of reconstruction and encoding. The quantization error, $e(t)$, is expressed as a function of the bit rate B.

ital representation. According to the given characteristics, for good error-rate performance of oversampled A/D conversion, efficient coding is more important than optimal reconstruction.

§5 Quantization of Weyl-Heisenberg expansions

Error analysis in the case of quantization of coefficients of Weyl- Heisenberg frame expansions is an immediate generalization of the results on oversampled A/D conversion from the previous section. The two cases which are considered first are: 1) frames derived from bandlimited window functions without restrictions on input signals, except that they are in $L^2(\mathbb{R})$; 2) time-limited input signals, with no restrictions on the window function except for the requirement that it is in $L^2(\mathbb{R})$.

CASE 1. σ-bandlimited window function. Let

$$\{\varphi_{m,n}(t) : \varphi_{m,n}(t) = \varphi(t - nt_0)e^{jm\omega_0 t}\} \tag{5.1}$$

be a Weyl-Heisenberg frame in $L^2(\mathbb{R})$, with the bounds

$$A\|f\|^2 \leq \sum_{m,n \in \mathbb{Z}^2} |\langle \varphi_{m,n}, f \rangle|^2 \leq B\|f\|^2. \tag{5.2}$$

Frame coefficients $\{c_{m,n} : c_{m,n} = \langle \varphi_{m,n}, f \rangle\}$ of a signal f can be expressed in the Fourier domain as

$$c_{m,n} = \int_{-\infty}^{\infty} \hat{f}(\omega - m\omega_0)\hat{\varphi}^*(\omega)e^{j\omega n t_0} d\omega. \tag{5.3}$$

The system which for an input signal gives these coefficients can be viewed as a multichannel system, containing a separate channel for each frequency shift $m\omega_0$, such that the m-th channel performs modulation of an input signal with $e^{jm\omega_0 t}$, then linear filtering with $\varphi(-t)$ and finally, sampling at points $\{nt_0\}$. Such a system is shown in Figure 11. For a fixed m, coefficients $c_{m,n}$ are samples of the signal

$$f_m(t) = \left(f(t)e^{jm\omega_0 t}\right) * \varphi(-t), \tag{5.4}$$

which will be called the *m-th subband component* of $f(t)$. In the sequel, the m-th subband component of a signal $s(t)$ will be denoted by $s_m(t)$ and the sequence obtained by sampling $s_m(t)$ with the interval t_0 will be denoted by $S_m[n]$, $S_m[n] = s_m(nt_0)$. Using this notation, frame coefficients of $f(t)$ are given by $c_{m,n} = F_m[n]$.

Such an interpretation of Weyl-Heisenberg frame coefficients of the signal $f(t)$ means that their quantization amounts to simple A/D conversion

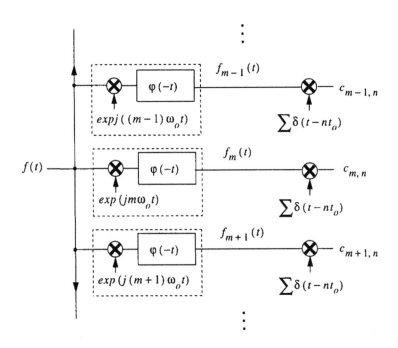

Figure 11. Evaluation of Weyl-Heisenberg frame coefficients. For a fixed m, coefficients $c_{m,n}$ are obtained as samples, with t_0 sampling interval of the signal $f_m(t)$, which is the result of modulating the input signal with $e^{jm\omega_0 t}$, followed by filtering with $\varphi(-t)$.

of the subband components of $f(t)$. Note that these coefficients are in general complex, and it is assumed here that real and imaginary parts are quantized separately. If the frame window $\varphi(t)$ is a σ-bandlimited function, each of the subband components is also a σ-bandlimited signal. In this context, a signal $g(t)$ is said to be a consistent estimate of $f(t)$ if they have the same quantized values of the frame coefficients and each subband component of $g(t)$ is continuously differentiable, $g_m(t) \in C^1$ (note that the subband signals, being bandlimited, are continuously differentiable a.e.). According to the results from the previous section, this indicates that if the frame redundancy is increased by decreasing the time step t_0 for a fixed ω_0, the quantization error of consistent reconstruction should decay as $O(t_0^2)$. This result is established by the following corollary of Theorem 1, and can be expressed in terms of the oversampling ratio, $r = \frac{2\pi}{\omega_0 t_0}$, as $\|e\|^2 = O(1/r^2)$.

Corollary 2. *Let $\{\varphi_{m,n}(t)\}$ be a Weyl-Heisenberg frame in $L^2(\mathbb{R})$, with time step t_0 and frequency step ω_0, derived from a σ-bandlimited window function $\varphi(t)$. Consider quantization of the frame coefficients of a signal $f(t) \in L^2(\mathbb{R})$ and suppose that for a certain ω_0 the following holds:*

i) *quantization threshold crossings of both real and imaginary parts of all the subband components $f_m(t) = \left(f(t)e^{jm\omega_0 t} \right) * \varphi(-t)$ form frame sequences for the space of σ-bandlimited signals, with frame bounds $0 < \alpha_m^r \le \beta_m^r < \infty$ and $0 < \alpha_m^i \le \beta_m^i < \infty$;*

ii)

$$\sup_{m \in \mathbf{Z}} \max \left(\frac{\beta_m^r}{\alpha_m^r}, \frac{\beta_m^i}{\alpha_m^i} \right) = M < \infty.$$

Then there exists a constant δ such that if $t_0 < \delta$ for any consistent estimate $g(t)$ of $f(t)$, the reconstruction error satisfies

$$\|f(t) - g(t)\|^2 \le k\|f(t)\|^2 t_0^2, \tag{5.5}$$

where k is a constant which does not depend on t_0.

Proof: Let $f(t)$ be reconstructed from its quantized coefficients as $g(t)$. Suppose that $g(t)$ is a consistent estimate of $f(t)$, that is, frame coefficients of $g(t)$ are quantized to the same values as those of $f(t)$. Under the bandlimitedness condition on $\varphi(t)$, all subband signals $f_m(t)$ are also σ-bandlimited, and each $g_m(t)$ is a consistent estimate of the corresponding $f_m(t)$, in the sense discussed in Section 4. Under assumption i), and as a consequence of Corollary 1, for each m there is a δ_m such that the m-th subband error component, $f_m(t) - g_m(t)$, satisfies

$$\|f_m(t) - g_m(t)\|^2 \le 9\sigma^2 \mu_m \|f_m(t)\|^2 t_0^2, \quad \mu_m = \max\left(\left(\frac{\beta_m^r}{\alpha_m^r}\right)^2, \left(\frac{\beta_m^i}{\alpha_m^i}\right)^2\right).$$
(5.6)

For a sampling interval $t_0 \le \pi/\sigma$ and any $s \in L^2(\mathbb{R})$, norms of subband signals s_m and their sampled versions satisfy

$$\|s_m\|^2 = t_0 \|S_m\|^2.$$
(5.7)

The frame condition (5.2) then implies

$$\|s\|^2 \le \frac{1}{A} \sum_m \|S_m\|^2$$
$$= \frac{1}{At_0} \sum_m \|s_m\|^2,$$
(5.8)

and

$$\|s\|^2 \ge \frac{1}{B} \sum_m \|S_m\|^2$$
$$= \frac{1}{Bt_0} \sum_m \|s_m\|^2.$$
(5.9)

From assumption ii) and Lemma 2, it follows that $\delta = \inf_{m \in \mathbb{Z}} \delta_m$ is strictly greater than zero, $\delta > 0$. For $t_0 \le \delta$, we have the following as a consequence of relations (5.6), (5.8), and (5.9):

$$\|f(t) - g(t)\|^2 \le \frac{1}{At_0} \sum_m \|f_m(t) - g_m(t)\|^2$$
$$\le \frac{1}{At_0} \sum_m 9\sigma^2 \mu_m \|f_m(t)\|^2 t_0^2$$
$$\le 9\sigma^2 M^2 \frac{B}{A} \|f\|^2 t_0^2.$$
(5.10)

This finally gives

$$\|f(t) - g(t)\|^2 \le k \|f(t)\|^2 t_0^2,$$
(5.11)

which for a constant ω_0 can be expressed as $\|f(t) - g(t)\|^2 = O(1/r^2)$. ∎

CASE 2. T-time-limited signals. Another way to interpret expansion coefficients of $f(t)$ with respect to the frame (5.1) is to consider them as samples of signals

$$\hat{\varphi}_n(\omega) = \left(\hat{\varphi}(\omega)e^{-jnt_0\omega}\right) * \hat{f}(\omega), \tag{5.12}$$

with ω_0 sampling interval. This is illustrated in Figure 12. Suppose that $f(t)$ is a T-time-limited signal. Then $\mathcal{F}\{\hat{\varphi}_n(\omega)\}$ is also T-time limited for each n, which makes all subband signals $\hat{\varphi}_n(\omega)$ bandlimited. From an argument completely analogous to the one in the previous case, it can be concluded that if for some fixed t_0 sequences of quantization threshold crossings of subband signals $\hat{\varphi}_n(\omega)$ satisfy certain frame properties, the quantization error of consistent reconstruction can be bounded as $\|e(t)\|^2 \leq k\|f(t)\|^2\omega_0^2$. Since we consider the case when $t_0 = \text{const}$ and $\omega_0 \to 0$, this can also be expressed as $\|e\|^2 = O(1/r^2)$. The precise formulation of this result is as follows.

Corollary 3. *Let $\{\varphi_{m,n}(t)\}$ be a Weyl-Heisenberg frame in $L^2(\mathbb{R})$, with time step t_0 and frequency step ω_0, derived from a window function $\varphi(t)$. Consider quantization of the frame coefficients of a T-time-limited signal $f(t) \in L^2(\mathbb{R})$. Suppose that for a certain t_0 the following hold:*

i)   *quantization threshold crossings of both real and imaginary parts of all the subband components $\hat{\varphi}_n(\omega) = \left(\hat{\varphi}(\omega)e^{-jnt_0\omega}\right) * \hat{f}(\omega)$ form frames for the space of T-time-limited signals, with frame bounds $0 < \alpha_n^r \leq \beta_n^r < \infty$ and $0 < \alpha_n^i \leq \beta_n^i < \infty$;*

ii)

$$\sup_{n \in \mathbf{Z}} \max\left(\frac{\beta_n^r}{\alpha_n^r}, \frac{\beta_n^i}{\alpha_n^i}\right) = M < \infty.$$

Then there exists a constant δ such that if $\omega_0 < \delta$ for any consistent estimate $g(t)$ of $f(t)$, the reconstruction error satisfies

$$\|f(t) - g(t)\|^2 \leq k\|f(t)\|^2\omega_0^2, \tag{5.13}$$

where k is a constant which does not depend on ω_0.

If the Weyl-Heisenberg frame is redefined as

$$\left\{\varphi_{m,n}(t) : \varphi_{m,n}(t) = \varphi(t - nt_0)e^{jm\omega_0(t - nt_0)}\right\},$$

then analogous results hold for the cases when input signals are bandlimited or the window function has a compact support in time.

The assumptions on bounded support of frame window functions or input signals in either time or frequency, introduced in the above considerations, are natural assumptions of time-frequency localized signal analysis.

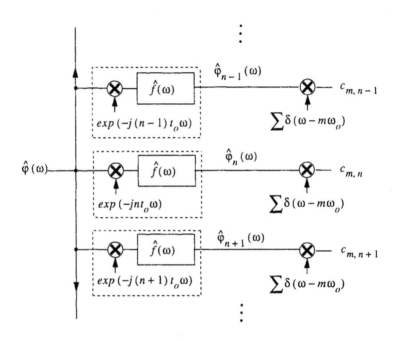

Figure 12. Evaluation of Weyl-Heisenberg frame coefficients. For a fixed n, coefficients $c_{m,n}$ are obtained as samples with ω_0 sampling interval of the signal $\hat{\varphi}_n(\omega)$, which is itself the result of modulating $\hat{\varphi}(\omega)$ by $e^{-jnt_0\omega}$, followed by filtering with $\hat{f}(\omega)$.

A question that arises is whether in the case when both the window function is bandlimited and the considered signals have finite support, the error decays as $\|e\|^2 = O(\omega_0^2 t_0^2)$. Another interesting case is when none of these assumptions is introduced. Is it then also possible to exploit frame redundancy for quantization error reduction so that the error norm tends to zero as the redundancy is increased, or even more $\|e\|^2 = O(1/r^2)$? These are still open problems.

§6 Appendix

6.1 Two lemmas on frames of complex exponentials

Estimates of bounds on the quantization error in Subsection 4.3 are derived from the next two lemmas [5].

Lemma 1. *Let* $\{e^{j\lambda_n \omega}\}$ *be a frame in* $L^2[-\sigma, \sigma]$. *If* M *is any constant and* $\{\mu_n\}$ *is a sequence satisfying* $|\mu_n - \lambda_n| \leq M$, *for all* n, *then there is a number* $C = C(M, \sigma, \{\lambda_n\})$ *such that*

$$\frac{\sum_n |f(\mu_n)|^2}{\sum_n |f(\lambda_n)|^2} \leq C \tag{6.1.1}$$

for every σ*-bandlimited signal* $f(x)$.

Lemma 2. *Let* $\{e^{j\lambda_n \omega}\}$ *be a frame in* $L^2[-\sigma, \sigma]$, *with bounds* $0 < A \leq B < \infty$ *and* δ *a given positive number. If a sequence* $\{\mu_n\}$ *satisfies* $|\lambda_n - \mu_n| \leq \delta$ *for all* n, *then for every* σ*-bandlimited signal* $f(x)$

$$A(1 - \sqrt{C})^2 \|f\|^2 \leq \sum_n |f(\mu_n)|^2 \leq B(1 + \sqrt{C})^2 \|f\|^2, \tag{6.1.2}$$

where

$$C = \frac{B}{A} \left(e^{\gamma\delta} - 1\right)^2. \tag{6.1.3}$$

Remark 1. If δ in the statement of Lemma 2 is chosen small enough, so that C is less then 1, then $\{e^{j\mu_n\omega}\}$ is also a frame in $L^2[-\sigma, \sigma]$. Moreover, there exists some $\delta_{1/4}(\{\lambda_n\}, \sigma)$, such that whenever $\delta < \delta_{1/4}(\{\lambda_n\}, \sigma)$, $\{e^{j\mu_n\omega}\}$ is a frame with frame bounds $A/4$ and $9B/4$.

6.2 An alternative proof of theorem 1

This proof is derived under the assumption that the quantization threshold crossings of the signal form a sequence of uniform density $d > \sigma/\pi$.

Proof of Theorem 1:   Recall that in order to have a sequence of quantization threshold crossings of a σ-bandlimited signal $f(t)$ with uniform density $d > \sigma/\pi$, the quantization step q must change in time following the decay of the signal. Since $f(t)$ is square integrable, $q = q(t)$ has to decay at least as fast as $1/t$, although it can be fixed on a given set of time segments.

Let A and B again denote the lower and the upper bounds, respectively, of the frame determined by the quantization threshold crossings, $\{x_n\}$, of $f(t)$, and let $g(t)$ be a consistent estimate of $f(t)$. Following the discussion in the proof of Theorem 1 (see (4.3.6)), the error norm is bounded as

$$\|e(t)\|^2 < \frac{\tau^2}{A} \sum_n |g'(\epsilon_n)|^2. \tag{6.2.1}$$

Since τ is smaller then the Nyquist sampling interval, $|x_n - \epsilon_n| < \pi/\sigma$ for all n. According to Lemma 1, there exists a constant C which does not depend on the sampling interval, such that for all $\tau < \pi/\sigma$

$$\sum_n |g'(\epsilon_n)|^2 \le C \sum_n |g'(x_n)|^2. \tag{6.2.2}$$

This gives

$$\|e(t)\|^2 \le \frac{\sigma^2 BC\tau^2}{A} \|g(t)\|^2. \tag{6.2.3}$$

Being a consistent estimate of $f(t)$, $g(t)$ can not differ from $f(t)$ by more than $q(n\tau)$, at time instants $\{n\tau\}$. The energy of $g(t)$ can be bounded by considering its samples at these points as

$$\|g(t)\|^2 = \tau \sum_n |g(n\tau)|^2 \tag{6.2.4}$$

$$\le \tau \sum_n \left(|f(n\tau)|^2 + 2|q(n\tau)f(n\tau)| + |q(n\tau)|^2 \right) \tag{6.2.5}$$

$$= \|f(t)\|^2 + E_s(\tau). \tag{6.2.6}$$

Note that

$$E_s(\tau) = \tau \sum_n \left(2|q(n\tau)g(n\tau)| + |q(n\tau)|^2 \right)$$

converges to

$$\lim_{\tau \to 0} E_s(\tau) = \int_{-\infty}^{+\infty} \left(2|q(t)g(t)| + |q(t)|^2 \right) dt, \tag{6.2.7}$$

which has to be finite since $q(t) = O\left(f(t)\right)$ and $f(t)$ is square integrable. Therefore, $E_s(\tau)$ has to be bounded by some E which does not depend on τ. This gives the following error bound

$$\|e(t)\|^2 \le \frac{\sigma^2 BC}{A} \left(\|f(t)\|^2 + E \right) \tau^2. \quad \blacksquare \tag{6.2.8}$$

Acknowledgment. The authors are grateful to N. Thao and T. Kalker for fruitful discussions on this topic, and to V. Goyal for his comments which improved the final version of this manuscript.

References

[1] Balian, R., Un principe d'incertitude fort en théorie du signal on mécanique quantique, *C. R. Acad. Sc. Paris* **292**, (1981).

[2] Cvetković, Z. and M. Vetterli, Deterministic analysis of errors in over-sampled A/D conversion and quantization of Weyl-Heisenberg frame expansions, submitted to *IEEE Trans. Inform. Theory*, April 1995.

[3] Daubechies, I., The wavelet transform, time-frequency localization and signal analysis, *IEEE Trans. Inform. Theory* **36** (1990), 961–1005.

[4] Daubechies, I., *Ten Lectures on Wavelets*, SIAM, Philadelphia, 1992.

[5] Duffin, R. J. and A. C. Schaeffer, A class of nonharmonic Fourier series, *Trans. Amer. Math. Soc.* **72** (1952), 341–366.

[6] Gabor, D., Theory of communications, *J. IEE* **93** (1946), 429–457.

[7] Goyal, V. K., M. Vetterli, and N. T. Thao, Quantization of overcomplete expansions, in *Proc. Data Compression Conference*, 1995, pp. 13–22.

[8] Goyal, V. K., Quantized overcomplete expansions: analysis, synthesis and algorithms, M.S. Thesis, University of California at Berkeley, July 1995.

[9] Hein, S. and A. Zakhor, Reconstruction of oversampled band-limited signals from $\Sigma\Delta$ encoded binary sequences, in *Proc. 25th Asilomar Conf. Signals Syst.*, November 1991, pp. 241–248.

[10] Hein, S. and A. Zakhor, Reconstruction of oversampled band-limited signals from $\Sigma\Delta$ encoded binary sequences, *IEEE Trans. Signal Processing* **42** (1994), 799–811.

[11] Kadec, I. M., The exact value of the Paley-Wiener constant, *Sov. Math. Dokl* (1964), 559–561.

[12] Low, F., Complete sets of wave packets, in *A Passion for Physics - Essays in Honor of Geoffrey Chew*, World Scientific, Singapore, 1985, pp. 17–22.

[13] Munch, N. J., Noise reduction in Weyl-Heisenberg frames, *IEEE Trans. Inform. Theory* **38** (1992), 608–616.

[14] Thao, N. T. and M. Vetterli, Oversampled A/D conversion using alternate projections, in *Proc. Conference on Information Sciences and Systems*, March 1991, pp. 241–248.

[15] Thao, N. T. and M. Vetterli, Deterministic analysis of oversampled A/D conversion and decoding improvement based on consistent estimates, *IEEE Trans. Signal Processing* **42** (1994), 519–531.

[16] Thao, N. T. and M. Vetterli, Reduction of the MSE in R-times oversampled A/D conversion from $O(1/R)$ to $O(1/R^2)$, *IEEE Trans. Signal Processing* **42** (1994), 200–203.

[17] Vetterli, M. and J. Kovacević, *Wavelets and Subband Coding*, Prentice-Hall, Englewood Cliffs, New Jersey, 1995.

[18] Young, R. M., *An Introduction to Nonharmonic Fourier Series*, Academic Press, New York, 1980.

Zoran Cvetković
Department of Electrical Engineering and Computer Sciences
University of California at Berkeley
Berkeley, CA 94720-1772
zoran@eecs.berkeley.edu

Martin Vetterli
Department d'Electricite EPFL
CH-1015 Lausanne
Switzerland
martin@eecs.berkeley.edu

IV.
Numerical Compression and Applications

Multiscale Inversion of Elliptic Operators

Amir Averbuch, Gregory Beylkin,
Ronald Coifman, and Moshe Israeli

Abstract. A fast adaptive algorithm for the solution of elliptic partial differential equations is presented. It is applied here to the Poisson equation with periodic boundary conditions. The extension to more complicated equations and boundary conditions is outlined.

The purpose is to develop algorithms requiring a number of operations proportional to the number of significant coefficients in the representation of the r.h.s. of the equation. This number is related to the specified accuracy, but independent of the resolution. The wavelet decomposition and the conjugate gradient iteration serve as the basic elements of the present approach.

The main difficulty in solving such equations stems from the inherently large condition number of the matrix representing the linear system that result from the discretization. However, it is known that periodized differential operators have an effective diagonal preconditioner in the wavelet system of coordinates. The condition number of the preconditioned matrix is $O(1)$ and, thus, depends only weakly on the size of the linear system.

The nonstandard form (nsf) is preferable in multiple dimensions since it requires $O(1)$ elements to represent the operator on all scales. Unfortunately, the preconditioned nsf turns out to be dense. This obstacle can be avoided if in the process of solving the linear system, the preconditioner is applied separately before and after the operator (to maintain sparsity).

A constrained version of the preconditioned conjugate gradient algorithm is developed in wavelet coordinates. Only those entries of the conjugate directions which are in the set of significant indices are used.

The combination of the above-mentioned elements yields an algorithm where the number of operations at each iteration is proportional to the number of elements. At the same time, the number of iterations is bounded by a constant.

§1 Introduction

In this paper we describe the components of a fast adaptive method for solving elliptic equations with periodic boundary conditions as well as de-

Signal and Image Representation in Combined Spaces
Y. Y. Zeevi and R. R. Coifman (Eds.), pp. 341–359.

velop a framework for solving problems with general boundary conditions. Let us consider the partial differential equation

$$\mathcal{L}u = f \quad x \in \mathbf{D} \subset \mathbf{R}^d, \tag{1.1}$$

with the boundary condition

$$\mathcal{B}u|_{\partial\mathbf{D}} = g, \tag{1.2}$$

where $\mathcal{L}$ is an elliptic operator,

$$\mathcal{L}u = - \sum_{i,j=1,\dots,d} (a_{ij}(x)\, u_{x_i})_{x_j} + b(x)\, u, \tag{1.3}$$

and $\mathcal{B}$ is the boundary operator,

$$\mathcal{B}u = \alpha u + \beta \frac{\partial u}{\partial N}. \tag{1.4}$$

We assume that the boundary $\partial\mathbf{D}$ is "complicated." As a practical matter we are interested in dimensions $d = 1, 2, 3$ though our considerations are valid in higher dimensions as well.

We adopt a classical approach to this problem which, until now, was not practical from the numerical point of view. We consider the following steps for solving the problem in (1.1) and (1.2):

1. We generate a function f_{ext}, a smooth extension of f outside the domain $\mathbf{D}$, such that f_{ext} is compactly supported in a rectangular box $\mathbf{B}$, $\mathbf{D} \subset \mathbf{B} \subset \mathbf{R}^d$, and $f = f_{\text{ext}}$ for $x \in \mathbf{D}$.

2. We solve the problem

$$\mathcal{L}u = f_{\text{ext}} \quad x \in \mathbf{B} \tag{1.5}$$

   with periodic boundary conditions.

3. Given the solution u_{ext} of (1.5), we look for the solution of (1.1) as

$$u = u_{\text{ext}} + v \tag{1.6}$$

   which yields the homogeneous equation

$$\mathcal{L}v = 0 \quad x \in \mathbf{D} \tag{1.7}$$

   with the boundary condition

$$\mathcal{B}v|_{\partial\mathbf{D}} = g - \mathcal{B}u_{\text{ext}}|_{\partial\mathbf{D}}, \tag{1.8}$$

   which we solve by boundary integral methods.

In order to realize the preceding steps we need to develop:

1. An algorithm for extending the function f outside the domain **D**.

2. An efficient method for solving (1.5).

3. An efficient method for solving the boundary integral equation derived from (1.7) and (1.8).

4. An effective algorithm for generating a representation of the solution v of (1.7) and (1.8) once the boundary integral equation is solved.

It is only recently that fast methods for solving the problem (1.7) and (1.8) using boundary integral equations have been developed, namely, Fast Multipole Method (FMM) in [9, 8, 7] and BCR algorithm in [6]. It is our understanding that 1 and 4 are solvable and the main difficulty resides with 2. We will address the algorithmic issues of 1, 3, and 4 elsewhere; and in this paper, we will concentrate on constructing an algorithm for solving (1.5).

§2 Approach

Our goal in solving (1.5) is to develop an adaptive algorithm where the number of operations will be proportional to the number of significant coefficients in the representation of f_{ext}. The usual solution procedure by current numerical methods requires discretization of the r.h.s. and of the solution in terms of a grid or a basis, such that the representations will resolve all features of interest. This might require a large number of grid points or elements not only near the singularities of the functions involved but also in the regions of smooth behavior, thus requiring proportionally large number of operations. Current adaptive procedures (for example, adaptive grids or irregular elements) are cumbersome, especially in higher dimensions and imply a considerable overhead both on the algorithmical and programming levels.

Our approach is based on using properties of representations of functions in wavelet bases and allows us to obtain a simple adaptive algorithm.

Let us illustrate it by considering Poisson's equation

$$\Delta u = f \quad x \in \mathbf{B} \tag{2.1}$$

with periodic boundary conditions where (with a slight abuse of notation) we used f instead of f_{ext} to denote the source term. The source term f may be discontinuous in the domain **B**.

Let us consider an MRA of $\mathbf{L}^2(\mathbf{R}^d)$,

$$\cdots \subset \mathbf{V}_2 \subset \mathbf{V}_1 \subset \mathbf{V}_0 \subset \mathbf{V}_{-1} \subset \mathbf{V}_{-2} \subset \cdots \tag{2.2}$$

where V_j is a subspace of an MRA spanned by translations of the scaling function,

$$\phi_{j,\mathbf{k}}(x) = 2^{-jd/2}\phi(2^{-j}x_1 - k_1)\,\phi(2^{-j}x_2 - k_2)\dots\phi(2^{-j}x_d - k_d), \quad (2.3)$$

where $x = (x_1,\dots,x_d)$ and $k = (k_1,\dots,k_d) \in \mathbf{Z}^d$. The function ϕ is the scaling function of MRA of $L^2(\mathbf{R})$.

Let us define the subspaces W_j as orthogonal complements of V_j in V_{j-1},

$$V_{j-1} = V_j \oplus W_j, \quad (2.4)$$

and represent the space $L^2(\mathbf{R}^d)$ as a direct sum

$$L^2(\mathbf{R}^d) = V_n \bigoplus_{j \leq n} W_j \quad (2.5)$$

where V_n is the subspace corresponding to the coarsest scale n. Let us define wavelets $\psi^\sigma_{j,\mathbf{k}}(x)$ which form an orthonormal basis of the subspaces W_j, $j \leq n$. We consider

$$\varphi^\rho(x) = \begin{cases} \psi(x) & \text{for } \rho = 1, \\ \phi(x) & \text{for } \rho = 0, \end{cases} \quad (2.6)$$

where ϕ and ψ are the scaling function and the wavelet of the MRA of $L^2(\mathbf{R})$. The wavelet ψ typically has several vanishing moments. Then it follows that

$$\psi^\sigma_{j,\mathbf{k}}(x) = 2^{-jd/2}\varphi^{\rho_1}(2^{-j}x_1 - k_1)\,\varphi^{\rho_2}(2^{-j}x_2 - k_2)\dots\varphi^{\rho_d}(2^{-j}x_d - k_d), \quad (2.7)$$

where the multi-index $\sigma = (\rho_1, \rho_2, \dots, \rho_d)$, $\sigma \neq \mathbf{0}$ and ρ_k, $k = 1, 2, \dots, d$, take values of either one or zero.

We also define projection operators on the subspaces V_j, $j \leq n$,

$$P_j : L^2(\mathbf{R}) \rightarrow V_j, \quad (2.8)$$

as follows

$$(P_j f)(x) = \sum_{\mathbf{k}} \langle f, \phi_{j,\mathbf{k}}\rangle \phi_{j,\mathbf{k}}(x) \quad (2.9)$$

while on the subspaces W_j, $j \leq n$,

$$Q_j : L^2(\mathbf{R}) \rightarrow W_j, \quad (2.10)$$

we define

$$Q_j = P_{j-1} - P_j = \sum_{\mathbf{k}}\sum_\sigma \langle f, \psi^\sigma_{j,\mathbf{k}}\rangle \psi^\sigma_{j,\mathbf{k}}(x), \quad (2.11)$$

where

$$\langle f, g \rangle = \int_{-\infty}^{\infty} f(x) \, g(x) dx. \tag{2.12}$$

The sum over σ is finite and the number of terms is $2^d - 1$ for each $\mathbf{k}$.

Let us represent the source term f and the solution u in (2.1) in the wavelet basis,

$$f(x) = \sum_{j \leq n} \sum_{\mathbf{k}} \sum_{\sigma} f_{j,\mathbf{k}}^{\sigma} \psi_{j,\mathbf{k}}^{\sigma}(x) + \sum_{\mathbf{k}} s_{n,\mathbf{k}}^{f} \phi_{n,\mathbf{k}}(x), \tag{2.13}$$

$$u(x) = \sum_{j \leq n} \sum_{\mathbf{k}} \sum_{\sigma} u_{j,\mathbf{k}}^{\sigma} \psi_{j,\mathbf{k}}^{\sigma}(x) + \sum_{\mathbf{k}} s_{n,\mathbf{k}}^{u} \phi_{n,\mathbf{k}}(x), \tag{2.14}$$

where

$$f_{j,\mathbf{k}}^{\sigma} = \langle f, \psi_{j,\mathbf{k}}^{\sigma} \rangle, \quad u_{j,\mathbf{k}}^{\sigma} = \langle u, \psi_{j,\mathbf{k}}^{\sigma} \rangle, \quad s_{n,\mathbf{k}}^{f} = \langle f, \phi_{n,\mathbf{k}} \rangle, \text{ and } s_{n,\mathbf{k}}^{u} = \langle u, \phi_{n,\mathbf{k}} \rangle. \tag{2.15}$$

We now define the ϵ-accuracy subspace for f to be the subspace on which f may be represented with accuracy ϵ, namely,

$$M_{\text{r.h.s}}^{\epsilon} = \mathbf{V}_n \bigcup \{\text{span}\{\psi_{j,\mathbf{k}}^{\sigma}\} \mid (j, \mathbf{k}, \sigma) : |f_{j,\mathbf{k}}^{\sigma}| > \epsilon\}, \tag{2.16}$$

and observe that the ϵ-accuracy subspace for the solution

$$M_{\text{sol}}^{\epsilon} = \mathbf{V}_n \bigcup \{\text{span}\{\psi_{j,\mathbf{k}}^{\sigma}\} \mid (j, \mathbf{k}, \sigma) : |u_{j,\mathbf{k}}^{\sigma}| > \epsilon\} \tag{2.17}$$

may be estimated given $M_{\text{r.h.s}}^{\epsilon}$. It may be verified that

Proposition 1. *Let*

$$u(x) = \sum_{j} \sum_{\mathbf{k}} \sum_{\sigma} u_{j,\mathbf{k}}^{\sigma} \psi_{j,\mathbf{k}}^{\sigma}(x) + \text{constant} \tag{2.18}$$

be the solution of

$$\Delta u = \psi_{j',\mathbf{k}'}^{\sigma'} \quad x \in \mathbf{B} \tag{2.19}$$

with periodic boundary conditions. For any $\epsilon > 0$ there exist $\lambda > 0$ and $\mu > 0$ such that all indices $(j, \mathbf{k}, \sigma)$ corresponding to the significant coefficients of the solution, $|u_{j,\mathbf{k}}^{\sigma}| \geq \epsilon$, satisfy $|\mathbf{k} - \mathbf{k}'| \leq \lambda$ and $|j - j'| \leq \mu$.

The size of $\mu > 0$ and $\lambda > 0$ depends on the particular choice of basis and, of course, on ϵ. Given $M_{\text{r.h.s}}^{\epsilon}$, we may construct the set $M_{\lambda,\mu}$ as a (λ, μ)-neighborhood of $M_{\text{r.h.s}}^{\epsilon}$. According to Proposition 1, $M_{\text{sol}}^{\epsilon} \subset M_{\lambda,\mu}$. We note that estimating the subspace amounts to constructing a mask which contains indices of significant coefficients.

Instead of estimating $M_{\lambda,\mu}$ directly, we may use an iterative approach. For example, solving directly on $M_{\text{r.h.s}}^{\epsilon}$ produces a solution $\tilde{u}$ with accuracy $\tilde{\epsilon} > \epsilon$. Applying the Laplacian to $\tilde{u}$, we generate $\tilde{f}$. Estimating the ϵ-accuracy subspace for $\tilde{f}$, we may use it to continue the iteration to improve the accuracy of the solution. In other words, the mask for $M_{\text{sol}}^{\epsilon}$ may be generated iteratively.

There are three main features in our approach to solve (2.1):

1. **Estimation of the ϵ-accuracy subspace for the solution.** Our first step is to explicitly estimate the subspace $M_{\text{sol}}^{\epsilon}$ given $M_{\text{r.h.s}}^{\epsilon}$. For elliptic operators, the dimension of $M_{\text{sol}}^{\epsilon}$ is proportional to that of $M_{\text{r.h.s}}^{\epsilon}$.

2. **Preconditioning of the operator.** A simple diagonal preconditioner is available for periodized differential operators in the wavelet bases [4, 5] which yields a condition number of $O(1)$. We will show below how to construct simple preconditioners in wavelet bases for more general operators.

3. **Constrained iterative solver.** We use the preconditioned Conjugate Gradient (CG) method which we constrain to the subspace estimated at Step 1, *e.g.* $M_{\lambda,\mu}$. The CG method requires only a constant number of iterations due to preconditioning at Step 2, whereas the cost of each iteration is proportional to the dimension of $M_{\text{sol}}^{\epsilon}$ provided we succeed to limit the number of operations required for the application of the operator (matrix) in the CG method (see below).

Steps 1–3 constitute an adaptive algorithm for solving Poisson's equation.

§3 Outline of the algorithm

Let us consider the projection L_0 of the periodized operator Δ on $\mathbf{V}_0$, the finest scale under consideration,

$$L_0 = P_0 \, \Delta \, P_0, \tag{3.1}$$

and L_s and L_{ns}, its standard (s-form) and non-standard forms (ns-form) [6].

One of the difficulties in solving (1.5) stems from the inherently large condition number of the linear system resulting from the discretization of (1.5). As was shown in [4] and [5], using a diagonal preconditioner in the wavelet system of coordinates yields a linear system with the condition number typically less than 10, independently of its size. Let $\mathcal{P}$ denote such a diagonal preconditioner.

In [5] the s-form is used to solve the two-point boundary-value problem. Alternatively, we may use the ns-form. Some care is required at this point since the preconditioned ns-form is dense unlike the s-form, which remains sparse. Thus, in the process of solving the linear system, it is necessary to apply the preconditioner and the ns-form sequentially in order to maintain sparsity. The ns-form is preferable in multiple dimensions since, for example, differential operators require $O(1)$ elements for representation on all scales (see *e.g.* [4]).

We develop a constrained (see below) preconditioned CG algorithm for solving (1.5) in an adaptive manner. Both the s-form and the ns-form may be used for this purpose but it appears that using the ns-form is more efficient, especially if compactly supported wavelets are used and high accuracy is required.

Let us consider (1.5) in the wavelet system of coordinates

$$L_{ns} u_w = f_w, \tag{3.2}$$

where f_w and u_w are representations of f and u in the wavelet system of coordinates. This equation should be understood to include the rules for applying the ns-form (see [6]).

Let us rewrite (3.2) using the preconditioner $\mathcal{P}$ as

$$\mathcal{P} L_{ns} \mathcal{P} v = \mathcal{P} f_w, \tag{3.3}$$

where $\mathcal{P} v = u$. For example, for the second derivative the preconditioner $\mathcal{P}$ is as follows:

$$\mathcal{P}_{il} = \delta_{il} 2^j \tag{3.4}$$

where $1 \le j \le n$ is chosen depending on i, l so that $n - n/2^{j-1} + 1 \le i, l \le n - n/2^j$, and $\mathcal{P}_{nn} = 2^n$.

The periodized operator Δ has the null space of dimension one which contains constants. If we use the full decomposition (over all n scales) in the construction of the ns-form, then the null space coincides with the subspace $\mathbf{V}_n$, which in this case has dimension one (see [5]). This allows us to solve (3.3) on the range of the operator,

$$\bigoplus_{1 \le j \le n} \mathbf{W}_j \tag{3.5}$$

where the linear system (3.3) is well conditioned.

Remark. Operators with variable coefficients. As in the case of the Laplacian, the ϵ-accuracy subspace for the solution may be estimated using corresponding subspaces for the r.h.s and the coefficients. Essentially, we consider the union of such subspaces as a starting point for constructing $M_{\lambda,\mu}$. These estimates may be revised in the process of iteration.

348   A. Averbuch et al.

§4 Preconditioner for the operator $-\Delta + $ Const

An "efficient" preconditioner is an essential element in the present approach. In a more restricted sense, "efficient" means insensitive to the size
of the problem.

Let us demonstrate how to construct a diagonal preconditioner for the
sum of operators $-\Delta + $ Const in the wavelet bases. We observe that if
A and B are diagonal operators with diagonal entries a_i and b_i, then the
diagonal operator with entries $1/(a_i + b_i)$ (provided $a_i + b_i \neq 0$) is an ideal
preconditioner.

In our case, the operator $-\Delta$ is not diagonal but we know a good
diagonal preconditioner for it in wavelet bases (3.4). Let us use this preconditioner instead of $-\Delta$ for the purpose of constructing a preconditioner
for $-\Delta + $ Const, where Const > 0. We note that in wavelet bases the
identity operator remains unchanged. We restrict Const $\cdot I$, where I is the
identity operator, to the subspace

$$\bigoplus_{1 \leq j \leq n} \mathbf{W}_j \tag{4.1}$$

and construct a preconditioner on this subspace.

We obtain

$$\mathcal{P}_{il} = \frac{\delta_{il}}{\sqrt{2^{-2j} + \text{Const}}} \tag{4.2}$$

where $1 \leq j \leq n$ is chosen depending on i,l so that $n - n/2^{j-1} + 1 \leq i,l \leq
n - n/2^j$, and $\mathcal{P}_{nn} = 1/\sqrt{2^{-2n} + \text{Const}}$. The square root appears in (4.2)
in order to symmetrize the application of the preconditioner as shown in
the previous section. In Table 1 we illustrate the effect of preconditioning
of the operator $-\frac{d^2}{dx^2} + $ Const by the diagonal matrix (4.2).

Remark. If we consider an operator $-\Delta + V$, where V is an operator of
multiplication by a function $V(x)$, a similar construction may be obtained
on fine scales. On fine scales where the function $V(x)$ does not change
significantly over the support of wavelets, we may consider the diagonal
operator V^{diag},

$$\left(V^{\text{diag}} \psi_{j,\mathbf{k}}^\sigma \right) = V(x_{j,\mathbf{k}}) \psi_{j,\mathbf{k}}^\sigma, \tag{4.3}$$

where $x_{j,\mathbf{k}}$ is a point within the support of the wavelet $\psi_{j,\mathbf{k}}^\sigma$. Using V^{diag}
instead of V, we obtain the preconditioner in a manner outlined above. We
will address the problem of constructing preconditioners for operators of
the form $-\Delta + V$ elsewhere.

Const	κ	κ_p
$7.1 \cdot 10^{-0}$	$2.4 \cdot 10^0$	2.1
$7.1 \cdot 10^{-1}$	$1.5 \cdot 10^1$	6.3
$7.1 \cdot 10^{-2}$	$1.4 \cdot 10^2$	9.4
$7.1 \cdot 10^{-3}$	$1.3 \cdot 10^3$	9.5
$7.1 \cdot 10^{-4}$	$6.7 \cdot 10^3$	7.5
$7.1 \cdot 10^{-5}$	$1.5 \cdot 10^4$	5.0

Table 1. Condition numbers κ before and κ_p after preconditioning of the operator $-\frac{d^2}{dx^2} + \text{Const}$ in the basis of Daubechies' wavelets with six vanishing moments. There are 8 scales and the matrix size is 256×256.

§5 Initial numerical experiments

Several numerical experiments have been performed in one dimension to verify the main points of the procedure outlined in this paper. A two- and three-dimensional version of an adaptive Poisson solver is being implemented and the results will be reported elsewhere.

In our one-dimensional experiment we set out to verify that using the CG method we can: (1) maintain the sparsity of the conjugate directions (and all other auxiliary vectors) used in the method, this is essential for establishing the sought for operation count, and: (2) solve the problem with a number of iterations that do not depend on the size of the problem.

In order to be able to make a convergence test independently of discretization errors, we first construct the solution and then apply the operator to obtain the right-hand side of the equation. We then solve the equation with a given accuracy ϵ and compare the results with the exact solution to verify the performance of the algorithm.

As the solution we chose initially the vector

$$u_i = \sin(2\pi i)/N, \tag{5.1}$$

where $i = 0, \ldots, N - 1$, the size $N = 2^n$ represents the size of the vector in the ordinary system of coordinate and n indicates the maximum number of scales in the problem. A more interesting problem is given by the solution vector

$$u_i = \tanh(\omega \sin(2\pi i)/N), \tag{5.2}$$

where ω is a parameter. By choosing ω we control the steepness of the slope of the function and, thus, the necessary number of scales to represent this vector in a wavelet basis up to some accuracy ϵ.

The initial experiments with very smooth solution vectors demonstrated a severe problem in the algorithm which is exhibited in Figures 1–5. For example, consider the wavelet coefficients of the solution vector for $\omega = 0.8$, $N = 1024$, analyzed with Daubechies' basis with 10 vanishing moments. Obviously, both the solution and the initial approximation have a very small number of significant coefficients concentrated in the coarse scales. (The wavelet coefficients are usually exhibited by plotting the magnitude against the number of the coefficient in the following sequence: the fine-scale wavelet coefficients numbered (in the present case) 1 to 512 appear first in the plot of the wavelet decomposition, immediately followed by the coarser scale with 256 coefficients, 513 to 768 followed by the third level with 128, from 769 to 896 then the levels containing the next 64, 32, 16, 8, 4, 2, 1 coefficients respectively, bringing us to 1023; the last coefficient, 1024, is the mean.)

Figure 1 shows only the largest coefficients (1010–1024), other coefficients are not visible on this scale. The other four figures show the behavior of the 64 coefficients in level four (897–960). The first iteration result, ans_1_iter, vanishes for this level, iteration 3 shows a significant growth, iteration five is even larger, iteration seven drops by an order of magnitude, iteration nine has again vanishingly small coefficients on this scale. We can summarize this behavior and our solution in the following way:

Constrained iterative solver. In order to solve (3.3) we apply the Conjugate Gradient method constrained to the subspace $M_{\lambda,\mu}$. Without such constraint the conjugate directions become "dense" at early stages of the iteration only to become small outside the subspace $M_{\lambda,\mu}$ later. Thus, constraining the solution to a subspace is critical for an adaptive algorithm.

In applying this "constrained" conjugate gradient method in the wavelet coordinates, we generate only those entries of conjugate directions which are in the set of significant indices which define the subspace $M_{\lambda,\mu}$. This yields an algorithm where the number of operations at each iteration is proportional to the number of elements of $M_{\lambda,\mu}$. The number of iterations is $O(1)$ and, thus, the overall number of operations is proportional to the number of significant coefficients of f, $i.e.$ the dimension of $M_{\text{r.h.s}}^{\epsilon}$.

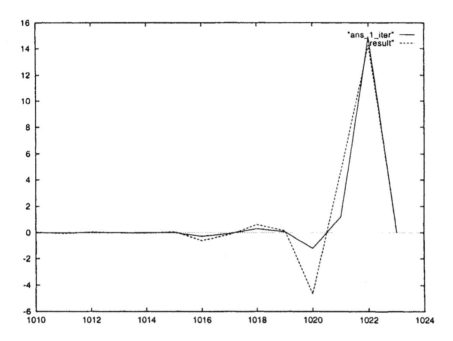

Figure 1. Compare the final result with the result after 1 iteration for $\omega = 0.8$, 1024 points, Daubechies 20, no skip.

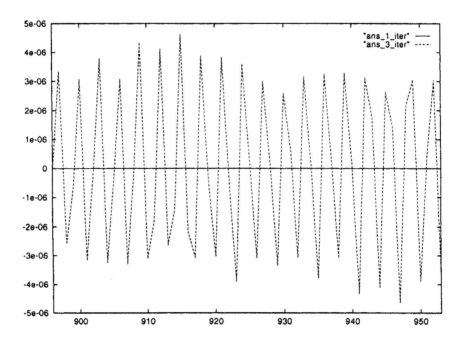

Figure 2. Compare the results after 1 and 3 iterations for $\omega = 0.8$, 1024 points, Daubechies 20, no skip.

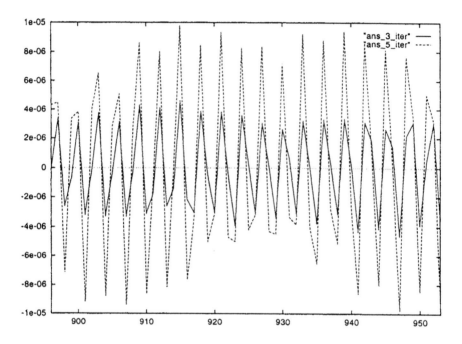

Figure 3. Compare the results after 3 and 5 iterations for $\omega = 0.8$, 1024 points, Daubechies 20, no skip.

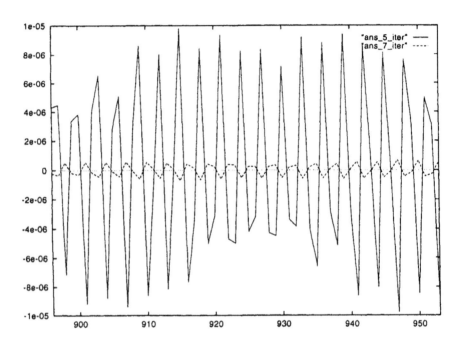

Figure 4. Compare the results after 5 and 7 iterations for $\omega = 0.8$, 1024 points, Daubechies 20, no skip.

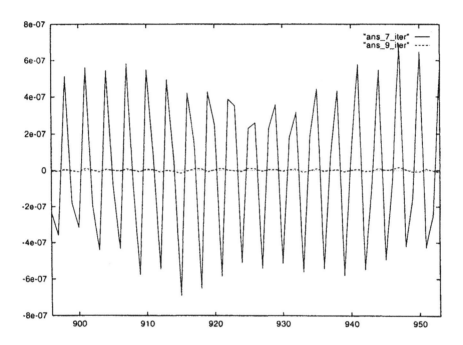

Figure 5. Compare the results after 7 and 9 iterations for $\omega = 0.8$, 1024 points, Daubechies 20, no skip.

# scales to skip	# iterations	$\| \ \|_\infty$	$\| \ \|_2$
0	36	10^{-13}	10^{-14}
1	36	0.45×10^{-9}	0.32×10^{-9}
2	34	0.96×10^{-8}	0.52×10^{-8}
3	32	0.17×10^{-6}	0.83×10^{-7}
4	27	0.30×10^{-5}	0.13×10^{-5}
5	17	0.48×10^{-4}	0.21×10^{-4}
6	9	0.80×10^{-3}	0.34×10^{-3}

Table 2. Coiflet 12, $2^{10} \times 2^{10}$ matrix.

§6   Further numerical experiments

In the second set of one-dimensional experiments we verify that using the "constrained CG" method we can maintain the sparsity of the conjugate directions. At the same time we must check that the convergence rate of the method was not damaged by the restriction of the search directions to a small subspace. In general we cannot expect success, however in the present case there are a number of special "mitigating circumstances."

For the smooth problems considered in the previous set of experiments the coefficients are concentrated in the coarse scales, so a very effective "mask" is obtained by setting to zero a number of fine scales. We demonstrate the results in Tables 2, 3, and 4 where we show the number of iterations of the constrained CG method as a function of the number of scales skipped (where 0 means all levels are used, 1 means the fine level was skipped, 2 means the two finest levels were skipped, etc.) and the accuracy achieved as measured against the exact solution. In the present approach skipping some levels means that the answer itself is represented with a small number of scales and the accuracy will be affected adversely. A more elaborate mask is used in our later experiments [1].

We observe that the number of iterations did not increase at all as a consequence of the constraint. Furthermore, in all cases checked it was a monotonically decreasing function of the number of scales skipped. We can conclude that at least for the case of smooth-forcing functions and constant coefficient operators the CPCG (constrained preconditioned conjugate gradient approach) results in a number of operations proportional to the number of significant coefficients of the forcing. More elaborate masks appropriate for non-smooth functions are explored in the next paper in this sequence [1].

# scales to skip	# iterations	$\| \ \|_\infty$	$\| \ \|_2$
0	29	10^{-14}	10^{-15}
1	28	0.20×10^{-13}	0.13×10^{-13}
2	28	0.17×10^{-11}	0.81×10^{-12}
3	26	0.97×10^{-10}	0.52×10^{-10}
4	23	0.64×10^{-8}	0.33×10^{-8}
5	16	0.40×10^{-6}	0.20×10^{-6}
6	9	0.25×10^{-4}	0.12×10^{-4}
7	5	0.14×10^{-2}	0.77×10^{-3}
8	3	0.65×10^{-1}	0.38×10^{-1}

Table 3. Daubechies 12, $2^{10} \times 2^{10}$ matrix.

# scales to skip	# iterations	$\| \ \|_\infty$	$\| \ \|_2$
0	22	10^{-14}	10^{-14}
1	22	10^{-14}	10^{-14}
2	22	10^{-14}	10^{-14}
3	21	10^{-14}	10^{-14}
4	19	0.36×10^{-13}	0.17×10^{-13}
5	14	0.34×10^{-10}	0.17×10^{-10}
6	9	0.33×10^{-7}	0.17×10^{-7}
7	5	0.27×10^{-4}	0.15×10^{-4}
8	3	0.13×10^{-1}	0.77×10^{-2}

Table 4. Daubechies 20, $2^{10} \times 2^{10}$ matrix.

Acknowledgment. The research of A. Averbuch, R. Coifman, and M. Israeli was supported by U.S-Israel Binational Science Foundation Grant # 92-00269/1. The research of M. Israeli was supported also by the Fund for the Promotion of Research at the Technion. The research of G. Beylkin was partially supported by ARPA grant F49620-93-1-0474 and ONR grant N00014-91-J4037.

References

[1] Averbuch, A., G. Beylkin, R. Coifman, and M. Israeli, Fast adaptive algorithm for elliptic equations with periodic boundary conditions, to appear.

[2] Averbuch, A., M. Israeli, and L. Vozovoi, Spectral multidomain technique with local Fourier basis, *J. Scientific Computing* **8**(2) (1993), 135–149.

[3] Aharoni, G., A. Averbuch, R. Coifman, and M. Israeli, Local cosine transform – A method for the reduction of the blocking effect in JPEG, *J. Math. Imaging and Vision*, Special Issue on Wavelets, **3** (1993), 7–38.

[4] Beylkin, G., On the representation of operators in bases of compactly supported wavelets, *SIAM J. Numer. Anal.* **29**(6) (1992), 1716–1740.

[5] Beylkin, G., On wavelet-based algorithms for solving differential equations, in *Wavelets: Mathematics and Applications*, J. J. Benedetto and M. W. Frazier (eds.), CRC Press, 1994, 449–466.

[6] Beylkin, G., R. R. Coifman, and V. Rokhlin, Fast wavelet transforms and numerical algorithms I, *Comm. Pure and Appl. Math.*, **44** (1993), 141–183. Yale University Technical Report YALEU/DCS/RR-696, August 1989.

[7] Carrier, J., L. Greengard, and V. Rokhlin. A fast adaptive multipole algorithm for particle simulations, *SIAM J. Scientific and Statistical Computing* **9**(4), 1988. Yale University Technical Report, YALEU/DCS/RR-496 (1986).

[8] Greengard, L. and V. Rokhlin, A fast algorithm for particle simulations, *J. Comp. Phys.* **73**(1) (1987), 325–348.

[9] Rokhlin, V., Rapid solution of integral equations of classical potential theory, *J. Comp. Phys.* **60**(2) (1985).

Amir Averbuch
School of Mathematical Sciences
Tel Aviv University
Tel Aviv 69978, Israel
amir@math.tau.ac.il

Gregory Beylkin
Program in Applied Mathematics
University of Colorado at Boulder
Boulder, CO 80309-0526
beylkin@julia.colorado.edu

Ronald Coifman
Department of Mathematics
P.O. Box 208283
Yale University
New Haven, CT 06520-8283
coifman@jules.math.yale.edu

Moshe Israeli
Faculty of Computer Science
Technion-Israel Institute of Technology
Haifa 32000, Israel
israeli@cs.technion.ac.il

Multiresolution Representation of Cell-Averaged Data: A Promotional Review

Ami Harten

Abstract. In this paper we review recent developments in techniques to represent data in terms of its *local* scale components. These techniques enable us to obtain data compression by eliminating scale-coefficients which are sufficiently small. This capability for data compression can also be used to reduce the cost of many numerical solution algorithms. The purpose of this paper is to promote the use of multiresolution representation schemes which are based on cell-average discretization as the "Standard Method" for data compression.

§1   Introduction

Fourier analysis, which provides a way to represent square-integrable functions in terms of their sinusoidal scale-components, has contributed greatly to all fields of science. The main drawback of Fourier analysis is in its globality – a single irregularity in the function dominates the behavior of the scale-coefficients and prevents us from getting immediate information about the behavior of the function elsewhere.

The development of the theory of wavelets (see [21] and [20]) was a great step towards local scale decomposition, and has already had great impact on several fields of science. In numerical analysis representation by compactly supported wavelets (see [6] and [5]) is used to reduce the cost of many numerical solution algorithms by either applying it to the numerical solution operator to obtain an approximate sparse form (see [4]), or by applying it to the numerical solution itself to obtain an approximate reduced representation in order to solve for less quantities (see [13]–[14]). The main drawback of the theory of wavelets is that it attempts to decompose any square integrable function into scale-components which are translates and dilates of a single function. Consequently, there are conceptual difficulties in extending wavelets to bounded domains and general geometries. Furthermore, the uniformity of the underlying wavelet approximation makes it

Signal and Image Representation in Combined Spaces
Y. Y. Zeevi and R. R. Coifman (Eds.), pp. 361–391.

impossible to obtain an adaptive (data-dependent) multiresolution *representation* which fits the approximation to the local nature of the data. The only adaptivity which is possible within the theory of wavelets is through redundant "dictionaries."

In a series of works [10]–[12] we have studied the question of how to represent discrete data which originates from unstructured grids in bounded domains in terms of scale decomposition. Combining ideas from multigrid methods, numerical solution of conservation laws, hierarchical bases of finite element spaces, subdivision schemes of Computer-Aided Design and of course – the theory of wavelets, we came up with the more general concept of "nested sequence of discretization." We say that a sequence of linear operators $\{D_k\}_{k=0}^{\infty}$ is a nested sequence of discretization if

(i)

$$\mathcal{D}_k : \mathcal{F} \xrightarrow{\text{onto}} V^k, \quad \dim V^k = J_k, \tag{1.1a}$$

(ii)

$$\mathcal{D}_k f = 0 \Rightarrow \mathcal{D}_{k-1} f = 0. \tag{1.1b}$$

Here $\mathcal{F}$ is a space of mappings and V^k is a linear space of dimension J_k.

Given any discrete data $v^L = \mathcal{D}_L f$ we show in [11]–[12] that it has a multiresolution representation, *i.e.* a one-to-one correspondence between the given data and its scale-decomposition:

$$v^L \xleftrightarrow{1:1} \{d^L, \dots, d^1, v^0\} =: \hat{v}. \tag{1.2a}$$

The k-th scale-coefficients $d^k = \{d_j^k\}_{j=1}^{J_k - J_{k-1}}$ represent the "difference in information" between $\mathcal{D}_k f$ and $\mathcal{D}_{k-1} f$, and $v^0 = \mathcal{D}_0 f$ is the discretization of f on the coarsest level. Observe that the number of components in $\hat{v}$ is the same as that of v^L because

$$\sum_{k=1}^{L} (J_k - J_{k-1}) + J_0 = J_L. \tag{1.2b}$$

A nested sequence of discretization comes equipped with a decimation operator

$$D_k^{k-1} : V^k \xrightarrow{\text{onto}} V^{k-1}, \tag{1.3a}$$

which is defined by assigning to $v^k = \mathcal{D}_k f \in V^k$ the value of $\mathcal{D}_{k-1} f \in V^{k-1}$, *i.e.*

$$D_k^{k-1}(\mathcal{D}_k f) =: \mathcal{D}_{k-1} f. \tag{1.3b}$$

To see that the decimation operator is well defined by the preceding observe that (1.1a) implies that

$$V^k = \mathcal{D}_k(\mathcal{F}), \quad V^{\|-\infty} = \mathcal{D}_{\|-\infty}(\mathcal{F})$$

and that if $v^k = \mathcal{D}_k f_1 = \mathcal{D}_k f_2$, then it follows from the nestedness (1.1b) that $\mathcal{D}_{k-1} f_1 = \mathcal{D}_{k-1} f_2$ because

$$0 = \mathcal{D}_k f_1 - \mathcal{D}_k f_2 = \mathcal{D}_k (f_1 - f_2) \Longrightarrow 0 = \mathcal{D}_{k-1}(f_1 - f_2) = \mathcal{D}_{k-1} f_1 - \mathcal{D}_{k-1} f_2;$$

thus the definition $D_k^{k-1} v^k = \mathcal{D}_{k-1} f$ is independent of the particular f. We refer to the assignment of $f \in \mathcal{F}$ to $v^k \in V^k$ such that $v^k = \mathcal{D}_k f$ as reconstruction, and denote it by $\mathcal{R}_k$

$$\mathcal{R}_k : V^k \to \mathcal{F}, \quad \mathcal{D}_k \mathcal{R}_k = I_k, \tag{1.4}$$

where I_k is the identity operator in V^k; note that $\mathcal{R}_k$ is a right-inverse of $\mathcal{D}_k$. We would like to point out that the reconstruction need not be a linear operator, *i.e.* it may depend on the data v^k.

Given any $v^L \in V^L$ we evaluate $\{v^k\}_{k=0}^{L-1}$ by repeated decimation

$$v^{k-1} = D_k^{k-1} v^k, \quad k = L, \dots, 1. \tag{1.5}$$

Starting from v^{k-1} in (1.5) we approximate v^k by

$$v^k \approx \mathcal{D}_k(\mathcal{R}_{k-1} v^{k-1}),$$

and denote

$$P_{k-1}^k = \mathcal{D}_k \mathcal{R}_{k-1} : V^{k-1} \to V^k. \tag{1.6a}$$

P_{k-1}^k, the prediction operator, is a right-inverse of the decimation D_k^{k-1}

$$D_k^{k-1} P_{k-1}^k = I_{k-1}. \tag{1.6b}$$

To see that take $f = \mathcal{R}_{k-1} v^{k-1}$ in (1.3b) and use (1.4). We observe that the prediction error e^k

$$e^k = v^k - P_{k-1}^k v^{k-1} = (I_k - P_{k-1}^k D_k^{k-1}) v^k \tag{1.7a}$$

satisfies the relation

$$D_k^{k-1} e^k = D_k^{k-1} v^k - (D_k^{k-1} P_{k-1}^k) v^{k-1} = v^{k-1} - v^{k-1} = 0$$

and therefore, it is in the null space of the decimation operator

$$e^k \in \mathcal{N}(D_k^{k-1}) = \{v| \quad v \in V^k, \ D_k^{k-1} v = 0\}. \tag{1.7b}$$

It follows from (1.3a) that

$$\dim \mathcal{N}(D_k^{k-1}) = J_k - J_{k-1} \tag{1.8a}$$

and therefore, the prediction error e^k, which is described in terms of J_k components in V^k, can be represented in $\mathcal{N}(D_k^{k-1})$ by $(J_k - J_{k-1})$ coefficients $d^k = \{d_j^k\}_{j=1}^{J_k - J_{k-1}}$, which we consider as a representation of the k-th scale. To be specific, let $\{\mu_j^k\}_{j=1}^{J_k - J_{k-1}}$ be any basis of $\mathcal{N}(D_k^{k-1})$,

$$\mathcal{N}(D_k^{k-1}) = \mathrm{span}\{\mu_j^k\}_{j=1}^{J_k - J_{k-1}}, \tag{1.8b}$$

and let d^k denote the coordinates of e^k in this basis

$$e^k = \sum_{j=1}^{J_k - J_{k-1}} d_j^k \mu_j^k =: E_k d^k, \quad d^k =: G_k e^k. \tag{1.9a}$$

Here G_k denotes the operator which assigns to $e^k \in \mathcal{N}(D_k^{k-1})$ its coordinates d^k in the basis $\{\mu_j^k\}_{j=1}^{J_k - J_{k-1}}$; observe that $E_k G_k$ is the identity operator in $\mathcal{N}(D_k^{k-1})$, *i.e.*

$$E_k G_k e^k = e^k \quad \text{for any } e^k \in \mathcal{N}(D_k^{k-1}). \tag{1.9b}$$

Using the above definition of scale-coefficients we obtain the multiresolution representation (MR) in (1.2a). The direct MR transform $\hat{v} = M \cdot v^L$ is given by the algorithm

$$\begin{cases} \mathrm{DO} \ \ k = L, \ldots, 1 \\[2mm] v^{k-1} = D_k^{k-1} v^k \\[2mm] d^k = G_k(I_k - P_{k-1}^k D_k^{k-1}) v^k =: G_k^D v^k. \end{cases} \tag{1.10}$$

The inverse MR transform $v^L = M^{-1} \cdot \hat{v}$ is given by

$$\begin{cases} \mathrm{DO} \ \ k = 1, \ldots, L \\[2mm] v^k = P_{k-1}^k v^{k-1} + E_k d^k. \end{cases} \tag{1.11}$$

In order to apply this multiresolution representation to real-life problems for purposes of analysis and data compression, we have to make sure that the direct MR transform and its inverse are stable with respect to perturbations. In [12] we present stability analysis for MR schemes and derive a sufficient condition which seems to be "close" to necessary; this condition also implies existence of a multiresolution basis for mappings in $\mathcal{F}$. In Appendix A we review some elements of this analysis and relate them to the particular examples of the present paper.

We remark that in multigrid terminology D_k^{k-1} is "restriction" and P_{k-1}^k is "prolongation." In signal processing D_k^{k-1} plays the role of "low-pass filter" while G_k^D, which is defined in (1.10), plays the role of "high-pass filter." This framework is a generalization of the theory of wavelets in the sense that under conditions of uniformity, its natural result is wavelets.

The selection of a nested sequence of discretization in (1.1a) constitutes a setting for the multiresolution representation (1.1b), *i.e.* it determines the operators D_k^{k-1}, G_k, and E_k. Once this is done, any choice of corresponding reconstruction operators $\{\mathcal{R}_k\}_{k=0}^L$ defines a MR scheme for discrete data v^L in V^L by taking $P_{k-1}^k = \mathcal{D}_k \mathcal{R}_{k-1}$ in the transforms (1.10)–(1.11). This opens up a tremendous number of possibilities for the design of MR schemes, where the primary consideration is the selection of an appropriate discretization. In this paper we claim that if we had to choose one of the many discretizations as a setting for the "standard data compression" scheme, then the best choice is that of cell-average discretization.

The considerations in selecting a particular discretization are : (1) Simplicity and flexibility; (2) Information contents; (3) Suitability to applications.

In Section 2 we show how to construct a nested sequence of cell-average discretization in any compact set in any number of space dimensions. In Section 3 we consider the one-dimensional case and show that any known method of interpolation gives rise to a corresponding technique for reconstruction from cell averages. Thus, the only competition to cell-average discretization with respect to simplicity and flexibility comes from discretization by point values (see [3]); however, cell-average discretization carries more information than point-value discretization.

In Section 6 we demonstrate the difference in information contents between these two discretizations by describing the subcell reconstruction technique by which a discontinuous piecewise-polynomial function can be recovered exactly from its cell averages; this is not true for point-value discretization. In [10] we consider discretizations by weighted averages and conclude that there are no major differences in regions of smoothness, but there are differences in the information about irregularities of the sampled function, which can be recovered from higher moments of the data. Thus "hat-averages" contain more information than cell averages, and one can exactly recover discontinuities as well as distributions in piecewise-polynomial functions from their "hat-averages" (see [2]). In this respect "spectral discretization," *e.g.* taking the first J_k Fourier coefficients, has the most information. However, cell-average discretization clearly wins over higher-moment discretizations in terms of simplicity and flexibility.

In Section 4 we describe the relation between the biorthogonal wavelets of [5] and our formulation. We show that biorthogonal wavelets can be thought of as the "uniform constant-coefficient" case of our framework, which corresponds to a choice of some weighted-average discretization in R. We study the case which corresponds to cell-average discretization and show that the corresponding biorthogonal wavelets are the hierarchical form of piecewise-polynomial reconstruction. The "competition" between Daubechies' orthonormal wavelets [6] to the biorthogonal wavelets of [5] is generally settled in favor of the latter, because in many applications the orthogonality doesn't buy you much, and it costs quite a bit.

In Section 5 we present stability analysis of data compression algorithms, which is based on the results of [12] and summarized in Appendix A. We also present our ideas on error-control algorithms, which enable us to prescribe an upper bound on the compression error and provides a meaningful and stable way to apply adaptive techniques. Because of the simplicity of cell-average discretization we can actually prescribe a strategy for truncation of scale coefficients and prove our contention about error control.

In Section 7 we present some numerical experiments and discuss some aspects of suitability of the discretization to various applications.

§2   MR schemes for cell-average discretization

Consider absolutely integrable functions $f \in \mathcal{F}$

$$f : \Omega \subset \mathrm{R}^m \longrightarrow \mathrm{R}, \qquad \mathcal{F} = L^1(\Omega) \tag{2.1a}$$

where Ω is a compact set, and let $C^k = \{c_i^k\}_{i=1}^{J_k}$ be a set of cells such that

$$\Omega^k =: \overline{\cup_{i=1}^{J_k} c_i^k} \subseteq \Omega, \qquad c_i^k \cap c_j^k = \emptyset \text{ for } i \neq j. \tag{2.1b}$$

We define the cell-average discretization by

$$(\mathcal{D}_k f)_i = \frac{1}{|c_i^k|} \int_{c_i^k} f(x)dx, \qquad |c_i^k| = \int_{c_i^k} dx, \tag{2.1c}$$

and consider a refinement sequence $\{C^k\}_{k=0}^{L}$ in which C^k is formed from C^{k-1} by dividing each cell c_i^{k-1} into, say q, disjoint cells $\{c_{i_\ell}^k\}_{\ell=1}^q$,

$$\overline{\cup_{\ell=1}^q c_{i_\ell}^k} = \overline{c_i^{k-1}}. \tag{2.1d}$$

Alternatively, we can consider (2.1d) to be a coarsening procedure in which we agglomerate every q cells of C^k into a larger cell of C^{k-1}; the only reason that we take here a fixed q is to simplify the notations. It follows from the additivity of the integral that

$$(\mathcal{D}_{k-1}f)_i = \frac{1}{|c_i^{k-1}|} \sum_{\ell=1}^{q} \int_{c_{i_\ell}^k} f(x)dx = \frac{1}{|c_i^{k-1}|} \sum_{\ell=1}^{q} |c_{i_\ell}^k|(\mathcal{D}_k f)_{i_\ell}. \qquad (2.2a)$$

This shows that condition (1.1b) is satisfied, and provides the definition of the decimation operator in (1.3)

$$v_i^{k-1} = (D_k^{k-1}v^k)_i =: \frac{1}{|c_i^{k-1}|} \sum_{\ell=1}^{q} |c_{i_\ell}^k| v_{i_\ell}^k. \qquad (2.2b)$$

Let e^k denote the prediction error in (1.7a), then

$$D_k^{k-1}e^k = 0 \Rightarrow \sum_{\ell=1}^{q} |c_{i_\ell}^k| e_{i_\ell}^k = 0. \qquad (2.3)$$

This relation shows that we can define the scale coefficients d^k by taking $(q-1)$ properly chosen linear combinations of the q prediction errors $\{e_{i_\ell}^k\}_{\ell=1}^{q}$ in each cell c_i^{k-1}. These linear combinations should be chosen so that together with (2.3) they constitute an invertible system of q linear equations for the prediction errors $\{e_{i_\ell}^k\}_{\ell=1}^{q}$ in the cell c_i^{k-1}. (See *e.g.* [18] for such combinations in representation of matrices.)

Using agglomeration as a coarsening technique may result in cells which are general polygons. In the following we describe a piecewise-polynomial reconstruction technique which is suitable for this purpose (see [17]). Let us denote by $\mathcal{S}_i^k$ a stencil of $s(r)$ cells in C^k which includes c_i^k, i.e.

$$\mathcal{S}_i^k = \{c_{i_m}^k\}_{m=1}^{s(r)}, \qquad c_i^k \in \mathcal{S}_i^k; \qquad (2.4a)$$

here $s(r)$ is the number of coefficients in a polynomial of degree $(r-1)$ in $\mathbb{R}^m$. Let $p_i^k(x; \mathcal{D}_k f)$ denote the unique polynomial of degree $(r-1)$ which attains the averages $(\mathcal{D}_k f)_{i_m}$ in $\mathcal{S}_i^k$, i.e. the one which satisfies the following system of $s(r)$ linear equations for its $s(r)$ coefficients:

$$\frac{1}{|c_{i_m}^k|} \int_{c_{i_m}^k} p_i^k(x; \mathcal{D}_k f)dx = (\mathcal{D}_k f)_{i_m}, \qquad m = 1, \ldots, s(r), \qquad (2.4b)$$

and define

$$(\mathcal{R}_k \mathcal{D}_k f)(x) = p_i^k(x; \mathcal{D}_k f) \quad \text{for } x \in c_i^k. \qquad (2.4c)$$

Clearly, (2.4) defines a reconstruction of $\mathcal{D}_k f$, which is exact for polynomial functions of degree less or equal $(r-1)$, and thus is r-th order accurate.

Note that for $r = 1$ in (2.4) we have $s(r) = 1$ and we get the piecewise-constant reconstruction

$$(\mathcal{R}_k \mathcal{D}_k f)(x) = \sum_i (\mathcal{D}_k f)_i \chi_{c_i^k}(x), \tag{2.5a}$$

where $\chi_C(x)$ denotes the characteristic function of the set C,

$$\chi_C(x) = \begin{cases} 1 & x \in C \\ 0 & \text{otherwise} \end{cases} \tag{2.5b}$$

In [17] we present a hierarchical algorithm for the selection of a "centered" stencil, which is applicable even to completely unstructured meshes C^k in $\mathbb{R}^m$. In this context the "centered" stencil is defined as the one which minimizes the reconstruction error for the one-higher degree polynomials (*i.e.* degree r). This algorithm is of "crystal growth" type: starting with the cell c_i^k we begin to add successively, one cell at a time, to the cluster of cells that we have at the beginning of each step. The cell which is being added is selected from the set of all side neighbors of the existing cluster by the requirement that it will minimize the reconstruction error of suitably chosen monomials.

In [17] we also present an adaptive "crystal growth" algorithm which is designed to assign a stencil S_i^k from the smooth part of $f(x)$, if available, to all cells c_i^k which are themselves in the smooth part of $f(x)$. This way a Gibbs-like phenomenon is avoided, and the resulting approximation is r-th order accurate everywhere, except at cells which contain a discontinuity. This is accomplished by selecting the cell from the set of side neighbors which minimizes the derivatives of the so-defined reconstruction. We refer to this adaptive technique as Essentially Non-Oscillatory (ENO) reconstruction, and we shall describe a one-dimensional version of it in Section 6 of this paper.

We refer the reader to [1] for details of special ENO reconstruction techniques for triangulated meshes.

§3   MR schemes in [0,1]

In this section we take in (2.1) $\Omega = [0,1]$ and let X^L be an arbitrary partition of $[0,1]$

$$X^L = \{x_i^L\}_{i=0}^{J_L}, \ x_0^L = 0, \ x_{J_L}^L = 1, \ J_L = 2^L J_0 \tag{3.1a}$$

where the sequence above is strictly increasing and J_0 is some integer. We define the grids $X^k = \{x_i^k\}_{i=0}^{J_k}, \ k = L-1, \ldots, 1$ by the coarsening

$$x_i^{k-1} = x_{2i}^k, \quad i = 0, \ldots, J_{k-1} =: J_k/2, \tag{3.1b}$$

in which we delete from X^k all the points with odd indices. (The only reason that we remove *every* other point is to simplify the notations.) We consider the covering of $[0,1]$ by cells

$$C^k = \{c_i^k\}_{i=1}^{J_k}, \quad c_i^k = (x_{i-1}^k, x_i^k), \tag{3.2a}$$

where $\{x_i^k\}$ are the gridpoints of X^k in (3.1); observe that

$$\overline{c_i^{k-1}} = \overline{c_{2i-1}^k \cup c_{2i}^k}. \tag{3.2b}$$

Let $v^k = \mathcal{D}_k f$ denote the cell-averages in (2.1c); v^{k-1} is obtained from v^k by the decimation (2.2b), *i.e.*

$$v_i^{k-1} = \frac{1}{|c_i^{k-1}|}(|c_{2i-1}^k|v_{2i-1}^k + |c_{2i}^k|v_{2i}^k), \quad i = 1, \dots, J_{k-1}. \tag{3.3}$$

From (1.7b) we get that the prediction error e^k satisfies the relation

$$|c_{2i-1}^k|e_{2i-1}^k + |c_{2i}^k|e_{2i}^k = 0 \quad \text{for} \quad i = 1, \dots, J_{k-1}. \tag{3.4a}$$

Therefore, if we define $d^k = G_k e^k$ in (1.9a) by

$$d_j^k = \frac{1}{|c_j^{k-1}|}(|c_{2j-1}^k|e_{2j-1}^k - |c_{2j}^k|e_{2j}^k) \quad \text{for} \quad j = 1, \dots, J_{k-1} \tag{3.4b}$$

we recover the prediction error by $e^k = E_k d^k$, which is defined as follows:

$$\begin{cases} e_{2i-1}^k = d_i^k \cdot \dfrac{|c_i^{k-1}|}{2|c_{2i-1}^k|} \\[2mm] e_{2i}^k = -d_i^k \cdot \dfrac{|c_i^{k-1}|}{2|c_{2i}^k|}. \end{cases} \tag{3.4c}$$

The direct MR transform $\hat{v} = M \cdot v^L$ is given in this case by the algorithm

$$\begin{cases} \text{DO } k = L, \dots, 1 \\[2mm] v_i^{k-1} = \dfrac{1}{|c_i^{k-1}|}(|c_{2i-1}^k|v_{2i-1}^k + |c_{2i}^k|v_{2i}^k), \quad i = 1, \dots, J_{k-1} \\[3mm] d_j^k = \dfrac{2|c_{2j-1}^k|}{|c_j^{k-1}|}[v_{2j-1}^k - (P_{k-1}^k v^k - 1)_{2j-1}], \quad j = 1, \dots, J_{k-1}. \end{cases} \tag{3.5}$$

The inverse MR transform $v^L = M^{-1} \cdot \hat{v}$ is given by

$$
\begin{cases}
\text{DO} \quad k = 1, \ldots, L \\[2mm]
\text{DO} \quad i = 1, \ldots, J_{k-1} \\[2mm]
v_{2i-1}^k = (P_{k-1}^k v^k - 1)_{2i-1} + \dfrac{|c_i^{k-1}|}{2|c_{2i-1}^k|} d_i^k \\[4mm]
v_{2i}^k = (|c_i^{k-1}| v_i^{k-1} - |c_{2i-1}^k| v_{2i-1}^k) / |c_{2i}^k| \, .
\end{cases}
\tag{3.6}
$$

Observe that the last statement of (3.6) is obtained from (3.3), and it is the "inverse" of the first step in algorithm (3.5).

In [19] we showed that any interpolation method gives rise to a corresponding method for reconstruction from cell averages by the following "reconstruction via primitive function" technique: Given cell averages $v^k = \mathcal{D}_k f$ we calculate the point values of the "primitive function"

$$
F_i^k = F(x_i^k), \quad F(x) = \int_0^x f(y) \, dy
$$

by

$$
F_0^k = 0, \quad F_i^k = \sum_{j=1}^{i} |c_j^k| v_j^k, \quad 1 \le i \le J_k,
\tag{3.7a}
$$

and define

$$
(\mathcal{R}_k v^k)(x) = \frac{d}{dx} \mathcal{I}_k(x; F^k),
\tag{3.7b}
$$

where $\mathcal{I}_k(x; F^k)$ is *any* interpolation of the values $F^k = \{F_i^k\}_{i=0}^{J_k}$ at the grid points of X^k in (3.2).

The prediction operator (1.6a) can be expressed in terms of the interpolation of the primitive values by

$$
(P_{k-1}^k v^k - 1)_{2i-1} = \frac{1}{|c_{2i-1}^k|} [\mathcal{I}_{k-1}(x_{2i-1}^k; F^{k-1}) - F_{i-1}^{k-1}].
\tag{3.8}
$$

The piecewise-polynomial reconstruction in (2.4) can be obtained via (3.7) from the following piecewise-polynomial interpolation: Let S_i^k denote a stencil of r consecutive *cells* of C^k which includes c_i^k and let $q_i^k(x; F^k)$ denote the unique polynomial of degree r which interpolates F^k at the $r+1$ endpoints of the cells in this stencil. We define the piecewise-polynomial reconstruction $(\mathcal{R}_k v^k)(x)$ by

$$
(\mathcal{R}_k v^k)(x) = p_i^k(x; v^k) =: \frac{d}{dx} q_i^k(x; F^k) \quad \text{for } x \in c_i^k;
\tag{3.9}
$$

observe that $p_i^k(x; v^k)$ is of polynomial degree $(r-1)$. For $r=1$ we have $S_i^k = \{c_i^k\}$ and $q_i^k(x; F^k)$ is the linear interpolation

$$q_i^k(x; F^k) = F_{i-1}^k + \frac{x - x_{i-1}^k}{x_i^k - x_{i-1}^k}(F_i^k - F_{i-1}^k) \quad \text{for} \quad x_{i-1}^k \leq x \leq x_i^k; \quad (3.10a)$$

in this case $\mathcal{R}_k v^k$ is the piecewise-constant reconstruction

$$(\mathcal{R}_k v^k)(x) = v_i^k \quad \text{for} \quad x \in c_i^k. \quad (3.10b)$$

Up to now we have not specified the stencil S_i^k of r consecutive cells of C^k that we assign to c_i^k. Clearly, if we choose S_i^k independently of the data v^k then the most accurate choice is that of a centered stencil (away from the boundaries), *i.e.* for $r = 2s + 1$ we take

$$S_i^k = \{c_{i-s}^k, \dots, c_{i+s}^k\} \quad \text{for} \quad s+1 \leq i \leq J_k - s, \quad (3.11a)$$

and near the boundaries

$$S_i^k = \{c_1^k, \dots, c_r^k\} \quad \text{for} \quad 1 \leq i \leq s$$
$$S_i^k = \{c_{J_k-r+1}^k, \dots, c_{J_k}^k\} \quad \text{for} \quad J_k - s + 1 \leq i \leq J_k. \quad (3.11b)$$

When the grid X^k is uniform we get a particularly simple expression for the prediction $P_{k-1}^k = \mathcal{D}_k \mathcal{R}_{k-1}$ in (3.10)–(3.11). For the centered stencil (3.11a) we get that the evaluation of the predicted value requires only s multiplications:

$$(P_{k-1}^k v^{k-1})_{2i-1} = v_i^{k-1} - \sum_{\ell=1}^{s} \gamma_\ell (v_{i+\ell}^{k-1} - v_{i-\ell}^{k-1}), \quad (3.12a)$$

where

$$\begin{cases} r = 3 \Longrightarrow \gamma_1 = -1/8 \\ \\ r = 5 \Longrightarrow \gamma_1 = -22/128, \ \gamma_2 = 3/128. \end{cases} \quad (3.12b)$$

In Figures 1 and 2 we show the subdivision limit of the piecewise-polynomial reconstruction for $r = 3$. This is done by applying the inverse MR transform with $L = 7$ levels and $J_0 = 8$ to $\hat{v} = \{0, \dots, 0, \eta_i^0\}$, where $(\eta_i^0)_j = \delta_{i,j}$ for $j = 1, \dots, J_0$, to obtain an approximation to

$$\varphi_i^0 = \lim_{L \to \infty} \prod_{\ell=0}^{L} (\mathcal{R}_\ell \mathcal{D}_\ell) \cdot (\mathcal{R}_0 \eta_i^0), \quad (3.13a)$$

and to $\hat{v} = \{0, \ldots, 0, \eta_i^0, 0\}$ to obtain an approximation to

$$\psi_i^1 = \lim_{L \to \infty} \prod_{\ell=1}^{L} (\mathcal{R}_\ell \mathcal{D}_\ell) \cdot (\mathcal{R}_1 \mu_i^1) \qquad (3.13b)$$

(see Appendix A). The subscripts a,b,c,d in Figures 1 and 2 stand for $i = 1, 2, 3, 4$, respectively. The limit functions for $i = 5, 6, 7, 8$ can be obtained from the former ones by an appropriate reflection.

In the next section we shall relate these limit functions to biorthogonal wavelets.

§4   Biorthogonal wavelets

In this section we derive the MR schemes which correspond to the bases of biorthogonal wavelets in [5]. These MR schemes are obtained from nested discretization of functions in $L^2_{loc}(\mathbf{R})$ by taking weighted-averages on a nested dyadic sequence of uniform grids of $\mathbf{R}$, as follows:

$$(\mathcal{D}_k f)_i = \frac{1}{h_k} \int_{-\infty}^{\infty} f(x) w \left(\frac{x - x_i^k}{h_k} \right) dx, \quad -\infty < i < \infty, \qquad (4.1a)$$

where $w \in L^2(\mathbf{R})$ is a weight-function

$$\int_{-\infty}^{\infty} w(x) dx = 1 \qquad (4.1b)$$

of compact support and

$$x_i^k = i h_k, \quad -\infty < i < \infty, \quad h_k = 2^{-k} h_0. \qquad (4.1c)$$

In [11] we show that if $w(x)$ is a solution of a dilation equation

$$w(x) = 2 \sum_{\ell=0}^{N} \alpha_\ell w(2x - \ell), \qquad (4.2a)$$

with coefficients that satisfy

$$\sum_{\ell} \alpha_{2\ell} = \sum_{\ell} \alpha_{2\ell-1} = 1/2, \qquad (4.2b)$$

then $\{\mathcal{D}_k\}$ is a nested sequence of operators and its decimation operator is given by

$$(D_k^{k-1} v)_i =: \sum_{\ell} \alpha_\ell v_{2i-\ell}. \qquad (4.2c)$$

In [6] and [12] it is shown that $\{\mu_j^k\}_{j=-\infty}^{\infty}$,

$$(\mu_j^k)_i = (-1)^{i+1}\alpha_{2j-i-1}, \quad -\infty < i < \infty \tag{4.3}$$

is a basis of the null space of the decimation operator $\mathcal{N}(D_k^{k-1})$ in (1.8b).

In order to have a "wavelet basis" of $L^2(\mathbb{R})$ which consists of dilates and translates of a single function, we consider reconstruction of the form

$$(\mathcal{R}_k v)(x) = \sum_i v_i \varphi\left(\frac{x - x_i^k}{h_k}\right), \tag{4.4a}$$

where $\varphi(x)$ is a solution of the dilation equation

$$\varphi(x) = \sum_\ell \beta_\ell \varphi(2x - \ell) \tag{4.4b}$$

which is normalized by

$$\int \varphi(x) w(x) dx = 1. \tag{4.4c}$$

In [11] we show that the coefficients $\{\beta_\ell\}$ have to satisfy the following relations

$$\sum_\ell \alpha_\ell \beta_{\ell+2m} = \delta_{m,0} \tag{4.5a}$$

$$\sum_\ell (-1)^\ell (\ell)^q \beta_\ell = 0 \quad \text{for} \quad q = 0, \ldots, (r-1), \quad r \geq 1. \tag{4.5b}$$

Relation (4.5a) is derived from the requirement that $\mathcal{R}_k$ in (4.4a) is indeed a reconstruction of the discretization (4.1), *i.e.* that $\mathcal{D}_k \mathcal{R}_k = I_k$. Relation (4.5b) is derived from the requirement that the reconstruction is exact for polynomial functions p of degree less or equal $(r-1)$, *i.e.* that $\mathcal{R}_k \mathcal{D}_k p = p$, $\deg(p) \leq (r-1)$.

It is easy to see that the corresponding prediction operator $P_{k-1}^k = \mathcal{D}_k \mathcal{R}_{k-1}$ is given by

$$(P_{k-1}^k v)_i = \sum_m \beta_{i-2m} v_m \tag{4.6}$$

and that the reconstruction $\mathcal{R}_k$ is hierarchical (see (A.4a) in Appendix A). In this case we show in [12] that for any $f \in \mathcal{F}$

$$\mathcal{R}_L \mathcal{D}_L f = \mathcal{R}_0 \mathcal{D}_0 f + \sum_{k=1}^{L} \sum_{j=-\infty}^{\infty} d_j^k \psi_j^k, \tag{4.7a}$$

where

$$\psi_j^k = \mathcal{R}_k \mu_j^k, \tag{4.7b}$$

and $d_j^k = d_j^k(f)$ are the scale coefficients in (1.10) :

$$d^k(f) = G_k e^k(f) = G_k \mathcal{D}_k (I - \mathcal{R}_{k-1} \mathcal{D}_{k-1}) f. \qquad (4.7c)$$

Consequently, $\{\psi_j^k\}_{j,k}$ is a wavelet basis of any Banach space $\mathcal{F}$ in which $\{(\mathcal{R}_k \mathcal{D}_k)\}_{k=0}^{\infty}$ is a sequence of approximation (see Appendix A).

Following the guidelines of the present paper, we first select a sequence of nested discretization and thus create a setting for MR schemes; in the framework of this section this amounts to a choice of coefficients $\{\alpha_\ell\}$ subject to conditions (4.2b). Once this is done, relations (4.5) become a system of *linear* equations for the coefficients $\{\beta_\ell\}$. For

$$\alpha_\ell = \frac{1}{2}(\delta_{\ell,0} + \delta_{\ell,-1}) \qquad (4.8a)$$

we get

$$w(x) = \begin{cases} 1 & -1 < x < 0 \\ 0 & \text{otherwise} \end{cases} \qquad (4.8b)$$

which is square integrable, and the discretization (4.1a) becomes the cell-average discretization (2.1c) in the uniform grid (4.1c). It is easy to verify that if we take $2r$ nonzero coefficients $\{\beta_\ell\}_{\ell=-r}^{r-1}$ then we get r linear equations from (4.5a), and another set of r linear equations from (4.5b). The solution for this system of $2r$ equations is unique and the resulting prediction operator (4.6) is identical to that of the centered stencil in (3.12a). We conclude that the two MR schemes are the same in the case of uniform grids in R, and that the reconstruction (4.4a) of the biorthogonal wavelets is the hierarchical form of the piecewise-polynomial reconstruction in (3.9) (see Appendix A). Thus, in Figure 1d $\varphi_4^0(x) = \varphi(\frac{x}{h_0} - 4)$, where $\varphi(x)$ is the solution of the dilation equation (4.4b), and in Figure 2d $\psi_4^1(x) = \psi(\frac{x}{h_1} - 4)$, where $\psi(x)$ is the "mother wavelet function" in the expansion (4.7). The boundary limit functions in Figures 1 and 2 show how to augment the scaling functions and wavelets of the infinite domain in order to get a "wavelet basis" for the interval $[0,1]$.

In [2] we take $w(x)$ in (4.1b) to be the "hat-function" and show how to derive MR schemes in this case.

Remark 1. Daubechies' orthonormal wavelets are obtained from the preceding formulation by adding the requirement $\beta_\ell = 2\alpha_\ell$ which couples the systems of equations for $\{\alpha_\ell\}$ and $\{\beta_\ell\}$, and then the biorthogonality condition (4.5a) becomes a *nonlinear* orthogonality condition. Observe that as we change the order of accuracy r, we necessarily change $\{\alpha_\ell\}$ and consequently, the nature of the discretization. In this respect Daubechies' orthonormal wavelets deviate from our setup in which the discretization is fixed, and we change the order of the reconstruction if desired. Also observe that we did not use the notion of orthogonality or biorthogonality in our formulation. The latter is an automatic consequence of the relation $\mathcal{D}_k \mathcal{R}_k = I_k$ and it does not impose an independent requirement.

§5 Data compression and error control

In this section we consider strategies for data compression and apriori bounds on the compression error.

We can obtain data compression by setting to zero all scale coefficients which fall below a prescribed tolerance. Let us denote

$$\tilde{d}_j^k = tr(d_j^k; \varepsilon_k) =: \begin{cases} 0 & \text{if } |d_j^k| \leq \varepsilon_k \\ \\ d_j^k & \text{if } |d_j^k| > \varepsilon_k, \end{cases} \tag{5.1a}$$

and refer to this operation as truncation. This type of data compression is used primarily to reduce the "dimensionality" of the data. Another aspect of data compression is to reduce the digital representation of the data for purposes of storage or transmission. In this case we use "quantization" which we model by

$$\tilde{d}_j^k = qu(d_j^k; \varepsilon_k) =: 2\varepsilon_k \cdot \text{ROUND} \left[\frac{d_j^k}{2\varepsilon_k} \right], \tag{5.1b}$$

where ROUND $[\,\cdot\,]$ denotes the integer which is obtained by rounding the number. For example, if $|d_j^k| \leq 256$ and $\varepsilon_k = 4$ then we can represent $|d_j^k|$ by an integer which is not larger than 32 and commit a maximal error of 4. Observe that $|d_j^k| < \varepsilon_k \implies qu(d_j^k; \varepsilon_k) = 0$ and that in both cases

$$|d^k - tr(d_j^k; \varepsilon_k)| \leq \varepsilon_k, \tag{5.2a}$$

$$|d^k - qu(d_j^k; \varepsilon_k)| \leq \varepsilon_k. \tag{5.2b}$$

Let

$$\tilde{v}^L = M^{-1} \cdot \{\tilde{d}^L, \ldots, \tilde{d}^1, v^0\}, \tag{5.3a}$$

denote the "decompressed" data; then, under the circumstances which are described in Appendix A, we typically get the following bound on the compression error

$$|\tilde{v}^L - v^L|_L \leq C \cdot \sum_{k=1}^{L} \varepsilon_k. \tag{5.3b}$$

Here $|\,\cdot\,|_L$ is the discrete norm which is defined in (A.2b) in Appendix A and C is a constant independent of L. It should be noted that we assume here that M^{-1} is a linear operator. Given any $\varepsilon > 0$ we can take for example

$$\varepsilon_k = (1-q)q^{L-k}\varepsilon \quad \text{for some} \quad 0 < q < 1, \qquad (5.4a)$$

in which case we get from (5.3b) that

$$|\tilde{v}^L - v^L|_L \le C \cdot \varepsilon. \qquad (5.4b)$$

Only in very simple situations we can choose a sequence of tolerances ε_k which ensures that for a prescribed ε

$$\max_{0 \le i \le J_L} |v_i^L - \tilde{v}_i^L| < \varepsilon. \qquad (5.5)$$

Thus, the typical situation in this setup is that we can specify the rate of compression, but we cannot specify an upper bound on the compression error.

In the following we describe a technique which enables us to specify a desired level of accuracy in the decompressed signal, independent of the particular form of the prediction operator; most importantly, it also applies to nonlinear (adaptive) prediction methods. This is accomplished by using a modification of the encoding algorithm (1.10) which keeps track of the cumulative compression error in the predetermined decoding procedure (1.11) and compresses accordingly. As is to be expected, we cannot specify compression rate at the same time. This algorithm applies to both the truncation (5.1a) and the quantization (5.1b), and we denote their operation by the generic name $cm(d^k; \varepsilon_k)$, where cm stands for compression.

We denote our error control decoding by M_ε and describe its operation on the input v^L by

$$\hat{v}_\varepsilon =: M_\varepsilon \cdot v^L =: \{\tilde{d}^L, \ldots, \tilde{d}^1, v^0\}. \qquad (5.6)$$

First we apply successive decimation to the input v^L

$$\begin{cases} \text{DO} \quad k = L, \ldots, 1 \\[2mm] v^{k-1} = D_k^{k-1} v^k . \end{cases} \qquad (5.7a)$$

These values of v^k are used in the following to monitor and control the accumulation of the compression error: Set

$$\tilde{v}^0 = v^0 \qquad (5.7b)$$

$$\begin{cases} \text{DO} \quad k = 1, \ldots, L \\[2mm] v^P = P_{k-1}^k \tilde{v}^{k-1} \\[2mm] \tilde{d}^k = cm[G_k(v^k - v^P); \varepsilon_k] \\[2mm] \tilde{v}^k = v^P + E_k \tilde{d}^k. \end{cases} \qquad (5.7c)$$

Clearly, if we apply the decoding M^{-1} :

$$\begin{cases} \text{DO} \quad k = 1, \ldots, L \\[2mm] v^P = P^k_{k-1} v^{k-1} \\[2mm] v^k = v^P + E_k d^k \end{cases} \tag{5.8a}$$

to the encoded data $\hat{v}_\varepsilon$ we get that

$$M^{-1} \cdot \{\tilde{d}^L, \ldots, \tilde{d}^1, v^0\} = \tilde{v}^L \tag{5.8b}$$

where $\tilde{v}^L$ is the quantity which is computed in (5.7c) for $k = L$. Our task is now to find G_k and ε_k such that (5.5) holds.

Next we apply this program to the case of cell-average discretization in a uniform grid of $[0, 1]$; here because of the simplicity of the expressions we can precisely monitor the cumulative compression error in the decoding algorithm and show how to control it. For simplicity we consider compression by truncation (5.1a). The predetermined decoding procedure is (3.6), which we now rewrite in the form

$$\begin{cases} \text{DO} \quad k = 1, \ldots, L \\[2mm] \text{DO} \quad i = 1, \ldots, J_{k-1} \\[2mm] v^P_{2i-1} = (P^k_{k-1} v^{k-1})_{2i-1} \\[2mm] v^k_{2i-1} = v^P_{2i-1} + d^k_i \\[2mm] v^k_{2i} = 2v^{k-1}_i - v^k_{2i-1} . \end{cases} \tag{5.9}$$

The modified encoding procedure is described algorithmically by the following:

(i) Apply decimation to the input v^L

$$\begin{cases} \text{DO} \quad k = L, \ldots, 1 \\[2mm] \text{DO} \quad i = 1, \ldots, J_{k-1} \\[2mm] v^{k-1}_i = \frac{1}{2}(v^k_{2i-1} + v^k_{2i}). \end{cases} \tag{5.10a}$$

(ii) Set

$$\tilde{v}^0 = v^0. \tag{5.10b}$$

(iii) Calculate

$$\begin{cases} \text{DO } k = 1,\ldots,L \\[2mm] \text{DO } j = 1,\ldots,J_{k-1} \\[2mm] v_{2j-1}^P = (P_{k-1}^k \tilde{v}^{k-1})_{2j-1} \\[2mm] \tilde{d}_j^k = tr(v_{2j-1}^k - v_{2j-1}^P - (v_j^{k-1} - \tilde{v}_j^{k-1}); \varepsilon_k) \\[2mm] \tilde{v}_{2j-1}^k = v_{2j-1}^P + \tilde{d}_j^k \\[2mm] \tilde{v}_{2j}^k = 2\tilde{v}_j^{k-1} - \tilde{v}_{2j-1}^k. \end{cases} \qquad (5.10c)$$

Note that we used the relation

$$v_{2j-1}^P + v_{2j}^P = 2\tilde{v}_j^{k-1}$$

in order to rewrite the expression for $\tilde{d}_j^k$ in (5.7b) as

$$[G_k(v^k - v^P)]_j =: \frac{1}{2}[(v_{2j-1}^k - v_{2j-1}^P) - (v_{2j}^k - v_{2j}^P)]$$

$$= v_{2j-1}^k - v_{2j-1}^P - (v_j^{k-1} - \tilde{v}_j^{k-1}).$$

Let us denote the cumulative compression error by

$$\mathcal{E}_j^k = v_j^k - \tilde{v}_j^k \qquad (5.11a)$$

and the prediction error

$$e_{2j-1}^P = v_{2j-1}^k - v_{2j-1}^P. \qquad (5.11b)$$

With this notation we get from (5.10c) that

$$\mathcal{E}_{\in|-\infty}^{\|} = 1_{\in|-\infty}^P - \sqcup\nabla(1_{\in|-\infty}^P - \mathcal{E}_{|}^{\|-\infty}; \varepsilon_{\|}), \qquad (5.11c)$$

$$\frac{1}{2}(\mathcal{E}_{\in|-\infty}^{\|} + \mathcal{E}_{\in|}^{\|}) = \mathcal{E}_{|}^{\|-\infty}. \qquad (5.11d)$$

Subtracting (5.11c) from (5.11d) we get

$$\frac{1}{2}(\mathcal{E}_{\in|}^{\|} - \mathcal{E}_{\in|-\infty}^{\|}) = \mathcal{E}_{|}^{\|-\infty} - 1_{\in|-\infty}^P + \sqcup\nabla(1_{\in|-\infty}^P - \mathcal{E}_{|}^{\|-\infty}; \varepsilon_{\|}). \qquad (5.12)$$

Let us now examine the two possibilities in (5.12):

$$|\mathcal{E}_{|}^{\|-\infty} - 1_{\in|-\infty}^P| \geq \varepsilon_{\|} \Rightarrow \frac{\infty}{\in}(\mathcal{E}_{\in|-\infty}^{\|} - \mathcal{E}_{\in|}^{\|}) = / \Rightarrow \mathcal{E}_{\in|-\infty}^{\|} = \mathcal{E}_{\in|}^{\|} = \mathcal{E}_{|}^{\|-\infty},$$

$$(5.13a)$$

$$|\mathcal{E}_|^{\|-\infty} - 1^P_{\in|-\infty}| < \varepsilon_\| \Rightarrow \begin{cases} \frac{1}{2}(\mathcal{E}^\|_{\in|} - \mathcal{E}^\|_{\in|-\infty}) = \mathcal{E}_|^{\|-\infty} - 1^P_{\in|-\infty} \\ \\ \frac{1}{2}(\mathcal{E}^\|_{\in|} + \mathcal{E}^\|_{\in|-\infty}) = \mathcal{E}_|^{\|-\infty}. \end{cases} \tag{5.13b}$$

From (5.13) we get the following inequalities

$$\max(|\mathcal{E}^\|_{\in|-\infty}|, |\mathcal{E}^\|_{\in|}|) = \overset{\infty}{\underset{\in}{}}|\mathcal{E}^\|_{\in|} + \mathcal{E}^\|_{\in|-\infty}| + \overset{\infty}{\underset{\in}{}}|\mathcal{E}^\|_{\in|} - \mathcal{E}^\|_{\in|-\infty}| \le |\mathcal{E}_|^{\|-\infty}| + \varepsilon_\|, \tag{5.14a}$$

$$(|\mathcal{E}^\|_{\in|-\infty}| + |\mathcal{E}^\|_{\in|}|) = \max(|\mathcal{E}^\|_{\in|} + \mathcal{E}^\|_{\in|-\infty}|, |\mathcal{E}^\|_{\in|} - \mathcal{E}^\|_{\in|-\infty}|) \le \in \max(|\mathcal{E}_|^{\|-\infty}|, \varepsilon_\|). \tag{5.14b}$$

Recalling that $\mathcal{E}' = \iota$ we get from (5.14a)

$$\|\mathcal{E}^\|\|_\infty \le \|\mathcal{E}^{\|-\infty}\|_\infty + \varepsilon_\| \le \cdots \le \sum_{\ell=\infty}^{\|} \varepsilon_\ell. \tag{5.15a}$$

Recalling that here $|c_i^k| = h_k = \frac{1}{2}h_{k-1}$ we get from (5.14b)

$$\|\mathcal{E}^\|\|_{\ell_\infty} = \langle_{\|}\sum_{)=\infty}^{\mathcal{J}_\|}|\mathcal{E}_)^\|| = \frac{1}{2}h_{k-1}\sum_{j=1}^{J_k-1}(|\mathcal{E}^\|_{\in|-\infty}| + |\mathcal{E}^\|_{\in|}|) \tag{5.15b}$$

$$\le h_{k-1}\sum_{j=1}^{J_k-1}\max(|\mathcal{E}_|^{\|-\infty}|, \varepsilon_\|).$$

It follows from (5.15) that

$$\|\mathcal{E}^L\|_\infty \le \sum_{\ell=\infty}^{L} \varepsilon_\ell, \tag{5.16a}$$

and if we choose $\{\varepsilon_\ell\}_{\ell=1}^{L}$ such that for all $\ell \ge 2$

$$\varepsilon_\ell \ge \sum_{m=1}^{\ell-1} \varepsilon_m, \tag{5.16b}$$

then the ℓ_1-error is

$$\|\bar{E}^L\|_{\ell_1} \le \varepsilon_L. \tag{5.16c}$$

Given ε it makes good sense to choose the tolerance-levels ε_k to be

$$\varepsilon_k = (1-q)q^{L-k}\varepsilon \quad \text{for some} \quad 0 < q \le \frac{1}{2}; \tag{5.17a}$$

in this case we get

$$\|v^L - \tilde{v}^L\|_\infty = \|\mathcal{E}^L\|_\infty \le \varepsilon, \tag{5.17b}$$

$$\|v^L - \tilde{v}^L\|_{\ell_1} = \|\mathcal{E}^L\|_{\ell_\infty} \le (\infty - \mathrm{II})\varepsilon. \tag{5.17c}$$

§6   ENO reconstruction and subcell resolution

In [19] we presented a data-dependent piecewise-polynomial reconstruction technique which avoids the Gibbs phenomenon by an adaptive selection of stencil $S_y^{\parallel}$ in (3.9); we refer to this technique as Essentially Non-Oscillatory (ENO) reconstruction. The basic idea of ENO reconstruction is to assign to a cell c_i^k which is in the smooth part of the sampled function, a stencil $S_y^{\parallel} = \{J_{y,}^{\parallel}, \ldots, J_{y,+\nabla-\infty}^{\parallel}\}$ with $i_0 = i_0(i)$, which is likewise in the smooth part of the function (provided that this is possible, *i.e.* that discontinuities are well separated and are far enough from the boundaries). This is done by choosing $S_y^{\parallel}$ to be the stencil for which the reconstruction polynomial $p_i^k(x; v^k)$ in (3.9) is the "smoothest" among all candidate stencils, *i.e.* those of r consecutive cells of C^k (starting with $c_{i_0}^k$) which contain the cell c_i^k, *e.g.* by taking $i_0(i)$ to be the index for which

$$\min_{i_0} \left| \frac{d^{r-1}}{dx^{r-1}} p_i^k(x; v^k) \right| \tag{6.1}$$

is attained among all candidate stencils. This enables us to get a good approximation everywhere except in the cells which contain a discontinuity.

Next we show that cell-average discretization enables us to get a good approximation even in cells which contain discontinuity by using "subcell resolution" (see [15]). Let $\mathcal{I}_k(x; F^k)$ denote the piecewise-polynomial interpolation of the primitive function in (3.9). Since it has formal order of accuracy $r + 1$, we get in regions of smoothness of f that

$$\mathcal{I}_{\parallel}(\S; \mathcal{F}^{\parallel}) = \mathcal{F}(\S) + \mathcal{O}((\langle_{\parallel})^{\nabla+\infty} \|\mathcal{F}^{(\nabla+\infty)}\|) \tag{6.2a}$$

then

$$(\mathcal{R}_k v^k)(x) = \frac{d}{dx} I_k(x; F^k) \quad = \frac{d}{dx} F(x) + O((h_k)^r \|F^{(r+1)}\|)$$
$$= f(x) + O((h_k)^r \|f^{(r)}\|). \tag{6.2b}$$

Assume now that $f(x)$ has $(p-1)$ continuous derivatives and that $f^{(p)}(x)$ is discontinuous but bounded. It is clear from relations (6.2) that the maximal accuracy that can be achieved from either point values or cell averages is $O(h^p \|f^{(p)}\|)$: Using cell-averages we gain one order of smoothness in the primitive function (3.5a) but we lose it in the differentiation (3.5b). Consequently there is no advantage in using cell averages rather than point values of $f(x)$ for smooth data.

There is a significant advantage however in using cell averages rather than point values of f when $f(x)$ is discontinuous in a finite number of points. To see that, let us assume that $f(x)$ is discontinuous at $x_d \in$

(x_{j-1}^k, x_j^k) and that in $[a, x_d) \cup (x_d, b]$, $0 \le a < x_d < b \le 1$, f has at least r continuous derivatives. Let $\mathcal{I}^{\mathcal{L}}$ and $\mathcal{I}^{\mathcal{R}}$ denote interpolation of either $f(x)$ or $F(x)$ at grid points in $[a, x_d)$ and $(x_d, b]$, respectively. We note that $F(x)$ is continuous in $[a, b]$, but has a discontinuous derivative at x_d. Consequently, if $F(x)$ is properly resolved on the k-th grid $\mathcal{I}^{\mathcal{L}}(\S; \mathcal{F}^{\parallel})$ and $\mathcal{I}^{\mathcal{R}}(\S; \mathcal{F}^{\parallel})$ will intersect at some point $\tilde{x}_d \in c_j^k$. From (6.2) we get that this intersection point is a good approximation to the location of the discontinuity within the cell c_j^k, *i.e.*

$$\tilde{x}_d - x_d = O((h_k)^r \| f^{(r)} \|). \tag{6.3}$$

On the other hand, having knowledge of point-values $\{f(x_i^k)\}$ in $[a, b]$, there is nothing much we can say about the location of the discontinuity *within* the cell c_j^k.

We describe now how to use the subcell-resolution technique of [15] in the prediction (3.8) in order to get an accurate prediction in a cell c_j^{k-1} which contains a discontinuity. Let us denote by $\tilde{F}_{2j-1}^k$ our approximation to $F(x_{2j-1}^k)$, and let $\mathcal{I}^{\mathcal{L}}(\S; \mathcal{F}^{\parallel - \infty})$ and $\mathcal{I}^{\mathcal{R}}(\S; \mathcal{F}^{\parallel - \infty})$ denote the ENO interpolation in the neighboring cells c_{j-1}^{k-1} and c_{j+1}^{k-1} on the left and right, respectively, and define

$$D(x) = \mathcal{I}^{\mathcal{R}}(\S; \mathcal{F}^{\parallel - \infty}) - \mathcal{I}^{\mathcal{L}}(\S; \mathcal{F}^{\parallel - \infty}). \tag{6.4a}$$

Since $D(\tilde{x}_d) = 0$ we assume that

$$D(x_{j-1}^{k-1}) \cdot D(x_j^{k-1}) < 0. \tag{6.4b}$$

$\tilde{F}_{2j-1}^k$ is now computed as follows

$$\tilde{F}_{2j-1}^k = \begin{cases} \mathcal{I}^{\mathcal{L}}(\S_{\in | -\infty}^{\parallel}; \mathcal{F}^{\parallel - \infty}) & \text{if } D(x_{2j-1}^k) \cdot D(x_j^{k-1}) \le 0 \\ \mathcal{I}^{\mathcal{R}}(\S_{\in | -\infty}^{\parallel}; \mathcal{F}^{\parallel - \infty}) & \text{otherwise}. \end{cases} \tag{6.5}$$

It is easy to see that if $f(x)$ is a piecewise-polynomial function

$$f(x) = \begin{cases} P_L(x) & a \le x < x_d \\ P_R(x) & x_d < x \le b \end{cases} \tag{6.6a}$$

with

$$\deg(P_L) \le r - 1, \quad \deg(P_R) \le r - 1, \tag{6.6b}$$

then

$$\tilde{F}_{2j-1}^k = F(x_{2j-1}^k), \tag{6.6c}$$

i.e. the procedure (6.5) is exact. More generally, if $f(x)$ has p continuous derivatives to the left and the right of the discontinuity, we get that

$$\tilde{F}^k_{2j-1} = F(x^k_{2j-1}) + O((h_k)^{\bar{p}+1}\|f^{(\bar{p})}\|), \qquad \bar{p} = \min(p, r). \qquad (6.7)$$

Remark 2. If we know that $f(x)$ has $q - 1$ continuous derivatives and a discontinuity of the q-th derivative in x_d, $x^{k-1}_{j-1} < x_d < x^{k-1}_j$ we can extend the subcell resolution technique of (6.4)–(6.5) to this case as follows: $\frac{d^q}{dx^q} F(x)$ has a discontinuous first derivative at x_d. If it is sufficiently resolved on the grid, we expect $\frac{d^q}{dx^q} I_L(x; F^{k-1})$ and $\frac{d^q}{dx^q} I_k(x; F^{k-1})$ to intersect at $\tilde{x}_d$ in c^{k-1}_j,

$$\tilde{x}_d - x_d = O(h^{r-q}). \qquad (6.8a)$$

It follows, therefore, that if we replace $D(x)$ in (6.4) by

$$D(x) = \frac{d^q}{dx^q} I^R(x; F^{k-1}) - \frac{d^q}{dx^q} I^L(x; F^{k-1}), \qquad (6.8b)$$

we get a subcell-resolution technique which is exact for the corresponding piecewise-polynomial problem (6.6); this implies (6.7).

Remark 3. Extrapolating the analysis of the information contents in cell averages vs. point values, we get that weighted averages with respect to the hat-function contain information that will enable us to obtain subcell resolution of δ-distributions; this may be useful for compression of digital images and propagation of singularities (see [2]).

§7   Numerical experiments and conclusions

In Figure 3 and Table 1 we present results of data compression for the MR schemes which are described in this paper. We generate the discrete data v^L for $J_L = 512$ by

$$v^L_i = f(\xi^L_i), \quad \xi^L_i = -1 + (i - \frac{1}{2})h_L, \quad h_L = 2/J_L, \quad i = 1, \ldots, J_L, \quad (7.1a)$$

where

$$f(x) = \begin{cases} -x\sin(\frac{\pi}{2}x^2) & -1 < x \leq -\frac{1}{3} \\ |\sin(2\pi x)| & |x| < \frac{1}{3} \\ 2x - 1 + \sin(3\pi x)/6 & \frac{1}{3} \leq x < 1. \end{cases} \qquad (7.1b)$$

This input data is displayed in Figure 3a by drawing a circle around the points (ξ^L_i, v^L_i) for $i = 1, \ldots, J_L$.

The MR schemes are used with 6 levels of resolution, *i.e.* $L = 6$ and $J_0 = 8$. The scale coefficients are truncated by (5.1a) with $\varepsilon = 10^{-3}$ and $q = \frac{1}{2}$ in (5.4a). The result of this truncation is displayed in the $x - k$ plane by drawing a circle around (ξ_j^k, k), where $\xi_j^k = -1 + (j - \frac{1}{2})h_k$, $h_k = 2^{L-k}h_L$ for each d_j^k which is above the tolerance ε_k.

In Figure 3b we show the data compression of the MR scheme which is based on Daubechies' orthonormal wavelets with $r = 5$, *i.e.* five vanishing moments.

In Figure 3c we show the results corresponding to the biorthogonal wavelets of [5] with $r = 5$. In both Figures 3b and 3c we used periodic extension at the boundaries.

In Figure 3d we show the results of the piecewise-polynomial reconstruction (3.11)–(3.12) with $r = 5$, where we used one-sided stencils near the boundaries.

In Figures 3e and 3f we used the error control algorithm (5.10) with the same choice of $\{\varepsilon_k\}$ as above. In Figure 3e we show the results of the ENO reconstruction with $r = 6$. The particular technique that we used here is described in [19]; it is the one-dimensional version of the "crystal-growth" algorithm in [17].

In Figure 3f we append this ENO reconstruction with subcell resolution (6.5). The subcell resolution is applied to cells c_j^k for which $\mathcal{S}_{j-1}^k \cap \mathcal{S}_{j+1}^k = \emptyset$, *i.e.* where the selection of ENO stencils "shies away" from c_j^k.

Method	Order r	Compression	$L_\infty - error$	$L_1 - error$	$L_2 - error$
Orthonormal wavelets	5	4.163	3.524×10^{-4}	1.081×10^{-5}	3.574×10^{-5}
Biorthogonal wavelets	5	4.923	7.659×10^{-5}	1.114×10^{-5}	2.141×10^{-5}
Piecewise-polynomial	5	5.626	2.967×10^{-4}	1.455×10^{-5}	2.824×10^{-5}
ENO reconstruction	6	9.143	2.525×10^{-4}	2.002×10^{-5}	3.514×10^{-5}
ENO +Subcell resol.	6	10.039	2.525×10^{-4}	2.088×10^{-5}	3.611×10^{-5}

Table 1. Data compression

The corresponding rates of compression and compression errors are listed in Table 1. We would like to make the following observations regarding these numerical results:

1. The compression error is about the same for all the MR schemes in this experiment.

2. The "smooth part" of the data is resolved to the prescribed tolerance at level 3; thus the significant scale-coefficients of levels 4 to 6 are the "signatures" of the various prediction operators of the two discontinuities in the function (7.1b) and the discontinuity in the derivative. In the case of wavelets where we used periodic extension, we also get the "signature" of the discontinuity, which is so introduced at the boundaries.

3. Biorthogonal wavelets perform better than orthonormal wavelets; their "signature" is narrower because of the smaller stencil and it is symmetric. We remark that the lack of symmetry in the coefficients of Daubechies' orthonormal wavelets increases the number of operations in various applications (see [18, 16, 9]).

4. As to be expected, the piecewise polynomial reconstruction in Figure 3d is identical to the biorthogonal wavelets in the interior; the only improvement is the removal of the "signature" of the boundary discontinuity which is accomplished by using one-sided stencils near the boundaries.

5. Since the main difference between the various MR schemes in this experiment is in their "signature" of discontinuities, there is a considerable improvement in using the ENO reconstruction, because then the signature is typically 1 point. This signature can be completely eliminated at times by using subcell resolution (after all the only information in a discontinuity is its location). Observe that the particular version of ENO reconstruction that is used here misses a bit at the discontinuous derivative which is located at $x = 0$. This can be fixed by using special procedures which are described in [7] (see also Remark 2).

In building the case to promote the use of cell-average discretization we have not yet discussed the issue of suitability to applications. As we have mentioned earlier, for smooth input data the performance of all MR schemes is about the same, and then pointwise discretization is usually the best choice because of its simplicity. However, when the input data is only piecewise smooth there is a clear advantage to using cell-average discretization – this is true in the numerical solution of hyperbolic conservation laws where the solution typically contains shock waves (see [13]–[14]), and in the solution of integral equations where the kernel is usually integrably singular (see [4, 18, 16] and [9]). Another important feature of cell-average discretization is that if the input data contains noise or local high-frequency components, then these are averaged out in the coarser levels. This enables us to to develop techniques to eliminate noise (see [22]) and to design special techniques to handle piecewise-smooth data which carries some local high frequency components (see [8]).

These characteristics of cell-average discretization make it ideal for the purpose of image compression, where the coarser levels of discretization behave like those of piecewise-smooth functions. Furthermore, the strategy of truncation in (5.4a) and (5.17a) dictates increases in accuracy as we go to larger cells. This is suitable to the averaging properties of the human eye and makes the decompressed image look "pleasant;" in some respects this works like cosmetic retouching, *e.g.* in compressing an image of a face it may eliminate freckles, but at the same time keep the larger features sharp.

We refer the reader to [18] and [16] where we describe tenser-product extension of the one-dimensional algorithms of the present paper to rectangular two-dimensional grids, and to [3] for details and experiments of data compression in unstructured meshes.

Acknowledgments. Many thanks to my collaborators Remi Abgrall, Paco Arandiga, Barna Bihari, Rosa Donat, and Itai Yad-Shalom.

This research was supported in UCLA by Grants ONR-N00014-92-J-1890 and NSF-DMS91-03104.

Appendix A. Stability analysis and existence of MR bases

We assume that sequence $\{(\mathcal{R}_k \mathcal{D}_k)\}_{k=0}^{\infty}$,

$$(\mathcal{R}_k \mathcal{D}_k) : \mathcal{F} \to \mathcal{F} , \tag{A.1a}$$

is a sequence of (discrete) approximation in the Banach space $\mathcal{F}$, *i.e.* that for any $f \in \mathcal{F}$

(i)

$$\|\mathcal{R}_k \mathcal{D}_k f\| \le C_A^k \|f\| \tag{A.1b}$$

(ii)

$$\|\mathcal{R}_k \mathcal{D}_k f - f\| \to 0 \quad \text{as} \quad k \to \infty . \tag{A.1c}$$

Using the principle of uniform boundedness we conclude that there exists a constant C_A such that for all k

$$C_A^k \le C_A. \tag{A.1d}$$

If (A.1a) is a *nested* sequence of discretization, we get that the direct MR transform (2.11) is stable with respect to perturbations in the input data v^L, and that

$$\langle \delta(d^k) \rangle_k = |\delta(e^k)|_k \le C_A(1 + C_A)|\delta(v^L)|_L$$

$$|\delta(v^0)|_0 \le C_A|\delta(v^L)|_L , \tag{A.2a}$$

where $\delta(\cdot)$ denotes the perturbation, and the discrete norms above are defined as follows:

$$|v^k|_k = \|\mathcal{R}_k v^k\| ; \tag{A.2b}$$

$$\langle d^k \rangle_k = |E_k d^k|_k . \tag{A.2c}$$

The "natural" function space for cell-average discretization is $\mathcal{F} = L_1(\Omega)$, and there (A.1b)–(A.1c) take the following form:

$$\int_\Omega |(\mathcal{R}_k \mathcal{D}_k f)(x)| dx \leq C_A^k \int_\Omega |f(x)| dx, \tag{A.3a}$$

$$\lim_{k \to \infty} \int_\Omega |(\mathcal{R}_k \mathcal{D}_k f)(x) - f(x)| dx = 0. \tag{A.3b}$$

In [12] we investigate the stability of the inverse MR transform (2.12) and the related question of existence of MR bases for mappings in $\mathcal{F}$. In the following we present a summary of the results:

Case 1. We assume that $\{(\mathcal{R}_k \mathcal{D}_k)\}_{k=0}^\infty$ is a hierarchic sequence of approximation, *i.e.* in addition to (A.1) it satisfies the following for all $k > 0$

$$(\mathcal{R}_k \mathcal{D}_k)\mathcal{R}_{k-1} = \mathcal{R}_{k-1}; \tag{A.4a}$$

note that another way to express (A.4a) is

$$\mathcal{R}_k P_{k-1}^k = \mathcal{R}_{k-1}. \tag{A.4b}$$

In this case we show in [12] that the inverse MR transform is stable with respect to perturbations

$$|\delta(v^L)|_L \leq |\delta(v^0)|_0 + \sum_{k=1}^L \langle \delta(d^k) \rangle_k , \tag{A.5}$$

and that for for any $f \in \mathcal{F}$

$$f = \mathcal{R}_0 \mathcal{D}_0 f + \sum_{k=1}^\infty \sum_{j=1}^{J_k - J_{k-1}} d_j^k \psi_j^k , \tag{A.6a}$$

where the coefficients $d_j^k = d_j^k(f)$ are the scale coefficients in (1.10)

$$d^k(f) = G_k e^k(f) = G_k \mathcal{D}_k(I - \mathcal{R}_{k-1}\mathcal{D}_{k-1})f. \tag{A.6b}$$

The functions of the MR basis of $\mathcal{F}$ are obtained by reconstruction of the basis (1.8b) of the null space of the decimation operator

$$\psi_j^k = \mathcal{R}_k \mu_j^k. \tag{A.6c}$$

Case 2. We assume that $\{\mathcal{R}_k \mathcal{D}_k\}_{k=0}^\infty$ is a sequence of approximation which is σ-contractive, *i.e.* that there exist $0 < q < 1$ and a convergent series $\sum_{\ell=0}^\infty \Delta_\ell < \infty$ of positive numbers such that for all $k \geq 0$ and any $f \in \mathcal{F}$

$$\sigma_{k+1}(\mathcal{R}_k \mathcal{D}_k f) \leq q \cdot (1 + \Delta_k) \cdot \sigma_k(f), \tag{A.7a}$$

where

$$\sigma_\ell(f) =: \|(I - \mathcal{R}_{\ell+1}\mathcal{D}_{\ell+1})\mathcal{R}_\ell\mathcal{D}_\ell f\|. \tag{A.7b}$$

In this case we show in [12] that the following limit exists

$$\mathcal{R}_k^H \mathcal{D}_k f =: \lim_{L \to \infty} \prod_{\ell=k}^{L} (\mathcal{R}_\ell \mathcal{D}_\ell) \cdot f, \tag{A.8}$$

that $\{\mathcal{R}_k^H \mathcal{D}_k\}$ is a hierarchic sequence of approximation, and that its MR scheme is exactly the same as that of $\{(\mathcal{R}_k\mathcal{D}_k)\}$. Hence, we can use the results of case 1 to conclude that the MR scheme is stable with respect to perturbations, and that the compression error is bounded by (A.2a) where now the discrete norms in (A.2b)–(A.2c) are defined with the hierarchic form $\mathcal{R}_k^H$. Furthermore, any $f \in \mathcal{F}$ has the expansion (A.6a), where $\mathcal{R}_0$ is replaced by $\mathcal{R}_0^H$ and

$$\psi_j^k =: \lim_{L \to \infty} \prod_{\ell=k}^{L} (\mathcal{R}_\ell \mathcal{D}_\ell) \cdot (\mathcal{R}_k \mu_j^k). \tag{A.9}$$

Remark 4. Let $\{\eta_i^k\}_{i=1}^{J_k}$ denote any basis of the linear space V^k in (1.1)

$$V^k = \text{span}\{\eta_i^k\}, \tag{A.10a}$$

and denote

$$\varphi_i^k =: \mathcal{R}_k^H \eta_i^k = \lim_{L \to \infty} \prod_{\ell=k}^{L} (\mathcal{R}_\ell \mathcal{D}_\ell) \cdot (\mathcal{R}_k \eta_i^k). \tag{A.10b}$$

In [11]–[12] we show that

$$\varphi_\ell^{k-1} = \sum_i (\hat{P}_{k-1}^k)_{i,\ell} \varphi_i^k \tag{A.10c}$$

where $\hat{P}_{k-1}^k$ is the matrix representation of the *original* prediction operator (1.6a). The sequence $\{\varphi_i^k\}$ is related to $\{\psi_j^k\}$ in (A.9) by

$$\psi_j^k = \sum_i (\hat{E}_k)_{i,j} \varphi_i^k, \tag{A.11}$$

where $\hat{E}_k$ is the matrix representation of the operator E^k in (1.9a).

Note that when $\hat{P}_{k-1}^k$ is the Töplitz-like matrix in (4.6), then all $\varphi_i^k(x)$ are generated from a single function $\varphi(x)$, which is the solution of the dilation equation (4.4b). From this point of view $\{\psi_j^k\}$ in (A.9) can be thought of as "generalized wavelets."

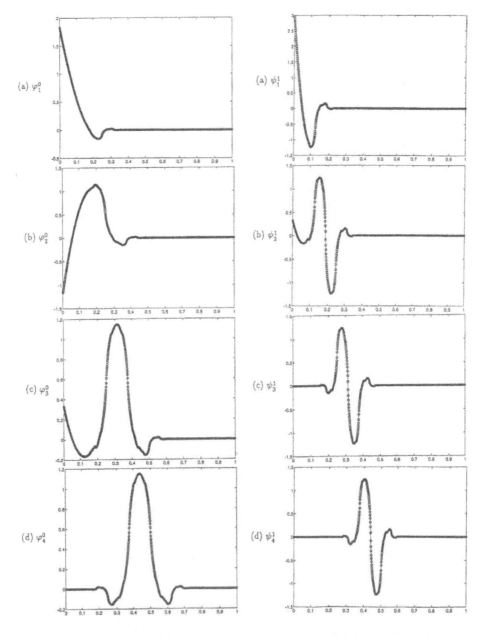

Figure 1. Subdivision limit for
$\{\varphi_i^0\}_{i=1}^4$.

Figure 2. Subdivision limit for
$\{\psi_j^1\}_{j=1}^4$.

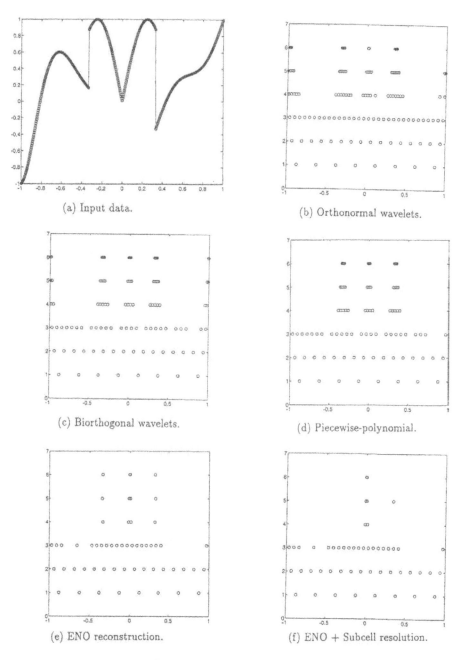

(a) Input data.

(b) Orthonormal wavelets.

(c) Biorthogonal wavelets.

(d) Piecewise-polynomial.

(e) ENO reconstruction.

(f) ENO + Subcell resolution.

Figure 3. Data compression.

Remark 5. Observe that the functionals σ_ℓ vanish if the sequence is hierarchic, and thus their size measures the deviation of the sequence from being hierarchic. We remark that wherever condition (A.7) has been verified, this was done with "hard" analysis; nevertheless, the design problem seems to be "soft." Although there are known examples of divergence of the above limit, most "natural" approximations give rise to stable MR schemes (as is evident from the experiments in this paper and others not reported here).

References

[1] Abgrall, R., Design of an essentially nonoscillatory reconstruction procedure on finite element type meshes, ICASE Report 91-84, (1991), in revised form INRIA Report # 1592 (1992), submitted to *Math. Comp.*

[2] Arandiga, F., R. Donat, and A. Harten, Multiresolution based on weighted averages of the hat function, UCLA CAM Report 93-34, September 1993.

[3] Abgrall, R. and A. Harten, Multiresolution representation in unstructured meshes. I. UCLA CAM Report 94-20, July 1994.

[4] Beylkin, G., R. Coifman, and V. Rokhlin, Fast wavelet transform and numerical algorithms. I, *Comm. Pure Appl. Math.* **44** (1991), 141–183.

[5] Cohen, A., I. Daubechies, and J.-C. Feauveau, Biorthogonal bases of compactly supported wavelets, *Comm. Pure Appl. Math.* **45** (1992), 485–560.

[6] Daubechies, I., Orthonormal bases of compactly supported wavelets, *Comm. Pure Appl. Math.* **41** (1988), 909–996.

[7] Donat, R., Studies on error propagation for certain nonlinear approximations to hyperbolic equations: Discontinuities in derivatives, *SIAM J. Num. Anal.* **31** (1994), 655–679.

[8] Donat, R. and A. Harten, Data compression algorithms for locally oscillatory data: A preliminary report, UCLA, October 1992, preprint.

[9] Guru Prasad, K., D. E. Keyes, and J. H. Kane, Generalized wavelet bases for boundary element matrices, 1994, preprint.

[10] Harten, A., Discrete multiresolution analysis and generalized wavelets, *J. Appl. Num. Math.* **12** (1993) 153–193; also UCLA CAM Report 92-08, February 1992.

[11] Harten, A., Multiresolution representation of data. I. UCLA CAM Report No. 93-13, June 1993.

[12] Harten, A., Multiresolution representation of data. II. General Framework, UCLA CAM Report No. 94-10, April 1994; also Technical Report 3-94, Dept. of Applied Mathematics, Tel Aviv University, April 1994.

[13] Harten, A., Multiresolution algorithms for the numerical solution of hyperbolic conservation laws, UCLA CAM Report 93-03, March 1993; *Comm. Pure Appl. Math.*, to appear.

[14] Harten, A., Adaptive multiresolution schemes for shock computations, UCLA CAM Report 93-06, April 1993; *J. Comput. Phys.*, to appear.

[15] Harten, A., ENO schemes with subcell resolution, *J. Comp. Phys.* **83** (1989), 148–184; also ICASE Report No. 87-56.

[16] Harten, A., Multiresolution representation and numerical algorithms: A brief review, UCLA CAM Report 94-12, June 1994.

[17] Harten, A., and S. R. Chakravarthy, Multi-dimensional ENO schemes for general geometries, ICASE Report 91-76, September 1991; also submitted to *J. Comput. Phys.*

[18] Harten, A. and I. Yad-Shalom, Fast multiresolution algorithms for matrix-vector multiplication, ICASE Report 92-55, October 1992; *SIAM J. Num. Anal.*, to appear.

[19] Harten, A., B. Engquist, S. Osher, and S. Chakravarthy, Uniformly high order accurate essentially non-oscillatory schemes, III, *J. Comput. Phys.* **71** (1987), 231–303.

[20] Mallat, S., Multiresolution approximation and wavelets orthonormal bases of $L^2(\mathbf{R})$, *Trans. Amer. Math. Soc.* **315** (1989), 69–87.

[21] Meyer, Y., *Ondelettes et Opérateurs*, Hermann, Paris, 1990.

[22] Mizrachi, D., Removing noise from discontinuous data, dissertation, Tel-Aviv University, (1991).

Ami Harten
School of Mathematical Sciences
Tel-Aviv University
Tel-Aviv, 69978 Israel
and
Department of Mathematics,
UCLA, Los Angeles, CA 90095

V.

Analysis of Waveform Representations

Characterization of Smoothness via Functional Wavelet Transforms

Charles K. Chui and Chun Li

Abstract. Characterization of the smoothness of signals (or functions) by using interpolatory wavelets is studied in this paper. The dual of an interpolatory wavelet, also called a distributional analyzing wavelet, plays an important role in interpolatory wavelet decomposition. The coefficients of the interpolatory wavelet series expansion are values of some functionals induced by this dual. In addition, while this series representation of a given signal models the so-called waveform of the signal, the vanishing moment property of the dual, rather than the interpolatory wavelet itself, gives rise to localized time information of the changes of the signal.

§1 Introduction

It is well known that the coefficient sequence of the Fourier series expansion of a periodic signal (or function) is often used to characterize the order of smoothness of the signal itself. This so-called Littlewood-Paley approach to wavelet series expansions (often called discrete wavelet transforms, DWT) is also well documented in the wavelet literature (see, for instance, the monograph [10] of Y. Meyer). This study, however, is within the L^p theory of wavelet analysis. Based on the pioneer work of Donoho [5], we recently developed in [2] a parallel theory of the DWT for the space C_u of bounded and uniformly continuous functions on $\mathbb{R}$. The basic functions that generate the local series expansions are interpolatory wavelets, such as those of Micchelli (see [2]) and Donoho [5]; and the coefficients are functional wavelet transforms, FnWT, of the functions in C_u under investigation. The importance of the FnWT is that it reveals the local details, just as the DWT does, of the functions under investigation by using their discrete function values directly, by means of Lagrange interpolation.

This paper may be considered a continuation of our work developed in [2]. The objective here is to characterize the smoothness of functions (or signals) in $C_u \cap L^p$ $(1 \le p \le \infty)$ in terms of the wavelet coefficients of the interpolatory wavelet series representation that are used in the process of interpolatory wavelet decompositions. This characterization will involve

Signal and Image Representation in Combined Spaces
Y. Y. Zeevi and R. R. Coifman (Eds.), pp. 395–414.

Besov spaces as well as Sobolev spaces. We emphasize that the important role of the duals of the interpolatory wavelets is mainly due to the property of vanishing moments to be described in Section 2. In this respect, our point of view and mathematical analysis are quite different from those in Donoho [5].

This paper is organized as follows. In Section 2 we will discuss the notion of interpolatory scaling functions, wavelets, and their duals as introduced in [2], and will describe some of their basic properties that are needed for stating and establishing our main results. Unlike the L^2 situation, the duals here are not functions in the same space as the wavelets and the original functions to be analyzed. In fact, they are distributions generated by the Dirac functional. The main results of this paper will be established in Sections 3 and 4.

§2   Interpolatory wavelets and their duals

Let ϕ be a compactly supported continuous function which satisfies some two-scale relation

$$\phi(x) = \sum_k p_k \phi(2x - k), \qquad x \in \mathbb{R}, \tag{2.1}$$

where $\{p_k\} \in \ell^\infty$; and assume that ϕ is so chosen that the z-transform of its values on the set of integers $\mathbb{Z}$ does not vanish on the unit circle, namely:

$$\Phi(z) := \sum_k \phi(k) z^k \neq 0, \quad \text{for all} \quad |z| = 1. \tag{2.2}$$

Set

$$V_j := \left\{ \sum_k c_k \phi(2^j \cdot - k) \colon \{c_k\} \in \ell^\infty \right\}, \qquad j \in \mathbb{Z}, \tag{2.3}$$

and consider the space

$$C_u = C_u(\mathbb{R}) \tag{2.4}$$

of uniformly continuous and bounded functions on $\mathbb{R}$. Note that each V_j is a subspace of C_u, but not of L^2. Also, we do not require $\{\phi(\cdot - k) : k \in \mathbb{Z}\}$ to be an L^2 Riesz basis. Instead, we assume that ϕ satisfies (2.2), and call (2.2) the interpolating condition of ϕ. The reason is that under this condition, we have, for each $j \in \mathbb{Z}$, a unique cardinal interpolation projection operator P_j from C_u to V_j in the sense that

$$(P_j f)\left(\frac{k}{2^j}\right) = f\left(\frac{k}{2^j}\right), \quad k \in \mathbb{Z}, \quad \text{for all } f \in C_u. \tag{2.5}$$

As pointed out in [8] (see also [2]), since ϕ is assumed to be a compactly supported continuous function, the two-scale sequence $\{p_k\}$ in (2.1) must have exponential decay. Hence, it is clear that (2.1) is equivalent to

$$\cdots V_{-1} \subset V_0 \subset V_1 \subset \cdots. \tag{2.6}$$

In [2] we observed, analogous to the L^2 situation, that this *interpolating multiresolution analysis*, IMRA, $\{V_j\}$ of C_u satisfies the following properties:

$$\mathrm{clos}_{C_u} \left(\bigcup_{j \in \mathbf{Z}} V_j \right) = C_u, \tag{2.7}$$

$$\bigcap_{j \in \mathbf{Z}} V_j = \mathbf{C}, \text{ the set of all complex numbers,} \tag{2.8}$$

$$f \in V_j \Leftrightarrow f(2\cdot) \in V_{j+1}, \quad j \in \mathbf{Z}, \tag{2.9}$$

$$f \in V_j \Rightarrow f(\cdot - 2^{-j}k) \in V_j, \quad k \in \mathbf{Z}, \quad \text{for all } j \in \mathbf{Z}. \tag{2.10}$$

Later, to facilitate the formulation of our main results, we will further assume that

$$\phi \text{ has } r\text{-regularity and } d\text{-polynomial span,} \tag{2.11}$$

meaning that ϕ is Hölder continuous of order r for some $r > 0$ and that the collection of formal sums $\sum_k c_k \phi(x - k)$ locally reproduces all polynomials of degree up to d for some integer $d \geq 0$.

Next, from the cardinal interpolation operators P_j in (2.5), we define the projections

$$\Delta_j := P_{j+1} - P_j, \quad j \in \mathbf{Z}, \tag{2.12}$$

and the corresponding subspaces

$$W_j := \{\Delta_j f : f \in V_{j+1}\} = \{f - P_j f : f \in V_{j+1}\}, \quad j \in \mathbf{Z}, \tag{2.13}$$

of C_u. It is clear that

$$V_{j+1} = V_j + W_j \quad \text{and} \quad V_j \cap W_j = \{0\}, \quad j \in \mathbf{Z}, \tag{2.14}$$

and we will use the direct-sum notation

$$V_{j+1} = V_j \dotplus W_j, \quad j \in \mathbf{Z}, \tag{2.15}$$

to describe (2.14).

From the condition (2.2), we may write

$$\frac{1}{\Phi(z)} = \sum_{k \in \mathbf{Z}} r_k z^k, \quad |z| = 1, \tag{2.16}$$

and note that the coefficient sequence $\{r_k\}$ has exponential decay. Let

$$\phi_L(x) := \sum_k r_k \phi(x-k). \tag{2.17}$$

Then it is clear that $\phi_L \in V_0$ satisfies $\phi_L(k) = \delta_{k,0},\ k \in \mathbf{Z}$, where $\delta_{k,0}$ denotes, as usual, the Kronecker delta symbol, and the interpolation operators P_j have the Langrange interpolating representation

$$(P_j f)(x) := \sum_{k \in \mathbf{Z}} f\left(\frac{k}{2^j}\right) \phi_L(2^j x - k), \qquad f \in C_u. \tag{2.18}$$

The function ϕ_L is called the *fundamental function* corresponding to ϕ.

Note that according to (2.3), (2.17), and the identity

$$\phi(x) = \sum_k \phi(k)\phi_L(x-k), \tag{2.19}$$

both $\{\phi(2^j \cdot -k) :\ k \in \mathbf{Z}\}$ and $\{\phi_L(2^j \cdot -k) :\ k \in \mathbf{Z}\}$ are bases of V_j. Hence, as usual, we will also refer to ϕ and ϕ_L as generators of the IMRA $\{V_j\}$. To construct generators of the complementary subspaces W_j, we consider any $g \in W_0$. Then by (2.13), there is an $f \in V_1$ such that $g(x) = f(x) - (P_0 f)(x)$. Since P_1 is a projection from C_u to V_1, we can write

$$f(x) = (P_1 f)(x) = \sum_k f\left(\frac{k}{2}\right)\phi_L(2x - k). \tag{2.20}$$

Also, as a consequence of (2.6), since the fundamental function ϕ_L has the two-scale relation

$$\phi_L(x) = \sum_k \phi_L\left(\frac{k}{2}\right)\phi_L(2x - k) \tag{2.21}$$

$$= \phi_L(2x) + \sum_k \phi_L\left(k + \frac{1}{2}\right)\phi_L(2x - 2k - 1),$$

we see, from (2.20), (2.18), and (2.21), that

$$g(x) = \sum_k f\left(\frac{k}{2}\right)\phi_L(2x - k) - \sum_\ell f(\ell)\phi_L(x - \ell) \tag{2.22}$$

$$= \sum_k f\left(\frac{k}{2}\right)\phi_L(2x - k) - \sum_\ell f(\ell)$$

$$\times \left[\phi_L(2x - 2\ell) + \sum_k \phi_L\left(k + \frac{1}{2}\right)\phi_L(2x - 2\ell - 2k - 1)\right]$$

$$= \sum_k f\left(k + \frac{1}{2}\right) \phi_L(2x - 2k + 1)$$

$$- \sum_k \left[\sum_\ell f(\ell)\phi_L\left(k - \ell + \frac{1}{2}\right)\right] \phi_L(2x - 2k - 1)$$

$$= \sum_k (f - P_0 f)\left(k + \frac{1}{2}\right) \phi_L(2x - 2k - 1)$$

$$= \sum_k c_k \psi_L(\cdot - k),$$

by introducing

$$\psi_L(x) := \phi_L(2x - 1), \tag{2.23}$$

and setting

$$c_k := (f - P_0 f)\left(k + \frac{1}{2}\right) = f\left(k + \frac{1}{2}\right) - \sum_\ell f(\ell)\phi_L\left(k - \ell + \frac{1}{2}\right). \tag{2.24}$$

Since $f \in C_u$, the sequence $\{c_k\}$ is clearly in ℓ^∞. Hence, by (2.22), we have

$$W_0 \subset \left\{\sum_k c_k \psi_L(\cdot - k): \{c_k\} \in \ell^\infty\right\}. \tag{2.25}$$

On the other hand, since $\psi_L := \phi_L(2\cdot -1) \in V_1$ and $\psi_L(\ell) = \phi_L(2\ell - 1) = 0$ for all $\ell \in \mathbf{Z}$, any $g(x) = \sum_k c_k \psi_L(x - k)$ with $\{c_k\} \in \ell^\infty$, satisfies $g \in V_1$ and $g(\ell) = 0$, so that $g = g - P_0 g \in W_0$. This proves that

$$W_0 \supset \left\{\sum_k c_k \psi_L(\cdot - k): \{c_k\} \in \ell^\infty\right\}. \tag{2.26}$$

Combining (2.25) and (2.26) yields

$$W_j = \left\{\sum_k c_k \psi_L(2^j \cdot -k): \{c_k\} \in \ell^\infty\right\} \tag{2.27}$$

for $j = 0$. By a scaling transform we see that (2.27) holds for all $j \in \mathbf{Z}$. Thus, we have shown that $\{\psi_L(2^j \cdot -k): k \in \mathbf{Z}\}$ is a basis of W_j; i.e. ψ_L generates $\{W_j\}$. This derives a result of Donoho given in [5]. Moreover, following [2] we consider another generator

$$\psi := \sum_k (-1)^{k-1} \phi(k - 1)\phi(2 \cdot -k), \tag{2.28}$$

for $\{W_j\}$ by verifying that

$$W_j = \left\{ \sum_k c_k \psi(2^j \cdot -k) : \{c_k\} \in \ell^\infty \right\}. \qquad (2.29)$$

Note that if $\phi \equiv \phi_L$, then $\psi \equiv \psi_L$. We call any generator of $\{W_j\}$ an *interpolatory wavelet*. Later we will introduce another interpolatory wavelet. In [2] we proved that $\{\phi(\cdot - k) : k \in \mathbf{Z}\}$ and $\{\psi(\cdot - k) : k \in \mathbf{Z}\}$ together constitute an ℓ^∞-stable basis of V_1. Thus, by a result of Jia and Micchelli [8] on stability, they are also ℓ^p-stable; *i.e.* for any p, $1 \le p \le \infty$, there are constants $0 < A \le B < \infty$, such that

$$A(\|\mathbf{c}\|_{\ell^p} + \|\mathbf{d}\|_{\ell^p}) \le \left\| \sum_k c_k \phi(\cdot - k) + \sum_k d_k \psi(\cdot - k) \right\|_p \le B(\|\mathbf{c}\|_{\ell^p} + \|\mathbf{d}\|_{\ell^p})$$

$$(2.30)$$

for all $\mathbf{c} = \{c_k\}_{k\in\mathbf{Z}} \in \ell^p$ and $\mathbf{d} = \{d_k\}_{k\in\mathbf{Z}} \in \ell^p$, where, as usual,

$$\|\mathbf{c}\|_{\ell^p} := \left(\sum_k |c_k|^p \right)^{1/p}, \text{ if } 1 \le p < \infty; \quad \|\mathbf{c}\|_{\ell^\infty} := \max_{k\in\mathbf{Z}} |c_k|. \qquad (2.31)$$

Assertion (2.30) is also valid for other scaling functions and interpolatory wavelets, such as ϕ_L and ψ_L.

We next turn to the discussion of duals. Let C_u^* be the space of all the continuous linear functionals on C_u. By $\langle f, \tilde{f} \rangle$, we mean the value of $\tilde{f} \in C_u^*$ evaluated at $f \in C_u$. We call $\tilde{f} \in C_u^*$ the dual of $f \in C_u$ if

$$\langle f(\cdot + k), \tilde{f} \rangle = \delta_{k,0}, \qquad k \in \mathbf{Z}. \qquad (2.32)$$

By using the classical Dirac delta function (distribution) δ, namely:

$$\langle f, \delta \rangle = f(0), \quad f \in C_u, \qquad (2.33)$$

and in view of

$$\langle \phi_L(\cdot + k), \delta \rangle = \phi_L(k) = \delta_{k,0}, \qquad k \in \mathbf{Z}, \qquad (2.34)$$

we see that δ is a dual of ϕ_L. In the following, we give a precise notion of dual functionals and describe what we mean by dilations, translations, and convergence of functionals.

Definition 1. *Let $X = X(\mathbb{R})$ be a normed linear space of functions over $\mathbb{R}$, and $X^* = X^*(\mathbb{R})$ be its dual space, consisting of all continuous linear functionals on X.*

(i) For an element $f^* \in X^*$, the *a-dilation and b-translation (shift)* $f^*(a \cdot -b)$, where $a, b \in \mathbb{R}$, $a \neq 0$, of f^* is defined by

$$\langle f, f^*(a \cdot -b) \rangle := \frac{1}{a} \left\langle f \left(\frac{\cdot + b}{a} \right), f^* \right\rangle, \quad f \in X. \qquad (2.35)$$

(ii) A pair (f, f^*), where $f \in X$ and $f^* \in X^*$, is said to be a *dual pair*, if

$$\langle f(\cdot + k), f^* \rangle = \langle f, f^*(\cdot - k) \rangle = \delta_{k,0} \quad \text{for all} \quad k \in \mathbb{Z}. \qquad (2.36)$$

(iii) Let $\{\tilde{f}_k\}$ be a sequence in X^*. The series $\sum_k \tilde{f}_k$ is said to be convergent in X^*, if the series $\sum_k \langle f, \tilde{f}_k \rangle$ is convergent for all $f \in X$, and its limit satisfies

$$\left| \sum_k \langle f, \tilde{f}_k \rangle \right| \leq C \|f\|, \quad f \in X,$$

for some positive constant C independent of f. Consequently, the series $\sum_k \tilde{f}_k$ can be considered as an element of X^*, in the sense that

$$\left\langle f, \sum_k \tilde{f}_k \right\rangle := \sum_k \langle f, \tilde{f}_k \rangle, \quad f \in X. \qquad (2.37)$$

Now, corresponding to the coefficient sequence $\{r_k\}$ of the Laurent expansion of $\Phi^{-1}(z)$ in (2.16), we consider the functional

$$\tilde{\phi} := \sum_k r_k \delta(\cdot + k). \qquad (2.38)$$

Since $\{r_k\} \in \ell^1$, we see that $\tilde{\phi}$ is a continuous linear functional on C_u in the sense of Definition (iii) above. It now follows from (2.34) that

$$\langle \phi, \tilde{\phi}(\cdot - j) \rangle = \langle \phi(\cdot + j), \tilde{\phi} \rangle \qquad (2.39)$$

$$= \sum_k r_k \langle \phi(\cdot + j), \delta(\cdot + k) \rangle$$

$$= \sum_k r_k \phi(j - k) = \phi_L(j) = \delta_{j,0}, \quad j \in \mathbb{Z},$$

so that $\tilde{\phi}$ is a dual of ϕ.

Note that although a function $f \in X$ may have more than one dual in different subspaces of X^*, yet the dual f^* of f in the subspace generated by the integer translates of f^* is unique. Before going into any further details, let us introduce the following "dual subspaces" generated by the Dirac delta functional δ:

$$\tilde{V}_j := \left\{ \sum_j d_k \delta(2^j \cdot -k) : \{d_k\} \in \ell^1 \right\}, \qquad j \in \mathbf{Z}. \qquad (2.40)$$

It is clear that δ satisfies the two-scale relation

$$\delta = 2\delta(2\cdot), \qquad (2.41)$$

so that

$$\tilde{V}_j \subset \tilde{V}_{j+1}, \quad j \in \mathbf{Z}. \qquad (2.42)$$

In addition, it follows from (2.27) and (2.40) that

$$W_j \perp \tilde{V}_j, \qquad j \in \mathbf{Z}, \qquad (2.43)$$

in the sense that

$$\langle g, \tilde{f} \rangle = 0 \quad \text{for all} \quad g \in W_j, \quad \tilde{f} \in V_j. \qquad (2.44)$$

Indeed,

$$\langle \psi_L(\cdot + k), \delta \rangle = \psi_L(k) = \phi_L(2k - 1) = 0, \quad k \in \mathbf{Z}. \qquad (2.45)$$

The "orthogonality" property (2.43) is one of the main reasons that the dual spaces $\tilde{V}_j$ are worth investigating. Next, following Cohen, Daubechies, and Feauveau [4], we will look for a subspace $\widetilde{W}_j$ of $\tilde{V}_{j+1}$ that satisfies

$$\tilde{V}_{j+1} = \tilde{V}_j \dot{+} \widetilde{W}_j, \qquad j \in \mathbf{Z}, \qquad (2.46)$$

$$V_j \perp \widetilde{W}_j, \qquad j \in \mathbf{Z}, \qquad (2.47)$$

so that a generator of $\widetilde{W}_j$ is a dual of a given generator of W_j. For this purpose, set

$$\tilde{\psi}_L := 2 \sum_k (-1)^{k-1} \phi_L \left(\frac{1-k}{2} \right) \delta(2 \cdot -k). \qquad (2.48)$$

Then we have

$$\langle \psi_L, \tilde{\psi}_L(\cdot - j) \rangle = \langle \psi_L(\cdot + j), \tilde{\psi}_L \rangle \qquad (2.49)$$

$$= \sum_k (-1)^{k-1} \phi_L \left(\frac{1-k}{2} \right) 2 \langle \phi_L(2 \cdot +2j - 1), \delta(2 \cdot -k) \rangle$$

$$= \sum_k (-1)^{k-1} \phi_L \left(\frac{1-k}{2} \right) \phi_L(k + 2j - 1)$$

$$= \sum_k (-1)^{k-1} \phi_L \left(\frac{1-k}{2} \right) \delta_{k+2j-1,0}$$

$$= \phi_L(j) = \delta_{j,0}, \qquad j \in \mathbf{Z},$$

so that $\tilde{\psi}_L$ is indeed a dual of ψ_L. Furthermore, observe that

$$\langle \phi_L, \tilde{\psi}_L(\cdot - j) \rangle = \langle \phi_L(\cdot + j), \tilde{\psi}_L \rangle$$

$$= \sum_k (-1)^{k-1} \phi_L \left(\frac{1-k}{2} \right) 2 \langle \phi_L(\cdot + j), \delta(2 \cdot - k) \rangle$$

$$= \sum_k (-1)^{k-1} \phi_L \left(\frac{1-k}{2} \right) \phi_L \left(\frac{k}{2} + j \right)$$

$$= \sum_\ell (-1)^\ell \phi_L \left(\frac{\ell}{2} + j \right) \phi_L \left(\frac{1-\ell}{2} \right)$$

$$= -\langle \phi_L, \tilde{\psi}_L(\cdot - j) \rangle, \qquad j \in \mathbf{Z},$$

so that

$$\langle \phi_L, \tilde{\psi}_L(\cdot - j) \rangle = 0, \qquad j \in \mathbf{Z}. \tag{2.50}$$

Hence, by setting

$$\widetilde{W}_j := \left\{ \sum_k d_k \tilde{\psi}_L(2^j \cdot - k) : \{d_k\} \in \ell^1 \right\}, \qquad j \in \mathbf{Z}, \tag{2.51}$$

we see that (2.47) holds. In [2], we already proved that (2.46) is also valid. From (2.38), (2.16), and (2.2), we also have

$$\delta = \sum_j \phi(j) \tilde{\phi}(\cdot + j). \tag{2.52}$$

Thus, $\tilde{\phi}(\cdot + j)$ is also a generator of $\{\tilde{V}_j\}$. Now let us define

$$\tilde{\psi} := 2 \sum_k h_k \tilde{\phi}(2 \cdot + k), \tag{2.53}$$

by using the coefficient sequence $\{h_k\}$ uniquely determined by

$$H(z) = \sum_k h_k z^k := 2z^{-1} P(-z) / \Phi(z^2). \tag{2.54}$$

We have shown in [2] that the above $\widetilde{\psi}$ also generates the sequence of subspaces $\widetilde{W}_j$ while it is a dual of the interpolatory wavelet ψ, as given by (2.28).

Note that both generators $\widetilde{\psi}_L$ and $\widetilde{\psi}$ of $\{\widetilde{W}_j\}$ are not compactly supported in general. So, it is natural to ask if there is a compactly supported generator of $\{W_j\}$. Fortunately, the answer is positive, and a choice is given by

$$\tilde{\eta} := 2\sum_k (-1)^{k-1}\phi\left(\frac{1-k}{2}\right)\delta(2\cdot -k). \tag{2.55}$$

Clearly, $\tilde{\eta}$ has compact support whenever ϕ is a compactly supported scaling function. The relation between $\tilde{\eta}$ and $\widetilde{\psi}_L$ is given by

$$\tilde{\eta} = \sum_\ell \phi(\ell)\widetilde{\psi}_L(\cdot + \ell) \quad \text{and} \quad \widetilde{\psi}_L = \sum_k r_k \tilde{\eta}(\cdot + k), \tag{2.56}$$

so that $\tilde{\eta}$ is also a generator of $\{W_j\}$. To see this, we have, by (2.48), (2.19), and (2.55),

$$\sum_\ell \phi(\ell)\widetilde{\psi}_L(\cdot + \ell) = \sum_\ell 2\sum_k (-1)^{k-1}\phi(\ell)\phi_L\left(\frac{1-k}{2}\right)\delta(2\cdot +2\ell - k)$$

$$= 2\sum_k\sum_\ell (-1)^{2\ell+k-1}\phi(\ell)\phi_L\left(\frac{1-k-2\ell}{2}\right)\delta(2\cdot -k)$$

$$= 2\sum_k (-1)^{k-1}\sum_\ell \phi(\ell)\phi_L\left(\frac{1-k}{2}-\ell\right)\delta(2\cdot +2\ell - k)$$

$$= 2\sum_k (-1)^{k-1}\phi\left(\frac{1-k}{2}\right)\delta(2\cdot -k) = \tilde{\eta}.$$

This proves the first equality in (2.56). Moreover, from (2.16), we see that the second relation in (2.56) also holds. Next, we observe that the dual of $\tilde{\eta}$ in W_0 is given by

$$\eta := \sum_k r_k \psi_L(\cdot - k). \tag{2.57}$$

Indeed, by (2.56), (2.57), (2.49), and (2.16), we have

$$\langle \eta, \tilde{\eta}(\cdot - j)\rangle = \left\langle \sum_k r_k \psi_L(\cdot - k), \sum_\ell \phi(\ell)\widetilde{\psi}_L(\cdot - j + \ell)\right\rangle \tag{2.58}$$

$$= \sum_k\sum_\ell r_k\phi(\ell)\langle \psi_L(\cdot - k), \widetilde{\psi}_L(\cdot - j + \ell)\rangle$$

$$= \sum_{k} \sum_{\ell} r_k \phi(\ell) \delta_{k,j-\ell}$$

$$= \sum_{k} r_k \phi(j-k) = \delta_{j,0}, \qquad j \in \mathbf{Z},$$

so that η and $\tilde{\eta}$ are duals to each other. In addition, in view of (2.57) and the relation

$$\psi_L = \sum_{k} \phi(k) \eta(\cdot - k), \qquad (2.59)$$

it follows that η is another generator of $\{W_j\}$. Hence, η is also another interpolatory wavelet.

To analyze the smoothness of functions (or signals), we actually need to use the dual of an interpolatory wavelet in $\widetilde{W}_0$. For this reason we usually call a dual interpolatory wavelet an "analyzing wavelet." Hence, any generator of $\{\widetilde{W}_j\}$ is an analyzing wavelet. These wavelets are not functions, but only functionals, being linear combinations of dilations and integer translations of the Dirac delta functional. We end this section by deriving the following property of vanishing moments of these analyzing wavelets.

Theorem 1. *Suppose that the scaling function ϕ locally reproduces all polynomials of degree up to d, or equivalently,*

$$x^\ell = \sum_{k} k^\ell \phi_L(x-k), \qquad \ell = 0, 1, \dots, d. \qquad (2.60)$$

Then

$$\langle p, \widetilde{\psi}_L \rangle = 0, \quad \text{for all} \quad p \in \pi_d, \qquad (2.61)$$

where π_d denotes the space of polynomials of degree not exceeding d.

Remark 1. Since a polynomial p of degree ≥ 1 is not in the space C_u, the meaning of $\langle p, \widetilde{\psi}_L \rangle$ must be clarified. This will be clear from the following proof of Theorem 1.

Proof: We first need to make sure that $\langle p, \widetilde{\psi}_L \rangle$ is well defined for any polynomial p. Observe that since ϕ_L has exponential decay, the series

$$2 \sum_{k} (-1)^{k-1} \phi_L \left(\frac{1-k}{2} \right) \langle p, \delta(2\cdot -k) \rangle = \sum_{k} (-1)^{k-1} \phi_L \left(\frac{1-k}{2} \right) p \left(\frac{k}{2} \right)$$

$$(2.62)$$

is absolutely convergent for any polynomial p. Hence, according to (2.48) and (2.62), we can define

$$\langle p, \widetilde{\psi}_L \rangle := \sum_{k} (-1)^{k-1} \phi_L \left(\frac{1-k}{2} \right) p \left(\frac{k}{2} \right), \qquad p \in \pi_d. \qquad (2.63)$$

This is similar to but not completely the same as Definition 1 (iii).

Now, for $p(x) = x^\ell$, $0 \le \ell \le d$, it follows from (2.63) and (2.34), and (2.60), that

$$\langle p, \widetilde{\psi}_L \rangle = \sum_k (-1)^{k-1} \phi_L \left(\frac{1-k}{2} \right) \left(\frac{k}{2} \right)^\ell = \left(\frac{1}{2} \right)^\ell - \sum_k k^\ell \phi_L \left(\frac{1}{2} - k \right) = 0.$$

(2.64)

Hence, it is clear that (2.61) holds. This completes the proof of the theorem. ∎

Since the polynomial space π_d is invariant under the operations of dilation and translation, (2.61) implies that

$$\langle p, \widetilde{\psi}_L(2^j - k) \rangle = 0, \quad j \in \mathbf{Z}, \ k \in \mathbf{Z}, \quad \text{for all } p \in \pi_d.$$

(2.65)

This can be written as

$$\pi_d \perp \widetilde{W}_j, \quad j \in \mathbf{Z}.$$

(2.66)

§3   Characterizing smoothness via interpolatory wavelets

The characterization of smoothness of functions is often done in Lipschitz spaces or Sobolev spaces. If we want to give a finer characterization of smoothness, we naturally require Besov spaces which have a unified framework that includes both the Sobolev spaces and the Lipschitz spaces. In this section we will give a characterization of functions in the Besov spaces via interpolatory wavelets introduced in the previous section.

To state and prove our main result, we need to use a proper definition of Besov spaces. For inhomogeneous Besov spaces $B_p^{s,q} := B_p^{s,q}(\mathbb{R})$ on the real line $\mathbb{R}$, there are several equivalent definitions. For our purpose, we adopt the one given in Meyer's book [10], as follows. We say that $f \in B_p^{s,q}$, if there exist $f_0, g_0, g_1, \ldots$ in the Sobolev space W_p^m and a sequence $\{\varepsilon_j\}_{j \in \mathbb{N}} \in \ell^q(\mathbb{N})$ such that

$$f = f_0 + g_0 + g_1 + \cdots, \quad \text{in } L^p,$$

(3.1)

$$\|g_j\|_p \le \varepsilon_j 2^{-sj}, \qquad j \in \mathbb{N}, \quad \text{and}$$

(3.2)

$$\|g_j^{(m)}\|_p \le C \varepsilon_j 2^{(m-s)j}, \qquad j \in \mathbb{N},$$

(3.3)

where $\mathbb{N} := \{0, 1, 2, \ldots\}$, $C > 0$ is some constant, and m can be any integer $> s$. The norm of f in $B_p^{s,q}$ can be defined as

$$\|f\|_{B_p^{s,q}} := \begin{cases} \|f_0\|_p + (\sum_{j \ge 0} (2^{sj} \|g_j\|_p)^q)^{1/q}, & \text{if } 1 \le p < \infty; \\ \|f_0\|_p + \max\{2^{sj} \|g_j\|_p : j \ge 0\}, & \text{if } p = \infty. \end{cases}$$

(3.4)

The other two equivalent definitions of Besov spaces involve modulus of continuity and Fourier transform, respectively. The interested reader is referred to the monographs [4] and [7]. From the K-functional theory, one can easily verify that the definition of Besov spaces in terms of the modulus of continuity is equivalent to the one given previously in (3.1)–(3.3). The reader is referred to [4, pp. 55, 177] for more detail.

In [2], (see also [5]), we proved that every $f \in C_u$ has a unique decomposition

$$f = P_0 f + \sum_{j \geq 0} \Delta_j f = \sum_k c_k \phi(\cdot - k) + \sum_{j \geq 0} \sum_k d_k^j \psi(2^j \cdot - k) \qquad \text{in} \quad C_u, \ (3.5)$$

where P_0 and $\Delta_j f$ are defined in (2.5) and (2.12), and they have been represented in terms of the scaling function ϕ and the interpolatory wavelet ψ, respectively. Based on this decomposition, we are now in a position to state the main results of this paper as follows. Here and throughout, we will always assume that the conditions (2.1), (2.2), and (2.11) are satisfied.

Theorem 2. *Let* $\min\{\lfloor r \rfloor, d\} > s > \frac{1}{p}, p, q \in [1, \infty]$, *and* $f \in C_u \cap L^p$. *Then* f *belongs to the Besov space* $B_p^{s,q}$ *if and only if* $\{2^{js} \|\Delta_j f\|_p\}_{j \in \mathbb{N}} \in \ell^q(\mathbb{N})$. *In addition, the norm of* f *in* $B_p^{s,q}$ *can be defined as*

$$\|f\|_{B_p^{s,q}} := \begin{cases} \|P_0 f\|_p + (\sum_{j \geq 0} (2^{sj} \|\Delta_j f\|_p)^q)^{1/q}, & \text{if} \quad 1 \leq q < \infty; \\ \|P_0 f\|_p + \max\{2^{sj} \|\Delta_j f\|_p : j \geq 0\}, & \text{if} \quad q = \infty. \end{cases}$$
$$(3.6)$$

This result was essentially given by Donoho [5], but he only considered the case when the scaling function ϕ is compactly supported and satisfies $\phi(k) = \delta_{k,0}, \ k \in \mathbf{Z}$ (*i.e.* $\phi = \phi_L$). Here, we will apply the duality analysis to give a *direct* proof of Theorem 2, which is different from the *indirect* proof given by Donoho [5], where the concept of almost-diagonal operators and other techniques are used.

Since ϕ and ψ are ℓ^p-stable in the sense of (2.30), we have the following.

Corollary 1. *Under the assumptions stated in Theorem 2, a function* $f \in C_u \cap L^p$ *is in* $B_p^{s,q}$ *if and only if its decomposition in (3.5) satisfies*

$$\{2^{j(s-1/p)} (\sum_k |d_k^j|^p)^{1/p}\}_{j \in \mathbb{N}} \in \ell^q(\mathbb{N}). \qquad (3.7)$$

Moreover, the quantity

$$\left(\sum_k |c_k|^p \right)^{1/p} + \left(\sum_{j \geq 0} |2^{j(s-1/p)} \left(\sum_k |d_k^j|^p \right)^{1/p}|^q \right)^{1/q} \qquad (3.8)$$

is an equivalent norm of f for the Besov space $B_p^{s,q}$. When $p = \infty$ or $q = \infty$, the usual change from summation to supremum should be made in (3.7) and (3.8).

Observe that the projection operators P_0 and Δ_j can also be represented by using the bases $\{\phi_L(\cdot - k) : k \in \mathbf{Z}\}$ and $\{\psi_L(2^j \cdot -k) : k \in \mathbf{Z}\}$ or $\{\eta(2^j \cdot -k) : k \in \mathbf{Z}\}$, respectively. In other words, each $f \in C_u \cap L^p$ can also be written either as

$$f = P_0 f + \sum_{j \geq 0} \Delta_j f = \sum_k c_k \phi_L(\cdot - k) + \sum_{j \geq 0} \sum_k d_k^j \psi_L(2^j \cdot -k) \qquad \text{in} \quad C_u,$$

(3.9)

or as

$$f = P_0 f + \sum_{j \geq 0} \Delta_j f = \sum_k c_k \phi_L(\cdot - k) + \sum_{j \geq 0} \sum_k d_k^j \eta(2^j \cdot -k) \qquad \text{in} \quad C_u,$$

(3.10)

where ϕ_L, ψ_L, and η are given in (2.17), (2.23) and (2.57), respectively. Hence, due to the ℓ^p-stability of ϕ_L, ψ_L, and η as mentioned earlier, the preceding corollary is also valid for the representations in (3.9) and (3.10). This is quite important, since the coefficients c_k and d_k^j in (3.10) are values of the dual functionals with compact support.

§4   Auxiliary results and proof of the theorem

According to the definition of Besov spaces given in the above section, the sufficiency direction of the theorem is almost obvious. Indeed, by taking $\varepsilon_j := 2^{js} \|\Delta_j f\|_p$, $g_j = \Delta_j f$, $j \geq 0$, $f_0 = P_0 f$, we see that (3.2) holds. Since $\{\varepsilon_j\}_{j \in \mathbf{N}} \in \ell^q(\mathbf{N})$, we also see that (3.5) is valid in L^p; that is, (3.1) holds. Also, by taking $m = \lfloor r \rfloor$, since $g_j = \Delta_j f = P_{j+1} f - P_j f \in V_{j+1}$, it follows from Bernstein's inequalities (see [10])

$$\|g^{(m)}\|_p \leq C 2^{mj} \|g\|_p, \quad \text{for all} \quad g \in V_j,$$

(4.1)

that (3.3) is valid. Hence, by the definition of Besov spaces, we have $f \in B_p^{s,q}$, and the quantity defined in (3.4) can be used as an equivalent norm for $B_p^{s,q}$.

The proof of the necessity direction involves a little more work. Let $f \in B_p^{s,q}$ and take $m = 1 + \min\{\lfloor r \rfloor, d\}$. Then there exist f_0, $g_j \in W_p^m$, $j \in \mathbf{N}$, such that (3.1), (3.2), and (3.3) hold. Since $m = 1 + \min\{\lfloor r \rfloor, d\} > s > \frac{1}{p}$, we see that W_p^m can be embedded into $C^{m-\frac{1}{p}} \subset C_u$, so that $\Delta_j f_0 = P_{j+1} f_0 - P_j f_0$ and $\Delta_j g_k = P_{j+1} g_k - P_j g_k$, $j, k \in \mathbf{N}$, are well defined. Moreover, it is also easy to see that (3.1) holds not only in L^p, but also in C_u. Hence, we have

$$\Delta_j f = \Delta_j f_0 + \sum_{k=0}^{\infty} \Delta_j g_k. \tag{4.2}$$

We need to derive some suitable estimates for $\Delta_j f_0$ and $\Delta_j g_k$. First, since $\Delta_j = P_{j+1} - P_j$, it is sufficient to estimate $P_j g$ for $g \in W_p^1$. To this end, recall that

$$(P_j g)(x) = \sum_{k \in \mathbf{Z}} g\left(\frac{k}{2^j}\right) \phi_L(2^j x - k), \tag{4.3}$$

with $\phi_L \in V_0$ being the fundamental function defined in (2.17) and having exponential decay. From the Lemma 3.3 in Li [9], we have

$$\|P_0 g\|_p \le C \left(\sum_k |g(k)|^p\right)^{1/p}, \tag{4.4}$$

with the constant given by $C := \left(\int_0^1 \left(\sum_k |\phi_L(x+k)|\right)^p dx\right)^{1/p} < \infty$. Moreover, the proof of Lemma 3.1 in [9] also shows that

$$\sum_k |g(k)|^p \le \|g\|_p^p + p\|g\|_p^{p/p'} \cdot \|g'\|_p \tag{4.5}$$

$$= \|g\|_p^p + p\|g\|_p^{p-1} \cdot \|g'\|_p,$$

where $\frac{1}{p'} + \frac{1}{p} = 1$, so that combining (4.4) and (4.5) yields

$$\|P_0 g\|_p \le C_1 (\|g\|_p^p + \|g\|_p^{p-1} \cdot \|g'\|_p)^{1/p}. \tag{4.6}$$

Now, for any $j \in \mathbf{N}$, since

$$(P_j g)(x) = (P_0 h)(2^j x), \quad \text{where} \quad h(x) = g\left(\frac{x}{2^j}\right), \tag{4.7}$$

a simple scaling operation and an application of (4.6) give us the following.

Lemma 1. For $g \in W_p^1$, $j \in \mathbf{Z}$,

$$\|P_j g\|_p \le C_1 (\|g\|_p^p + 2^{-j} \|g\|_p^{p-1} \cdot \|g'\|_p)^{1/p}. \tag{4.8}$$

We will use Lemma 1 to estimate $\|\Delta_j g_k\|$ for $k \ge j$. However, to establish a proper estimate for $\|\Delta_j f_0\|_p$ and $\|\Delta_j g_k\|_p$ when $0 \le k < j$, we need to adopt other techniques.

From the duality relation between ψ and $\widetilde{\psi}$, we can write, as in [2],

$$(\Delta_j g)(x) = \sum_k 2^j \langle g, \widetilde{\psi}(2^j \cdot -k)\rangle \psi(2^j x - k), \qquad (4.9)$$

where the interpolatory wavelet ψ and its dual $\widetilde{\psi}$ are given in (2.28) and (2.53), respectively. Moreover, from (2.53) and (2.38), we can write

$$\widetilde{\psi} = 2 \sum_n t_n \delta(2 \cdot -n), \qquad (4.10)$$

for some coefficients sequence $\{t_n\}_{n \in \mathbf{Z}}$ with exponential decay. Since both $\widetilde{\psi}_L$ and $\widetilde{\psi}$ are generators of $\{\widetilde{W}_j\}$, it follows from the vanishing moment property of Theorem 1 (see (2.61) or (2.66)) that

$$\langle p, \widetilde{\psi} \rangle = 0, \qquad p \in \pi_d. \qquad (4.11)$$

Lemma 2. Let ℓ be an integer satisfying $1 \leq \ell \leq d+1$, and $N_\ell(x)$ denote the cardinal B-spline of order ℓ with support $[0, \ell]$ (see [1, pp.85–86]). Then, there exists a function

$$\psi_\ell(x) = \sum_n c_n^{(\ell)} N_\ell(2x - n) \qquad (4.12)$$

where the sequence $\{c_n^{(\ell)}\}_{n \in \mathbf{Z}}$ has exponential decay as $|n| \to \infty$, such that

$$\psi_\ell^{(\ell)} = \widetilde{\psi}. \qquad (4.13)$$

Here, the distributional derivative is used.

Proof: Let us first consider the case $\ell = 1$. Since $N_1(x) = [0,1](\cdot - x)_+^0 = (1-x)_+^0 - (-x)_+^0$, we have

$$N_1'(x) = \delta(-x) - \delta(1 - x) = \delta(x) - \delta(x - 1).$$

Writing $\psi_1(x) := \sum_n c_n^{(1)} N_1(2x - n)$, we obtain

$$\psi_1'(x) = \sum_n 2 c_n^{(1)} N_1'(2x - n)$$

$$= \sum_n 2 c_n^{(1)} [\delta(2x - n) - \delta(2x - n - 1)]$$

$$= \sum_n 2 (c_n^{(1)} - c_{n-1}^{(1)}) \delta(2x - n).$$

Let $c_n^{(1)} := \sum_{k=-\infty}^{n} t_k$, where $\{t_n\}$ is given as in (4.10). We then obtain (4.13) for $\ell = 1$. Moreover, from $\langle 1, \tilde{\psi} \rangle = 0$ we have $\sum_{k \in \mathbb{Z}} t_k = 0$, so that $c_n^{(1)} = \sum_{k \le n} t_k = -\sum_{k > n} t_k$, and therefore, $\{c_n^{(1)}\}$ as well as $\{t_n\}$ have exponential decay. Furthermore, the vanishing moment property (4.11) implies that

$$\langle p, \psi_1 \rangle = 0, \quad \text{for all} \quad p \in \pi_{d-1}. \tag{4.14}$$

By applying this fact and the B-spline recurrence formula

$$N_j'(x) = N_{j-1}(x) - N_{j-1}(x-1), \qquad 1 \le j \le d+1, \tag{4.15}$$

(see, *e.g.* [1, p.86]) we may conclude, by a similar argument, that (4.12) and (4.13) hold for any integer ℓ satisfying $1 \le \ell \le d+1$. This proves Lemma 2. ∎

To complete the proof of Theorem 2, we need the following lemma concerning the boundedness of linear operators on L^p. This lemma can be found, at least, implicitly in literature such as [10] or [6]. For completeness, we state this lemma and include a simple proof.

Lemma 3. *Let K be a linear operator on L^p, $1 \le p \le \infty$, defined by*

$$(Kg)(x) := \int_{\mathbb{R}} k(x,y)g(y)dy, \quad g \in L^p, \tag{4.16}$$

where the kernel $k(x,y)$ satisfies

$$|k(x,y)| \le C_1(1+|x-y|)^{-2}, \qquad x, y \in \mathbb{R}. \tag{4.17}$$

Then there is a constant $C > 0$, such that

$$\|Kg\|_p \le C\|g\|_p \qquad g \in L^p. \tag{4.18}$$

Proof: By the definition in (4.16) and the condition in (4.17), it is easy to see that (4.18) holds for $p = 1$ and $p = \infty$. Thus, by the Riesz-Thorin interpolation theorem on linear operators (see *e.g.* [6, p.193]), (4.18) holds for all p, $1 \le p \le \infty$. This proves the lemma. ∎

Proof of Theorem 2. As we have already seen at the beginning of this section, the sufficiency direction of the theorem is clear. Let us continue the proof of the other direction. Since $1 \le m = 1 + \min\{\lfloor r \rfloor, d\} \le d+1$, it follows from (4.9) and Lemma 2 and by integration by parts that

$$(\Delta_j g)(x) = \sum_k 2^j \langle g, \psi_m^{(m)}(2^j \cdot -k) \rangle \psi(2^j x - k)$$

$$= (-1)^m 2^j \cdot 2^{-jm} \sum_k \langle g^{(m)}, \psi_m(2^j \cdot -k) \rangle \psi(2^j x - k).$$

Here, $\langle g^{(m)}, \psi_m(2^j \cdot -k) \rangle$ is nothing else but the integral $\int_{\mathbf{R}} g^{(m)}(y) \psi_m(2^j y - k) dy$, so that

$$(\Delta_j g)(x) = (-1)^m 2^{-jm} \cdot 2^j \int_{\mathbf{R}} \sum_k \psi(2^j x - k) \psi_m(2^j y - k) g^{(m)}(y) dy \tag{4.19}$$

$$= (-1)^m 2^{-jm} 2^j \int_{\mathbf{R}} k(2^j x, 2^j y) g^{(m)}(y) dy,$$

where

$$k(x, y) := \sum_k \psi(x - k) \psi_m(y - k). \tag{4.20}$$

Since ψ as given in (2.28) is compactly supported and ψ_m as derived in (4.12) has exponential decay, one can easily see that $k(x, y)$ as defined in (4.20) satisfies (4.17) (see, *e.g.* [3]). Thus, by Lemma 3 and (4.19), we have

$$\|\Delta_0 g\|_p \le C \|g^{(m)}\|_p. \tag{4.21}$$

Moreover, since $(\Delta_j g)(x) = (-1)^m 2^{-jm} (\Delta_0 h)(2^j x)$ for $g^{(m)}\left(\frac{z}{2^j}\right) = h^{(m)}(z)$, we obtain

$$\|\Delta_j g\|_p \le C 2^{-jm} \|g^{(m)}\|_p. \tag{4.22}$$

Applying (4.22) to $g = f_0$, $g = g_k$, $k \le j$, we now have

$$\|\Delta_j f_0\|_p + \sum_{k=0}^{j} \|\Delta_j g_k\|_p \le C \sum_{k=0}^{j} 2^{-jm} \|g_k^{(m)}\|_p \tag{4.23}$$

$$\le C \sum_{k=0}^{j} 2^{-jm} \varepsilon_k 2^{(m-s)k}, \quad k \le j,$$

where the last inequality follows from (3.3). On the other hand, by Lemma 1, we also have

$$\|P_j g_k\|_p \le C_1 \|g_k\|_p^{1-\frac{1}{p}} (\|g_k\|_p + 2^{-j} \|g_k'\|_p)^{1/p}. \tag{4.24}$$

In addition, Stein's inequalities (see [11]), (3.2), and (3.3) together imply that

$$\|g_k'\|_p \le C \|g_k\|_p^{1-\frac{1}{m}} \|g_k^{(m)}\|_p^{\frac{1}{m}} \tag{4.25}$$

$$\le C \varepsilon_k (2^{-sk})^{1-\frac{1}{m}} (2^{(m-s)k})^{1/m} = C \varepsilon_k 2^{(1-s)k}.$$

Hence, (4.24), (4.25), and (3.2) together yield, for $k > j$,

$$\|P_j g_k\|_p \le C_2 \varepsilon_k (2^{-sk})^{1-\frac{1}{p}} (2^{-sk} + 2^{-j} 2^{(1-s)k})^{1/p} \le C_2 \varepsilon_k 2^{-sk+(k-j+1)/p}.$$

$$(4.26)$$

Consequently, we have

$$\|\Delta_j g_k\|_p \le \|P_j g_k\|_p + \|P_{j+1} g_k\|_p \le C \varepsilon_k 2^{(\frac{1}{p}-s)k-\frac{j}{p}}, \quad k > j. \qquad (4.27)$$

Now, applying (4.23) for $k \le j$ and (4.27) for $k > j$ to (4.2), we conclude that

$$\|\Delta_j f\|_p \le \|\Delta_j f_0\|_p + \sum_{k \ge 0} \|\Delta_j g_k\|_p$$

$$(4.28)$$

$$\le C 2^{-jm} \sum_{k=0}^{j} \varepsilon_k 2^{(m-s)k} + C 2^{-j/p} \sum_{k=j+1}^{\infty} \varepsilon_k 2^{(\frac{1}{p}-s)k}$$

$$\le C \tilde{\varepsilon}_j 2^{-sj},$$

where

$$\tilde{\varepsilon}_j := \sum_{k=0}^{j} \varepsilon_k 2^{(s-m)(j-k)} + \sum_{k=j+1}^{\infty} \varepsilon_k 2^{(s-\frac{1}{p})(j-k)} = \sum_{k=0}^{\infty} \varepsilon_k b_{j-k}, \qquad (4.29)$$

with

$$b_l := \begin{cases} 2^{(s-m)l}, & l \ge 0; \\ 2^{(s-\frac{1}{p})l}, & l < 0. \end{cases} \qquad (4.30)$$

We extend the sequence $\varepsilon = \{\varepsilon_k\}_{k \in \mathbb{N}}$ from $\ell^q(\mathbb{N})$ to $\ell^q(\mathbf{Z})$ by setting $\varepsilon_k = 0$ for $k < 0$, so that the sequence $\{\tilde{\varepsilon}_j\}_{j \in \mathbf{Z}}$ given in (4.29) is a convolution of $\{\varepsilon_k\}_{k \in \mathbf{Z}}$ with $\{b_l\}_{l \in \mathbf{Z}}$. Since $m > s > 1/p$, we see from (4.30) that $\{b_l\}_{l \in \mathbf{Z}} \in \ell^1(\mathbf{Z})$, and hence, $\{\tilde{\varepsilon}_j\}_{j \in \mathbf{Z}}$ is also in $\ell^q(\mathbf{Z})$. From (4.28), we conclude that $\{2^{js}\|\Delta_j f\|_p\}_{j \in \mathbb{N}} \in \ell^q(\mathbb{N})$. This completes the proof of the theorem. ∎

Acknowledgments. This research was supported by NSF Grant DMS-92-06928 and ARO Contract DAAH 04-93-G-0047.

References

[1] Chui, C. K., *An Introduction to Wavelets*, Academic Press, Boston, 1992.

[2] Chui, C. K. and Chun Li, Dyadic affine decompositions and functional wavelet transforms, *SIAM J. Math. Analysis* **27**(3) (1996), 865–890.

[3] Chui, C. K. and X. L. Shi, On L^p-boundness of affine frame operators, *Indag. Math., New Series* **4** (1993), 431–438.

[4] DeVore, R. A. and G. G. Lorentz, *Constructive Approximation*, Springer-Verlag, Berlin, 1993.

[5] Donoho, D. L., Interpolating wavelet transforms, preprint.

[6] Folland, G. B., *Real Analysis*, John Wiley & Sons, New York, 1984.

[7] Frazier, M., B. Jawerth, and G. Weiss, *Littlewood-Paley Theory and the Study of Function Spaces*, CBMS Regional Conference Series in Mathematics, #79, Amer. Math. Soc., Providence, R.I., 1991.

[8] Jia, R. Q. and C. A. Micchelli, Using the refinement equation for the construction of pre-wavelets II: power of two, in *Curves and Surfaces*, P. J. Laurent, A. Le Méhauté, and L. L. Schumaker (eds.), Academic Press, Boston, 1991, 209–246.

[9] Li, Chun, Infinite-dimensional widths in the spaces of functions II, *J. Approx. Theory* **69** (1992), 15–34.

[10] Meyer, Y., *Ondelettes et Opérateurs I: Ondelettes*, Hermann Publ., Paris, 1990.

[11] Stein, E. M., Functions of exponential type, *Ann. of Math.* **65** (1957), 582–592.

Charles K. Chui
Center for Approximation Theory
Texas A&M University
College Station, TX 77843
cchui@tamu.edu

Chun Li
Institute of Mathematics
Academia Sinica
Beijing 100080, China.

Characterizing Convergence Rates for Multiresolution Approximations

Mark A. Kon and Louise A. Raphael

"Although this may seem a paradox, all exact science is dominated by the idea of approximation." Bertrand Russell

Abstract. Characterizations of convergence rates of multiresolution and wavelet approximations with respect to the supremum norm are given for functions in the L^2-Sobolev spaces H^s. Under certain assumptions on the homogeneous Sobolev space H_h^s to which the basic wavelet ψ or scaling function ϕ belongs, it is shown that the error at level n for $f \in H^s$ is given by

$$\sup_x |f(x) - P_n f(x)| \le C 2^{-ns} \|f\|_{H^s},$$

where P_n is the projection onto the scaled space at resolution level n. In other cases, rates of convergence depend on the homogeneous Sobolev class of ϕ or ψ and not on f. Necessary and sufficient conditions on the wavelet, scaling function, and operator $I - P_n$ are provided for a given rate of convergence. Finally, these conditions are related to the Strang-Fix conditions.

Introduction – previous research

The effectiveness of wavelets is due to their accurate and efficient approximation of given functions or signals. In this paper we show how fast the sequence of errors associated with a multiresolution approximation converges to zero under the worst-case scenario, using the supremum norm.

Let P_n be a projection onto a scaled space V_n and $\| \ \|$ a norm reasonable for measuring distance. A classical problem in approximation theory is to estimate the rate of convergence of the error $E_n = \|f - P_n f\|$ as $n \to \infty$ given the smoothness of f, and to derive lower limits on the smoothness order of f given the convergence rate of E_n [9].

In the theory of wavelets, Y. Meyer and S. Mallat [17] were the first to sharply measure the decay of the L^2-approximation error of a function f and its multiresolution approximation under mild smoothness conditions

Signal and Image Representation in Combined Spaces   415
Y. Y. Zeevi and R. R. Coifman (Eds.), pp. 415–437.
Copyright ©1998 by Academic Press.
All rights of reproduction in any form reserved.
ISBN 0-12-777830-6

on the scaling function associated with a multiresolution analysis. Mallat characterized the Sobolev spaces H^s in terms of L^2 decay rates of the approximation errors. Under rapid decay of the scaling function ϕ, its partial derivatives and moments $\int x^k \phi(x) dx = 0$, Mallat proved

$$f \text{ is in } H^s \Leftrightarrow \sum_{n=-\infty}^{\infty} 2^{2ns} \|f - P_n f\|_2^2 < \infty.$$

Recently, Walter [23] proved that under conditions on the scaling function involving the Zak transform, r-regular wavelet expansions converge uniformly for continuous functions f at a rate $O(2^{-n(s-1/2)})$ when f belongs to H^s.

Meyer [18], among the many results in Chapters 2 and 6 of *Wavelets and Operators*, provides an elegant treatise for r-regular multiresolution approximations of $L^2(\mathbb{R}^d)$ functions in the settings of Sobolev, Hölder, and Besov spaces. Bertoluzza [1], using Meyer's approach, found error estimates for wavelet expansions with respect to Besov and L^p norms. She also found Sobolev error estimates by adopting the numerical procedure of [11] to approximate the wavelet expansion of a function f with respect to the Daubechies wavelet basis.

In the theory of approximation, de Boor, De Vore, and Ron [7] and more recently Jetter and Zhou [12], provide L^2-approximation errors using sequences of dilations of shift-invariant spaces. Multiresolution analysis is a specialization of their theory. Also, de Boor and Ron [8] have studied order s convergence in supremum norm for principal shift-invariant subspaces. Fisher [10] has derived rates of approximation for uniformly bounded sequences of linear operators $\{T_n\}$ such that $T_n f = f$ on the intersection of the scaling space V_n and the spaces of continuous bounded functions on $\mathbb{R}^d$. She applied her theory to projection operators and sampling spaces

Our techniques differ from the above. We use the L^1-convolution kernel bounds on the kernels associated with the projection operators P_n, developed in [14] together with Fourier analysis.

In Section 1 we define basic concepts, properties, and state theorems from [14] needed for the proof of the principal theorem. In Section 2 we state the main theorem, and in Sections 3 and 4 we sketch the outline of the proof for necessary and sufficient conditions on the wavelet and scaling function, respectively. Section 5 closes with a discussion of the relation between these conditions and the Strang-Fix conditions.

§1   Basic concepts

The easiest way to construct a wavelet for $L^2(\mathbb{R}^d)$ is by using multiresolution analysis as developed by Mallat [17] and Meyer [18]. We recall that

$L^p(\mathbb{R}^d)$ denotes the space of all measurable functions defined on $\mathbb{R}^d$ such that $\int_{\mathbb{R}^d} |f(x)|^p dx < \infty$ when $0 < p < \infty$, and $\|f\|_\infty = \sup_x |f(x)|$, $x \in \mathbb{R}^d$, where sup denotes supremum modulo sets of measure 0. A multiresolution analysis (MRA) on $\mathbb{R}^d$, $d \geq 1$, decomposes the space $L^2(\mathbb{R}^d)$ into an increasing sequence of closed subspaces V_n

$$\ldots, V_{-2} \subset V_{-1} \subset V_0 \subset V_1 \subset V_2 \ldots, \qquad (1.1\text{a})$$

with the property that the space V_i is a "rescaling" of the space V_{i-1}, that is,

$$\phi(x) \in V_i \text{ if and only if } \phi(2x) \in V_{i+1} \text{ for all } i. \qquad (1.1\text{b})$$

It is also required that

$$\cap_{j\in\mathbf{Z}}V_j = \{0\}; \qquad (1.1\text{c})$$

$$\overline{\cup_{j\in\mathbf{Z}}V_j} = L^2(\mathbb{R}^d), \qquad (1.1\text{d})$$

where the overline denotes closure. Moreover, V_0 is closed under integer translations, *i.e.*

$$\phi(x) \in V_0 \quad \Rightarrow \quad \phi(x - k) \in V_0 \qquad (1.1\text{e})$$

for all $k \in \mathbf{Z}^d$. Finally, it is assumed that there exists a function $\phi \in L^2(\mathbb{R}^d)$ such that

$$\{\phi_k(x) \equiv \phi(x - k)\}_{k\in\mathbf{Z}^d} \text{ form an orthonormal basis for } V_0. \qquad (1.1\text{f})$$

Such a function ϕ is called a *scaling function*. Let the space W_i denote the orthogonal complement of V_i in V_{i+1}, *i.e.* $W_i = V_{i+1} \ominus V_i$, so that $V_{i+1} = V_i \oplus W_i$. From existence of ϕ it follows (see *e.g.* [5, 18]) that there is a set of basic wavelets $\{\psi^\lambda(x)\}_{\lambda\in\Lambda}$ (with Λ a finite index set) such that $\psi_{jk}^\lambda(x) \equiv 2^{jd/2}\psi^\lambda(2^j x - k)(j \in \mathbf{Z}, k \in \mathbf{Z}^d)$ which form an orthonormal basis for W_j for fixed j, and form an orthonormal basis for $L^2(\mathbb{R}^d)$ as j, k, λ vary.

By (1.1a) and (1.1b), it follows that there exist constants $\{h_k\}$ such that the scaling equation

$$\phi(x) = 2^d \sum_{k=0}^{\infty} h_k \phi(2x - k) \qquad (1.1\text{g})$$

is satisfied. The function

$$m_0(\xi) \equiv \sum_{k=0}^{\infty} h_k e^{-k\xi} \qquad (1.1\text{h})$$

is called the *symbol* of the sequence $\{h_k\}$, and satisfies

$$\hat{\phi}(\xi) = m_0(\xi/2)\hat{\phi}(\xi/2)$$

via Fourier transforming (1.1g), where $\hat{\phi}(\xi) = (2\pi)^{-d} \int_{\mathbf{R}^d} \phi(x) e^{-i\xi \cdot x} dx$.

Our results hold in arbitrary dimension. The most direct construction of multidimensional wavelets is through tensor products of one-dimensional multiresolution analyses (see *e.g.* [4, 5, 18]). For example, in two dimensions a basis for W_j is of the form

$$\{\psi_{jk}^h, \psi_{jk}^v, \psi_{jk}^d\}_{k \in \mathbf{Z}^2} = \{\psi_{jk}^\lambda\}_{k \in \mathbf{Z}^2, \lambda \in \{h,v,d\}},$$

where

$$\psi^h(x,y) = \phi(x)\psi(y); \quad \psi^v(x,y) = \psi(x)\phi(y); \quad \psi^d(x,y) = \psi(x)\psi(y) \quad (1.2)$$

and $\psi_{jk}^\lambda(x,y) = 2^j \psi_0^\lambda(2^j x - k_1, 2^j y - k_2)$ (here $(k_1, k_2) = k$). This orthonormal basis respects the decomposition, and analogous bases can be constructed in higher dimensions. Thus, in general we will write as a wavelet basis for $L^2(\mathbf{R}^d)$, the collection $\{\psi_{jk}^\lambda\}_{j \in \mathbf{Z}, k \in \mathbf{Z}^d, \lambda \in \Lambda}$, with Λ an indexing set containing $2^d - 1$ elements.

A more general construction of wavelet bases in multiple dimension is given in [18]. It is assumed there that a multiresolution analysis is given in which the scaling function is r-regular; that is, ϕ and its partial derivatives decay faster than the inverse of any polynomial when $|x| \to \infty$.

Our results hold for any set of $\{\phi_{jk}\}, \{\psi_{jk}^\lambda\}$ whose translations and dilations form an orthonormal basis for $L^2(\mathbf{R}^d)$ as constructed by any of the above algorithms. In addition, they also can be stated in cases where ϕ or ψ fail to exist.

The following concepts are essential for our formulation of the sharp conditions for order s sup-norm convergence of multiresolution approximations.

Definition 1. *A function $f(x)$ is in the class $\mathcal{RB}$ if it is absolutely bounded by an L^1 radial decreasing function $\eta(x)$, i.e. with $\eta(x_1) = \eta(x_2)$ whenever $|x_1| = |x_2|$, and with $\eta(x_1) \leq \eta(x_2)$ whenever $|x_1| \geq |x_2|$, and $\eta(x) \in L^1(\mathbf{R}^d)$. (note that since η is defined at the origin, ϕ must be bounded). We define P_n and Q_n, respectively, to be the orthogonal projections onto the spaces V_n and W_n, with kernels $P_n(x,y)$ and $Q_n(x,y)$.*

We will say that given $f \in L^2$,

(i) the *multiresolution approximation* of f is defined by the sequence $\{P_n f\}_n$.

(ii) the *wavelet expansion* of f is

$$\sum_{j;k,\lambda} a_{jk}^\lambda \psi_{jk}^\lambda(x) \sim f, \quad (1.3)$$

where the a_{jk}^λ are the L^2-expansion coefficients of f.

(iii) the *scaling expansion* of f

$$\sum_k b_k \phi_k(x) + \sum_{j \geq 0; k; \lambda} a_{jk}^\lambda \psi_{jk}^\lambda(x) \sim f, \qquad (1.4)$$

where the b_k, a_{jk}^λ are L^2-expansion coefficients of f defined by

$$b_k = \int_{\mathbf{R}^d} \phi_k(x) f(x) dx \quad \text{and} \quad a_{jk}^\lambda = \int_{\mathbf{R}^d} \psi_{jk}^\lambda(x) f(x) dx.$$

We say that such expansion sums (1.3) and (1.4) converge in any given sense (*e.g.* pointwise, in L^p, etc.), if they do so in a semi-order independent way. Specifically, it is required that the sums be calculated in such a way that the range (largest minus smallest) of j values for which the sum over k is incomplete, remains uniformly bounded.

Most importantly, as shown in [14], partial sums of wavelet and scaling expansions as well as projections $P_n f$ are almost everywhere equal to each other at each level N of summation under our assumptions.

Definition 2. *Letting $\hat{f}(\xi)$ be the Fourier transform of f, we define the L^2 Sobolev space H^s by:*

$$H^s \equiv \{ f \in L^2(\mathbf{R}^d) : \|f\|_s \equiv \sqrt{\int |\hat{f}(\xi)|^2 (1 + |\xi|^2)^s d\xi} < \infty \}. \qquad (1.5)$$

The L^2 homogeneous Sobolev space is defined by:

$$H_h^s \equiv \{ f \in L^2(\mathbf{R}^d) : \|f\|_{h,s} \equiv \sqrt{\int |\hat{f}(\xi)|^2 |\xi|^{2s} d\xi} < \infty \}. \qquad (1.6)$$

We note that H^s is contained in H_h^s, but H_h^s contains functions f whose Fourier transform may be quite singular at the origin. For the purposes of studying multiscale expansions, the space H_h^s is advantageous over H^s in that H_h^s norms scale nicely with changes in scales of functions.

We now relate the two Sobolev norms $\| \ \|_s$ and $\| \ \|_{h,s}$. Note that since for positive a, b, we have $a^s + b^s \leq (a + b)^s \leq 2^{s+1}(a^s + b^s)$, it follows that

$$\begin{aligned}
\|f\|_s &= \left(\int (1 + |\xi|^2)^s |\hat{f}(\xi)|^2 d\xi \right)^{1/2} \\
&\sim \left(\int |\xi|^{2s} |\hat{f}(\xi)|^2 d\xi \right)^{1/2} + \left(\int |\hat{f}(\xi)|^2 d\xi \right)^{1/2} \qquad (1.7) \\
&= \|f\|_{h,s} + \|f\|_0,
\end{aligned}$$

where in the above equivalences the expressions are considered as functions of the argument $f \in H^s$, so that $A(f) \sim B(f)$ means that there exist constants K_1 and K_2, such that $K_1 A(f) \leq B(f) \leq K_2 A(f)$ for all f for which both sides are defined. We note the h, s norm scales conveniently under dilations.

If $\hat{f}$ and $\mathcal{F}(f)$ denote the Fourier transform of f, then note $\mathcal{F}(f(ax)) = a^{-d} \hat{f}(\xi/a))$ letting

$$f_n(x) = 2^{nd/2} f(2^n x),$$

we have

$$\|f_n\|_{h,s} = 2^{ns} \|f\|_{h,s}. \tag{1.8}$$

Next, we cite the scaling property of the kernels $P_n(x, y)$ (see [18]), given by

$$P_n(x, y) = 2^{nd} P_0(2^n x, 2^n y). \tag{1.9}$$

Finally, we define the pointwise order of convergence for a family of wavelets.

Definition 3. *An MRA or family of wavelets ψ^λ yields pointwise order of convergence (or pointwise order of approximation) s in H^r if for any function $f \in H^r$, the nth level multiresolution approximation $P_n f$ satisfies*

$$\|f - P_n f\|_\infty \leq C 2^{-ns} \|f\|_s \tag{1.10}$$

where C may depend on f. The MRA or family of wavelets yields maximal order of convergence s in H^r if s is the largest number such that (1.10) holds for all f. If the supremum s of the numbers for which (1.10) holds is not attained, then we denote the maximal order of convergence by s^-.

1.1 Basic lemmas

It is shown in [23] that wavelet expansions of functions which are sufficiently smooth converge at rates commensurate with the differentiability of the wavelet used in the expansion. We will give sharpened versions of these results in necessary and sufficient form. We begin by reviewing some needed results from [13, 14]. Below ϕ and ψ denote, respectively, the scaling function and basic wavelet of a multiresolution expansion. All results in this section are in $\mathbb{R}^d$ unless stated otherwise.

Lemma 1. [14]

(i) *If the scaling function $\phi \in \mathcal{RB}$, then the kernel*

$$P_0(x, y) \equiv \sum_{k \in \mathbb{Z}^d} \phi(x - k) \overline{\phi}(y - k)$$

satisfies

$$|P_0(x, y)| \leq H_0(|x - y|), \tag{1.11}$$

where $H_0(x)$ is a bounded radial decreasing L^1 function. Further, the convergence of this sum is uniform on $\mathbb{R}^{2d}$. This sum forms the L^2 kernel of the projection P_0 onto V_0.

(ii) If $\psi^\lambda(x) \in \mathcal{RB}$ for each λ, then $Q_0(x, y) \equiv \sum_{k \in \mathbb{Z}^d; \lambda} \psi^\lambda(x-k)\overline{\psi}^\lambda(y-k)$ converges uniformly and absolutely on $\mathbb{R}^{2d}$, and is bounded. Further, if $\psi^\lambda(x) \ln(2 + |x|) \in \mathcal{RB}$ for each λ, then

$$|Q_0(x, y)| \leq H_1(|x - y|)/\ln(2 + |x - y|) \tag{1.12}$$

where $H_1(x)$ is a bounded radial L^1 decreasing function. This sum $Q_0(x, y)$ is the kernel of the orthogonal projection Q_0 onto W_0.

Let $D = \{(x, y) : x = y\} \subset \mathbb{R}^d \times \mathbb{R}^d$ denote the diagonal hyperplane. Then we also have

Lemma 2. [14] *If $\psi^\lambda(x) \cdot \ln(2 + |x|) \in \mathcal{RB}$ for each λ, then in any set $F \subset \mathbb{R}^d \times \mathbb{R}^d$ with $F \cap D = \emptyset$, the kernel $M_n(x, y) \equiv \sum_{0 \leq j < n; k; \lambda} \psi_{jk}^\lambda(x)\overline{\psi}_{jk}^\lambda(y)$ satisfies $M_n(x, y) \underset{n \to \infty}{\longrightarrow} M(x, y)$, where $M(x, y)$ satisfies*

$$|M(x, y)| \leq H_1(|x - y|), \tag{1.13}$$

with $H_1(x)$ a radial decreasing L^1 function (possibly infinite at 0). Further, the convergence of M_n to M is uniform and absolute if F has a positive distance from the diagonal $D = \{(x, y) : x = y\}$, and the absolute sum satisfies

$$\sum_{0 \leq j \leq n; k; \lambda} |\psi_{jk}^\lambda(x)\overline{\psi}_{jk}^\lambda(y)| \leq H_2(|x - y|),$$

uniformly in n, for some other function H_2 with the same properties as H_1 above.

Let Q_j be the projection onto the wavelet space W_j, which has kernel $Q_j(x, y) = \sum_{k, \lambda} \psi_{jk}^\lambda(x)\overline{\psi}_{jk}^\lambda(y)$.

Lemma 3. [14] *Under only the assumption $\psi^\lambda(x) \in \mathcal{RB}$ for all λ, the large scale part of the kernel $N_{-n}(x, y) = \sum_{-n \leq j < 0} Q_j(x, y)$ converges uniformly and absolutely to a bounded kernel $P_0(x, y)$ as $n \to \infty$, uniformly on $\mathbb{R}^d \times \mathbb{R}^d$. The limit $P_0(x, y)$ is the kernel of P_0, the projection onto V_0.*

§2 The main results

Conditions (a) through (d''') in the following theorems are equivalent. The wavelet and scaling function results are independent, in that the results on wavelets do not require the existence of a scaling function, and vice-versa.

We assume that our wavelets satisfy $\psi(x)\ln(2+|x|) \in \mathcal{RB}$, and/or that our scaling functions are in $\mathcal{RB}$. These assumptions are needed for various L^p and almost everywhere pointwise convergence properties of wavelet expansions (1.3) and (1.4) [14].

In Theorems 1, 2, and 3 we give necessary and sufficient conditions on the basic wavelet ψ or the scaling function ϕ for order s convergence of expansions in a given Sobolev space.

We emphasize that the statements in the theorems apply to all three types of expansions and approximations of Definition 1: wavelet expansions, scaling expansions, and the projections $P_n f$. This follows as the partial sums of (1.3), (1.4) and projections $P_n f$ are almost equal everywhere to each other at each level of summation [14].

Theorem 1. *Given a multiresolution analysis with either*

(i) *a scaling function $\phi \in \mathcal{RB}$,*

(ii) *basic wavelets satisfying $\psi^\lambda(x)\ln(2+|x|) \in \mathcal{RB}$, or*

(iii) *a kernel for the basic projection P satisfying $|P(x,y)| \le H(|x-y|)$ with $H(x) \in \mathcal{RB}$, the following conditions (a to b) are equivalent for $s > d/2$:*

(a) *The multiresolution approximation yields pointwise order of convergence $s - d/2$ in H^s.*

(a′) *The multiresolution approximation yields pointwise order of convergence $r - d/2$ in H^r for all $r \le s$.*

(a″) *The multiresolution approximation yields maximal pointwise order of convergence $s - d/2$ in H^s.*

(a‴) *The multiresolution approximation yields maximal pointwise order of convergence $r - d/2$ in H^r for all $r \le s$.*

(b) *The projection $I - P_n : H_h^r \to L^\infty$ is bounded, where I is the identity and d denotes dimension.*

Theorem 2. *Under the assumptions of Theorem 1, if there exists a family $\{\psi^\lambda\}$ of basic wavelets corresponding to $\{P_n\}$ with $\psi^\lambda(x)\ln(2+x) \in \mathcal{RB}$, then the following conditions are equivalent to those in Theorem 1.*

(c) *For every such family of basic wavelets and each $\lambda, \psi^\lambda \in H_h^{-s}$.*

(c′) For every such family of basic wavelets and for each λ,

$$\int_{|\xi|<\delta} |\hat{\psi}^\lambda(\xi)|^2 |\xi|^{-2s} d\xi < \infty \tag{2.1}$$

for some (or for all) $\delta > 0$ (including $\delta = \infty$).

(c″) For some such family of basic wavelets, (2.1) holds.

Theorem 3. *If there exists a scaling function ϕ corresponding to $\{P_n\}$ (with or without a family of wavelets), then the following conditions are equivalent to those in Theorems 1 and 2.*

(d) For every such scaling function, $1 - (2\pi)^{d/2}|\phi| \in H_h^{-s}$.

(d′) For every scaling function $\phi \in \mathcal{RB}$ corresponding to $\{P_n\}$

$$\int_{|\xi|<\delta} (1 - (2\pi)^{d/2}|\hat{\phi}(\xi)|)|\xi|^{-2s} d\xi < \infty \tag{2.2}$$

for some (or all) $\delta > 0$ (including $\delta = \infty$).

(d″) For some scaling function ϕ corresponding to $\{P_n\}$, (2.2) holds.

(d‴) For every scaling function ϕ corresponding to $\{P_n\}$ and every $\delta > 0$,

$$\int_{|\xi|<\delta} \sum_{\ell\neq 0} |\hat{\phi}(\xi + 2\pi\ell)|^2 |\xi|^{-2s} d\xi < \infty.$$

We remark that condition (b) can refer to any given n, or to all n, since boundedness of $I - P_n$ for one n is equivalent to that for all n.

Note that statements (a)–(b) of the preceding theorem imply that the set A of $s > d/2$ for which the preceding statements hold has the property that if $s_1 \in A$ then for all $s_2 < s_1, s_2 \in A$. It therefore follows that there is a "largest" s in A, *i.e.* that for some $S > 0$,

$$A = (d/2, S] \quad \text{or} \quad A = (d/2, S).$$

It also follows from this theorem that for $s < S$, the maximal rate of convergence in $H^{s+d/2}$ is s. In addition, depending on whether $S \in A$ or not, the maximal rate of convergence in $H^{S+d/2}$ is S or S^-, where the latter indicates that all rates of convergence *less* than S are valid, but that convergence of order S itself fails. Thus we have:

Corollary 1. *Let* $S \equiv \sup\{s : I - P_n : H_h^s \to L^\infty \text{ is bounded}\}$. *Then*

(i) *For* $s < S$, *the maximal rate of convergence for the wavelet expansion in* H^s *is* $s - d/2$.

(ii) *The maximal rate of convergence for the wavelet expansion in* H^S *is* $S - d/2$ *or* $(S - d/2)^-$, *according to whether* $S \in \sup\{s : I - P_n : H_h^s \to L^\infty \text{ is bounded}\}$ *or not.*

See equation (1.1h) for the definition of the symbol $m_0(\xi)$. The next three statements for the symbol $m_0(\xi)$ are equivalent to those of Theorems 1, 2, and 3.

Theorem 4. *If* $m_0(\xi)$ *denotes a symbol of a multiresolution expansion corresponding to a sequence of projections as in Theorem 1, then the following conditions are equivalent to those in Theorems 1, 2, and 3:*

(i) *For every symbol* $m_0(\xi)$ *corresponding to* $\{P_n\}$,

$$\int_{|\xi| < \delta} (1 - |m_0(\xi)|^2)|\xi|^{-2s} d\xi < \infty \tag{2.3}$$

for some (or all) $\delta > 0$ *(including* $\delta = \infty$).

(ii) *For some symbol* $m_0(\xi)$ *corresponding to* $\{P_n\}$, *(2.3) holds.*

(iii) *Every (or some) symbol* $m_0(\xi)$ *corresponding to* $\{P_n\}$ *satisfies*

$$\int_{|\xi - \pi\epsilon| < \delta} |m_0(\xi)|^2 |\xi - \pi\epsilon|^{-2s} d\xi < \infty$$

for some (or all) $\delta > 0$ *(including* $\delta = \infty$) *and for every non-zero* ϵ *in* $\mathbb{R}^d$ *with components from the set* $\{0, 1\}$.

The preceding conditions on P_0 give the essential information we will need regarding orders of approximation in various Sobolev spaces. In fact, $H^{s+d/2}$ is the "critical" space for order of approximation s, in that it is the lowest-order Sobolev space in which order of approximation s can possibly occur. This theorem states that whether in fact order s convergence does occur in $H^{s+d/2}$ depends on the boundedness of $I - P_0$ from $H^{s+d/2}$ to L^∞. Equivalently, if there exist wavelets ψ^λ and a scaling function ϕ which implement the multiresolution analysis, we have order s convergence in the critical space $H^{s+d/2}$ if and only if ψ^λ or $1 - (2\pi)^{d/2}|\hat\phi| \in H_h^{(-s+d/2)}$.

The key idea of the proof is to essentially examine the distance between the kernel $P_n(x, y)$ (or P_n) and the delta function in a negative Sobolev space (this viewpoint is taken in [23]). We will view this difference as an operator into L^∞, and study the error

$$\|E_n\| = \|\delta(x - y) - P_n(x, y)\|_{H^s \to L^\infty},$$

where $\delta(x - y)$ is the delta distribution (in the negative Sobolev space H^{-s}, $s > d/2$) centered at y, as a convolution "kernel" applicable to functions in sufficiently smooth Sobolev spaces. Strictly speaking, of course, the latter norm is just of the operator $I - P_n$, where I denotes the identity, since there is no kernel for the above.

§3 Sketch of the proof for wavelets

Recalling the definitions of the Sobolev space H^s and homogeneous Sobolev space H^s_h, we let $s > 0$ and define $E_n = I - P_n$ to be the error operator which gives the difference between a function f and its projection $P_n f$ into V_n. Then in terms of the preceding definitions, the error E_n satisfies:

Proposition 1.

(a) For $s > 0$, the error $E_n = I - P_n$ satisfies the scaling identity

$$\|E_n\|_{H^s \to L^\infty} \sim 2^{n(d/2-s)} \|E_0\|_{H^s_h \to L^\infty}, \qquad (3.1)$$

with the equivalence uniform over all values of n, if the right side is finite.

(b) If the right side of (3.1) is infinite, then

$$\frac{\|E_n\|_{H^s \to L^\infty}}{2^{n(d/2-s)}} \xrightarrow[n \to \infty]{} \infty. \qquad (3.2)$$

Thus, if the right side of (3.1) is finite for some n, (a) of Proposition 1 implies that there are constants such that for all n,

$$C_1 2^{n(d/2-s)} \|E_0\|_{H^s_h \to L^\infty} < \|E_n\|_{H^s \to L^\infty} < C_2 2^{n(d/2-s)} \|E_0\|_{H^s_h \to L^\infty}.$$

Taking (3.1), replacing s by $s + d/2$, and noting (3.2) if $\|E_0\|_{H^s_h \to L^\infty} = \infty$, we get the following, which proves the equivalence of (a) and (b) of Theorem 1.

Proposition 2. ((a)⟺(b) of Theorem 1) *Consider a given multiresolution analysis with projections P_n onto the subspace V_n, and let $s > 0$. A necessary and sufficient condition that this multiresolution analysis yield pointwise approximations of order $s - d/2$ in H^s is that $H^s_h \to L^\infty$ be bounded, where P_0 is the projection onto V_0, and I is the identity.*

Remark.

1. Since P_0 is an operator in L^2, the statement of Proposition 1 should technically be interpreted as the assertion that $I - P_0$ maps that part of $H_h^{s+d/2}$ which is in L^2 into L^∞. However, since $L^2 \cap H_h^s$ is dense in H_h^s for all real s, these statements are equivalent and will be considered interchangeable.

2. Note that since $s < d/2$ is allowed in Proposition 1 above, this theorem also formally describes situations in which pointwise approximations are guaranteed to diverge at the specified rate $O(2^{-nd/2})$ for some functions f. This may be applicable to some multiresolution-type expansions. However, in the context of wavelet expansions the cases where $s < d/2$ are vacuous, since it is never true that in this case $I - P_0 : H_h^s \to L^\infty$ is bounded. Indeed, since we assume that ψ is bounded, it follows easily that $P_0 f \in L^\infty$ for $f \in L^2$. On the other hand, there exist unbounded functions in H^s and hence, in H_h^s for $s < d/2$. Hence, for $s < d/2$, it is impossible that $(I - P)f$ is bounded for all $f \in H_h^s$, and so in this case, the hypothesis of the theorem is never satisfied in our context for $s < d/2$.

We will next consider what properties of the basic wavelet ψ imply convergence of given order s. We will translate the above conditions into one on the basic wavelet ψ, and will see that the correct condition is that ψ be in the appropriate dual Sobolev space.

Recall that by Lemma 1, under the hypothesis that $\psi \in \mathcal{RB}$, the projection kernel $Q_j(x, y)$ is equal pointwise to the uniformly convergent sum $\Sigma_{k,\lambda} \psi_{jk}^\lambda(x) \overline{\psi_{jk}^\lambda(y)}$. Before proving the equivalence of (b) and (c) of Theorems 1 and 2, we observe that the Fourier transform of ψ,

$$(2\pi)^{d/2}(\hat{\psi}_{jk}(x)) = \int 2^{jd/2} \psi(2^j x - k) e^{-ix\xi} dx$$

$$= 2^{-jd/2} e^{-i\xi 2^{-j}k} (2\pi)^{d/2} \hat{\psi}(2^{-j}\xi). \qquad (3.3)$$

In Theorem 5 below, note that the condition $\psi(x) \in H_h^{s^*} = H_h^{-s}$ is a condition on the function ψ, viewed as a linear functional. This condition is equivalent to bounds on the Fourier transform of ψ, which dictate the approach of the proof. Recall that P_0 denotes the orthogonal projection onto V_0 in L^2.

Theorem 5. ((b)⇔(c) of Theorems 1 and 2) *Let $s > d/2$ and assume that $\psi^\lambda \ln(2 + |x|) \in \mathcal{RB}$ for all λ. Then $I - P_0 : H_h^s \to L^\infty$ is bounded if and only if $\psi(x) \in H_h^{s^*}$, the dual space of H_h^s.*

Sketch of Proof: Assume first that $\psi \in H_h^{s^*}$. We note from Lemma 2 that the series $P_n(x,y) \equiv \sum_{j<n} Q_j(x,y) = \sum_{j<n;k;\lambda} \psi_{jk}^\lambda(x)\overline{\psi_{jk}^\lambda}(y)$ converges uniformly and absolutely as $n \to \infty$ on any compact set not intersecting the diagonal $D \equiv \{(x,y): x = y\}$. Further, the series $(I - P_n)(x,y) \equiv \sum_{j\geq n} Q_j(x,y) = \sum_{j\geq n;k\in\mathbf{Z}^d;\lambda} \psi_{jk}^\lambda(x)\overline{\psi_{jk}^\lambda}(y)$ converges uniformly off the diagonal $\{x = y\}$ by Lemma 2, and its sum is in $\mathcal{RB}$. Thus, if $\mathcal{D}$ denotes C^∞ functions with compact support (which we will need to use below in order to interchange summation and integration in integrals), we have the following:

$$\|I - P_n\|_{H_h^s \to L^\infty} = \sup_{f \in \mathcal{D}} \frac{\|\sum_{j\geq n,k\in\mathbf{Z}^d;\lambda} \psi_{jk}^\lambda(x) \int \overline{\psi_{jk}^\lambda}(y)f(y)dy\|_\infty}{\|f\|_{H_h^s}}$$

$$\leq \sup_{f \in \mathcal{D}} \| \sum_{j\geq n,k\in\mathbf{Z}^d;\lambda} |\psi_{jk}^\lambda(x)| \frac{|\int \overline{\psi_{jk}^\lambda}(y)f(y)dy|}{\|f\|_{H_h^s}} \|_\infty, \quad (3.4)$$

where we have used that (Theorem 1 in [14]) under our hypotheses, wavelet expansions (1.3) converge almost everywhere, and the partial sums can be represented in terms of the projections P_n.

By standard calculations, one can show that

$$\|I - P_n\|_{H_h^s \to L^\infty} \leq \sum_{j\geq n;\lambda} \|\psi^\lambda\|_{h,-s} 2^{-js} \| \sum_{k\in\mathbf{Z}^d} |\psi_{jk}^\lambda(x)| \|_\infty,$$

where $\| \sum_{k\in\mathbf{Z}^d} |\psi_{jk}^\lambda(x)| \|_\infty$ for $\psi \in \mathcal{RB}$ can be easily bounded as a discrete approximation of

$$2^{jd/2} \int_{\mathbf{R}^d} |\psi^\lambda(2^j x - k)| dk = 2^{jd/2} \|\psi^\lambda\|_{L^1}.$$

Thus, we have

$$\|I - P_n\|_{H_h^s \to L^\infty} \leq \sum_\lambda C' \|\psi^\lambda\|_{h,-s} \sum_{j\geq n} 2^{-j(s-d/2)} < \infty$$

if $s > d/2$, since the sum over λ is finite.

Conversely, assume that $\psi^{\lambda_1} \notin H_h^{s^*} = H_h^{-s}$ for some $\lambda_1 \in \Lambda$. Without loss we will assume that $\lambda_1 = 1$. Thus, there exists a sequence of functions $f_n \in H_h^s$ of uniformly bounded H_h^s norm such that

$$\int \overline{\psi^1}(-\xi)\hat{f}_n(\xi)d\xi \xrightarrow[n\to\infty]{} +\infty.$$

We assume without loss that $\|f_n\|_{h,s} = 1$. Let $\hat{Q}_0(x,\xi)$ denote the Fourier transform of $Q_0(x,y)$ in the variable y. This exists by virtue of the bounds in Lemma 1. Using (3.3), we have (L^∞ norms are with respect to the x variable)

$$\left\| \int Q_0(x,y) f_n(y) \right\|_\infty = \left\| \int \hat{Q}_0(x,\xi) \hat{f}_n(\xi) d\xi \right\|_\infty$$

$$= \left\| \int \hat{f}_n(\xi) \sum_\lambda \overline{\hat{\psi}^\lambda}(-\xi) \right.$$

$$\left. \times \sum_k e^{-i\xi k} \psi_{0k}^\lambda(x) d\xi \right\|_\infty, \tag{3.5}$$

where all integrals and sums converge absolutely and uniformly.

Through rather lengthy, constructive, technical arguments, we can show that, for a particular choice of $\{f_n\}$, the supremum norm above, associated with a subsequence $\{f_{n_k}\}$ becomes infinite as $k \to \infty$. For a detailed proof, see [16]. That is,

$$\left\| \int Q_0(x,y) f_{n_k}(y) dy \right\|_\infty \xrightarrow[k \to \infty]{} \infty$$

where $\|f_{n_k}\|_{hs} = 1$ for all n_k. By scaling, we therefore conclude that for all n,

$$\|Q_n\|_{H_h^s \to L^\infty} = \infty.$$

Now consider the projection $P_1 = P_0 + Q_0$. We have

$$\|(I - P_1) f_{n_k}\|_\infty = \|(I - P)_0 f_{n_k} - Q_0 f_{n_k}\|_\infty, \tag{3.6}$$

so that if $\|(I - P_0) f_{n_k}\|_\infty$ is bounded in k, then by (3.6) $\|(I - P_1) f_{n_k}\|_\infty$ is unbounded in k. Therefore either $\|I - P_0\|_{H_h^s \to L^\infty}$ or $\|I - P_1\|_{H_h^s \to L^\infty}$ is infinite. Thus, by scaling invariance, it follows that $\|I - P_0\|_{H_h^s \to L^\infty} = \|I - P_n\|_{H_h^s \to L^\infty} = \infty$ (the above two norms are infinite or finite together). This completes the proof. ∎

Next we remark that (c) and (c') of Theorem 2 are equivalent. This follows because for $\psi \in L^2$,

$$\int_{|\xi| < \delta} |\hat{\psi}^\lambda(\xi)|^2 |\xi|^{-2s} d\xi < \infty \tag{3.7}$$

characterizes the functions $\psi \in H_h^{-s}$.

Remark.

1. Note that this result makes intuitive sense, since order $s - d/2$ convergence in a natural way requires $s > d/2$. For if $f \in H^r$ for $r < d/2$, then not only does convergence above fail, but by standard theorems

about Sobolev spaces, the function f need no longer be continuous and can take on an infinite essential supremum. Therefore, it would be impossible for such functions to have wavelet expansions which converge uniformly for this independent reason.

2. We note here that the order s convergence implied by this Theorem (as well as by the other conditions stated in Theorems 1, 2, and 3) in fact applies to any order of summation in which scale j wavelets are added before scale $j + 1$ wavelets. Thus, the rates of convergence apply to partial sums which include all wavelets at scale j and any subset of those at the next scale.

§4 Sketch of proof for scaling functions

We now state a condition analogous to (3.7) for the scaling function ϕ, if it exists. We first note that under our hypotheses,

$$\int_{\mathbf{R}^d} \phi(x)dx = 1, \tag{4.1}$$

if the function ϕ is multiplied by an appropriate phase constant. Let $\mathcal{G} = \{0,1\}^d$ denote all vectors with entries consisting of $\{0,1\}$. We have then the following basic properties [5].

Lemma 4. *If ϕ is a scaling function for a multiresolution analysis, then*

$$\sum_{\ell \in \mathbf{Z}^d} |\hat{\phi}(\xi + 2\pi\ell)|^2 = (2\pi)^{-d}.$$

Furthermore, for any $\epsilon \in \mathcal{G}$,

$$\sum_{\ell \in 2\mathbf{Z}^d + \epsilon} |\hat{\phi}(\xi + \pi\ell)|^2 = (2\pi)^{-d}.$$

Lemma 5. *If ϕ is a scaling function for a multiresolution analysis, then $\hat{\phi}(0) = (2\pi)^{-d/2}$, and $\hat{\phi}(2\pi\ell) = 0$ for $\ell \neq 0$, and*

$$\sum_{\ell \in \mathbf{Z}^d; \ell \neq 0} |\hat{\phi}(\xi + 2\pi\ell)|^2 \leq C((2\pi)^{-d/2} - |\hat{\phi}(\xi)|)$$

for some constant C.

Also $\sum_{\epsilon \in \mathcal{G}} |m_0(\xi + \pi\epsilon)|^2 = 1$ almost everywhere for $\xi \in \mathbf{R}^d$.

We also state for the purpose of the following theorem a general version
of the Poisson summation formula. Given a function $f(x) \in L^1(\mathbb{R}^d)$, this
theorem states (using the present conventions in defining the Fourier transform)

$$\sum_{k \in \mathbf{Z}^d} f(x+k) = (2\pi)^{d/2} \sum_{l \in \mathbf{Z}^d} \hat{f}(2\pi l)e^{2\pi i x l}, \qquad (4.2)$$

where $\hat{f}$ is the Fourier transform of f.

The detailed proof of Theorem 6 can be found in [16].

Theorem 6. ((a) ⇔ (d′) of Theorems 1 and 3) *Let ϕ be the scaling function of a multiresolution analysis and assume $\phi \in \mathcal{RB}$. Then, a necessary
and sufficient condition for order $s - d/2$ ($s > d/2$) convergence of the
multiresolution expansion with scaling function ϕ, is that ϕ satisfy*

$$\int_{|\xi|<\delta} (1 - (2\pi)^{d/2}|\hat{\phi}(\xi)|)|\xi|^{-2}d\xi < \infty \qquad (4.3)$$

for some (or for all) $\delta > 0$.

Proof: Assume first that (4.3) is satisfied. Then, according to Theorem 5,
we need to show the projection $I - P_0$ is bounded from H_h^s to L^∞.

Thus, assume $f(x) \in \mathcal{S}(\mathbb{R}^d) \cap H_h^s(\mathbb{R}^d)$, where $\mathcal{S}$ is the Schwartz space of
rapidly decreasing infinitely differentiable functions, and let $g(x) \in L^1(\mathbb{R}^d)$.
Then (here ˇ denotes the inverse Fourier transform) applying Parseval's
relation in just the y variable, we have

$$\begin{aligned}
(I - P_0)f(x) &= \left(f(x) - \int \sum_k \phi(x-k)\overline{\phi}(y-k)f(y)dy\right) \\
&= f(x) - \int \sum_k \phi(x-k)\overline{e^{ik\xi}\hat{\phi}(-\xi)}\check{f}(\xi)d\xi \qquad (4.4) \\
&= \int \left((2\pi)^{-d/2}e^{-ix\xi} - Z(x,\xi)\overline{\hat{\phi}}(-\xi)\right)\check{f}(\xi)d\xi,
\end{aligned}$$

where $Z(x,\xi) \equiv \Sigma_k\phi(x-k)e^{-i\xi k} = \Sigma_k\phi(x+k)e^{i\xi k}$ is the Zak transform of
ϕ.

After some calculations using the symbol $m_0(\xi)$ of $\phi(x)$ and a form of
Poisson summation formula, we derive a sequence of inequalities which lead
to

$$\int \|(2\pi)^{-d/2}e^{-ix\xi} - Z(x,\xi)\overline{\hat{\phi}}(-\xi)\|_\infty|\xi|^{-2s-d}d\xi < \infty.$$

Thus,

$$\|(I - P_0 f)(x)\|_\infty \le \int \|(2\pi)^{-d/2} e^{-ix\xi} - Z(x,\xi)\overline{\hat{\phi}}(-\xi)\|_\infty |\check{f}(\xi)| d\xi$$

$$\le \left(\int \|(2\pi)^{-d/2} e^{-ix\xi} - Z(x,\xi)\overline{\hat{\phi}}(-\xi)\|_\infty^2 |\xi|^{-2s} d\xi \right)^{1/2}$$

$$\times \left(\int |\check{f}(\xi)|^2 |\xi|^{2s} d\xi \right)^{1/2},$$

where L^∞ norms refer to x. Thus, $(I - P_0)f(x) \in L^\infty$ with norm bounded independently of f for $\|f\|_{s,h} \le 1$, so that $I - P_0$ is bounded from $H_h^s(\mathbb{R}^d)$ to $L^\infty(\mathbb{R}^d)$, and thus, by the part of Theorem 1 which has already been established, the multiresolution expansion with scaling function ϕ has order of convergence s.

To show the converse, assume

$$\int_{|\xi|<\delta} (1 - (2\pi)^{d/2} |\hat{\phi}(\xi)|) |\xi|^{-2s} d\xi = \infty.$$

By a long technical argument we can prove that there exists a sequence $\{\check{f}_n^*(\xi)\}$ with $\|f_n^*\|_{h,s} = 1$ such that

$$\|(I-P_0)f_n^*(x)\|_\infty \ge C\| \int_A e^{-i\xi x} \left((2\pi)^{-d/2} - e^{i\xi x} Z(x,\xi)\overline{\hat{\phi}}(-\xi) \right) \check{f}_n^*(\xi) d\xi\|_\infty$$

can be made arbitrarily large, where A is a set of ξ such that the arguments of the functions in the integrand differ by at most $\pi/2$ from each other. It follows that $\|I - P_0\|_{H^{s+d/2} \to L^\infty} = \infty$, as desired, which yields the converse statement.

§5 Strang-Fix conditions

The conditions of our main Theorems 1, 2, and 3 on $\hat{\phi}$ can be related to differentiability and then moment conditions on ψ, though the latter conditions can be stated only for integer derivatives or moments. That is, the above conditions are related to the so-called Strang-Fix conditions in approximation theory [22], which relate H^s and L^∞ order of convergence of spline expansions to such conditions. These results can be viewed as an extension of the so-called Condition A in [21].

We note that it is in fact possible to have fractional maximal orders of convergence for multiresolution expansions. To see this, note that in Theorems 2 and 3 it is shown that the order of convergence of a wavelet expansion is determined essentially by the asymptotics of the Fourier transform of the wavelet or that of the scaling function near the origin. Thus, to show any order of convergence is possible, it suffices to show one can find scaling functions with arbitrary asymptotics near the origin. This can be done in one dimension as follows. Observe that for finding a valid scaling function for a multiresolution expansion, it is sufficient (see *e.g.* [5]) to find a 2π-periodic function $m_0(\xi)$ such that

$$|m_0^2(\xi)| + |m_0^2(\xi + \pi)| = 1, \tag{5.1}$$

with $m(0) = 1$, and such that $\hat{\phi}(\xi) = (2\pi)^{-1} \prod_{k=1}^{\infty} m_0(\xi/2^k)$ is continuous at the origin and in L^2. Indeed, if $m_0(\xi) = \sum_{k=-\infty}^{\infty} h_k e^{-k\xi}$, then (5.1) implies that the integer translates of $\phi(x)$ are orthonormal, while the fact that $\hat{\phi}$ is continuous at the origin and equal to $(2\pi)^{-1}$ guarantees that the family of scaling spaces V_j satisfies $\overline{\cup_j V_j} = L^2$ and $\cap_j V_j = \{0\}$. Finally, ϕ satisfies a scaling equation with coefficients h_k, so that we also have $V_j = SV_{j-1}$, where S satisfies $Sf(x) = 2^{d/2}f(2x)$. Thus, such a ϕ corresponds to a full multiresolution analysis $\{V_j\}$. That $\hat{\phi}(\xi)$ can have arbitrary asymptotics near the origin follows essentially from the fact that $m_0(\xi)$ can have arbitrary asymptotics there.

Theorems 1, 2, and 3 are related to the Strang-Fix conditions [22] as follows. Using L^2 and L^∞ error criteria in (1.10), the Strang-Fix conditions state conditions for order s convergence of a wavelet expansion. Let $m_0(\xi)$ denote the symbol of the scaling equation defined in (1.1h).

Then, under mild hypotheses, a sufficient condition in one dimension for order s convergence (for s an integer and for sufficiently smooth f) in L^2 is that $m_0(\xi)$ have a zero of order s at $\xi = \pi$. There are some other ways of stating this condition (see [21]):

1. All polynomials of order $s-1$ are linear combinations of integer translates $\phi(x - k)$.

2. The first s moments of the wavelet $\psi(x)$ are zero:

$$\int x^k \psi(x) dx = 0 \qquad (k = 0, \ldots, s - 1).$$

3. The wavelet coefficients of a smooth function decay like

$$\int f(x) \psi_{jk}(x) dx \le C2^{-js}.$$

We now consider the relation of our conditions for order s convergence with these conditions.

First note that condition (c) of Theorem 2 is equivalent to $(-\Delta)^{-s/2}\psi^\lambda \in L^2(\mathbb{R}^d)$, where $-\Delta$ is the positive Laplace operator, defined by the usual operator calculus. This in turn can be expressed as a singular integral condition on ψ itself (see *e.g.* [20]). Equivalently, condition (c) can be stated as

$$\psi^\lambda = (-\Delta)^{s/2}\eta^\lambda,$$

for some $\eta^\lambda(x) \in L^2(\mathbb{R}^d)$.

We now specialize to dimension $d = 1$. In this case, condition (c) is equivalent to $(\frac{d}{dx})^{-s}\psi \in L^2(\mathbb{R})$. When, further, $s = k$ is an integer, it is equivalent to

$$\psi = \frac{d^k}{dx^k}\eta, \tag{5.2}$$

for some $\eta \in L^2(\mathbb{R})$ (note since we know that ψ is also in L^2, this implies $(d^\ell/dx^\ell)\eta \in L^2$ for all $0 \le \ell \le k$).

Let us translate this latter statement into better known moment conditions on ψ [17, 18, 5, 22]. We begin by making a growth assumption on the wavelet, namely that

$$|\psi(x)| \le C(1 + |x|)^{-k-\epsilon}, \tag{5.3}$$

which implies that the ℓth derivative $\eta^{(\ell)}$ satisfies $|\eta^{(\ell)}(x)| \le C(1+|x|)^{-\ell-\epsilon}$ for integers $0 \le \ell \le k$ and some $\epsilon > 0$. Then, when (5.2) holds, we have by integration by parts,

$$\int x\psi dx = \int x\eta^{(k)} dx = x\eta^{(k-1)}\Big|_{-\infty}^{\infty} - \int \eta^{(k-1)} dx = 0 - \eta^{(k-2)}(x)\Big|_{-\infty}^{\infty} = 0$$

and generally for $\ell \le k - 1$

$$\int x^\ell \psi dx = C_\ell \eta^{(k-\ell-1)}(x)\Big|_{-\infty}^{\infty} = 0,$$

so that all moments of order $k - 1$ or less vanish.

Conversely, if

$$\int x^\ell \psi(x) dx = 0 \quad \text{for} \quad 0 \le \ell \le k - 1, \tag{5.4}$$

then, still under assumption (5.3), we have ($\psi \equiv \psi_k$)

$$\int \psi(x) dx = 0 \quad \Rightarrow \quad \psi_{k-1} \equiv \int_{-\infty}^x \psi_k(x) dx \le C(1 + |x|)^{1-k-\epsilon};$$

$$\int x\psi(x) dx = 0 \quad \Rightarrow \quad \psi_{k-2} \equiv \int_{-\infty}^x \psi_{k-1}(x) dx$$

$$\equiv x\psi_{k-1}\Big|_{-\infty}^x - \int_{-\infty}^x x\psi_k(x) dx \le C(1 + |x|)^{2-k-\epsilon};$$

$$\int x^2\psi(x)dx = 0 \quad \Rightarrow \quad \psi_{k-3} \equiv \int_{-\infty}^{x} \psi_{k-2}(x)dx \leq C(1+|x|)^{3-k-\epsilon};$$

$$\vdots$$

$$\int x^{k-1}\psi(x)dx = 0 \quad \Rightarrow \quad \psi_0 \equiv \int_{-\infty}^{x} \psi_1(x)dx \leq C(1+|x|)^{-\epsilon},$$

where $\psi_{k-\ell} = \int_{-\infty}^{x} \psi_{k-(\ell-1)}(x)dx$. Letting $\eta(x) \equiv \psi_0(x)$, we conclude (5.2) holds for some $\eta(x) \in L^2(\mathbb{R})$. Thus, under assumption (5.3), conditions (5.2) and (5.4) are equivalent.

Thus, condition (c) in Theorem 1 where $s = k$ is an integer is equivalent to the moment condition (5.4). This in turn is equivalent to other versions of the Strang-Fix conditions [22]; see above. Thus, our conditions are in any dimension extensions of the Strang-Fix conditions to non-integer s.

We close with two remarks.

Remark 1. Condition (2.2) is akin to conditions in [18] which are consequences of so-called r-regularity. Meyer proves (Proposition 7 in Chapter 2) that r-regular scaling functions in d dimensions satisfy

$$|\hat{\phi}(\xi)|^2 = (2\pi)^{-d} + O(|\xi|^{2r+2}) \quad (\xi \to 0).$$

This of course implies $\int (|\hat{\phi}|^2 - (2\pi)^{-d})|\xi|^{-2r-3+\epsilon}d\xi < \infty$, so setting the exponent in (2.2) equal to that here,

$$-2s = -2r - 3 + \epsilon \quad \Rightarrow \quad s = r + 3/2 - \epsilon'.$$

Therefore, according to the above Proposition in [18], we conclude pointwise convergence of order $s - d/2 = r + 1 - \epsilon'$ for r-regular expansion of functions in H^s in one dimension, for all $\epsilon' > 0$

Remark 2. The extendability of our results to rates of convergence for errors of best L^2-approximations by splines is a consequence of the fact that such approximations can be framed in the context of multiresolution expansions. This has been proved in [16]. The following proposition applies to splines of arbitrary order on uniform meshes.

Proposition 3. In d dimensions, best L^2 spline approximations using a B-spline ϕ^1 on a uniform mesh yields maximal pointwise order of convergence $s - d/2$ in H^s for best L^2 spline approximations ($s > d/2$) if and only if $\sqrt{\sum_{\ell\in\mathbb{Z}^d;\ell\neq 0} |\hat{\phi}^1(\xi + 2\pi\ell)|^2} \in H_h^{-s}$, or equivalently

$$\int \sum_{\ell\neq 0} |\hat{\phi}^1(\xi + 2\pi\ell)|^2 |\xi|^{-2s}d\xi < \infty.$$

For further results on splines, pre-wavelets, and wavelets see [2, 3, 6].

Acknowledgments. We thank Ron De Vore and Amos Ron for helpful conversations which led us to formulate a more complete theory whose proofs will appear elsewhere [15]. The second author wishes to thank Professor Joshua Zeevi for extending an invitation to participate in the Neaman Workshop. This research was partly supported by the Air Force Office of Scientific Research, the National Security Agency, and the National Science Foundation.

References

[1] Bertoluzza, S., Some error estimates for wavelet expansions, *Math. Models and Methods in Applied Sciences* **2**(4) (1992), 489–906.

[2] Buhmann, M. and C. Micchelli, Spline prewavelets for non-uniform knots, *Numer. Math.* **61** (1992), 455–474.

[3] Chui, C. K., *An Introduction to Wavelets*, Academic Press, Boston, 1992.

[4] Daubechies, I., Orthonormal bases of compactly supported wavelets, *Comm. Pure Appl. Math.* **41** (1988), 909–996.

[5] Daubechies, I., Ten Lectures on Wavelets, in *CBMS-NSF Series in Applied Mathematics*, #61, SIAM, Philadelphia, 1992.

[6] de Boor, C., On max-norm bound for the least-squares splines approximant, in *Approximation and Function Spaces*, Z. Ciesielski (ed.), Warsaw, 1981, pp. 163–175.

[7] de Boor, C., R. De Vore, and A. Ron, Approximation from shift-invariant subspaces of $L^2(\mathbb{R}^d)$, *Trans. Amer. Math. Soc.*, to appear.

[8] de Boor, C. and A. Ron, Fourier analysis of principal shift-invariant space, *Const. Approx.* **8** (1992), 427–462.

[9] De Vore, R., *The approximation of continuous functions by positive linear operators*, Springer, Lecture Notes in Math., **293**, 1972.

[10] Fischer, A., Multiresolution analysis and multivariate approximation of smooth signals in $C_B(\mathbb{R}^d)$, *J. Fourier Anal. and Appl.*, to appear.

[11] Glowinski, R., W. Lawton, M. Ravachol, and E. Tenenbaum, Wavelet solution of linear and nonlinear elliptic, parabolic and hyperbolic problems in one space dimension, in *Proc. of the 9th International Conf. on Numerical Methods in Applied Science and Engineering*, SIAM, 1990.

[12] Jetter, K. and D.-X Zhou, Order of linear approximation in shift-invariant spaces, Duisberg, 1994, preprint.

[13] Kelly, S., M. Kon, and L. Raphael, Pointwise convergence of wavelet expansions, *Bull. AMS* **30** (1994), 87–94.

[14] Kelly, S., M. Kon, and L. Raphael, Local convergence of wavelet expansions, *J. Functional Analysis* **126** (1994), 102–138.

[15] Kon, M. and L. Raphael, Convergence rates of multiscale and wavelet expansions, in preparation.

[16] Kon, M. and L. Raphael, Applications of Generalized Multiresolution Analysis of Convergence of Spline Expansions on $\mathbb{R}^d$, in *Proceedings of Signal Processing and Mathematical Analysis*, Cairo, 1994, to appear.

[17] Mallat, S., Multiresolution approximation and wavelets, *Trans. Amer. Math. Soc.* **315** (1989), 69–88.

[18] Meyer, Y., *Wavelets and Operators* (translated by D. H. Salinger), Cambridge University Press, 1992.

[19] Ruskai, M. B., G. Beylkin, R. Coifman, I. Daubechies, S. Mallat, Y. Meyer, and L. Raphael (eds.), *Wavelets and their Applications*, Jones and Barlett, Boston, 1992.

[20] Stein, E., *Singular Integrals and Differentiability Properties of Functions*, Princeton University Press, Princeton, 1970.

[21] Strang, G., Wavelets and dilation equations, a brief introduction, *SIAM Rev.* **31** (1989), 614–627.

[22] Strang, G. and G. Fix, A Fourier analysis of the finite element variational method, in *Constructive Aspects of Functional Analysis*, Edizioni Cremonese, Rome, 1973.

[23] Walter, G., Approximation of the delta function by wavelets, *J. Approx. Theor.* **71** (1992), 329–343.

Mark A. Kon
Department of Mathematics
Boston University
Boston, MA 02215

Louise A. Raphael
Department of Mathematics
Howard University
Washington, DC 20059

On Calderón's Reproducing Formula

Boris Rubin

Abstract. A generalization of Calderón's reproducing formula involving a finite Borel measure is considered. This generalization gives rise to new representations of singular integral operators with constant coefficients on a real line and clarifies conditions under which a reproducing formula holds. Convergence of new representations in L^p-norm and almost everywhere is studied. An algorithm of approximating initial function ϕ by a Calderón-type construction with a perturbed data $\tilde{\phi}$ is suggested.

§1 Introduction

A classical Calderón's reproducing formula has the form

$$\phi = \frac{1}{c_{u,v}} \int\limits_0^\infty \frac{\phi * u_t * v_t}{t} dt \tag{1.1}$$

where $u_t(x) = \frac{1}{t}u(\frac{x}{t})$, $v_t(x) = \frac{1}{t}v(\frac{x}{t})$, $x \in \mathbb{R}$, $c_{u,v}$ is a constant depending on functions u, v which satisfy suitable conditions. (1.1) plays an important role in the wavelet theory and in many areas of analysis. After a pioneering paper by A.P. Calderón [1], (1.1) was studied by many authors (see, *e.g.* [4-7], [6, 10, 12, 13]). In mentioned publications, functions u and v are quite good (almost every author imposes his own conditions) and satisfy certain relations in the Fourier terms. The problem is to relax such conditions as much as possible and to present them without using a Fourier transform. Since the properties of the right-hand side of (1.1) are defined not by u and v separately but by their convolution $\mu = u * v$, it is natural to investigate an operator

$$I(\mu, \phi) = \int\limits_0^\infty \frac{\phi * \mu_t}{t} dt \tag{1.2}$$

Signal and Image Representation in Combined Spaces
Y. Y. Zeevi and R. R. Coifman (Eds.), pp. 439–455.

439

where μ is an arbitrary complex-valued finite Borel measure (the set of all such measures will be denoted by $\mathcal{M}$),

$$(\phi * \mu_t)(x) = \int_{-\infty}^{\infty} \phi(x - ty)d\mu(y). \tag{1.3}$$

The equality $I(\mu, \phi) = \phi$ represents a reproducing formula which gives rise to modified wavelet transform $(V_\mu\phi)(x,t) = (\phi * \mu_t)(x)$. Such transforms were considered by A. Grossmann and J. Morlet in [9] and called "voice transforms." One should mention the paper [2] by R. R. Coifman and R. Rochberg in which the reader may find reproducing formulae of different types.

Notation

$$I_{\varepsilon,\rho}(\mu, \phi) = \int_{\varepsilon}^{\rho} \frac{\phi * \mu_t}{t} dt, \quad 0 < \varepsilon < \rho < \infty. \tag{1.4}$$

From here on, for $\mu \in \mathcal{M}$ the values $\mu(\{\pm\infty\})$ are assumed to be zero,

$$\hat{\mu}(\xi) = \int_{-\infty}^{\infty} e^{ix\xi}d\mu(x), \quad (H\phi)(x) = p.v.\frac{1}{\pi i} \int_{-\infty}^{\infty} \frac{\phi(y)}{x - y}dy \text{ (the Hilbert transform)}.$$

In this notation, $(H\phi)^\wedge(\xi) = \hat{\phi}(\xi)\text{sgn } \xi$.

§2   Main results

Given $\mu \in \mathcal{M}$, denote

$$\hat{\mu}_{\pm}(\eta) = \frac{\hat{\mu}(\eta) \pm \hat{\mu}(-\eta)}{2}, \quad \alpha_{\pm} = \int_{0}^{\infty} \frac{\hat{\mu}_{\pm}(\eta)}{\eta} d\eta, \tag{2.1}$$

the integrals $\alpha_{\pm}$ being interpreted as improper ones. Note that

$$\alpha_- < \infty \ \forall \mu \in \mathcal{M} \text{ and } \alpha_- = \frac{\pi i}{2} \int_{-\infty}^{\infty} \text{sgn } x \, d\mu'(x) \tag{2.2}$$

where $\mu' = \mu - \mu(\{0\})\delta_0$, δ_0 being a Dirac measure at the point $x = 0$. Indeed,

$$
\begin{aligned}
\alpha_- &= \lim_{\substack{\varepsilon \to 0 \\ \rho \to \infty}} i \int_\varepsilon^\rho \frac{d\eta}{\eta} \int_{-\infty}^\infty \sin(x\eta) d\mu(x) \\
&= \lim_{\substack{\varepsilon \to 0 \\ \rho \to \infty}} i \int_{-\infty}^\infty \operatorname{sgn} x \, d\mu'(x) \int_{\varepsilon x}^{\rho x} \frac{\sin \eta}{\eta} d\eta = \frac{\pi i}{2} \int_{-\infty}^\infty \operatorname{sgn} x \, d\mu'(x).
\end{aligned}
$$

Theorem 1. *Let $\phi \in L^2$, $\mu \in \mathcal{M}$. The limit*

$$
I(\mu, \phi) = \overset{(L^2)}{\underset{\substack{\varepsilon \to 0 \\ \rho \to \infty}}{\lim}} \int_\varepsilon^\rho \frac{\phi * \mu_t}{t} dt \tag{2.3}
$$

exists iff $\alpha_+ < \infty$. If $\alpha_+ < \infty$, then

$$
I(\mu, \phi) = \alpha_+ \phi + \alpha_- H\phi. \tag{2.4}
$$

This theorem gives rise to the following:

Problem. To characterize the set of all measures $\mu \in \mathcal{M}$ for which $\alpha_+ < \infty$.

This problem seems to be open. Obviously, μ must enjoy a cancellation property $\hat{\mu}(0) = 0$ *i.e.* $\mu(\mathbb{R}) = 0$, but of course this is not enough. The following theorem contains some sufficient conditions which are very close to be necessary, and extends the statement of Theorem 1 to the case $\phi \in L^p$.

Theorem 2. *Let $\mu \in \mathcal{M}$ be such that*

$$
\mu(\mathbb{R}) = 0, \quad \int_{-\infty}^\infty |\log|x|| \, d|\mu|(x) < \infty \tag{2.5}
$$

or

$$
d\mu(x) = g(x)dx, \quad g \in H^1(\text{a Hardy space}). \tag{2.6}
$$

Then the following statements hold:
(i) $\alpha_+ < \infty$ and may be evaluated as follows:

$$
\alpha_+ = \int_{-\infty}^\infty \log \frac{1}{|x|} d\mu(x) \quad \text{in the case (2.5)}, \tag{2.7}
$$

and

$$\alpha_+ = \frac{\pi i}{2} \int\limits_{-\infty}^{\infty} (Hg)(x)\text{sgn } x \, dx \quad \text{in the case (2.6).} \tag{2.8}$$

(ii) If $\phi \in L^p$, $\quad 1 < p < \infty$, then

$$I_{\varepsilon,\rho}(\mu, \phi) \to \alpha_+\phi + \alpha_-H\phi \tag{2.9}$$

in L^p-norm as $\varepsilon \to 0$, $\quad \rho \to \infty$. Furthermore, if

$$\int\limits_{|x|<1} |x|^{-\delta}d|\mu|(x) < \infty \quad \text{and} \quad \int\limits_{|x|>1} |x|^{\delta}d|\mu|(x) < \infty \text{ for some } \quad \delta > 0,$$

$$\tag{2.10}$$

then (2.9) holds almost everywhere on $\mathbb{R}$.

Theorem 2 enables us to write out natural sufficient conditions under which a reproducing formula holds. Namely, we have

Corollary 1. *Assume that* $\phi(\not\equiv 0) \in L^p$, $\quad 1 < p < \infty$, *and* (2.5) *holds. Then,* $I(\mu, \phi) = \phi$ *if and only if*

$$\int\limits_{-\infty}^{\infty} \log \frac{1}{|x|}d\mu(x) = 1 \quad \text{and} \quad \mu\big((-\infty,0)\big) = \mu\big((0,+\infty)\big).$$

The main point in these results is that not only is equation (1.1) generalized by our considering a finite measure instead of usual functions u, v, but also represent conditions for μ in terms of only μ itself without using a Fourier transform $\hat{\mu}(\eta)$ as was done in previous publications. Here, one should mention a paper of M. Holschneider [11] in which either u or v may be a distribution, but nevertheless the convolution $u * v$ is regular.

In the case $\alpha_- = 0$ one can add the following:

Remark. Theorem 2 may be extended to the case $p = \infty$ provided that the space L^∞ is replaced by the space C^0 of continuous functions vanishing at infinity. In this case, we have a uniform convergence in (2.9). If $p = 1$ and μ satisfies (2.10), then a.e.-convergence in (2.9) also takes place.

Note that for $\mu \in H^1$ the integral in (2.7) may not be absolutely convergent (see [19], p. 178), but since $\log|x| \in BMO$, one can interpret it in the sense of duality between BMO and H^1 [19].

Proof of Theorem 1:

The argument to be used here is quite standard (see [7]). Assume first that $\phi \in L^1 \cap L^2$. Then, $I_{\varepsilon,\rho}(\mu, \phi) \in L^1 \cap L^2$ and $[I_{\varepsilon,\rho}(\mu, \phi)]^{\wedge}(\xi) = \hat{k}_{\varepsilon,\rho}(\xi)\hat{\phi}(\xi)$ where

$$\hat{k}_{\varepsilon,\rho}(\xi) = \int\limits_{\varepsilon}^{\rho} \frac{\hat{\mu}(t\xi)}{t}dt = \int\limits_{\varepsilon}^{\rho} \frac{\hat{\mu}_+(t\xi)}{t}dt + \int\limits_{\varepsilon}^{\rho} \frac{\hat{\mu}_-(t\xi)}{t}dt = \hat{k}_{\varepsilon,\rho}^+(\xi) + \hat{k}_{\varepsilon,\rho}^-(\xi).$$

For $\xi \neq 0$ we have

$$\hat{k}_{\varepsilon,\rho}^+(\xi) = \int\limits_{\varepsilon|\xi|}^{\rho|\xi|} \frac{\hat{\mu}_+(\eta)}{\eta}d\eta, \quad \hat{k}_{\varepsilon,\rho}^-(\xi) = \text{sgn } \xi \int\limits_{\varepsilon|\xi|}^{\rho|\xi|} \frac{\hat{\mu}_-(\eta)}{\eta}d\eta.$$

Thus,

$$(I_{\varepsilon,\rho}(\mu, \phi))^{\wedge}(\xi) = \hat{\phi}(\xi)\left(\int\limits_{\varepsilon|\xi|}^{\rho|\xi|} \frac{\hat{\mu}_+(\eta)}{\eta}d\eta + \text{sgn } \xi \int\limits_{\varepsilon|\xi|}^{\rho|\xi|} \frac{\hat{\mu}_-(\eta)}{\eta}d\eta\right), \quad \xi \neq 0.$$

$$(2.11)$$

Since integrals $\alpha_{\pm}$ from (2.1) are finite, the functions

$$\psi_{\pm}(t) = \int\limits_{0}^{t} \frac{\hat{\mu}_{\pm}(\eta)}{\eta}d\eta$$

are continuous on $[0, \infty]$. It follows that there exist constants $A_{\pm} > 0$ for which $|\hat{k}_{\varepsilon,\rho}^{\pm}(\xi)| \leq A_{\pm} \quad \forall \rho > \varepsilon > 0, \quad \xi \in \mathbb{R}$, and

$$\|I_{\varepsilon,\rho}(\mu, \phi)\|_2 = \|\hat{\phi}(\xi)\hat{k}_{\varepsilon,\rho}(\xi)\|_2 \leq A\|\phi\|_2, \quad A = A_+ + A_-. \quad (2.12)$$

Now the Lebesgue theorem on dominated convergence yields

$$\|I_{\varepsilon,\rho}(\mu, \phi) - \alpha_+\phi - \alpha_-H\phi\|_2 \to 0 \text{ as } \varepsilon \to 0, \ \rho \to \infty. \quad (2.13)$$

The result for arbitrary $\phi \in L^2$ may be easily derived from (2.12) and (2.13) by using the inequality

$$\|I_{\varepsilon,\rho}(\mu, \phi)\|_2 \leq V_{\mu}\|\phi\|_2 \log \frac{\rho}{\varepsilon}$$

where V_{μ} is a total variation of μ. ∎

Proof of Theorem 2:

STEP 1. Represent $I_{\varepsilon,\rho}(\mu,\phi)$ in the form

$$I_{\varepsilon,\rho}(\mu,\phi) = k_{\varepsilon}*\phi - k_{\rho}*\phi, \quad k_{\varepsilon}(x) := \frac{1}{\varepsilon}k\left(\frac{x}{\varepsilon}\right), \quad k_{\rho}(x) := \frac{1}{\rho}k\left(\frac{x}{\rho}\right) \quad (2.14)$$

with

$$k(x) = \frac{1}{x}\int_0^x d\mu(y) = \frac{1}{|x|}\left\{ \begin{array}{ll} \mu((0,x)) & \text{if } x > 0, \\ \mu((x,0)) & \text{if } x < 0. \end{array} \right. \quad (2.15)$$

In order to obtain (2.14), for $\phi \in L^p$, $1 \le p \le \infty$, we have

$$I_{\varepsilon,\rho}(\mu,\phi)(x) = \int_\varepsilon^\rho \frac{dt}{t} \int_{-\infty}^\infty \phi(x-ty)d\mu(y)$$

$$= \int_{-\infty}^\infty d\mu'(y) \int_\varepsilon^\rho \phi(x-ty)\frac{dt}{t} + \mu(\{0\})\phi(x)\log\frac{\rho}{\varepsilon}, \quad \mu' = \mu - \mu(\{0\})\delta_0.$$

Since $\mu(\{0\}) = 0$ then the last term vanishes and we obtain

$$I_{\varepsilon,\rho}(\mu,\phi)(x) = \int_{-\infty}^\infty \phi(x-t)k_{\varepsilon,\rho}(t)dt \quad (2.16)$$

where $k_{\varepsilon,\rho} \in L^1$ and has the form

$$k_{\varepsilon,\rho}(t) = \frac{1}{|t|}\left\{ \begin{array}{ll} \mu([t/\rho, t/\varepsilon)) & \text{if } t > 0, \\ \mu((t/\varepsilon, t/\rho]) & \text{if } t < 0. \end{array} \right.$$

This gives (2.14).

The kernel $k(x)$ plays a key role in the sequel. In general, it does not belong to L^1. In order to manage it we proceed as follows. Divide $k(x)$ into its even part and the odd part by putting $k = k^+ + k^-$ where

$$k^+(x) = \frac{\mu((-|x|,|x|))}{2|x|}, \quad k^-(x) = \frac{\mu((0,|x|)) - \mu((-|x|,0))}{2x}. \quad (2.17)$$

Then introduce a conjugate Poisson kernel $q(x) = (1/\pi i)x(1+x^2)^{-1}$ for which $\hat{q}(\xi) = e^{-|\xi|}\text{sgn }\xi$ (see [14]), and put

$$h = k^- - \alpha_- q \quad (2.18)$$

with α_- from (2.2). This results

$$I_{\varepsilon,\rho}(\mu,\phi) = A(\varepsilon) - A(\rho) \quad \text{where} \quad A(t) = k_t^+*\phi + h_t*\phi + \alpha_- q_t*\phi, \quad t = \varepsilon, \rho. \quad (2.19)$$

STEP 2. Let us prove that (2.5) implies the following relations

$$k^+ \in L^1, \quad h \in L^1, \quad \int_{-\infty}^{\infty} k^+(x)dx = \int_{-\infty}^{\infty} \log\frac{1}{|x|}d\mu(x), \quad \int_{-\infty}^{\infty} h(x)dx = 0.$$

(2.20)

First we check that $k^+ \in L^1$. Since k^+ is even it suffices to prove that $k^+ \in L^1(0,1)$ and $k^+ \in L^1(1,\infty)$. We have

$$\int_0^1 |k^+(x)|dx \le \frac{1}{2}\int_0^1 \frac{dx}{x}\int_{-x}^x d|\mu||y| = \frac{1}{2}\left[\int_0^1 d|\mu|(y)\int_y^1 \frac{dx}{x} + \int_{-1}^0 d|\mu||y|\int_{-y}^1 \frac{dx}{x}\right]$$

$$= \frac{1}{2}\int_{-1}^1 \log\frac{1}{|y|}d|\mu||y| < \infty.$$

(2.21)

Since $\mu(\mathbb{R}) = \mu(0) = 0$, then for $x > 0$,

$$k^+(x) = -\frac{\mu((-\infty,-x)) + \mu((x,+\infty))}{2x}$$

(2.22)

and therefore

$$\int_1^\infty |k^+(x)|dx \le \frac{1}{2}\int_1^\infty \left(\int_{-\infty}^{-x} d|\mu|(y) + \int_x^\infty d|\mu|(y)\right)\frac{dx}{x}$$

(2.23)

$$= \frac{1}{2}\left[\int_{-\infty}^1 d|\mu||y|\int_1^{-y} \frac{dx}{x} + \int_1^\infty d|\mu|(y)\int_1^y \frac{dx}{x}\right] = \frac{1}{2}\int_{|y|>1} \log|y|d|\mu||y| < \infty.$$

Actually, the same evaluation gives representation to the integral of k^+ (see (2.20)).

Consider $h = k^- - \alpha_- q$. Since $h(x)$ is odd, it suffices to show that $h \in L^1(0,1)$ and $h \in L^1(1,\infty)$. For the first relation, the similar estimates as in (2.21) are used. For proving the second one we apply the equality $\mu(\mathbb{R}) = \mu(\{0\}) = 0$ which yields

$$h(x) = \frac{\mu((-\infty,-x)) + \mu((x,+\infty))}{2x} + \frac{\alpha_-}{\pi i}\left(\frac{1}{x} - \frac{x}{1+x^2}\right) \in L^1(1,\infty)$$

(2.24)

(see (2.23)). The last equality in (2.20) is obvious because h is odd.

STEP 3. Let us prove that the coefficient α_+, defined in (2.1) in the Fourier terms, may be evaluated by

$$\alpha_+ = \int_{-\infty}^\infty k^+(x)dx = \int_{-\infty}^\infty \log\frac{1}{|x|}d\mu(x).$$

(2.25)

By (2.19) and (2.11) for sufficiently good ϕ (one can take ϕ belonging to the Schwarz class S),

$$\hat{\phi}(\xi)\left(\int_{\varepsilon|\xi|}^{\rho|\xi|}\frac{\hat{\mu}_+(\eta)}{\eta}d\eta + \text{sgn } \xi \int_{\varepsilon|\xi|}^{\rho|\xi|}\frac{\hat{\mu}_-(\eta)}{\eta}d\eta\right) = \hat{\phi}(\xi)[w(\varepsilon\xi) - w(\rho\xi)] \quad (2.26)$$

where

$$w(\xi) = \hat{k}^+(\xi) + \hat{h}(\xi) + \alpha_- e^{-|\xi|}\text{sgn } \xi.$$

Fix $\xi \neq 0$ such that $\hat{\phi}(\xi) \neq 0$. Since

$$\lim_{\substack{\varepsilon \to 0 \\ \rho \to \infty}} [w(\varepsilon\xi) - w(\rho\xi)] = \hat{k}^+(0) + \alpha_-\text{sgn } \xi = \int_{-\infty}^{\infty} k^+(x)dx + \alpha_-\text{sgn } \xi,$$

then by STEP 2 the desired result follows immediately. Thus, the relation (2.7) in Theorem 2 is proved.

By the way, it is interesting to note that

$$\hat{k}^+(\xi) = \int_{|\xi|}^{\infty} \hat{\mu}_+(\eta)d\eta/\eta. \quad (2.27)$$

Really, put $\varepsilon = 1$ in (2.26) and let $\rho \to \infty$. Since ϕ may be an arbitrary function from S, by using a Riemann-Lebesgue theorem we obtain

$$\int_{|\xi|}^{\infty} \hat{\mu}_+(\eta)d\eta/\eta + \text{sgn } \xi \int_{|\xi|}^{\infty} \hat{\mu}_-(\eta)d\eta/\eta = \hat{k}^+(\xi) + \hat{h}(\xi) + \alpha_- e^{-|\xi|}\text{sgn } \xi. \quad (2.28)$$

This gives (2.27) by taking into account that $\hat{k}^+$ is even and $\hat{h}$ is odd.

STEP 4. Consider the situation arising when $d\mu(x) = g(x)dx$, $g \in H^1$, where H^1 is a real Hardy space (see (2.6)). We remind some basic facts related to the H^1-space (see, *e.g.* [3, 19]).

Definition 1. *A function $a(x)$ supported by an interval I is called an atom if*

$$\int_I a(x)dx = 0 \quad \text{and} \quad |a(x)| \leq 1/|I|.$$

Theorem 3. *A summable function g belongs to H^1 if and only if*

$$g = \sum_{k=0}^{\infty} \lambda_k a_k \qquad (2.29)$$

where a_k, $k \geq 0$ is an atom and $\sum_{k=0}^{\infty} |\lambda_k| < \infty$. If $g \in H^1$, then $||g||_{H^1} \sim$ inf $\sum_{k=0}^{\infty} |\lambda_k|$ where the infimum is taken over all decompositions (2.29).

We need the following:

Lemma 1. *An operator*

$$(K^+ g)(x) = \frac{1}{2|x|} \int\limits_{-|x|}^{|x|} g(y) dy (\equiv k^+(x)) \qquad (2.30)$$

is bounded from H^1 into L^1.

Proof: Let first $g(x) \equiv a(x)$ be an atom supported by I. Then $|(K^+ a)(x)| \leq 1/I$ and $||K^+ a||_1 \leq 4$ (we leave a simple verification of the last inequality to the reader). In the general case we have a representation (2.29). Since $||a_k||_1 \leq 1$ and $\sum_{k=0}^{\infty} |\lambda_k| < \infty$, then the series (2.29) converges in L^1-norm and therefore $K^+ g = \sum_{k=0}^{\infty} \lambda_k K^+ a_k$. It follows

$$||K^+ g||_1 \leq 4 \sum_{k=0}^{\infty} |\lambda_k| \leq \text{ const } \cdot ||g||_{H^1}. \qquad (2.31)$$

∎

Let us prove that for all $\phi \in L^p$, $1 < p < \infty$, and for all $g \in H^1$,

$$I_{\varepsilon,\rho}(g,\phi) = (K^+ g)_\varepsilon * \phi - (K^+ g)_\rho * \phi + (K^+ Hg)_\varepsilon * H\phi - (K^+ Hg)_\rho * H\phi \quad (2.32)$$

where $(\cdot)_\varepsilon$ and $(\cdot)_\rho$ denote corresponding dilations and H stands for the Hilbert transform. Since the operators $K^+ : H^1 \to L^1$, $H : H^1 \to H^1$, $H : L^p \to L^p$ ($1 < p < \infty$) are bounded, an L^p-norm of the right-hand side of (2.32) does not exceed $||\phi||_p ||g||_{H^1}$ up to a constant factor. At the same time

$$||I_{\varepsilon,\rho}(g,\phi)||_p \leq \int\limits_{\varepsilon}^{\rho} \frac{||\phi * g_t||_p}{t} dt \leq ||\phi||_p ||g||_1 \log \frac{\rho}{\varepsilon} \leq$$

$$\leq ||\phi||_p ||g||_{H^1} \log \frac{\rho}{\varepsilon}. \qquad (2.33)$$

Thus, it suffices to prove (2.25) for g, ϕ belonging to some set Φ which is dense in H^1 and L^p. One may take

$$\Phi = \{\omega \in S : \int_{-\infty}^{\infty} x^j \omega(x)dx = 0 \quad \forall j = 0, 1, 2, \dots\}$$

(see [19], p. 128). Assume that $g, \phi \in \Phi$ and write out the Fourier transform of the left-hand side of (2.32). By (2.11) and (2.27),

$$(I_{\varepsilon,\rho}(g,\phi))^\wedge(\xi) = \hat{\phi}(\xi) \left(\hat{k}^+(\varepsilon\xi) - \hat{k}^+(\rho\xi) + \operatorname{sgn} \xi \int_{\varepsilon|\xi|}^{\rho|\xi|} \frac{\hat{g}_-(\eta)}{\eta} d\eta \right)$$

where

$$\begin{aligned} k^+ &= K^+ g, \\ \hat{g}_-(\eta) &= \frac{\hat{g}(\eta) - \hat{g}(-\eta)}{2} = \frac{(Hg)^\wedge(\eta) + (Hg)^\wedge(-\eta)}{2} \\ &= (Hg)^\wedge_+(\eta) \quad \text{for} \quad \eta > 0. \end{aligned}$$

The relations (2.27) and (2.30) yield

$$\int_{\varepsilon|\xi|}^{\rho|\xi|} \frac{\hat{g}_-(\eta)}{\eta} d\eta = \int_{\varepsilon|\xi|}^{\rho|\xi|} \frac{(Hg)^\wedge_+(\eta)}{\eta} d\eta = (K^+ Hg)^\wedge(\varepsilon\xi) - (K^+ Hg)^\wedge(\rho\xi). \quad (2.34)$$

It follows

$$\begin{aligned} (I_{\varepsilon,\rho}(g,\phi))^\wedge(\xi) &= \hat{\phi}(\xi)[(K^+g)^\wedge(\varepsilon\xi) - (K^+g)^\wedge(\rho\xi)] \\ &+ (H\phi)^\wedge(\xi)[(K^+ Hg)^\wedge(\varepsilon\xi) - (K^+ Hg)^\wedge(\rho\xi)] \end{aligned}$$

which coincides with the Fourier transform of the right-hand side of (2.32). The proof of (2.32) is complete.

By putting

$$\zeta = K^+ Hg \quad (\in L^1) \quad (2.35)$$

it is convenient to write (2.32) in the form

$$I_{\varepsilon,\rho}(g,\phi) = k^+_\varepsilon * \phi - k^+_\rho * \phi + \zeta_\varepsilon * H\phi - \zeta_\rho * H\phi. \quad (2.36)$$

Now let us check the equalities

$$\int_{-\infty}^{\infty} k^+(x)dx = \int_{-\infty}^{\infty} (K^+ g)(x)dx = \frac{\pi i}{2} \int_{-\infty}^{\infty} (Hg)(x)\operatorname{sgn} x\, dx, \quad (2.37)$$

$$\int_{-\infty}^{\infty} \zeta(x)dx = \int_{-\infty}^{\infty} (K^+ Hg)(x)dx = \frac{\pi i}{2} \int_{-\infty}^{\infty} g(x)\operatorname{sgn} x\, dx. \quad (2.38)$$

Since the integrals above represent linear bounded functionals on H^1, it suffices to prove (2.37) and (2.38) for $g \in \Phi$. By (2.27), (2.2) for (2.37) we have

$$\int_{-\infty}^{\infty} k^+(x)dx = \hat{k}^+(0) = \int_0^{\infty} \frac{\hat{g}_+(\eta)}{\eta} d\eta = \int_0^{\infty} \frac{(Hg)_-^{\wedge}(\eta)}{\eta} d\eta$$

$$= \frac{\pi i}{2} \int_{-\infty}^{\infty} (Hg)(x) \operatorname{sgn} x \, dx.$$

The relation (2.38) is a consequence of (2.37) (one should replace g by Hg and take into account that $HHg = g$).

STEP 5. Now we are ready to prove the statement (2.9) of Theorem 2 with the convergence being intepreted in L^p-norm. Consider the case (2.5) and recall the relation (2.19):

$$I_{\varepsilon,\rho}(\mu, \phi) = k_\varepsilon^+ * \phi + h_\varepsilon * \phi + \alpha_- q_\varepsilon * \phi - k_\rho^+ * \phi - h_\rho * \phi - \alpha_- q_\rho * \phi. \quad (2.39)$$

By taking into account (2.20) and (2.25), one can apply well-known results on the approximation of the identity ([18], [14]) and related properties of the conjugate Poisson kernel ([14]). This gives

$$k_\varepsilon^+ * \phi \to \alpha_+ \phi, \quad h_\varepsilon * \phi \to 0, \quad q_\varepsilon * \phi \to H\phi \quad \text{as} \quad \varepsilon \to 0 \quad (2.40)$$

in L^p-norm. At the same time, since $k^+, h \in L^1$, then $k_\rho^+ * \phi \to 0$, $h_\rho * \phi \to 0$ as $\rho \to \infty$ (see, e.g. [17], p. 22). In order to prove that $\lim_{\rho \to \infty} \|q_\rho * \phi\|_p = 0$ we take the sequence of compactly supported smooth functions w_m such that $\lim_{m \to \infty} \|\phi - w_m\|_p = 0$ and make use of the uniform estimate $\|\phi * q_\rho\|_p \le A_p \|\phi\|_p$ (see [14]). Then

$$\|\phi * q_\rho\|_p \le \|(\phi - w_m) * q_\rho\|_p + \|w_m * q_\rho\|_p \le A_p \|\phi - w_m\|_p + \rho^{-1/p'} \|w_m\|_1 \|q\|_p$$

and the desired assertion follows. Similarly, in the case of (2.6) a desired statement follows from (2.36)–(2.38) by taking into account that $k^+, \zeta \in L^1$.

STEP 6. Let us prove an a.e.-convergence in (2.9) under the condition (2.10). First we note that by (2.10), k^+ and h have the order $O(|x|^{\delta-1})$ for $|x| < 1$ and the order $O(|x|^{-\delta-1})$ for $|x| > 1$. Really, in the case $|x| < 1$ we have

$$|k^+(x)| \le \frac{1}{2|x|} \int_{-|x|}^{|x|} d|\mu|(y) \le c|x|^{\delta-1}.$$

Similarly, by using (2.18) and (2.17) one can obtain $h(x) = O(|x|^{\delta-1}) + O(|x|)$. In the case $|x| > 1$ the functions $\mu(|x|, +\infty)$, $\mu(-\infty, -|x|)$ have the order $O(|x|^{-\delta})$. Therefore, according to (2.22) and (2.24), $k(x) = O(|x|^{-\delta-1}) + O(|x|^{-3})$. By taking into account these estimates one can see the a.e.-convergence in (2.40) to be held due to the classical results from [18] and [14]. Let us prove that

$$k_\rho * \phi \equiv (k_\rho^+ + h_\rho + \alpha_- q_\rho) * \phi \to 0$$

almost everywhere as $\rho \to \infty$. For any $\delta \in (0,1)$, by (2.15) we have

$$|(k_\rho * \phi)(x)| \leq \int_0^\delta |\phi(x-t)| \frac{|\mu|((0,t/\rho))}{t} dt + \int_{-\delta}^0 |\phi(x-t)| \frac{|\mu|((t/\rho,0))}{t} dt$$

$$+ \int_\delta^\infty |\phi(x-t)| \frac{|\mu((0,t/\rho))|}{t} dt + \int_{-\infty}^{-t} |\phi(x-t)| \frac{|\mu((t/\rho,0))|}{t} dt$$

$$= A_\delta^+(x,\rho) + A_\delta^-(x,\rho) + B_\delta^+(x,\rho) + B_\delta^-(x,\rho).$$

If $\rho \geq 1$, then

$$A_\delta^+(x,\rho) \leq \int_0^t |\phi(x-t)| \frac{|\mu|((0,t))}{t} dt.$$

By (2.10), $|\mu|((0,t))/t \in L^1(0,1)$ and hence, the function

$$\psi_x(t) := \begin{cases} |\phi(x-t)||\mu|((0,t))/t & \text{if } t < 1, \\ 0, & \text{if } t \geq 1, \end{cases}$$

is summable for almost all x. It follows that $A_\delta^+(x,\rho)$ may be done arbitrarily small for sufficiently small $\delta = \delta(x)$ uniformly with respect to $\rho \geq 1$. The same is true for $A_\delta^-(x,\rho)$. By taking $\delta = \delta(x)$ fixed, consider $B_\delta^+(x,\rho)$. If $1 \leq p < \infty$, then $t \to |\phi(x-t)|/t$ is a summable function for $t > \delta$ and therefore,

$$\lim_{\rho \to \infty} B_\delta^+(x,\rho) = \int_\delta^\infty \frac{|\phi(x-t)|}{t} \lim_{\rho \to \infty} |\mu((0,t/\rho))| dt = 0.$$

A similar argument holds for $B_\delta^-(x,\rho)$.   ∎

The proof of Theorem 2 is completed.

§3 Control of approximation in a reproducing formula

Here we follow the philosophy originally used in [8] and [15] for regularization of fractional derivatives.

Let $\phi \in L^p$ be a true function which is unknown, and $\tilde{\phi}$ be a given perturbed function such that

$$||\phi - \tilde{\phi}||_p < \varepsilon. \tag{3.1}$$

As we know,

$$\phi = \lim_{\substack{a \to 0 \\ b \to \infty}}^{(L^p)} \int_a^b \frac{\phi * \mu_t}{t} dt$$

for a suitable measure μ. The problem is to choose $a = a_\varepsilon$ and $b = b_\varepsilon$ in such a way that the error

$$\Delta(\varepsilon) = \left\| \phi - \int_{a_\varepsilon}^{b_\varepsilon} \frac{\tilde{\phi} * \mu_t}{t} dt \right\|_p \tag{3.2}$$

was minimal and to estimate an order of $\Delta(\varepsilon)$ for $\varepsilon \to 0$. Here we study this problem in the case when ϕ and $\tilde{\phi}$ are supported by a finite interval, say, $[c, d]$.

For simplicity, assume that μ is supported by a positive half-line $\mathbb{R}_+$ and satisfies the relations

$$\mu(\mathbb{R}_+) = 0, \qquad \int_0^\infty |\log x| \, d|\mu|(x) < \infty. \tag{3.3}$$

Assume also that μ is normalized by

$$\int_0^\infty \log \frac{1}{|x|} d\mu(x) = 1. \tag{3.4}$$

Denote

$$I_{a,b}\phi = \int_a^b \frac{\phi * \mu_t}{t} dt, \qquad k(x) = \frac{1}{x} \int_0^x d\mu(y). \tag{3.5}$$

From here on $||\cdot||_p$ stands for $||\cdot||_{L^p(c,d)}$.

We have

$$||I_{a,b}\tilde{\phi} - \phi||_p \leq ||I_{a,b}(\tilde{\phi} - \phi)||_p + ||I_{a,b}\phi - \phi||_p = \Delta_1 + \Delta_2. \tag{3.6}$$

Obviously,

$$\Delta_1 \leq c_1 ||\phi - \tilde{\phi}||_p \log \frac{b}{a} < c_1 \varepsilon \log \frac{b}{a} \quad \text{where} \quad c_1 = \int_0^\infty d|\mu|(x). \tag{3.7}$$

Since $I_{a,b}\phi = k_a * \phi - k_b * \phi$ (see (2.14)), then

$$\Delta_2 \leq ||k_a\phi - \phi||_p + ||k_b * \phi||_p. \tag{3.8}$$

Note that by (3.3) and (3.4), $k \in L^1(\mathbb{R}_+)$ and $\int_0^\infty k(x)dx = 1$. It follows that $\Delta_2 \to 0$ as $a \to 0$, $b \to \infty$ (e.g. for $1 < p < \infty$). This information is not enough to estimate an accuracy of the approximation. Assume additionally that ϕ belongs to the Lipschitz space $H_p^\lambda(c, d)$ for some $\lambda \in (0, 1]$, $1 \leq p \leq \infty$ (in the case $p = \infty$ we deal with continuous Lipschitz functions). This space consists of L^p-functions with a finite norm

$$||\phi||_{\lambda,p} = \max \left(||\phi||_p; \quad \sup_h \frac{||\phi(x - h) - \phi(x)||_p}{|h|^\lambda} \right).$$

Here ϕ is assumed to be extended by zero outside of $[c, d]$. If

$$||\phi||_{\lambda,p} \leq E \qquad \text{for some} \quad E > 0, \tag{3.9}$$

then

$$||k_a * \phi - \phi||_p = \left\| \int_0^\infty [\phi(x - ay) - \phi(x)]k(y)dy \right\|_p \leq c_1'a^\lambda, \tag{3.10}$$

where

$$c_1' = E \int_0^\infty y^\lambda|k(y)|dy \leq \frac{E}{\lambda} \int_0^\infty |1 - x^\lambda|d|\mu|(x) = c_2 \tag{3.11}$$

(the last inequality follows from the relation $\mu(\mathbb{R}_+) = 0$). In order to manage the second term in (3.8) we restrict ourselves by assuming $\mathrm{supp}\mu \subset [\delta, +\infty)$ for some $\delta > 0$. Then,

$$(k_b * \phi)(x) = \int_\delta^{(x-c)/b} \phi(x - by)k(y)dy \equiv 0 \quad \text{for } x < d, \quad b > \frac{d-c}{\delta}. \tag{3.12}$$

As a result, from (3.6)–(3.12) we have

$$||I_{a,b}\tilde{\phi} - \phi||_p < c_1\varepsilon \log\frac{b}{a} + c_2a^\lambda. \tag{3.13}$$

Our next goal is to minimize the right-hand side of (3.13). Since we are interested in a small a and a big b it is natural to link a and b by putting $b = a^{-\gamma}$ with some $\gamma > 0$. Then,

$$||I_{a,b}\phi - \phi||_p < c_2a^\lambda - c_1\varepsilon(1 + \gamma) \log a \overset{\text{def}}{=} f(a).$$

A function $f(a)$ achieves its minimum at the point

$$a = a_\varepsilon = \left(\frac{c_1 \varepsilon (1 + \gamma)}{c_2 \lambda} \right)^{1/\lambda}. \tag{3.14}$$

For such a,

$$f(a_\varepsilon) = \varepsilon \frac{c_1 (1 + \gamma)}{\lambda} \left[\log \frac{1}{\varepsilon} + \log \frac{ec_2 \lambda}{c_1 (1 + \gamma)} \right] \overset{\text{def}}{=} \tilde{\varepsilon}. \tag{3.15}$$

Put $b_\varepsilon = a_\varepsilon^{-\gamma}$. If $\varepsilon \to 0$, then $a_\varepsilon \to 0$ and $b_\varepsilon \to \infty$. In particular, for sufficiently small ε, namely, $\varepsilon < \varepsilon_0 = c_2 \lambda b_0^{-\lambda/\gamma} / c_1 (1 + \gamma)$, $b_0 = (d - c) / \delta$, b_ε becomes bigger than b_0 (see (3.12)). Thus, for $\varepsilon < \varepsilon_0$ we have

$$\left\| \phi - \int_{a_\varepsilon}^{b_\varepsilon} \frac{\tilde{\phi} * \mu_t}{t} dt \right\|_p < f(a_\varepsilon) = O\left(\varepsilon \log \frac{1}{\varepsilon} \right) \quad \text{as } \varepsilon \to 0.$$

Resume obtained results.

Theorem 4. *Let* $\mu \in \mathcal{M}$ *be such that* $\mu(\mathbb{R}) = 0$ *and* supp $\mu \subset [\delta, \infty)$ *for some* $\delta > 0$. *Let* $\phi \in H_p^\lambda(c, d)$, $0 < \lambda \leq 1$, $1 \leq p \leq \infty$, *and* $\|\phi\|_{\lambda, p} \leq E$. *Denote*

$$c_1 = \int_\delta^\infty d|\mu|(x), \quad c_2 = \frac{E}{\lambda} \int_0^\infty |1 - x^\lambda| d|\mu|(x), \quad b_0 = \frac{d - c}{\delta},$$

and put

$$a_\varepsilon = \left(\frac{c_1 \varepsilon (1 + \gamma)}{c_2 \lambda} \right)^{1/\lambda}, \quad b_\varepsilon = a_\varepsilon^{-\gamma}, \quad \varepsilon_0 = \frac{c_2 \lambda b_0^{-\lambda/\gamma}}{c_1 (1 + \gamma)}$$

for $\varepsilon > 0$ *and arbitrary fixed* $\gamma > 0$. *If* $\|\phi - \tilde{\phi}\|_p < \varepsilon$, *then for* $\varepsilon < \varepsilon_0$,

$$\left\| \phi - \int_{a_\varepsilon}^{b_\varepsilon} \frac{\tilde{\phi} * \mu_t}{t} dt \right\|_p < \tilde{\varepsilon}$$

where

$$\tilde{\varepsilon} = \frac{c_1 \varepsilon (1 + \gamma)}{\lambda} \left[\log \frac{1}{\varepsilon} + \log \frac{ec_2 \lambda}{c_1 (1 + \gamma)} \right] = O\left(\varepsilon \log \frac{1}{\varepsilon} \right) \quad \text{as } \varepsilon \to 0.$$

Acknowledgment. The author is deeply grateful to R. R. Coifman and Y. Meyer for the useful discussion, and to V. P. Havin for valuable advice. The investigation was supported in part by the National Council for Research and Development of Israel (grant no. 3418) and in part by the Edmund Landau Center for Research in Mathematical Analysis sponsored by the Minerva Foundation (Germany).

References

[1] Calderón, A. P., Intermediate spaces and interpolation, the complex method, *Studia Math.* **24** (1964), 113–190.

[2] Coifman, R. R. and R. Rochberg, Representation theorems for holomorphic and harmonic functions in L^p, *Astérisque* **77** (1980), 11–66.

[3] Coifman, R. R. and G. Weiss, Extensions of Hardy spaces and their use in analysis, *Bull. Amer. Math. Soc.* **83** (1977), 569–645.

[4] Daubechies, I., *Ten Lectures on Wavelets*, CBMS-NSF Series in Appl. Math. #61, SIAM Publ., Philadelphia, 1992.

[5] Dynkin, E. M., Methods of the theory of singular integrals, Littlewood-Paley theory and its applications, in *Commutative harmonic analysis* **IV**. V. P. Khavin and N. K. Nikol'ski (eds.), Springer, Berlin, 1992, pp. 97–194. (Encycl. Sci. Math., vol.42).

[6] Folland, G. B. and E. M. Stein, *Hardy Spaces on Homogeneous Groups*, Princeton Univ. Press, Princeton, NJ, 1982.

[7] Frazier, M., B. Jawerth, and G. Weiss, *Littlewood-Paley Theory and the Study of Function Spaces*, CBMS-Conf. Lect. Notes **79**, Amer. Math. Soc., Providence, RI, 1991.

[8] Gorenflo, R. and B. Rubin, Locally controllable regularization of fractional derivatives, *Inverse problems* **10** (1994), 881–893.

[9] Grossmann, A. and J. Morlet, Decomposition of functions into wavelets of constant shape, and related transforms, in *Mathematics + Physics, 1*, Ed. L. Streit (ed.), World Scientific Publ., Singapore, 1985, pp. 135–165.

[10] Han, Y. S., Calderón-type reproducing formula and Tb Theorem, *Rev. Mat. Iberoamericana* **10** (1994), 51–91.

[11] Holschneider, M., Inverse radon transforms through inverse wavelet transform, *Inverse Problems* **7** (1991), 853–861.

[12] Holschneider, M. and Ph. Tchamitchian, Pointwise analysis of Riemann's "nondifferentiable" function, *Invent. Math.* **105** (1991), 157–175.

[13] Meyer, Y., *Wavelets and Operators*, Cambridge Studies in Adv. Math. **37**, Cambridge Univ. Press, 1992.

[14] Neri, U., *Singular Integrals*, Lect. Notes in Math. 200, Springer-Verlag, Heidelberg, 1971.

[15] Rubin, B., Hypersingular integrals of Marchaud type and the inversion problem for potentials, *Math. Nachr.* **165** (1994), 245–321.

[16] Rubin, B. and E. Shamir, Calderón's reproducing formula and singular integral operators on a real line, *Integr. Equat. Oper. Th.* **21** (1995), 77–92.

[17] Samko, S. G., *Hypersingular Integrals and their Applications*, Rostov Univ. Publ., Rostov-on-Don, 1984.

[18] Stein, E. M., *Singular Integrals and Differentiability Properties of Functions*, Princeton Univ. Press, Princeton, NJ, 1970.

[19] Stein, E. M., *Harmonic Analysis, Real Variable Methods, Orthogonality, and Oscillation Integrals*, Princeton Univ. Press, Princeton, NJ, 1993.

Boris Rubin
Department of Mathematics
Hebrew University of Jerusalem
Givat Ram 91904
Jerusalem, Israel
boris@math.huji.ac.il

Continuous Wavelet Transforms on a Sphere

Boris Rubin

Abstract. A general method for constructing continuous wavelet transforms is suggested. Such transforms arise naturally as a result of special procedures of analytic continuation for analytic families of operators. As an example, the case of n-dimensional sphere in $\mathbb{R}^{n+1}$ is considered.

Introduction

During the last decade, a modern wavelet analysis has been extensively developed in the series of books and papers by C. Chui, R. Coifman, I. Daubechies, A. Grossmann, P. G. Lemarié, S. Mallat, Y. Meyer and many other authors who contributed greatly to this field. A theory of continuous wavelet transforms is one of the basic tools of this analysis (see, *e.g.* [3, 9, 18, 21, 24]). In mentioned publications, this theory has been built on the basis of Calderón's reproducing formula, or by making use of square integrable group representations ([18, 19]). Concerning Calderón's reproducing formula, see, *e.g.* [2, 3, 9, 12, 11, 20, 22, 24, 26]. In the following we present an alternative approach which may be helpful in the cases when methods mentioned above fail or their realization is difficult. Such a situation arises *e.g.* on a spherical surface. Here the classical Calderón's reproducing formula (and related wavelet transforms) may not be used in principle for geometrical reasons. On the other hand it is not clear how to use representations of the rotation group for setting up a spherical analogue of the Calderón's reproducing formula. A list of such examples may be continued.

Our approach is based on the following observation. Any reproducing formula may be interpreted as a representation of the identity operator I. Consider an analytic operator family $\{K^{\alpha}\}$, $\alpha \in \mathbb{C}$, for which $K^0 = I$ (for instance, one can take any analytic semigroup of operators). Then a reproducing formula is nothing else but the trivial relation

$$\varphi = (a.c.)K^{\alpha}\varphi|_{\alpha=0}$$

where $(a.c.)K^{\alpha}$ stands for some representation of analytic continuation of K^{α}. Now the problem is to choose a suitable operator family $\{K^{\alpha}\}$ and to

Signal and Image Representation in Combined Spaces
Y. Y. Zeevi and R. R. Coifman (Eds.), pp. 457–476.

write out $(a.c.)K^\alpha$ in appropriate form. If we want to get a reproducing formula, say, on a set $\mathcal{A}$, it is natural to choose operators K^α acting just on this set. A desired construction of analytic continuation may be obtained by using a generalization of A. Marchaud's idea ([23, 28]) that was originally used for constructing fractional derivatives (see Section 1 for the details). This construction gives rise to natural continuous wavelet transforms on $\mathcal{A}$ and to the corresponding reproducing formula. Note that the classical Calderón's reproducing formula on $\mathbb{R}^n$ can be obtained in the framework of our approach if we take K^α to be a semigroup of Riesz potentials for which $(\widehat{K^\alpha\varphi})(\xi) = |\xi|^{-\alpha}\hat{\varphi}(\xi)$ in the Fourier terms.

In Sections 2 and 3 a new method is illustrated for the case when $\mathcal{A}$ is an n-dimensional unit sphere $\Sigma_n \subset \mathbb{R}^{n+1}$ (one should mention a paper of Segman and Zeevi [33] in which the term "spherical wavelet transform" has another meaning). Various topics of the harmonic analysis on a sphere are presented in [1, 7], [4-6], [16, 17, 25, 30, 34, 35, 36]. This list of references may be essentially extended.

The content of this article is to a large extent originated by papers [27, 28].

Notation

$$\Sigma_n = \{x \in \mathbb{R}^{n+1} : |x| = 1\}, \quad \sigma_n = |\Sigma_n| = 2\pi^{(n+1)/2}/\Gamma((n+1)/2).$$

For $x, y \in \Sigma_n$ the notation xy is used for the scalar product of x and y; dx denotes a Lebesgue measure on Σ_n; $\mathcal{Y}(\Sigma_n) = \{Y_{k,j}(x)\}$ stands for the complete orthonormal system of spherical harmonics on Σ_n; $k = 0, 1, \ldots$; $j = 1, 2, \ldots, d_n(k)$, $d_n(k)$ being the dimension of the subspace of harmonics of the order k, $d_n(k) = (n + 2k - 1)\frac{(n+k-2)!}{k!(n-1)!}$ (see [35]).

$$f(x, r) = \frac{1}{\sigma_n} \int_{\Sigma_n} \frac{1 - r^2}{|y - rx|^{n+1}} f(y)dy, \quad 0 < r < 1,$$

is a Poisson integral of f.

Below $C(\Sigma_n)$ is a linear space of continuous functions on Σ_n. The notation $L^p(\Sigma_n)$ is standard for $1 \leq p < \infty$. For our convenience, a similar notation $L^\infty(\Sigma_n)$ is used for the space $C(\Sigma_n)$.

We use the symbol "$\leq\approx$" instead of "$\leq$" when corresponding relations hold up to a nonessential constant factor.

§1   A generalization of Marchaud's method

A great number of operators K^α which constitute an analytic family and satisfy an equality $K^0 = I$ can be represented in the form

$$(K^\alpha \varphi)(x) = \int\limits_0^\infty \frac{t^{\alpha-1}}{\Gamma(\alpha)} f_{\varphi,x}(t) dt, \quad \text{Re } \alpha > 0, \tag{1.1}$$

where $f_{\varphi,x}(t)$ is a certain expression depending on x, φ and being sufficiently good in t-variable (here φ is also assumed to be sufficiently good). For example, consider a Riesz potential

$$(I^\alpha \varphi)(x) = c_{n,\alpha} \int\limits_{\mathbf{R}^n} |y|^{\alpha-n} \varphi(x-y) dy, \quad c_{n,\alpha} = \frac{\Gamma(\frac{n-\alpha}{2}) 2^{-\alpha}}{\pi^{n/2} \Gamma(\alpha/2)}, \quad 0 < \text{Re } \alpha < n. \tag{1.2}$$

By passing to polar coordinates, we have

$$(I^\alpha \varphi)(x) = \frac{\Gamma(\frac{n-\alpha}{2}) 2^{-\alpha} \Gamma(\alpha)}{\pi^{n/2} \Gamma(\alpha/2)} \int\limits_0^\infty \frac{t^{\alpha-1}}{\Gamma(\alpha)} (S_x \varphi)(t) dt \tag{1.3}$$

$$= \frac{\Gamma(\frac{n-\alpha}{2})}{\pi^{n/2} 2^{1+\alpha}} \int\limits_0^\infty \frac{t^{\alpha/2-1}}{\Gamma(\alpha/2)} (S_x \varphi)(\sqrt{t}) dt, \tag{1.4}$$

where

$$(S_x \varphi)(t) = \int\limits_{\Sigma_{n-1}} \varphi(x - t\sigma) d\sigma.$$

Thus, it makes sense to investigate the following model situation. Given an infinitely differentiable rapidly decreasing function $f(t)$, $t \geq 0$, put

$$I_f(\alpha) = \int\limits_0^\infty \frac{t^{\alpha-1}}{\Gamma(\alpha)} f(t) dt, \quad \text{Re } \alpha > 0. \tag{1.5}$$

A function $I_f(\alpha)$ may be extended to all $\alpha \in \mathbb{C}$ as an entire function ([14]). There are many ways to realize this extension. We use the following approach based on the idea of Marchaud [23]. Write (1.5) in the form

$$I_f(\alpha) \int\limits_0^\infty t^{\alpha-1} e^{-t} dt = \int\limits_0^\infty t^{\alpha-1} f(t) dt, \quad \text{Re } \alpha > 0.$$

Substitute t for ty, $y > 0$, then integrate with respect to some measure $d\mu(y)$ and change the order of integration. This gives

$$I_f(\alpha) \int\limits_0^\infty t^{\alpha-1}dt \int\limits_0^\infty e^{-ty}d\mu(y) = \int\limits_0^\infty t^{\alpha-1}dt \int\limits_0^\infty f(ty)d\mu(y), \quad \mathrm{Re}\,\alpha > 0. \quad (1.6)$$

Here we do not care of justification of all operations because our present purpose is just to construct a heuristic model. The reader may think of μ to be finite and compactly supported away from the origin. Precise justification must be given in each concrete situation.

If

$$\int\limits_0^\infty y^j d\mu(y) = 0 \quad \forall j = 0, 1, \ldots, \ell - 1 \quad (\ell \in \mathbb{N}),$$

then (1.6) may be extended to $\mathrm{Re}\,\alpha > -\ell$ and we get

$$I_f(\alpha) = \frac{1}{c_{\alpha,\mu}} \int\limits_0^\infty \frac{(\mu_t, f)}{t^{1-\alpha}}dt, \quad \mathrm{Re}\,\alpha > -\ell, \quad (1.7)$$

where

$$c_{\alpha,\mu} = \int\limits_0^\infty \frac{\tilde{\mu}(t)}{t^{1-\alpha}}dt, \tilde{\mu}(t) \quad = \quad \int\limits_0^\infty e^{-ty}d\mu(y) \quad \text{(a Laplace transform of } \mu),$$

$$(\mu_t, f) \quad = \quad \int\limits_0^\infty f(ty)d\mu(y).$$

Here we assume that $c_{\alpha,\mu} \neq 0$.

If $f(t) \equiv f_{\varphi,x}(t)$ (see (1.1)) then (1.7) represents a reproducing formula (in the case $\alpha = 0$) and gives rise to wavelet type representations of potentials, singular, and hypersingular operators in the cases $\mathrm{Re}\,\alpha > 0$, $\mathrm{Re}\,\alpha = 0$, $(\alpha \neq 0)$ and $\mathrm{Re}\,\alpha < 0$, respectively. By (1.7), a natural definition of the continuous wavelet transform (generated by the family $\{K^\alpha\}$) is

$$(V\varphi)(x, t) = \int\limits_0^\infty f_{\varphi,x}(ty)d\mu(y) \quad (1.8)$$

with the inversion formula

$$\varphi(x) = \frac{1}{c_\mu} \int\limits_0^\infty (V\varphi)(x, t)\frac{dt}{t}, \quad c_\mu = \int\limits_0^\infty \frac{\tilde{\mu}(t)}{t}dt. \quad (1.9)$$

Similar transforms in the special case $x \in \mathbb{R}^1$, $d\mu(y) = g(y)dy$, $f_{\varphi,x}(ty) = \varphi(x - ty)$ are known as "voice tranforms" ([18]).

An interested reader may apply this algorithm to Riesz potentials as well as to many other potential type operators (see [28]), and to see what will be. Note that both representations (1.3) and (1.4) are acceptable ([28], Section 4).

Wavelet type representations of potentials, singular, and hypersingular integrals will be considered in future publications. Now let us go back to the sphere and show how our method works in the case $\alpha = 0$.

§2   Reproducing formula on a sphere

According to the general philosophy presented in Section 1 we begin by choosing an appropriate operator family on the sphere. Various fractional integrals on the sphere were investigated in [27]. It is reasonable to select the following two families that are the simplest:

$$(I^\alpha \varphi)(x) \;=\; c_{n,\alpha} \int\limits_{\Sigma_n} \frac{\varphi(y)dy}{|x - y|^{n-\alpha}} \sim \sum_{k,j} \frac{\Gamma(k + (n - \alpha)/2)}{\Gamma(k + (n + \alpha)/2)} \varphi_{k,j} Y_{k,j}(x),$$

$$0 < \operatorname{Re}\alpha < n, \qquad (2.1)$$

$$(J^\alpha \varphi)(x) \;=\; \frac{1}{\Gamma(\alpha)} \int\limits_0^1 \left(\log \frac{1}{\rho}\right)^{\alpha-1} \varphi(x, \rho)d\rho \sim \sum_{k,j} (k + 1)^{-\alpha} \varphi_{k,j} Y_{k,j}(x),$$

$$\operatorname{Re}\alpha > 0, \qquad (2.2)$$

where $c_{n,\alpha}$ is the same as in (1.2) and $\varphi(x, \rho)$ is the Poisson integral of φ. In this section we establish a reproducing formula and introduce corresponding wavelet transforms generated by potentials (2.1).

Let us pass to the "polar coordinates" on a sphere by making use of the formula

$$\int\limits_{\Sigma_n} a(xy)\varphi(y)dy = \sigma_{n-1} \int\limits_{-1}^1 a(\eta)(M_\eta \varphi)(x)(1 - \eta^2)^{n/2-1}d\eta \qquad (2.3)$$

where

$$(M_\eta \varphi)(x) = \frac{(1 - \eta^2)^{(1-n)/2}}{\sigma_{n-1}} \int\limits_{xy=\eta} \varphi(y)dy \qquad (2.4)$$

is a mean value of φ on a planar section $\{y \in \Sigma_n : xy = \eta\}$ (See [29],

p. 183). By virtue of (2.3),

$$(I^\alpha \varphi)(x) = d_{n,\alpha} \int_0^2 \frac{t^{\alpha/2-1}}{\Gamma(\alpha/2)} f_{\varphi,x}(t) dt, \quad 0 < \text{Re } \alpha < n, \tag{2.5}$$

$$d_{n,\alpha} = 2^{1-(\alpha+n)/2} \Gamma\left(\frac{n-\alpha}{2}\right) / \Gamma(n/2), \quad f_{\varphi,x}(t) = (2-t)^{n/2-1}(M_{1-t}\varphi)(x).$$

Thus, we have an integral of the form (1.1) (up to a factor $d_{n,\alpha}$). Take for simplicity $d\mu(\tau) = g(\tau)d\tau$ and put (see (1.8))

$$
\begin{aligned}
(V_g\varphi)(x,t) &= 2^{1-n/2} \int_0^{2/t} (2-ty)^{n/2-1}(M_{1-ty}\varphi)(x)g(y)dy \\
&= \frac{2^{1-n/2}}{t} \int_{-1}^1 (1+\eta)^{n/2-1}(M_\eta\varphi)(x)g\left(\frac{1-\eta}{t}\right)d\eta.
\end{aligned}
$$

By (2.3),

$$(V_g\varphi)(x,t) = \frac{2^{1-n/2}}{t\sigma_{n-1}} \int_{\Sigma_n} (1-xy)^{1-n/2} g\left(\frac{1-xy}{t}\right)\varphi(y)dy, \quad t > 0. \tag{2.6}$$

Definition 1. *Let g be a summable function on $(0,\infty)$ for which*

$$\int_0^\infty g(\tau)d\tau = 0 \quad \text{and} \quad \int_0^\infty |g(\tau)\log\tau|d\tau < \infty. \tag{2.7}$$

Then $(V_g\varphi)(x,t)$ defined by (2.6) will be called a spherical wavelet transform of φ.

In the case $n = 2$, $(V\varphi)(x,t)$ has an especially simple form

$$(V_g\varphi)(x,t) = \frac{1}{2\pi t} \int_{\Sigma_2} g\left(\frac{1-xy}{t}\right)\varphi(y)dy. \tag{2.8}$$

Remark 1. In [33] a term "spherical wavelet transform" has another meaning and is used for functions on $\mathbb{R}^n$. A "logarithmic" condition in (2.7) is quite natural (see [26] for the detailed discussion).

Let us justify the inversion formula

$$\varphi(x) = \frac{1}{c_g} \int_0^\infty (V_g\varphi)(x,t)\frac{dt}{t}, \quad c_g = \int_0^\infty \frac{\tilde{g}(t)}{t}dt \tag{2.9}$$

where $\tilde{g}(t)$ stands for the Laplace transform of g. Denote

$$(I_{\varepsilon,\rho}\varphi)(x) = \int_{\varepsilon}^{\rho} (V_g\varphi)(x,t)\frac{dt}{t}, \quad 0 < \varepsilon < \rho < \infty,$$

and assume $\varphi \in L^p(\Sigma_n)$, $1 \le p \le \infty$. We have

$$(I_{\varepsilon,\rho}\varphi)(x) = \frac{2^{1-n/2}}{\sigma_{n-1}} \int_{\Sigma_n} (1-xy)^{1-n/2}\varphi(y)dy \int_{\varepsilon}^{\rho} g\left(\frac{1-xy}{t}\right)\frac{dt}{t^2} = K_\varepsilon\varphi - K_\rho\varphi$$

$$(2.10)$$

where

$$(K_\varepsilon\varphi)(x) = \frac{2^{1-n/2}}{\sigma_{n-1}} \int_{\Sigma_n} \varphi(y)(1-xy)^{1-n/2}dy \int_{\varepsilon}^{\infty} g\left(\frac{1-xy}{t}\right)\frac{dt}{t^2}$$

$$= \frac{2^{1-n/2}}{\sigma_{n-1}} \int_{\Sigma_n} \varphi(y)k_\varepsilon(1-xy)dy,$$

$$k_\varepsilon(\tau) = \tau^{1-n/2}\int_{\varepsilon}^{\infty} g\left(\frac{\tau}{t}\right)\frac{dt}{t^2} = \frac{\tau^{1-n/2}}{\varepsilon}k\left(\frac{\tau}{\varepsilon}\right), \quad k(\tau) = \frac{1}{\tau}\int_0^\tau g(z)dz.$$

$K_\rho\varphi$ has the same form with ε replaced by ρ.

Thus, it is clear that we have to deal with approximations to the identity on the sphere. Let us present an auxiliary general result from [27] which will be used in the sequel for proving "almost everywhere" convergence. Consider a family of spherical convolutions

$$(A_\varepsilon\varphi)(x) = \int_{\Sigma_n} a_\varepsilon(1-xy)\varphi(y)dy$$

and a corresponding maximal operator $(A^*\varphi)(x) = \sup_{\varepsilon>0} |(A_\varepsilon\varphi)(x)|$.

Denote $\sigma_t(x) = \{y \in \Sigma_n : xy > t\}$, where $t \in (-1,1)$, $x \in \Sigma_n$;

$$\varphi^*(x) = \sup_{t\in(-1,1)} \frac{1}{(1-t)^{n/2}} \int_{\sigma_t(x)} |\varphi(y)|dy$$

$$= \sup_{t\in(-1,1)} \frac{\sigma_{n-1}}{(1-t)^{n/2}} \int_t^1 (1-\tau^2)^{n/2-1}(M_\tau|\varphi|)(x)d\tau, \quad (2.11)$$

$$\varphi^{**}(x) = \sup_{t\in(-1,1)} \frac{1}{\text{mes }\sigma_t(x)} \int_{\sigma_t(x)} |\varphi(y)|dy.$$

$\varphi^{**}(x)$ is a usual maximal function on Σ_n. It is easy to see that $c_1\varphi^*(x) \leq \varphi^{**}(x) \leq c_2\varphi^*(x)$ for some positive constants c_1, c_2 which depend only on n.

Lemma 1. *Let*

$$|a_\varepsilon(\tau)| \leq \frac{\tau^{1-n/2}}{\varepsilon}\lambda(\tau/\varepsilon), \tag{2.12}$$

$\lambda(\xi)$ *being a non-increasing integrable function on* $(0, \infty)$. *Then*

$$(A^*\varphi)(x) \leq Ac_n\varphi^*(x), \qquad A = \int_0^\infty \lambda(\xi)d\xi,$$

c_n *being a constant depending on* n.

Proof: We may assume $\varphi \geq 0$. Using the argument of Theorem 2 from [34, p. 64] we have

$$|(A_\varepsilon\varphi)(x) \leq \sigma_{n-1}\int_0^{2/\varepsilon} \lambda(\xi)(2 - \varepsilon\xi)^{n/2-1}(M_{1-\varepsilon\xi}\varphi)(x)d\xi$$

$$\leq A \sup_{0 < h < 2} \psi_{x,\varphi}(h),$$

where

$$\psi_{x,\varphi}(h) = \frac{\sigma_{n-1}}{h}\int_{1-h}^1 (1-\tau)^{1-n/2}[(1-\tau^2)^{n/2-1}(M_\tau\varphi)(x)]d\tau.$$

Let us estimate the last integral. We have

$$\psi_{x,\varphi}(h) = \frac{1}{h}\int_{1-h}^1 u(\tau)dv(\tau),$$

where

$$u(\tau) = (1-\tau)^{1-n/2}, \quad v(\tau) = -\sigma_{n-1}\int_\tau^1 (1-t^2)^{n/2-1}(M_t\varphi)(x)dt.$$

By (2.11), $|v(\tau)| \leq (1-\tau)^{n/2}\varphi^*(x)$. Hence,

$$\psi_{x,\varphi}(h) = \frac{1}{h}[uv|_{1-h}^1 + (1 - \frac{n}{2})\int_{1-h}^1 v(\tau)(1-\tau)^{-n/2}d\tau$$

$$= -h^{-n/2}v(1-h) - \frac{n/2-1}{h}\int_{1-h}^1 v(\tau)(1-\tau)^{-n/2}d\tau \leq c(n)\varphi^*(x). \quad \blacksquare$$

Corollary 1. Let $\varphi \in L^p(\Sigma_n)$, $1 \le p \le \infty$. If $a_\varepsilon(\tau)$ satisfies (2.12), then there exist constants c_1, c_2 independent from φ such that

$$\|A^*\varphi\|_p \le c_1\|\varphi\|_p \quad \text{if } 1 < p \le \infty,$$

and

$$\text{mes } \{x \in \Sigma_n : (A^*\varphi)(x) > a\} \le \frac{c_2}{a}\|\varphi\|_1 \quad \text{if } p = 1, \ a > 0.$$

This assertion follows from the similar one for $\varphi^{**}(x)$. The latter may be verified using the argument from [34] with insignificant modifications when proving a covering lemma (these modifications are caused by the compactness of Σ_n).

Now we are ready to prove the following:

Theorem 1. Let $\varphi \in L^p(\Sigma_n)$, $1 \le p \le \infty$.[1] Assume that a "wavelet function" g is the same as in Definition 1, and such

$$c'_g = \int_0^\infty g(\tau) \log \frac{1}{\tau} d\tau \ne 0. \tag{2.13}$$

Then

$$\varphi(x) = \lim_{\substack{\varepsilon \to 0 \\ \rho \to \infty}} \frac{1}{c'_g} \int_\varepsilon^\rho (V_g\varphi)(x, t)\frac{dt}{t} \tag{2.14}$$

in $L^p(\Sigma_n)$-norm. If, besides that,

$$\int_0^1 \tau^{-\delta}|g(\tau)|d\tau < \infty \quad \text{and} \quad \int_1^\infty \tau^\delta|g(\tau)|d\tau < \infty \text{ for some } \delta > 0, \tag{2.15}$$

then the convergence in (2.14) holds almost everywhere on Σ_n.

Proof: Let us turn to (2.10):

$$\int_\varepsilon^\rho (V_g\varphi)(x, t)\frac{dt}{t} = K_\varepsilon\varphi - K_\rho\varphi, \quad \left(K_s\varphi\right)(x) = \frac{2^{1-n/2}}{\sigma_{n-1}} \int_{\Sigma_n} \varphi(y)k_s(1 - xy)dy,$$

$$k_s(\tau) = (\tau^{1-n/2}/s)k(\tau/s), \quad k(\tau) = \frac{1}{\tau} \int_0^\tau g(z)dz; \quad s = \varepsilon, \rho.$$

---

[1] Remember our agreement $L^\infty(\Sigma_n) \equiv C(\Sigma_n)$ (see Notation).

One can readily check (see [26]) that by (2.7),

$$k \in L^1(0, \infty), \quad \int\limits_0^\infty k(\tau)d\tau = c'_g = c_g, \qquad (2.16)$$

c_g and c'_g being defined by (2.9) and (2.13), respectively.

Next we prove that

$$\sup_{0 < \varepsilon < 1} \|K_\varepsilon \varphi\|_p \lesssim \|\varphi\|_p, \quad 1 \le p \le \infty. \qquad (2.17)$$

By Young's inequality (see, e.g. [30]), $\|k_\varepsilon \varphi\|_p \le 2^{1-n/2} c \|\varphi\|_p$ where

$$c = \int\limits_{-1}^1 |k_\varepsilon(1-t)|(1-t^2)^{n/2-1}dt = \int\limits_0^{2/\varepsilon} (2-\varepsilon z)^{n/2-1}|k(z)|dz \lesssim \int\limits_0^\infty |k(z)|dz$$

if $n \ge 2$. Let $n = 1$. Then,

$$c = \int\limits_0^{2/\varepsilon} [(2-\varepsilon z)^{-1/2} - 2^{-1/2}]|k(z)|dz + 2^{-1/2} \int\limits_0^{2/\varepsilon} |k(z)|dz = c_1 + c_2.$$

The estimate for c_2 is obvious, and for c_1 we have

$$c_1 = \varepsilon \int\limits_0^{2/\varepsilon} \frac{z|k(z)|dz}{2^{1/2}(2-\varepsilon z)^{1/2}(2^{1/2} + (2-\varepsilon z)^{1/2})}$$

$$\lesssim \varepsilon^{1/2} \int\limits_0^{2/\varepsilon} \frac{z|k(z)|}{(2/\varepsilon - z)^{1/2}}dz = c_{11} + c_{12}.$$

Here

$$c_{11} = \varepsilon^{1/2} \int\limits_0^1 \frac{z|k(z)|}{(2/\varepsilon - z)^{1/2}}dz \le \varepsilon^{1/2} \left(\frac{2}{\varepsilon} - 1\right)^{-1/2} \int\limits_0^1 |k(z)|dz \le \int\limits_0^1 |k(z)|dz,$$

and (by (2.7))

$$c_{12} = \varepsilon^{1/2} \int\limits_1^{2/\varepsilon} \frac{dz}{(2/\varepsilon - z)^{1/2}} \int\limits_z^\infty |g(\tau)|d\tau \lesssim \varepsilon^{1/2} \left(\frac{2}{\varepsilon} - 1\right)^{1/2} \int\limits_1^\infty |g(\tau)|d\tau$$

$$\leq \approx \int\limits_{1}^{\infty} |g(\tau)| d\tau.$$

Thus, (2.17) is proved and we can pass to the next step. Let us check that

$$\lim_{\varepsilon \to 0} \|K_\varepsilon Y_{k,j} - c_g' Y_{k,j}\|_p = 0, \quad \lim_{\rho \to \infty} \|k_\rho Y_{k,j}\|_p = 0 \qquad (2.18)$$

for any spherical harmonic $Y_{k,j} \in \mathcal{Y}(\Sigma_n)$. By Funk-Hecke theorem ([10], p. 247), $K_\varepsilon Y_{k,j} = m_k^\varepsilon Y_{k,j}$ where

$$\begin{aligned} m_k^\varepsilon &= 2^{1-n/2} \int\limits_{-1}^{1} k_\varepsilon (1-t) P_k(t)(1-t^2)^{n/2-1} dt \\ &= 2^{1-n/2} \int\limits_{0}^{2/\varepsilon} k(t) P_k(1-t\varepsilon)(2-t\varepsilon)^{n/2-1} dt, \qquad (2.19) \end{aligned}$$

$P_k(t)$ being Legendre polynomials normalized by $P_k(1) = 1$. Relations (2.18) will be proved if we check that

$$\lim_{\varepsilon \to 0} m_k^\varepsilon = c_g', \quad \lim_{\rho \to \infty} m_k^\rho = 0. \qquad (2.20)$$

Consider the first equality. For $n \geq 2$ it follows immediately from (2.19) by Lebesgue theorem on the dominated convergence. For $n = 1$ we have

$$m_k^\varepsilon = \int\limits_{0}^{2/\varepsilon} k(t) P_k(1-t\varepsilon)\left[(1-t\varepsilon/2)^{1/2}-1\right] dt + \int\limits_{0}^{2/\varepsilon} k(t) P_k(1-t\varepsilon) dt = a_\varepsilon + b_\varepsilon.$$

$$(2.21)$$

The second term tends to c_g' as $\varepsilon \to 0$. The first one may be done arbitrarily small. Really,

$$|a_\varepsilon| \leq \left(\int\limits_{0}^{1} + \int\limits_{1}^{2/\varepsilon} \right) |k(t) P_k(1-t\varepsilon)| \frac{t\, dt}{(1-t\varepsilon/2)^{1/2}} = a_1 + a_2$$

where

$$a_1 \quad \lesssim \varepsilon^{1/2} \int\limits_0^1 \frac{|k(t)|dt}{(2/\varepsilon - t)^{1/2}} \lesssim \frac{\varepsilon^{1/2}}{(2/\varepsilon - 1)^{1/2}} \to 0, \quad \varepsilon \to 0,$$

$$a_2 \quad \lesssim \varepsilon^{1/2} \int\limits_1^{2/\varepsilon} \frac{dt}{(2/\varepsilon - t)^{1/2}} \int\limits_t^\infty |g(\tau)|d\tau$$

$$= \varepsilon^{1/2} \left\{ \int\limits_1^{2/\varepsilon} |g(\tau)|d\tau \int\limits_1^\tau \frac{dt}{(2/\varepsilon - t)^{1/2}} + 2\,(2/\varepsilon - 1)^{1/2} \int\limits_{2/\varepsilon}^\infty |g(\tau)|d\tau \right\}$$

$$\lesssim \int\limits_1^{2/\varepsilon} |g(\tau)|d\tau \int\limits_{1-\varepsilon\tau/2}^{1-\varepsilon/2} \frac{du}{u^{1/2}} + \int\limits_{2/\varepsilon}^\infty |g(\tau)|d\tau \to 0, \quad \varepsilon \to 0.$$

Let us check the second relation in (2.20). For $n \geq 2$ it is obvious from (2.19) (with ε replaced by ρ):

$$|m_k^\rho| \lesssim \int\limits_0^{2/\rho} |k(t)|dt \to 0 \quad \text{as } \rho \to \infty.$$

If $n = 1$, then, as in (2.21), we obtain $m_k^\rho = a_\rho + b_\rho$ where $\lim\limits_{\rho\to\infty} b_\rho = 0$ and

$$|a_\rho| \quad \lesssim \rho \int\limits_0^{2/\rho} \frac{|k(t)|t\,dt}{(1 - t\rho/2)^{1/2}} \lesssim \rho^{1/2} \int\limits_0^{2/\rho} \frac{dt}{(2/\rho - t)^{1/2}} \int\limits_0^t |g(\tau)|d\tau$$

$$= 2 \int\limits_0^{2/\rho} |g(\tau)|(2 - \rho\tau)^{1/2}d\tau \to 0$$

as $\rho \to \infty$.

Now all preparations have been done and we can complete the proof. Since span $\mathcal{Y}(\Sigma_n)$ is dense in $L^p(\Sigma_n)$, then (2.17) and (2.18) yield the relations

$$\lim_{\varepsilon\to 0} \|K_\varepsilon\varphi - c_g'\varphi\|_p = 0, \quad \lim_{\rho\to\infty} \|K_\rho\varphi\|_p = 0, \quad 1 \leq p \leq \infty, \qquad (2.22)$$

which imply (2.14). If (2.15) holds, then $k(\tau) = (1/\tau)\int_0^\tau g(z)dz$ has a non-increasing integrable majorant (the reader may easily check this statement by making use of (2.7); see also [26]). In this case, Corollary 1 together with (2.22) and (2.20) yields an "almost everywhere" convergence in (2.14) (see, e.g. an argument in [34, p. 8]). ∎

§3 Spherical wavelet transform of a Poisson type

In this section we introduce a spherical wavelet transform generated by the operator family

$$(J^\alpha\varphi)(x) = \frac{1}{\Gamma(\alpha)}\int_0^1 \left(\log\frac{1}{\rho}\right)^{\alpha-1}\varphi(x,\rho)d\rho \qquad (3.1)$$

where $\varphi(x,\rho)$ is the Poisson integral of φ. A new transform essentially differs from that in Section 2. Put $\rho = e^{-t}$. Then

$$(J^\alpha\varphi)(x) = \int_0^\infty \frac{t^{\alpha-1}}{\Gamma(\alpha)}\varphi(x,e^{-t})dt.$$

According to (1.8) (with $d\mu(y) = g(y)dy$, $g \in L^1(0,\infty)$) put

$$(V_g\varphi)(x,t) = \int_0^\infty \varphi(x,e^{-ty})e^{-ty}g(y)dy \qquad (3.2)$$

or

$$(V_g\varphi)(x,t) = \frac{1}{t}\int_0^1 \varphi(x,\rho)g\left(\frac{1}{t}\log\frac{1}{\rho}\right)d\rho. \qquad (3.3)$$

If g enjoys conditions of Definition 1, then $\varphi \to (V_g\varphi)(x,t)$ will be called a *spherical wavelet transform of a Poisson type*. By the contraction property of the Poisson integral ([35]),

$$\|(V_g\varphi)(\cdot,t)\|_p \le \|\varphi\|_p\|g\|_{L^1(0,\infty)}, \quad 1 \le p \le \infty. \qquad (3.4)$$

According to (1.9), $(V_g\varphi)(x,t)$ should admit the following inversion

$$\varphi(x) = \frac{1}{c_g}\int_0^\infty (V_g\varphi)(x,t)\frac{dt}{t}, \quad c_g = \int_0^\infty \frac{\tilde{g}(t)}{t}dt. \qquad (3.5)$$

Our goal is to justify this formula. Consider the corresponding truncated integral

$$\int_\varepsilon^\rho (V_g\varphi)(x,t)\frac{dt}{t} = M_\varepsilon\varphi - M_\rho\varphi, \quad 0 < \varepsilon < \rho < \infty, \qquad (3.6)$$

where

$$(M_\delta\varphi)(x) = \int_\delta^\infty (V_g\varphi)(x,t)\frac{dt}{t} = \int_\delta^\infty \frac{dt}{t}\int_0^\infty \varphi(x,\varepsilon^{-ty})e^{-ty}g(y)dy; \quad \delta = \varepsilon,\rho.$$

$$(3.7)$$

Lemma 2. *Let* g *satisfy conditions of Definition 1. Then the following statements hold:*

(i) *If* $\varphi \in L^p(\Sigma_n)$, $1 \le p \le \infty$, *then* $\sup\limits_{\delta > 0} \|M_\delta \varphi\|_p \le c\|\varphi\|_p$, *where* c *does not depend on* φ.

(ii) *For any spherical harmonic* $Y_{k,j} \in \mathcal{Y}(\Sigma_n)$,

$$\lim_{\varepsilon \to 0} (M_\varepsilon Y_{k,j})(x) = c'_g Y_{k,j}(x), \quad \lim_{\rho \to \infty} (M_\rho Y_{k,j})(x) = 0 \qquad (3.8)$$

where the limits are uniform in $x \in \Sigma_n$, *and* c'_g *is the same as in* (2.13) *and* (2.16).

(iii) *For* $\varphi \in L^p(\Sigma_n)$, *a maximal operator* $M^*\varphi = \sup\limits_{\delta > 0} |M_\delta \varphi|$ *enjoys the following estimates*

$$\|M^*\varphi\|_p \le c_1 \|\varphi\|_p \quad if \quad 1 < p \le \infty \qquad (3.9)$$

and

$$\text{mes } \{x \in \Sigma_n : (M^*\varphi)(x) > a\} \le \frac{c_2}{a} \|\varphi\|_1 \quad if \ p = 1, \ a > 0. \quad (3.10)$$

Proof: In order to prove (i) it suffices to check that

(a) $$\|M_\delta Y_{k,j}\|_p \le c\|Y_{k,j}\|_p \quad \forall Y_{k,j} \in \mathcal{Y}\Sigma_n)$$

and

(b) $$\|M_\delta \varphi\|_p \le c_\delta \|\varphi\|_p \quad \forall \varphi \in L^p(\Sigma_n)$$

with some constant c_δ depending on δ. Really, if (a) and (b) hold and φ_m is a linear combination of spherical harmonics such that $\lim_{m \to \infty} \|\varphi_m - \varphi\|_p = 0$ (in the case $p = \infty$ we take $\varphi \in C(\Sigma_n)$), then

$$\|M_\delta \varphi\|_p \le \|M_\delta(\varphi - \varphi_m)\|_p + \|M_\delta \varphi_m\|_p \le c_\delta \|\varphi - \varphi_m\|_p + c\|\varphi_m\|$$

$$\le (c_\delta + c)\|\varphi - \varphi_m\|_p + c\|\varphi\|_p,$$

and passage to the limit as $m \to \infty$ gives a desired result.

Let us prove (a). Since for any spherical harmonic $Y_{k,j}(x)$, its Poisson integral $Y_{k,j}(x, \rho)$ is just $\rho^k Y_{k,j}(x)$ [35], one can write $\|M_\delta Y_{k,j}\|_p = |\tilde{c}_\delta| \, \|Y_{k,j}\|_p$, where

$$\tilde{c}_\delta = \int_\delta^\infty \frac{dt}{t} \int_0^\infty e^{-ty(k+1)} g(y) dy.$$

By making use of (2.7) we have

$$
\begin{aligned}
|\tilde{c}_\delta| &= \left| \int\limits_\delta^\infty \frac{dt}{t} \int\limits_0^\infty \left[e^{-ty(k+1)} - e^{-t} \right] g(y)\,dy \right| \\
&\leq \int\limits_0^\infty |g(y)|dy \int\limits_0^\infty \frac{|e^{-ty(k+1)} - e^{-t}|}{t} dt \\
&= \int\limits_0^\infty |g(y)\log[y(k+1)]|dy \\
&< \infty \quad ([15],\ \text{Formula 3.434(2))}.
\end{aligned}
\tag{3.11}
$$

The inequality (b) is actually obvious due to the contraction property of the Poisson integral:

$$
\begin{aligned}
\|M_\delta\varphi\|_p &\leq \|\varphi\|_p \int\limits_0^\infty |g(y)|dy \int\limits_{\delta y}^\infty \frac{e^{-z}}{z} dz \\
&\leq \approx \|\varphi\|_p \left(\int\limits_0^{1/2\delta} |g(y)| \log\frac{1}{\delta y} dy + \int\limits_{1/2\delta}^\infty |g(y)|dy \int\limits_{1/2}^\infty \frac{e^{-z}}{z} dz \right) \\
&= c_\delta \|\varphi\|_p.
\end{aligned}
\tag{3.12}
$$

Let us prove (ii). As above we have

$$
M_\delta Y_{k,j} = \tilde{c}_\delta Y_{k,j} \quad \text{where } \tilde{c}_\delta = \int\limits_0^\infty g(y)dy \int\limits_\delta^\infty \frac{e^{-ty(k+1)} - e^{-t}}{t} dt.
\tag{3.13}
$$

If $\delta \to 0$, then

$$
\tilde{c}_\delta \to \int\limits_0^\infty g(y) \log\frac{1}{y(k+1)} dy \stackrel{(2.7)}{=} \int\limits_0^\infty g(y) \log\frac{1}{y} dy.
$$

If $\delta \to \infty$ one can pass to the limit under the integration sign (this is possible due to estimates in (3.11) and obtain $\lim_{\delta\to\infty} \tilde{c}_\delta = 0$.

It remains to prove (iii). First we verify an equality

$$
M_\delta\varphi = \int\limits_0^\infty \varphi(x, e^{-z}) e^{-z} k_\delta(z)dz, \quad \varphi \in L^p(\Sigma_n),\ 1 \leq p \leq \infty,
\tag{3.14}
$$

where $k_\delta(z) = (1/z)k(z/\delta)$, $k(z) = \frac{1}{z}\int_0^z g(y)dy \in L^1(0,\infty)$. Since operators in both sides of (3.14) are bounded in $L^p(\Sigma_n)$ it suffices to check (3.14) on spherical harmonics. For the left-hand side we have an equality (3.13). The right-hand side of (3.14) is just

$$Y_{k,j}\int\limits_0^\infty e^{-(k+1)z}\frac{dz}{z}\int\limits_0^{z/\delta} g(y) = Y_{k,j}\int\limits_0^\infty g(y)dy\int\limits_{\delta y}^\infty \frac{e^{-(k+1)z}}{z}dz = \tilde{c}_\delta Y_{k,j}$$

the interchange of integration is possible because

$$\int\limits_0^\infty |g(y)|dy\int\limits_{\delta y}^\infty \frac{e^{-(k+1)z}}{z}dz < \infty \quad (\text{see}(3.12)).$$

From (3.14), by taking into account a maximal estimate $\sup_{o<\rho<1}|\varphi(x,\rho)| \leq c_n\varphi^*(x)$ ([17], p. 169), we obtain

$$M^*\varphi \leq c_n\varphi^*(x)\int\limits_0^\infty |k(z)|dz$$

which implies (3.9) and (3.10).   ∎

Theorem 2. *Let $\varphi \in L^p(\Sigma_n)$, $1 \leq p \leq \infty$, and g satisfy Definition 1. If $c_g' \neq 0$ (see (2.13)), then*

$$\varphi(x) = \lim\limits_{\substack{\varepsilon\to 0 \\ \rho\to\infty}} \frac{1}{c_g'}\int\limits_\varepsilon^\rho (\mathcal{V}_g\varphi)(x,t)\frac{dt}{t} \tag{3.15}$$

where the limit is interpreted in L^p-norm and in "a.e."- sense.

Proof: Let us turn to (3.6):

$$\int\limits_\varepsilon^\rho (\mathcal{V}_g\varphi)(x,t)\frac{dt}{t} = M_\varepsilon\varphi - M_\rho\varphi.$$

By Lemma 2 (i, ii) the relation (3.15) holds with the limit being interpreted in L^p-norm. Convergence in a.e.-sense is a direct consequence of (3.9) and (3.10).   ∎

Remark 2. A reproducing formula (3.5) may be written in terms of "two convolutions" as the original formula of Calderón. Let $u(y)$ and $v(y)$ be $L^1(0, \infty)$-functions for which

$$g(y) \equiv \int_0^y u(\tau)v(y - \tau)d\tau$$

satisfies Definition 1. Redenote

$$u_t(\rho) = \frac{1}{t}u\left(\frac{1}{t}\log\frac{1}{\rho}\right), \quad \mathcal{V}_u\varphi = \varphi \bullet u_t \quad \text{(similarly for } v\text{)}.$$

Then, by making use of the multiplication property of a Poisson integral

$$\frac{1}{\sigma_n}\int_{\Sigma_n} \frac{(1 - r^2)\varphi(y, \rho)}{|y - rx|^{n+1}}dy = \varphi(x, r\rho),$$

one can readily see that $(\mathcal{V}_g\varphi)(x, t) = (\varphi \bullet u_t \bullet u_t)(x)$ and

$$\varphi = \frac{1}{c_g'}\int_0^\infty \frac{\varphi \bullet u_t \bullet v_t}{t}dt.$$

We leave this simple exercise to the reader.

Remark 3. The results of this paper appeared first in Preprint No. 14 of the Edmund Landau Center for Research in Mathematical Analysis (Inst. of Math., Hebrew University of Jerusalem, 1994). After that the author was aware of other publications related to spherical wavelet transforms (S. Dahlke and P. Maass [8], W. Freeden and U. Windheuser [13], P. Schröder and W. Sweldens [31, 32], B. Torresani [37]). The approaches of mentioned authors differ from that presented above.

Acknowledgment. The author is deeply grateful to R. R. Coifman and A. Grossmann for useful inspiring discussions. The investigation was supported in part by the National Council for Research and Development of Israel (grant no. 3418) and in part by the Edmund Landau Center for Research in Mathematical Analysis sponsored by the Minerva Foundation (Germany).

References

[1] Berens, H., P. L. Butzer, and S. Pawelke, Limitierung verfahren von Reihen mehrdimensionaler kugelfunktionen und deren Saturationsverhalten, *Publs. Res. Inst. Math. Sci. Ser. A*, **4** (1968), 201–268.

[2]   Calderón, A. P., Intermediate spaces and interpolation, the complex method, *Studia Math.* **24** (1964), 113–190.

[3]   Chui, C. K., *An Introduction to Wavelets*, Academic Press, Boston, 1992.

[4]   Coifman, R. R. and G. Weiss, Representations of compact groups and spherical harmonics, *L'Enseignement mathém.*, t. XIY (1968), fasc. **2**, 121–172.

[5]   Coifman, R. R. and G. Weiss, Transference methods in Analysis, *CBMS Reg. Conf. Ser. Math.* **31**, 1977.

[6]   Coifman, R. R. and G. Weiss, *Analyse Harmonique Non-Commutative sur Certains Espaces Homogenes*, Lecture Notes in Math. **242**, 1971.

[7]   Colzani, L., Hardy spaces on unit spheres, Bollettino U.M.I., *Analisi Functionale e Applicazioni*, Ser. YI, **IY - C.N.1** (1985), 219–244.

[8]   Dahlke, S. and P. Maass, Continuous wavelet transform on tangent bundles of spheres, The University of Potsdam, preprint.

[9]   Daubechies, I., *Ten Lectures on Wavelets*, CBMS-NSF Series in Appl. Math. #61, SIAM Publ., Philadelphia, 1992.

[10] Erdélyi, A. (ed.), *Higher Transcendental Functions*, vol. II, McGraw-Hill, New York, 1953.

[11] Folland, G. B. and E. M. Stein, *Hardy Spaces on Homogeneous Groups*, Princeton Univ. Press, Princeton, NJ, 1982.

[12] Frazier, M., B. Jawerth, and G. Weiss, Littlewood-Paley Theory and the Study of Function Spaces, *CBMS-Conf. Lect. Notes # 79*, Amer. Math. Soc., Providence, R.I. 1991.

[13] Freeden, W. and U. Windheuser, Spherical wavelet transform and its discretization, University of Kaiserslautern, Bericht Nr. 125, 1994.

[14] Gelfand, I. M. and G. E. Shilov, *Generalized Functions*, vol. 1, *Properties and Operations*, Academic Press, New York, 1964.

[15] Gradshteyn, I. S. and I. M. Ryzhik, *Table of Integrals, Series, and Products*, Academic Press, 1980.

[16] Greenwald, H. C., Lipschitz spaces on the surface of the unit sphere in Euclidean n-spaces, *Pacific J. Math.* **50** (1974), 63–80.

[17] Greenwald, H. C., Lipschitz spaces of distributions on the surface of unit sphere in Eucledian n-space, *Pacific J. Math.* **70** (1977), 163–176.

[18] Grossmann, A. and J. Morlet, Decomposition of functions into wavelets of constant shape, and related transforms, in *Mathematics and Physics* **1**, L. Streit (ed.), pp. 135–165, World Scientific Publ., Singapore, 1985.

[19] Grossmann, A., J. Morlet, and T. Paul, Transforms associated to square integrable group representations, I. General results, *J. Math. Phys.* **27**, 1985, 2473–2479; II. Examples, Ann. Inst. H.Poincaré **45**, 1986, 293–309.

[20] Han, Y. S., Calderón-type reproducing formula and Tb Theorem, *Rev. Mat. Iberoamericana* **10** (1994), 51–91.

[21] Heil, C. and D. Walnut, Continuous and discrete wavelet transforms, *SIAM Rev.* **31** (1989), 628–666.

[22] Holschneider, M. and Ph. Tchamitchian, Pointwise analysis of Riemann's "nondifferentiable" function, *Invent. math.* **105** (1991), 157–175.

[23] Marchaud, A., Sur les dériviées et sur les differences des fonctions de variables réelles, *J. Math. Pures el Appl.* **6** (1927), 337–425.

[24] Meyer, Y., *Wavelets and Operators*, Cambridge Studies in Adv. Math. **37**, Cambridge Univ. Press, 1992.

[25] Neri, U., *Singular Integrals*, Lect. Notes in Math. **200**, Springer-Verlag, Heidelberg, 1971.

[26] Rubin, B., On Calderón's reproducing formula. in *Signal and Image Representations in Combined Spaces*, Y. Y. Zeevi and R. Coifman (eds.), Academic Press, 1997.

[27] Rubin, B., The inversion of fractional integrals on a sphere, *Israel Journal of Math.* **79** (1992), 47–81.

[28] Rubin, B., Hypersingular integrals of Marchaud type and the inversion problem for potentials, *Math. Nachr.* **165** (1994), 245–321.

[29] Samko, S. G., Generalized Riesz potentials and hypersingular integrals with homogenous characteristics, their symbols and inversion, *Proceeding of the Steklov Inst. of Math.* **2** (1983), 173–243.

[30] Samko, S. G., Singular integrals over a sphere and the construction of the characteristic from the symbol, *Soviet Math.* (Iz. VUZ) **27** (1983), 35–52.

[31] Schröder, P. and W. Sweldens, Spherical wavelets: Efficiently representing functions on the sphere, University of South Carolina, preprint.

[32] Schröder, P. and W. Sweldens, Spherical wavelets: Texture processing, University of South Carolina, preprint.

[33] Segman, J. and Y. Y. Zeevi, Spherical wavelets and their applications to image representation, *Journal of Visual Communication and Image Representation* **4** (1993), 263–270.

[34] Stein, E. M., *Singular Integrals and Differentiability Properties of Functions*, Princeton Univ. Press, Princeton, NJ, 1970.

[35] Stein, E. M. and G. Weiss, *Introduction to Fourier Analysis on Euclidean Spaces*, Princeton Univ. Press, Princeton, NJ, 1971.

[36] Strichartz, R. S., Multipliers for spherical harmonic expansions, *Trans. Amer. Math. Soc.* **167** (1972), 115–124.

[37] Torresani, B., Phase space decompositions: Local Fourier analysis on spheres, preprint, March 1993.

Boris Rubin
Department of Mathematics
Hebrew University of Jerusalem
Givat Ram 91904
Jerusalem, Israel
boris@math.huji.ac.il

Periodic Splines, Harmonic Analysis, and Wavelets

Valery A. Zheludev

Abstract. We discuss here wavelets constructed from periodic spline functions based on a new computational technique called spline harmonic analysis (SHA). SHA is a version of harmonic analysis operating in the spaces of periodic splines of defect 1 with equidistant nodes. Discrete Fourier transform is a special case of SHA. The continuous Fourier analysis is the limit case of SHA as the degree of splines involved tends to infinity. Thus, SHA bridges the gap between the discrete and the continuous versions of Fourier analysis. SHA can be regarded as a computational version of the harmonic analysis of continuous periodic functions from discrete noised data. The SHA approach to wavelets yields a tool for constructing a diversity of spline wavelet bases, for a fast implementation of the decomposition of a function into a fitting wavelet representation and its reconstruction. Via this approach we are able to construct wavelet packet bases for refined frequency resolution of signals. In this paper we also present algorithms for digital signal processing by means of spline wavelets and wavelet packets. The algorithms established are embodied in a flexible multitasking software for digital signal processing.

§1 Introduction

The objective of this paper is the presentation of techniques of adaptive signal processing based on the spline wavelet analysis.

At present most popular wavelet schemes are based on the compactly supported orthonormal wavelet bases invented by Ingrid Daubechies [9]. Exploiting the so-called wavelet packets [12, 7], provides essential advantages because these generate a library of bases and provide opportunities for adaptive representation of signals. It should be pointed out that the compact support of basic wavelets and the orthonormality of corresponding wavelet bases are not compatible with the symmetry of the wavelets concerned. The lack of the symmetry is a noticeable handicap when using the wavelets by Daubechies for signal processing. To attain the symmetry one should sacrifice at least one of these properties. In [8] the authors constructed biorthogonal bases of compactly supported symmetric wavelets.

Signal and Image Representation in Combined Spaces
Y. Y. Zeevi and R. R. Coifman (Eds.), pp. 477–509.

However, a certain inconvenience of the construction lies in the fact that dual wavelets belong to different wavelet spaces.

Early examples of wavelets were based on spline functions [11, 1, 10]. Later, spline wavelets were shadowed by the wavelets by Daubechies. However, in numerous situations spline wavelets offer advantages over these latter ones. In recent years spline-wavelets were subjected to detailed study, mainly by C. Chui and collaborators [2, 3, 4, 5, 6, 7]. In particular, in [3] the authors succeeded in constructing compactly supported wavelets in the spaces of cardinal splines. The significant property of such spline wavelets, in addition to the symmetry, is that their dual wavelets belong to the same spaces as the original ones.

We discuss in this paper periodic spline wavelets. Our approach to the spline wavelet analysis is based on an original computational technique—the so-called spline harmonic analysis (SHA) which is a version of the harmonic analysis (HA) in spline spaces. SHA in some sense bridges the gap between the continuous and discrete HA. It is a rather universal technique applicable to a great variety of numerical problems, not necessarily to wavelet analysis [18, 19, 21]. Application of the SHA techniques to wavelet analysis is found to be remarkably fruitful. This approach led to the construction of a rich diversity of wavelet bases as well as wavelet packet ones. Moreover, we present an efficient scheme of decomposition into the wavelet representation and reconstruction of a signal, based on SHA.

In Section 2 we outline the properties of splines with equidistant nodes which will be of use for wavelet analysis, especially the properties of *B*-splines. Section 3 is devoted to the presentation of the SHA. In Section 4 we discuss the multiscale analysis of splines. We establish the two-scale relation, construct orthogonal bases in spline wavelet spaces, and present a spectral algorithm for decomposition of a spline into frequency multichannel representation and reconstruction from this representation. We introduce the high- and low-frequency wavelet spaces. In Section 5 we construct the families of *father* and *mother* wavelets as well as the spline wavelet packets. We discuss methods of the most informative digital representation of signals by means of spline wavelets and establish a useful quadrature formula.

Some results presented in the paper have been announced in the papers [23, 24, 22]. On the basis of algorithms established in this paper we have developed a flexible multitasking software for digital signal processing by means of spline wavelets and wavelet packets. The software allows processing of periodic signals as well as non-periodic ones in real-time mode.

§2 Splines with equidistant nodes

In this section we outline the properties of polynomial splines with equidistant nodes, most of which are known [23, 24]. A function $_pS(x)$ will be referred to as a spline of order p if

1) $_pS(x) \in C^{p-2}$

2) $_pS(x) = P_k(x)$ as $x \in (x_k, x_{k+1})$, $P_k(x) \in \Pi_{p-1}$

Π_{p-1} is the space of polynomials whose degree does not exceed $p - 1$. In what follows we deal exclusively with splines whose nodes $\{x_k\}$ are equidistant $x_k = hk$, $k = -\infty, ..., \infty$ and denote this space by $_pS$. These splines are referred to as *cardinal* ones. The most significant advantages of these splines over others originate from the fact that in the space $_pS$ there exist bases each of which consists of translates of a unique spline. The most important basis for our subject is $\mathcal{B}$-splines.

2.1 The $\mathcal{B}$-splines

We define the truncated powers as $x_+^k = \left(\frac{1}{2}(x + |x|) \right)^k$.

The following linear combination of truncated powers:

$$_pB_h(x) = \frac{h^{-p}}{(p-1)!} \sum_{l=0}^{p} (-1)^l \binom{p}{l} (x - lh)_+^{p-1}$$

is referred to as the $\mathcal{B}$-spline. It is a spline of order p with nodes in the points $\{hk\}_{-\infty}^{\infty}$.

Properties of $\mathcal{B}$-splines.

1. $\operatorname{supp} {}_pB_h(x) = (0, hp)$.

2. $_pB_h(x) > 0$ as $x \in (0, ph)$.

3. $_pB_h(x)$ is symmetric about $x = hp/2$ where it attains its unique maximum.

4. $\int_{-\infty}^{\infty} {}_pB_h(x)\, dx = \int_0^{ph} {}_pB_h(x)\, dx = 1$.

Point out a property concerned with discrete values of $\mathcal{B}$-splines. Define discrete and continuous moments of $\mathcal{B}$- splines as

$$\mu_s({}_pB_h)(t) = h \sum_{r=-\infty}^{\infty} \left(h\left(t + r - \frac{p}{2}\right) \right)^s {}_pB(h(t + r))$$

$$M_s({}_pB_h) = \int_0^{ph} (x - hp/2)^s {}_pB_h(x)\, dx.$$

Proposition 1. [([24, 22]] *Provided $s \leq p - 1$, the discrete moments $\mu_s({}_pB_h)(t)$ do not depend on t and coincide with the corresponding continuous moments $M_s({}_pB_h)$. The moment*

$$\mu_p({}_pB_h)(t) = (-1)^{p-1}\beta_h(t)h^p + M_p({}_pB_h) \quad \text{as } t \in [0,1]$$

and 1-periodic with respect to t; $\beta_p(t)$ is the Bernoulli polynomial of degree p.

This property will be of use in Section 5 to derive an important quadrature formula.

The $\mathcal{B}$-splines of any order can be computed immediately.

Define a function which will be of important concern in what follows

$$_pu_h(\omega) = h \sum_k e^{-iwhk} \, {}_pB((k+p/2)h). \tag{2.1.1}$$

Due to the symmetry of B-splines, the functions $_pu_h(\omega)$ are real-valued cosine polynomials. These functions were extensively studied in [13] and [15]. They are related to the Euler-Frobenius polynomials [14].

Proposition 2. *The functions $_pu_h(\omega)$ are strictly positive, moreover*

$$0 < K_p = {}_pu_h(\pi/h) \leq {}_pu_h(\omega) \leq {}_pu_h(0) = 1.$$

The constants K_p do not depend on h and $\lim_{p \to \infty} K_p = 0$.

The Fourier transform of the $\mathcal{B}$-spline is:

$$_p\widehat{B}_h(\omega) = \int_{-\infty}^{\infty} e^{-i\omega t} \, {}_pB_h(t) \, dt \tag{2.1.2}$$

$$= \left(\frac{1 - e^{-i\omega h}}{i\omega h}\right)^p = e^{-\frac{ip\omega h}{2}}\left(\frac{\sin \omega h/2}{\omega h/2}\right)^p. \tag{2.1.3}$$

2.2 The $\mathcal{B}$-spline representation of cardinal splines

Recall that $_p\mathcal{S}$ denotes the space of cardinal splines of order p with their nodes at the points $\{hk\}_{-\infty}^{\infty}$.

Proposition 3. [14] *Any spline $_pS(x) \in {}_p\mathcal{S}$ can be represented as follows*

$$_pS(x) = h \sum_k q_k \, {}_pB_h(x - hk). \tag{2.2.1}$$

Here and below $\sum_k$ stands for $\sum_{k=-\infty}^{\infty}$.

Remark 1. If x is any fixed value, then the series (2.2.1) contains only p non-zero addends. So, given a set of coefficients $\{q_k\}$, values of the spline $_pS(x)$ can be computed immediately.

Provided the coefficients $\{q_k\}_{k=-\infty}^{\infty} \in l_2$, the Fourier transform of the spline $_pS(x)$ is

$$
\begin{aligned}
{}_p\widehat{S}(x) &= \int_{-\infty}^{\infty} e^{-i\omega x} h \sum_{k=-\infty}^{\infty} q_k \; {}_pB_h(x - hk) \\
&= h \sum_k e^{-i\omega hk} q_k \; {}_p\widehat{B}_h(\omega) = \breve{q}(\omega) e^{-\frac{ip\omega h}{2}} \left(\frac{\sin \omega h/2}{\omega h/2} \right)^p, \\
\breve{q}(\omega) &= h \sum_k e^{-i\omega hk} q_k.
\end{aligned}
$$

So, we can say that the approximation of a signal by a spline is, as a matter of fact, a kind of low-pass filtering. Actually, the signal is being filtered into the band $[\frac{-1}{2h}, \frac{1}{2h}]$.

<h2 align="center">§3   Spline harmonic analysis</h2>

HA is a powerful mathematical tool for solving a great diversity of theoretical and computational problems. HA techniques are best suited to solving problems associated with the operators of convolution and of differentiation. It stems from the fact that the basic functions of the conventional HA — the exponential functions — are eigenvectors of these operators.

However, the conventional HA is not quite relevant for dealing with signals of finite order of smoothness determined by a finite set of functionals (may be noised) because of at least two reasons: 1) The basic functions of HA — the exponential functions — are infinitely differentiable. 2) Practical computing of the coordinates — the Fourier coefficients or the Fourier integrals — poses a lot of problems.

To circumvent these obstacles it would be attractive to have a version of HA which deals with discrete issue data and provides solutions of problems as immediately computable functions of the smoothness required.

We present here such a version of HA. It is based on periodic splines and we name it the Spline Harmonic Analysis (SHA).

3.1 Periodic splines

We introduce some notations. In what follows, we will assume the step of a mesh involved to be $h = 1/N$, $N = 2^j$. Throughout, $\sum_k^j$ stands for $\sum_{k=-2^{j-1}}^{2^{j-1}-1}$. If a sequence is furnished with the upper index j it will imply that it is 2^j-periodic (*e.g.* $\{u_k^j\}$). Throughout, we denote $\omega = e^{2\pi i/N}$. The direct and inverse Discrete Fourier Transform (DFT) of a vector $\vec{a} = \{a_k\}$ is

$$T_r^j(\vec{a}) = \frac{1}{N} \sum_k^j \omega^{-rk} a_k \qquad a_k = \sum_r^j \omega^{rk} T_r^j(\vec{a}). \qquad (3.1.1)$$

The discrete convolution of the vector $\vec{a}$ with a vector $\vec{b} = \{b_k\}$ and its DFT are

$$\vec{a} * \vec{b} = \left\{ 2^{-j} \sum_l^j a_{k-l} b_l \right\}_k^j \qquad T_r^j(\vec{a} * \vec{b}) = T_r^j(\vec{a}) \cdot T_r^j(\vec{b}).$$

To unify the notations in what follows we will denote the $\mathcal{B}$-spline $_pB_{1/N}(x)$ as $_pB_j(x)$.

Any 1-periodic cardinal spline of order p can be represented as follows

$$_pS^j(x) = \frac{1}{N} \sum_k^j q_k^j \, _pM^j(x - k/N) \qquad (3.1.2)$$

where

$$_pM^j(x) =: \sum_l \, _pB_j(x - l)$$

is a 1-periodic spline. We name it the periodic $\mathcal{B}$-spline. The sequence of the coefficients $\{q_k^j\}$ is N-periodic.

The properties of $\mathcal{B}$-splines $_pM^j$ are determined completely by the $\mathcal{B}$-splines $_pB_j$. It should be noted only that if $p > N$ then the supports of adjacent splines $_pB_j(x + l)$ and $_pB_j(x + l + 1)$ overlap and, therefore, the support of the periodic spline $_pM^j$ has no gaps in this case.

As for spectral properties, due to the periodicity, the $\mathcal{B}$-spline $_pM^j$ is being expanded into the Fourier series:

$$_pM^j(x) = \sum_n e^{2\pi i n x} e^{-\pi i n p / N} \left(\frac{\sin \pi n/N}{\pi n/N} \right)^p.$$

This relation implies that if a spline $_pS^j$ is represented as in (3.1.2) then its Fourier coefficients are

$$
\begin{aligned}
C_n(_pS^j) &= \int_0^1 e^{-2\pi i n x} \, _pS^j(x)\, dx \qquad (3.1.3) \\
&= e^{-\pi i n p/N} \left(\frac{\sin \pi n/N}{\pi n/N} \right)^p T_n^j(\vec{q}^{\,j}) = \left(\frac{1 - \omega^{-n}}{2\pi i n/N} \right)^p T_n^j(\vec{q}^{\,j}).
\end{aligned}
$$

Recall that DFT $\{T_n^j(\vec{q}^{\,j})\}$ forms an N-periodic sequence.

Denote by $_p\mathcal{V}^j$ the space of 1-periodic splines of order p with their nodes in the points $\{k2^{-j}\}_{-\infty}^\infty$. The relation (3.1.2) implies that the shifts $\{_pM^j(x - k/2^j)\}_k^j$ form a basis of $_p\mathcal{V}^j$.

3.2 Periodic ortsplines

To start constructing SHA we carry out a simple transformation. Let a spline $_pS^j \in {_p}\mathcal{V}^j$ be represented as follows

$$_pS^j(x) = \frac{1}{N} \sum_k^j q_k \; _pM^j(x - k/N). \tag{3.2.1}$$

Due to (3.1.1), we can write

$$q_k = \sum_r^j \omega^{rk} T_r^j(\vec{q})$$

where $\vec{q} = \{q_k\}_k^j$. Substituting it to (3.2.1), we come to the relation

$$\begin{aligned}
_pS^j(x) &= \frac{1}{N} \sum_k^j {_pM^j}(x - k/N) \sum_r^j \omega^{rk} T_r^j(\vec{q}) \\
&= \sum_r^j T_r^j(\vec{q}) \frac{1}{N} \sum_k^j \omega^{rk} \; _pM^j(x - k/N).
\end{aligned}$$

Setting

$$\xi_r = T_r^j(\vec{q}) = \frac{1}{N} \sum_k^j \omega^{-rk} q_k, \tag{3.2.2}$$

$$_pm_r^j(x) = \frac{1}{N} \sum_k^j \omega^{rk} \; _pM^j(x - k/N), \quad r = 0, ..., 2^j - 1, \tag{3.2.3}$$

we write finally

$$_pS^j(x) = \sum_r^j \xi_r \; _pm_r^j(x).$$

Point out at once the reciprocal relations

$$q_k = \sum_r^j \omega^{rk} \xi_r \tag{3.2.4}$$

$$_pM^j(x - k/N) = \sum_r^j \omega^{-rk} \; _pm_r^j(x), \quad _pM^j(x) = \sum_r^j \; _pm_r^j(x). \tag{3.2.5}$$

We take some time to discuss properties of the splines $_pm_r^j$ which are basic for our constructions.

Proposition 4. *The sequence* $\{ _pm_r^j(x)\}$ *is N-periodic with respect to r.*

It follows immediately from 1-periodicity of the $\mathcal{B}$-splines $_pM^j(x)$.
Eq. (3.2.5) implies:

Proposition 5. *The splines* $\{ _pm_r^j(x)\}_r^j$ *form a basis of the space* $_p\mathcal{V}^j$.

It will be of use to compute the Fourier coefficients. Keeping in mind
(3.1.3) we write

$$
\begin{aligned}
C_n(_pm_r^j) &= e^{-\pi inp/N}\left(\frac{\sin \pi n/N}{\pi n/N}\right)^p \cdot \frac{1}{N}\sum_k \omega^{-k(n-r)} \\
&= \delta_n^r(\mathrm{mod}N)\cdot e^{-\pi inp/N}\left(\frac{\sin \pi n/N}{\pi n/N}\right)^p \\
&= \delta_n^r(\mathrm{mod}N)(1-\omega^{-r})^p\frac{1}{(2\pi in/N)^p}. \quad (3.2.6)
\end{aligned}
$$

So,

$$
\begin{aligned}
_pm_r^j(x) &= (1-\omega^{-r})^p\sum_l e^{2\pi i(r+lN)x}\frac{1}{(2\pi i(r+lN)/N)^p} \quad (3.2.7) \\
&= \left(\frac{N(1-\omega^{-r})}{2\pi i}\right)^p e^{2\pi irx}\sum_l e^{2\pi ilNx}\frac{1}{(r+lN)^p} \\
&= \sum_l e^{2\pi i(r+lN)(x-p/2N)}\left(\frac{\sin \pi(r+lN)/N}{(\pi(r+lN))/N}\right)^p \\
&= e^{2\pi i(x-p/2N)r}(\sin \pi r/N)^p\sum_l e^{2\pi ilNx}(\pi(r+lN)/N)^{-p}.
\end{aligned}
$$

Let us derive some consequences from (3.2.6) and (3.2.7). First we
denote

$$
_pu_r^j =: {} _pm_r^j(p/2N) = \frac{1}{N}\sum_k^j \omega^{-rk} {} _pM((p/2+k)/N).
$$

Substituting the identity $_pM^j(x) = \sum_l {} _pB_j(x-l)$ into the latter relation, we obtain

$$
\begin{aligned}
_pu_r^j &= \frac{1}{N}\sum_k^j e^{-2\pi irk/N}\sum_{l=-\infty}^{\infty} {} _pB_j((p/2+k+lN)/N) \\
&= \sum_{n=-\infty}^{\infty} e^{-2\pi irn/N} {} _pB_j\left(\left(\frac{p}{2}+n\right)/N\right) = {} _pu_{1/N}(2\pi r).
\end{aligned}
$$

Recall that the function $_pu_h(\omega)$ was defined in (2.1.1). So, dealing with the N-periodic sequence $_pu_r^j$ we can compute it immediately and may refer to the properties $_pu_h(\omega)$ marked in Section 2. This sequence is of importance for us. Substitution $x = p/2N$ into (3.2.7) results in the identity

$$_pu_r^j = \sum_l \left(\frac{\sin\pi(r+lN)/N}{(\pi(r+lN)/N)}\right)^p = (N\sin\pi r/N)^p \sum_l \frac{(-1)^{lp}}{(\pi(r+lN))^p}.$$

The Parseval equality for the Fourier series entails the important relation:

Proposition 6. *The inner product is*

$$\langle _pm_r^j, \,_pm_s^j\rangle = \delta_r^s \,_{2p}u_r^j.$$

Corollary 1. *The splines* $\{_pm_r^j\}_r^j$ *form an orthogonal basis of* $_p\mathcal{V}^j$ *and the splines* $\left\{\frac{1}{p}m_r^j = \,_pm_r^j/\sqrt{_{2p}u_r^j}\right\}_r^j$ *form an orthonormal one.*

Proposition 7. *The splines* $_pm_r^j$ *are eigenvectors to the shift operator. To be specific,*

$$_pm_r^j(x+l/N) = \omega^{rl}\,_pm_r^j(x). \tag{3.2.8}$$

Proof: Eq. (3.2.3) implies

$$_pm_r^j(x+l/N) = \frac{1}{N}\sum_k^j \omega^{rk}\,_pM^j\left(x-\frac{(k-l)}{N}\right)$$

$$= \omega^{rl}\frac{1}{N}\sum_k^j \omega^{rk}\,_pM^j(x-k/N) = \omega^{rl}\,_pm_r^j(x). \quad\blacksquare$$

Corollary 2. *The splines* $_p^2m_r^j(x) = \,_pm_r^j(x+p/2N)/\,_pu_r^j$ *interpolate the exponential functions* $\mu_r(x) = e^{2\pi irx}$. *Namely,*

$$_p^2m_r^j(l/N) = \mu_r(l/N).$$

Indeed, (3.2.8) implies

$$_pm_r^j((p/2+l)/N) = e^{2\pi irl/N}\,_pm_r^j(p/2N) = e^{2\pi irl/N}\,_pu_r^j.$$

Let $\sigma_n^j(x) = \sum_r^j\,^n\xi_r\,_pm_r^j(x)$ be an orthogonal projection of the exponential function $\mu_n(x)$ onto the space $_p\mathcal{V}^j$. Then

$$^n\xi_r\langle _pm_r^j, \,_pm_s^j\rangle = \langle \mu_n, \,_pm_s^j\rangle \iff$$

$$^n\xi_r\delta_r^s\,_{2p}u_r^j = \overline{C_n}(_pm_s^j) \implies$$

$$^n\xi_r = \delta_n^r(\text{mod}N)\cdot e^{\pi inp/N}\left(\frac{\sin\pi n/N}{\pi n/N}\right)^p\frac{1}{_{2p}u_r^j}.$$

Hence, it follows

Proposition 8. *The spline*

$$_p\sigma_n^j(x) = \delta_n^r(\mathrm{mod}N) \cdot e^{\pi inp/N} \left(\frac{\sin \pi n/N}{\pi n/N} \right)^p \frac{_pm_r^j(x)}{2p\,u_r^j},$$

is an orthogonal projection of the exponential function $\mu_n(x)$ *onto the space* $_p\mathcal{V}^j$ *provided* $n = r(\mathrm{mod}N)$.

Denote $_p^3 m_r^j(x) = {}_p\sigma_r^j(x)$. The latter properties relate the splines $_p m_r^j$ to the orthogonal exponential functions $\mu_n(x) = e^{2\pi inx}$. We will see further that this relation is much more intimate, but for the moment we make a terminology remark.

As the splines $\{ _p m_r^j \}$ form an orthogonal basis, it is pertinent to call these *Ortsplines* (OS). The connection of OS with the operators of convolution and of differentiation is related to this $\mu_n(x)$. To be specific

Proposition 9. *There holds the relation*

$$_p m_r^j(x)^{(s)} = (N(1 - \omega^{-r}))^s {}_{p-s}m_r^j(x).$$

Proof: In accordance with (3.2.7)

$$
\begin{aligned}
_p m_r^j(x)^{(s)} &= (1 - \omega^{-r})^s (1 - \omega^{-r})^{p-s} \sum_l e^{2\pi i(r+lN)x} \frac{(2\pi i(r + lN))^s}{(2\pi i(r + lN)/N)^p} \\
&= [N(1 - \omega^{-r})]^s {}_{p-s}m_r^j(x). \quad \blacksquare
\end{aligned}
$$

Remark 2. Emphasize that OS $_{p-s}m_r^j(x)$ is a replica of OS $_p m_r^j(x)$ in the space $_{p-s}\mathcal{V}^j$.

Let us turn now to the convolution. Provided f, g are square integrable 1-periodic functions, we mean under the convolution $f * g$ the following integral

$$f * g(x) = \int_0^1 f(x - y)\, g(y)\, dy.$$

Recall the Fourier coefficients

$$C_n(f * g) = C_n(f) \cdot C_n(g). \qquad (3.2.9)$$

Proposition 10. *The convolution of the ortsplines is*

$$_q m_s^j * {}_p m_r^j(x) = {}_{q+p}m_r^j(x) \cdot \delta_s^r.$$

The relation results immediately from (3.2.6) and (3.2.9).

Corollary 3. *Let a spline $_q\sigma^j(x)$ be given as*

$$_q\sigma^j(x) = \sum_s \eta_s \; _qm_s^j(x).$$

Then the convolution is

$$_q\sigma^j \ast \; _pm_r^j(x) = \eta_r \cdot \; _{p+q}m_r^j(x).$$

Emphasize that OS $_{p+s}m_r^j(x)$ *is a replica of* OS $_pm_r^j(x)$ *in the space* $_{p-s}\mathcal{V}^j$.

Remark 3. We may interpret Proposition 9 and Corollary 3 as follows: OS $\{\, _pm_r^j\}$ are *generalized* eigenvectors of the operators of *convolution* and *differentiation* unlike the exponential functions which are the conventional eigenvectors of these operators.

The properties of OS established give rise to formulas related to corresponding ones of HA.

Proposition 11. *Given two splines of* $_p\mathcal{V}^j$

$$_pS^j(x) = \sum_r^j \xi_r \; _pm_r^j(x) \tag{3.2.10}$$

$$_p\widetilde{S}^j(x) = \sum_r^j \chi_r \; _pm_r^j(x), \tag{3.2.11}$$

there holds the Parseval equality

$$\langle \, _pS^j, \; _p\widetilde{S}^j \rangle = \sum_r^j \xi_r \overline{\chi_r} \; _{2p}u_r^j.$$

Proposition 12. *Let a spline $_pS^j$ be given by (3.2.10) and $_q\sigma^j(x) = \sum_s^j \eta_s \; _qm_r^j(x) \in \; _q\mathcal{V}^j$. Then the convolution is*

$$_pS^j \ast \; _q\sigma^j(x) = \sum_r^j \xi_r \eta_r \; _{p+q}m_r^j(x) \in \; _{p+q}\mathcal{V}^j.$$

Proposition 13. *Let a spline* $_pS^j$ *be given by* (3.2.10). *Then the derivative is*

$$_pS^j(x)^{(s)} = \sum_r^j (N(1 - \omega^r))^s \xi_r\ _{p-s}m_r^j(x) \in\ _{p-s}\mathcal{V}^j.$$

Remark 4. Propositions 11–13 enable us to look upon the expansion of a spline $_pS^j \in\ _p\mathcal{V}^j$ with respect to OS basis (3.2.10) as upon a peculiar HA of the spline. Moreover, OS $\{\ _pm_r^j\}$ acts as harmonics, and the coordinates $\{\xi_r\}$ as the Fourier coefficients or a *spectrum* of the spline.

We stress that, given the $\mathcal{B}$-spline representation of the spline (3.1.2), the OS representation can be derived at once by means of DFT (3.2.2) as well as the reciprocal change (3.2.4). It is natural to employ in the process the Fast Fourier transform (FFT) algorithms.

We call this HA in spline spaces the spline harmonic analysis (SHA). A great deal of operations with splines rise to remarkable simplicity by means of SHA. It provides a powerful and flexible tool for dealing with splines and, moreover, with functions of finite order of smoothness when discrete (may be noised) samples of these functions are available. Recently, a promising field of application of SHA has appeared—the *spline wavelet analysis*. We discuss it in further detail.

In [24] we have presented relations which allow one to assert that DFT is a special case of SHA whereas the continuous Fourier analysis is a limit case of SHA. In some sense, SHA bridges the gap between the discrete and the continuous versions of HA.

Remarks on SHA applications.   SHA techniques can be applied successfully for solving problems of spline functions (interpolation, smoothing, approximation) and also for problems associated with the operators of convolution and of differentiation (integral equations of convolution type, differential equations with constant coefficients). When solving inverse problems, the phenomena of ill-posedness occurs frequently. The SHA approach allows the implementation of efficient regularizing algorithms. These applications of SHA techniques are presented in literature.

§4   Multiscale analysis of a spline

4.1   Decomposition of a spline space into wavelet spaces

First we point out that the space $_p\mathcal{V}^{j-1}$ is the subspace of $_p\mathcal{V}^j$. The space $_p\mathcal{W}^{j-1}$ is the orthogonal complement of $_p\mathcal{V}^{j-1}$ in the space $_p\mathcal{V}^j$. So

$$_p\mathcal{V}^j = {}_p\mathcal{V}^{j-1} \oplus\ _p\mathcal{W}^{j-1}.$$

The space $_pW^{j-1}$ is usually called the *wavelet space*. The space $_p\mathcal{V}^{j-1}$ can in turn be decomposed as

$$_p\mathcal{V}^{j-1} = {_p\mathcal{V}^{j-2}} \oplus {_p\mathcal{W}^{j-2}}.$$

Correspondingly

$$_pS^{j-1}(x) = {_pS^{j-2}(x)} \oplus {_pW^{j-2}(x)}.$$

Iterating this procedure we obtain

$$
\begin{aligned}
_p\mathcal{V}^j &= {_p\mathcal{V}^{j-m}} \oplus {_p\mathcal{W}^{j-m}} \oplus {_p\mathcal{W}^{j-m+1}} \oplus \cdots \oplus {_p\mathcal{W}^{j-1}} && (4.1.1) \\
_pS^j(x) &= {_pS^{j-m}(x)} \oplus {_pW^{j-m}(x)} \oplus {_pW^{j-m+1}(x)} \oplus \cdots \oplus {_pW^{j-1}(x)}.
\end{aligned}
$$
$$(4.1.2)$$

The relation (4.1.2) represents a spline as the sum of its smoothed out, "blurred" version $_pS^{j-m}$ and "details" $\{_pW^{j-\nu}\}_{\nu=1}$. Emphasize that all addends in (4.1.2) are mutually orthogonal.

We will call the decomposition of a spline $_pS^j(x)$ the *multiscale analysis* (MSA) of this spline and, provided, the spline $_pS^j(x) = {_pS^j(f,x)}$ is a spline approximating a signal f, as MSA of the signal f.

Now, provided we are able to project a signal f onto the appropriate spline space $_p\mathcal{V}^j$, $f \to {_pS^j(f)}$ and to decompose the spline $_pS^j(f)$, in accordance with (4.1.1) we get an opportunity to process the signal in several frequency channels simultaneously. If need be, the channels obtain bandwidths arranged according to the logarithmic scale which can be subdivided into more narrow channels by means of the so-called *wavelet packets*. This subject will be discussed later.

After multichannel processing one needs to reconstruct the spline processed from its multiscale representation of type (4.1.2) into the conventional $\mathcal{B}$-spline representation where its values can be computed immediately. It is being implemented in accordance with the pyramidal diagram

$$
\begin{array}{ccccccccc}
_pS^{j-m} & \longrightarrow & S^{j-m+1} & \longrightarrow & \cdots & \longrightarrow & S^{j-2} & \longrightarrow & S^{j-1} & \longrightarrow & S^j \\
_pW^{j-m} & \nearrow & W^{j-m+1} & \nearrow & \cdots & \nearrow & W^{j-2} & \nearrow & W^{j-1} & \nearrow &
\end{array}
\quad (4.1.3)
$$

The algorithms of such reconstruction to be established are of high-rate efficiency as well.

SHA provides a powerful tool for developing the wavelet analysis. We start with MSA, *i.e.* we establish algorithms to decompose a spline into an orthogonal sum of type (4.1.2) and to reconstruct it.

4.2 Two-scale relations

We first discuss projecting a spline $_pS^j \in {_p\mathcal{V}^j}$ onto the subspace $_p\mathcal{V}^{j-1}$. Corresponding algorithms result from the so-called two-scale relations which correlate ortsplines of the spaces $_p\mathcal{V}^j$ and $_p\mathcal{V}^{j-1}$.

As a rule, in what follows we will omit the index $_p$. for splines belonging to the spaces $_p\mathcal{V}^\nu$. The term u_r^ν will stand for $_{2p}u_r^\nu$.

Theorem 1. *There hold the two-scale relations for* $r = 0, 1, ..., 2^{j-1} - 1$:

$$m_r^{j-1}(x) = b_r^j\, m_r^j(x) + b_{r-N/2}^j\, m_{r-N/2}^j(x), \qquad b_r^j = 2^{-p}(1+\omega^{-r})^p. \quad (4.2.1)$$

Proof: In accordance with (3.2.7) we have

$$
\begin{aligned}
m_r^{j-1}(x) &= \left(\frac{N(1-\omega^{-2r})}{4\pi i}\right)^p e^{2\pi i r x} \sum_{l=-\infty}^{\infty} e^{\pi i l N x}\frac{1}{(r+lN/2)^p} \\
&= 2^{-p}(1+\omega^{-r})^p\left(\frac{N(1-\omega^{-r})}{2\pi i}\right)^p e^{2\pi i r x} \sum_{\nu=-\infty}^{\infty} e^{2\pi i \nu N x}\frac{1}{(r+\nu N)^p} \\
&+ 2^{-p}(1+\omega^{-(r-N/2)})\left(\frac{N(1-\omega^{-(r-N/2)})}{2\pi i}\right)^p e^{2\pi i(r-N/2)x} \\
&\quad \times \sum_{\nu=-\infty}^{\infty} e^{2\pi i \nu N x}\frac{1}{(r-N/2+\nu N)^p} \\
&= 2^{-p}(1+\omega^{-r})^p m_r^j(x) + 2^{-p}(1+\omega^{-(r-N/2)})^p m_{r-N/2}^j(x). \quad \blacksquare
\end{aligned}
$$

Remark 5. The relation (4.2.1) implies an identity which will be employed repeatedly in what follows. Namely, writing this relation for the splines of order $2p$ with $x = 2p/N$, we have

$$
\begin{aligned}
_{2p}u_r^{j-1} &= {}_{2p}m_r^{j-1}(2p/N) = 4^{-p}(1+\omega^{-r})^{2p}\,{}_{2p}m_r^j(2p/N) \\
&+ 4^{-p}(1-\omega^{-r})^{2p}\,{}_{2p}m_{r-N/2}^j(2p/N).
\end{aligned}
$$

But

$$
_{2p}m_r^j(2p/N) = {}_{2p}m_r^j(p/N)\omega^{pr} = \omega^{rp}\,{}_{2p}u_r^j
$$

and we obtain finally

$$
\begin{aligned}
_{2p}u_r^{j-1} &= 4^{-p}\omega^{rp}\left[(1+\omega^{-r})^{2p}\,{}_{2p}u_r^j + (-1)^p(1-\omega^{-r})^{2p}\,{}_{2p}u_{r-N/2}^j\right] \\
&= |b_r^j|^2\,{}_{2p}u_r^j + |b_{r-N/2}^j|^2\,{}_{2p}u_{r-N/2}^j. \quad (4.2.2)
\end{aligned}
$$

Theorem 1 enables us to implement projecting a spline $S^j(x) \in {}_p\mathcal{V}^j$ onto the space $_p\mathcal{V}^{j-1}$.

Theorem 2. *Let the spline*

$$S^{j-1}(x) = \sum_r^{j-1} \xi_r^{j-1} m_r^{j-1}(x)$$

be the orthogonal projection of a spline

$$S^j(x) = \sum_l^j \xi_l^j m_l^j(x) \in {}_pV^j$$

onto the space ${}_pV^{j-1}$. *Then the coordinates are*

$$
\begin{aligned}
\xi_r^{j-1} &= \langle S^j, m_r^{j-1} \rangle / u_r^{j-1} = \langle S^j, (b_r^j\, m_r^j + b_{r-N/2}^j\, m_{r-N/2}^j) \rangle / u_r^{j-1} \\
&= \frac{1}{u_r^{j-1}} \left(\xi_r^j\, u_r^j\, \overline{b_r^j} + \xi_{r-N/2}^j\, u_{r-N/2}^j\, \overline{b}_{r-N/2}^j \right).
\end{aligned}
\tag{4.2.3}
$$

4.3   Ortwavelets

Now, we proceed to projecting a spline of ${}_pV^j$ onto the space ${}_pW^{j-1}$. To do it, first of all we need a basis of the space ${}_pW^{j-1}$. We now construct an orthogonal basis of ${}_pW^{j-1}$.

Theorem 3. *There exists an orthogonal basis* $\{w_r^{j-1}(x)\}_r^{j-1}$ *of* ${}_pW^{j-1} \subset {}_pV^j$

$$
\begin{aligned}
w_r^{j-1}(x) &= a_r^j\, m_r^j(x) + a_{r-N/2}^j\, m_{r-N/2}^j(x), \tag{4.3.1} \\
a_r^j &= \omega^r\, \overline{b}_{r-N/2}^j\, u_{r-N/2}^j = 2^{-P} \omega^r (1 - \omega^r)^P\, u_{r-N/2}^j. \tag{4.3.2}
\end{aligned}
$$

Moreover,

$$\|w_r^{j-1}\|^2 = \langle w_r^{j-1}, w_r^{j-1} \rangle = v_r^{j-1}$$

where $v_r^{j-1} = u_r^j\, u_{r-N/2}^j\, u_r^{j-1}$ *is a* 2^{j-1}-*periodic sequence.*

Proof: The orthogonality of a spline w_r^{j-1} to any w_l^{j-1}, m_l^{j-1}, $l \neq r$ is readily apparent from the orthogonality of the splines $\{m_r^j\}_r^j$ to each other. We should establish the orthogonality w_r^{j-1} to m_r^{j-1}. Due to (4.2.1), we write

$$
\begin{aligned}
\langle m_r^{j-1}, w_r^{j-1} \rangle &= b_r^j\, \overline{a}_r^j\, u_r^j + b_{r-N/2}^j\, \overline{a}_{r-N/2}^j\, u_{r-N/2}^j \\
&= \omega^{-r} b_r^j\, b_{r-N/2}^j\, u_{r-N/2}^j\, u_r^j - \omega^{-r} b_{r-N/2}^j\, b_r^j\, u_{r-N/2}^j\, u_r^j.
\end{aligned}
$$

We have employed here the periodicity of the sequence u_r^j, namely, the relation $u_{r-N/2-N/2}^j = u_r^j$. Moreover, $\omega^{-r-N/2} = \omega^{-r} \cdot e^{\pi i} = -\omega^{-r}$. Therefore,

$$\langle m_r^{j-1}, w_r^{j-1} \rangle = 0.$$

Similarly, in view of (4.2.2), we can write

$$
\begin{aligned}
\langle w_r^{j-1}, w_r^{j-1} \rangle &= |a_r^j|^2 u_r^j + |a_{r-N/2}|^2 u_{r-N/2}^j \\
&= u_r^j u_{r-N/2}^j (|b_r^j|^2 {}_{2p} u_r^j + |b_{r-N/2}^j|^2 {}_{2p} u_{r-N/2}^j) = v_r^{j-1}. \blacksquare
\end{aligned}
$$

We will call the splines w_r^{j-1} the *ortwavelets* (OW). Note that the OW, just as the OS, are eigenvectors of the operator of shift.

Proposition 14. *There holds the relation*

$$w_r^{j-1}(x + 2k/N) = \omega^{2kr} w_r^j(x). \tag{4.3.3}$$

Proof: Since $\omega^{2k(r-N/2)} = \omega^{2kr} e^{-2\pi i k} = \omega^{2kr}$, we have

$$
\begin{aligned}
w_r^{j-1}(x + 2k/N) &= a_r^j m_r^j(x + 2k/N) + a_{r-N/2}^j m_{r-N/2}^j(x + 2k/N) \\
&= \omega^{2kr}(a_r^j m_r^j(x) + a_{r-N/2}^j m_{r-N/2}(x)) = \omega^{2kr} w_r^{j-1}(x). \blacksquare
\end{aligned}
$$

Theorem 4. *Let the spline*

$$W^{j-1}(x) = \sum_r^{j-1} \eta_r^{j-1} w_r^{j-1}(x)$$

be the orthogonal projection of a spline

$$S^j(x) = \sum_l^j \xi_l^j m_l^j(x) \in {}_p\mathcal{V}^j$$

onto the space ${}_p\mathcal{W}^{j-1}$. *Then the coordinates are*

$$\eta_r^{j-1} = \langle S^j, w_r^{j-1} \rangle / v_r^{j-1} \tag{4.3.4}$$

$$= \frac{1}{v_r^{j-1}} (\xi_r^j u_r^j \bar{a}_r^j + \xi_{r-N/2}^j u_{r-N/2}^j \bar{a}_{r-N/2}) \tag{4.3.5}$$

$$= \frac{\omega^{-r}}{u_r^{j-1}} (\xi_r^j b_{r-N/2}^j - \xi_{r-N/2}^j b_r^j). \tag{4.3.6}$$

Now we have carried out the first step of the decomposition

$$S^j(x) = S^{j-1}(x) \oplus W^{j-1}(x).$$

In the spectral domain we have split the frequency band $[-N/2, N/2]$ into the subband $[-N/4, N/4]$ and two strips $[-N/2, -N/4]$, $[N/4, N/2]$. Subjecting the spline S^{j-1} to the procedures suggested, we carry out the second step and so on until we get (4.1.2). The frequency band will be split in accordance with the logarithmic scale.

4.4 High- and low-frequency wavelet subspaces

If a refined frequency resolution in the strips $[-N/2, -N/4]$, $[N/4, N/2]$ is wanted, we suggest projecting a spline $W^{j-1}(x) \in {}_pW^{j-1}$ onto two subspaces ${}_p^lW^{j-2}$, ${}_p^hW^{j-2}$ one of which is an orthogonal complement of the other in ${}_pW^{j-1}$ and such that ${}_p^lW^{j-2}$ is, loosely speaking, concentrated at the frequency strips $[-\frac{3N}{8}, -\frac{N}{4}]$, $[\frac{N}{4}, \frac{3N}{8}]$ and ${}_p^hW^{j-2}$ in the strips $[-\frac{N}{2}, -\frac{3N}{8}]$, $[\frac{3N}{8}, \frac{N}{2}]$.

To construct these subspaces we start with bases.

Let us call the splines

$${}^hw_r^{j-2}(x) = b_r^{j-1}\, w_r^{j-1}(x) + b_{r-N/4}^{j-1}\, w_{r-N/4}^{j-1}(x) \in {}_pW^{j-1},$$

$r = 0, 1, ..., 2^{j-2} - 1$, the *high-frequency* OW (HOW) and the splines

$$
\begin{aligned}
{}^lw_r^{j-2}(x) &= {}^la_r^{j-1}\, w_r^{j-1}(x) + {}^la_{r-N/4}^{j-1}\, w_{r-N/4}^{j-1}(x) \in {}_pW^{j-1}, \\
{}^la_r^{j-1} &= \omega^{2r} b_{r-N/4}^{j-1}\, v_{r-N/4}^{j-1} = 2^{-p}\omega^{2r}(1 - \omega^{2r})^p v_{r-N/4}^{j-1},
\end{aligned}
$$

$r = 0, 1, ..., 2^{j-2} - 1$, the *low-frequency* OW (LOW).

Theorem 5. *There hold the relations*

$$\langle {}^lw_r^{j-2}, {}^lw_s^{j-2} \rangle = \delta_r^s\, {}^lv_r^{j-2}, \qquad \langle {}^hw_r^{j-2}, {}^hw_s^{j-2} \rangle = \delta_r^s\, {}^hv_r^{j-2}$$

where

$$
\begin{aligned}
{}^hv_r^{j-2} &= |b_r^{j-1}|^2 v_r^{j-1} + |b_{r-N/4}^{j-1}|^2 v_{r-N/4}^{j-1} && (4.4.1) \\
&= (\cos(2\pi r/N))^{2p} v_r^{j-1} + (\sin(2\pi r/N))^{2p} v_{r-N/4}^{j-1}, \\
{}^lv_r^{j-2} &= |a_r^{j-1}|^2 v_r^{j-1} + |a_{r-N/4}^{j-1}|^2 v_{r-N/4}^{j-1} && (4.4.2) \\
&= (\sin(2\pi r/N))^{2p} (v_{r-N/4}^{j-1})^2 v_r^{j-1} \\
&\quad + (\cos(2\pi r/N))^{2p} (v_r^{j-1})^2 v_{r-N/4}^{j-1} (v_r^{j-1})^2.
\end{aligned}
$$

Moreover, $\langle {}^lw_r^{j-2}, {}^hw_s^{j-2} \rangle = 0 \ \forall\, r, s.$

Corollary 4. *The splines* $\{{}^{l}w_r^{j-2}(x), {}^{h}w_r^{j-2}(x)\}_r^{j-2}$ *form an orthogonal basis of the space* ${}_pW^{j-1}$.

Proof: The relations $\langle {}^{i}w_r^{j-2}, {}^{k}w_s^{j-2} \rangle = 0$ if $r \neq s$, $i = l, h$, $k = l, h$ are readily apparent due to the orthogonality of the OS $\{w_r^{j-1}\}$ to each other. The formulas (4.4.1) and (4.4.2) are evident also. Let us examine the inner product

$$\langle {}^{h}w_r^{j-2}, {}^{l}w_r^{j-2} \rangle = \omega^{-2r}(b_r^{j-1} b_{r-N/4}^{j-1} v_r^{j-1} v_{r-N/r}^{j-1} - b_{r-N/4}^{j-1} b_r^{j-1} v_{r-N/4}^{j-1} v_r^{j-1})$$

$$= 0.$$

We have exploited the fact that $v_{r-N/4-N/4}^{j-1} = v_r^{j-1}$, just as b_r^{j-1}. ∎

These results enable us to decompose the wavelet space ${}_pW^{j-1}$ into an orthogonal sum of spaces. To be specific, define the space ${}^{l}W^{j-2} \subset {}_pW^{j-1}$ as ${}^{l}W^{j-2} =: \text{span}\{{}^{l}w_r^{j-2}(x)\}_r^{j-2}$ and the space ${}^{h}W^{j-2} \subset {}_pW^{j-1}$ as ${}^{h}W^{j-2} =: \text{span}\{{}^{h}w_r^{j-2}(x)\}_r^{j-2}$. It can be verified immediately that

$$_pW^{j-1} = {}^{l}W^{j-2} \oplus {}^{h}W^{j-2}.$$

It is reasonable to refer to the space ${}^{l}W^{j-2}$ as the low-frequency wavelet subspace and the space ${}^{h}W^{j-2}$ as the high-frequency wavelet subspace. The space ${}^{l}W^{j-2}$ is "concentrated" at the bands $[-\frac{3}{8N}, -\frac{N}{4}], [\frac{N}{4}, \frac{3}{8N}]$, whereas ${}^{h}W^{j-2}$ at $[-\frac{1}{2N}, -\frac{3}{8N}], [\frac{3}{8N}, \frac{1}{2N}]$.

If necessary we can decompose in a similar manner one (or both) of the subspaces ${}^{l}W^{j-2}, {}^{h}W^{j-2}$ into orthogonal sums of subspaces ${}^{ll}W^{j-3} \oplus {}^{hl}W^{j-3}$ and ${}^{lh}W^{j-3} \oplus {}^{hh}W^{j-3}$, respectively and to iterate this process.

Proposition 14 entails the following fact.

Proposition 15. *There hold the relations*

$$^{i}w_r^{j-2}(x + 4k/N) = \omega^{4kr} \cdot {}^{i}w_r^{j-2}(x), \quad i = l, h.$$

Similar formulas hold for ${}^{ik}w_r^{j-3}$, $i = l, h; k = l, h.$

To project a spline

$$W^{j-1}(x) = \sum_r^{j-1} \eta_r^{j-1} w_r^{j-1}(x) \in {}_pW^{j-1}$$

onto the spaces ${}^{l}W^{j-2}$ and ${}^{h}W^{j-2}$, one should act in a way similar to that used for establishing (4.2.3) and (4.3.4). So, we have

$$W^{j-1}(x) = {}^{l}W^{j-2}(x) \oplus {}^{h}W^{j-2}(x),$$

$$^h W^{j-2}(x) = \sum_r^{j-2} {}^h\eta_r^{j-2} \, {}^h w_r^{j-2}(x),$$

$$^h\eta_r^{j-2} = \frac{1}{{}^h v_r^{j-2}} \left(\eta_r^{j-1} v_r^{j-1} \overline{b}_r^{j-1} + \eta_{r-N/4}^{j-1} v_{r-N/4}^{j-1} \overline{b}_{r-N/4}^{j-1} \right),$$

$$^l W^{j-2}(x) = \sum_r^{j-2} {}^l\eta_r^{j-2} \, {}^l w_r^{j-2}(x),$$

$$^l\eta_r^{j-2} = \frac{1}{{}^l v_r^{j-2}} \left(\eta_r^{j-1} v_r^{j-1} \, {}^l\overline{a}_r^{j-2} + \eta_{r-N/4}^{j-1} v_{r-N/4}^{j-1} \, {}^l\overline{a}_{r-N/4}^{j-2} \right).$$

4.5 Reconstruction of a spline

Let a spline $_pS^j(x)$ be given in the decomposed form. It is required to reconstruct it in the conventional form suited for computation. To be specific, suppose we have two splines

$$S^{j-1}(x) = \sum_r^{j-1} m_r^{j-1}(x)\xi_r^{j-1} \in {}_p\mathcal{V}^{j-1},$$

$$\mathcal{W}^{j-1}(x) = \sum_r^{j-1} w_r^{j-1}(x)\eta_r^{j-1} \in {}_p\mathcal{W}^{j-1}.$$

Let $S^j(x) = S^{j-1}(x) \oplus \mathcal{W}^{j-1}(x)$. We are able to come up with the following assertion.

Theorem 6. *There hold the relations*

$$S^j(x) = \frac{1}{N} \sum_k^j q_k^j M^j(x - k/N) = \sum_r^j \xi_r^j \, {}_p m_r^j(x), \qquad (4.5.1)$$

$$\xi_r^j = b_r^j \xi_r^{j-1} + a_r^j \eta_r^{j-1}, \qquad q_k^j = \sum_r^j \omega^{kr} \xi_r^j. \qquad (4.5.2)$$

Proof: Due to (4.2.1), we can write

$$S^{j-1}(x) = \sum_r^{j-1} \xi_r^{j-1} \left(b_r^j m_r^j(x) + b_{r-N/2} m_{r-N/2}^j \right) = \sum_r^j \xi_r^{j-1} b_r^j m_r^j(x).$$

Similarly, (4.3.1) entails

$$W^{j-1}(x) = \sum_r^j \eta_r^{j-1} a_r^j m_r^j(x).$$

These two relations imply (4.5.1) and (4.5.2). ∎

By this means, given the representation of a spline in the form (4.1.2), it is possible to reconstruct it into the conventional form (4.5.1) in line with the diagram (4.1.3).

The algorithm suggested allows a fast implementation.

Remark 6. To compute values and display graphically the spline

$$\mathcal{W}^{j-1}(x) = \sum_{r}^{j-1} w_r^{j-1}(x)\eta_r^{j-1}$$

one may carry out the suggested reconstruction procedure assuming $\xi_r^{j-1} = 0$.

§5   Wavelets and multichannel processing of a signal

In Section 4 we carried out decomposition of a spline belonging to $_p\mathcal{V}^j$ into the set of its projections onto the subspaces $_p\mathcal{V}^{j-m}$, $_p\mathcal{W}^{j-m}$. Now we are going to process the spline in these subspaces. To do it we need relevant bases of the subspaces $_p\mathcal{V}^{j-m}$, $_p\mathcal{W}^{j-m}$. We start with the space $_p\mathcal{V}^j$.

5.1   Father wavelets

Definition 1. *A spline* $^s\varphi^j(x) \in {_p\mathcal{V}^j}$ *will be referred to as the father wavelet (FW) if its shifts* $^s\varphi^j(x - k/2^j)$, $k = 0, 1, ..., 2^j - 1$ *form a basis of the space* $_p\mathcal{V}^j$. *Two FW are said to be the dual ones if* $\langle\, ^s\varphi^j(\cdot - k/2^j),\, ^\sigma\varphi^j(\cdot - l/2^j)\rangle = \delta_k^l$.

We establish conditions to a spline to be the FW and to two FWs to be the dual ones.

Theorem 7. *A spline*

$$^s\varphi^j(x) = 2^{-j/2}\sum_r^j {}^s\rho_r^j\, m_r^j(x) \tag{5.1.1}$$

is the FW if and only if $^s\rho_r^j \neq 0\ \forall\ r$. *Two FWs are dual to each other if and only if*

$$^s\rho_r^j\, {}^\sigma\bar{\rho}_r^j\, {}_{2p}u_r^j = 1. \tag{5.1.2}$$

Proof: Let a spline $^s\varphi^j(x)$ be written as in (5.1.1). Due to (3.2.8), we have

$$^s\varphi^j(x - k/2^j) = 2^{-j/2} \sum_r^j {}^s\rho_r^j \omega^{-kr} m_r^j(x). \qquad (5.1.3)$$

Hence, it follows

$$^s\rho_r^j m_r^j(x) = 2^{-j/2} \sum_k^j \omega^{kr} {}^s\varphi^j(x - k/2^j).$$

These two relations imply the first assertion. Indeed, if some of $\{\rho_r^j\}$ are zero, then the dimension of the span$\{{}^s\varphi^j(x - k/2^j)\}$ is less than 2^j; if all of $\{\rho_r^j\}$ are nonzero, then all of m_r^j belong to the span. To establish the second assertion, write the inner product keeping in mind (5.1.3):

$$\langle {}^s\varphi^j(\cdot - k/2^j),\ {}^\sigma\varphi^j(\cdot - l/2^j) \rangle = 2^{-j} \sum_r^j {}^s\rho_r^j\ {}^\sigma\overline{\rho}_r^j\ {}_{2p}u_r^j\ \omega^{(l-k)r}.$$

The latter sum is equal to δ_k^l if and only if (5.1.2) holds. ■

The following assertion relates the coordinates of a spline with respect to a FW basis with those in the OS one.

Theorem 8. *Let*

$$^s\varphi^j(x) = 2^{-j/2} \sum_r^j {}^s\rho_r^j m_r^j(x)$$

be a FW, and a spline $S^j(x)$ is expanded with respect to the two bases

$$S^j(x) = \sum_k^j {}^sq_k^j\ {}^s\varphi^j(x - k/N) = \sum_r^j \xi_r^j m_r^j(x).$$

Then

$$^sq_k^j = 2^{-j/2} \sum_r^j \omega^{rk} \xi_r^j / {}^s\rho_r^j, \qquad \xi_r^j = 2^{-j/2}\ {}^s\rho_r^j \sum_k^j {}^sq_k^j \omega^{-rk}. \qquad (5.1.4)$$

Proof: Let us employ (5.1.3) once more:

$$S^j(x) = \sum_k {}^s q_k^j \, {}^s \varphi^j(x - k/N)$$

$$= \sum_k {}^s q_k^j \, 2^{-j/2} \sum_r {}^s \rho_r^j \, \omega^{-rk} \, m_r^j(x)$$

$$= \sum_r m_r^j \, {}^s \rho_r^j \, 2^{-j/2} \sum_k \omega^{-rk} \, {}^s q_k^j.$$

Hence,

$$\xi_r^j = 2^{-j/2} \, {}^s \rho_r^j \sum_k {}^s q_k^j \, \omega^{-rk}.$$

The second relation of (5.1.4) can be obtained immediately by means of DFT. ■

Proposition 16. *If FW $^\sigma \varphi^j$ is dual to FW $^s \varphi^j$ then*

$$^s q_k^j = \langle S^j, \, ^\sigma \varphi^j(\cdot - k/2^j)\rangle.$$

Remark 7. Eq. (5.1.4) implies that to make the change from a FW basis to the OS one or the reciprocal change, one has to carry out DFT. Of course, it should employ a FFT algorithm.

We present some examples of FWs.

Examples

1. *B-spline.* Suppose $^1 \rho_r^j = 1$. Then immediately from (3.2.5), we can derive that $^1 \varphi^j(x) = 2^{-j/2} M^j(x)$.

2. *FW dual to $^1 \varphi^j(x)$.* Suppose $^2 \rho_r^j = 1/{}_{2p} u_r^j$. Then, in accordance with (5.1.2), the FW $^2 \varphi^j(x)$ is dual to $^1 \varphi^j(x)$.

   Emphasize that if $S^j(x) = \sum_k^j {}^2 q_k^j \, {}^2 \varphi^j(x - k/N)$ then

$$^2 q_k^j = 2^{-j/2} \int_0^{p/N} S^j(x - k/N) M^j(x) \, dx.$$

   Provided $S^j(x) = S^j(f, x)$ is an orthogonal projection of a function f onto the spline space $_p \mathcal{V}^j$, we have

$$^2 q_k^j = 2^{-j/2} \int_0^{p/N} f(x - k/N) M^j(x) \, dx.$$

3. Setting $^3\rho_r^j = (\,_{2p}u_r^j)^{-1/2}$ we obtain the *self-dual FW* $^3\varphi^j(x)$ whose shifts form an orthonormal basis of $_p\mathcal{V}^j$ [1, 10].

4. *Interpolating FW.* If we set $^4\rho_r^j = 1/\,_pu_r^j$ then FW $^1\varphi^j(x) = 2^{-j/2}\,_pL^j(x)$. $_pL^j(x)$ is the so-called fundamental spline, namely

$$_pL^j((k+p/2)/N) = \begin{cases} 1 & \text{if } k = 0 \\ 0 & \text{if } k = 1, ..., N-1 . \end{cases}$$

Therefore, the spline

$$S^j(x) = \sum_k^j z_k \,_pL^j(x - k/N)$$

interpolates the vector $\{z_k\}_k^j$. To be specific, $S^j(k/N + p/2N) = z_k \forall\ k$.

5.2 Mother wavelets

We present here a family of bases of the space $_p\mathcal{W}^{j-1}$. The contents of this section are related to that of Subsection 5.1 where we introduced FWs.

Definition 2. A spline $^s\psi^{j-1}(x) \in\ _p\mathcal{W}^{j-1}$ will be referred to as the *mother wavelet (MW)* if its shifts $^s\psi^{j-1}(x - k/2^{j-1})$, $k = 0, 1, ..., 2^{j-1} - 1$ form a basis of the space $_p\mathcal{W}^{j-1}$. Two MWs are said to be the *dual* ones if

$$\left\langle\, ^s\psi^{j-1}(\cdot - k/2^{j-1}),\ ^\sigma\psi^{j-1}(\cdot - l/2^{j-1})\right\rangle = \delta_k^l.$$

We will establish conditions to a spline to be MW and to two MWs to be dual ones.

Theorem 9. *A spline*

$$^s\psi^{j-1}(x) = 2^{(1-j)/2}\sum_r^{j-1}\, ^s\tau_r^{j-1}\, w_r^{j-1}(x) \tag{5.2.1}$$

is a MW if and only if $^s\tau_r^{j-1} \neq 0\ \forall\ r$. *Two MWs are dual to each other if and only if*

$$^s\tau_r^{j-1}\,\overline{^\sigma\tau_r^{j-1}}\, v_r^{j-1} = 1\,\forall\ r. \tag{5.2.2}$$

Proof: Let a spline $^s\psi^{j-1}(x)$ be written as in (5.2.1). Due to (4.3.3), we have

$$^s\psi^{j-1}(x - k/2^{j-1}) = 2^{(1-j)/2} \sum_{r}^{j-1} {}^sT_r^{j-1}\omega^{-2kr}w_r^{j-1}(x).$$

Hence, it follows

$$^sT_r^{j-1}w_r^{j-1} = 2^{(1-j)/2} \sum_{k}^{j-1} \omega^{2kr}\,{}^s\psi(x - k/2^{j-1}).$$

These two relations imply the first assertion of the theorem. To establish the second assertion, we write the inner product

$$\langle\,{}^s\psi^{j-1}(\cdot - k/2^{j-1}),\,{}^\sigma\psi^{j-1}(\cdot - l/2^{j-1})\rangle = 2^{1-j}\sum_{r}^{j-1} {}^sT_r^{j-1}\overline{{}^\sigma T_r^{j-1}}v_r^{j-1}\omega^{2(l-k)r}$$

$$= \delta_l^k$$

provided (5.2.2) holds. ∎

The following assertion relates the coordinates of a spline with respect to a MW basis with those in the OW one.

Theorem 10. *Let*

$$^s\psi^{j-1}(x) = 2^{(1-j)/2}\sum_{r}^{j-1} {}^sT_r^{j-1}w_r^{j-1}(x) \tag{5.2.3}$$

be a MW and a spline $W^{j-1}(x) \in {}_p\mathcal{W}^{j-1}$ is expanded with respect to the two bases

$$W^{j-1}(x) = \sum_{k}^{j-1} {}^sp_k^{j-1}\,{}^s\psi^{j-1}(x - 2k/N) = \sum_{r}^{j-1} \eta_r^{j-1}w_r^{j-1}(x). \tag{5.2.4}$$

Then

$$^sp_k^{j-1} = 2^{(1-j)/2}\sum_{r}^{j-1} \omega^{2rk}\eta_r^{j-1}/{}^2T_r^{j-1}, \tag{5.2.5}$$

$$\eta_r^{j-1} = {}^sT_r^{j-1}2^{(1-j)/2}\sum_{k}^{j-1} {}^sp_k^{j-1}\omega^{-2rk}. \tag{5.2.6}$$

Proof: Substituting (5.2.3) into (5.2.4), we obtain in view of (4.3.3)

$$
\begin{aligned}
W^{j-1}(x) &= \sum_k^{j-1} {}^sp_k^{j-1}\, 2^{(1-j)/2} \sum_r^{j-1} \omega^{-2kr}\, {}^s\tau_r^{j-1} w_r^{j-1}(x) \\
&= \sum_r^{j-1} w_r^{j-1}(x) \cdot {}^s\tau_r^{j-1}\, 2^{(1-j)/2} \sum_k^{j-1} {}^sp_k^{j-1}\, \omega^{-2kr}.
\end{aligned}
$$

This implies (5.2.6). Carrying out DFT, we hence derive (5.2.5). ∎

Remark 8. If MW ${}^\sigma\psi^{j-1}$ is dual to MW ${}^s\psi^{j-1}$, then for any spline $W^{j-1}(x)$ as given in (5.2.4)

$$
{}^sp_k^{j-1} = \langle W^{j-1}, {}^\sigma\psi^{j-1}(\cdot - k/2^{j-1})\rangle.
$$

Provided the spline $W^{j-1}(x)$ is an orthogonal projection of a spline $S^j(x)$ onto ${}_p\mathcal{W}^{j-1}$,

$$
{}^sp_k^{j-1} = \langle S^j, {}^\sigma\psi^{j-1}(\cdot - k/2^{j-1})\rangle.
$$

Remark 9. Theorem 10 implies that to make the change from a MW basis to the OW one or the reciprocal change, one has to carry out DFT.

We present some examples of MW.

Examples

1. *B-wavelet.* Suppose ${}^1\tau_r^{j-1} = 1\ \forall\ r$. The determining feature of wavelet ${}^1\psi^{j-1}(x)$ is the compactness (up to periodization) of its support. To be precise, $\mathrm{supp}\,{}^1\psi^{j-1}(x) \subseteq ((-2p)/N, (2p-2)/N)(\mathrm{mod}1)$. The wavelet $\psi^{j-1}(x) = 2^{(-1+j)/2} \cdot {}^1\psi^{j-1}(x)$ is a periodization of the B-wavelet invented by Chui and Wang [3].

2. *MW dual to* ${}^1\psi^{j-1}(x)$. Suppose ${}^2\tau_r^{j-1} = 1/v_r^{j-1}$. Then, in accordance with (5.2.2), the MW ${}^2\psi^{j-1}(x)$ is dual to ${}^1\psi^{j-1}(x)$.
   Emphasize that if $S^j(x) = S^{j-1}(x) \oplus W^{j-1}(x)$ and

$$
W^{j-1}(x) = \sum_k^{j-1} {}^2p_k^{j-1}\, {}^2\psi^{j-1}(x - 2k/N) \tag{5.2.7}
$$

   then,

$$
{}^2p_k^{j-1} = \int_{-2p/N}^{(2p-2)/N} S^j(x - 2k/N)\, {}^1\psi^{j-1}(x)\, dx.
$$

   Provided $S^j(x) = S^j(f, x)$ is an orthogonal projection of a function f onto the spline space ${}_p\mathcal{V}^j$, we have

$$
{}^2p_k^{j-1} = \int_{-2p/N}^{(2p-2)/N} f(x - 2k/N)\, {}^1\psi^{j-1}(x)\, dx. \tag{5.2.8}
$$

3. Setting $^3\tau_r^j = (v_r^{j-1})^{-1/2}$ we obtain the *self-dual MW* $^3\psi^{j-1}(x)$ whose shifts form an orthonormal basis of $_p\mathcal{W}^{j-1}$. This MW is a periodization of the Battle-Lemarié wavelet [9, 20].

4. *Cardinal MW.* If we set $^4\tau_r^{j-1} = 1/(u_r^j\, u_{r-N/2}^j)$ then we obtain MW $^4\psi^{j-1}(x) = 2^{(1-j)/2}\, _{2p}L^j(x + 1/N)^{(p)}$, where $_{2p}L^j(x)$ is the fundamental spline of the degree $2p - 1$ introduced in Section 7. MW $2^{(1-j)/2}\, ^4\psi^{j-1}(x)$ is a periodization of the cardinal wavelet suggested by Chui and Wang in [2].

5.3   Wavelet packets

Now we discuss briefly bases in the low- and high-frequency wavelet spaces $^l\mathcal{W}^{j-2}$ and $^h\mathcal{W}^{j-2}$.

Just as in previous sections, we can find splines whose shifts form bases of the subspaces $^l\mathcal{W}^{j-2}$ and $^h\mathcal{W}^{j-2}$. For example, a spline $^l_s\psi^{j-2}(x) \in {}^l\mathcal{W}^{j-2}$ will be referred to as the low-frequency MW (LMW) if its shifts $^l_s\psi^{j-2}(x - k/2^{j-2})$, $k = 0, 1, ..., 2^{j-2} - 1$ form a basis of the space. Two LMW are said to be the dual ones if

$$\langle\, ^l_s\psi^{j-2}(\cdot - k/2^{j-2}),\, ^l_\sigma\psi^{j-2}(\cdot - l/2^{j-2})\rangle = \delta_k^l.$$

Theorem 11. *A spline*

$$^l_s\psi^{j-2} = 2^{(2-j)/2}\sum_{r}^{j-2} {}^l_s\tau_r^{j-2}\, ^l_sw_r^{j-2}(x)$$

is the LMW if and only if $^l_s\tau_r^{j-2} \neq 0\ \forall\, r$. *Two MW are dual to each other if and only if*

$$^l_s\tau_r^{j-2}\,\overline{^l_\sigma\tau_r^{j-2}} \cdot {}^l v_r^{j-2} = 1\ \forall\, r.$$

There holds an assertion related to Theorem 8 and Theorem 10.

Setting $^l_1\tau_r^{j-2} = 1$, we obtain the LMW of minimal support, so to say, *B*-LMW.

Similar considerations can be conducted in the space $^h\mathcal{W}^{j-2}$. We are now able to construct a diversity of bases of the space $_p\mathcal{W}^{j-1}$ for refined frequency resolution of a certain signal f under processing. For example, one such base may be structured as follows:

$$\{^l_s\psi^{j-2}(x-k/2^{j-2})\}_k^{j-2},\ \{^{lh}_\sigma\psi^{j-3}(x-l/2^{j-3})\}_l^{j-3},\ \{^{hh}_\gamma\psi^{j-3}(x-\nu/2^{j-3})\}_\nu^{j-3}.$$

The MWs of type $\{^l_s\psi^{j-2}, ^{lh}_\sigma\psi^{j-3}, ^{hh}_\gamma\psi^{j-3}\}$ are called the *wavelet packets* (compare with [7]).

5.4 Digital processing a periodic signal by means of spline wavelets

We discuss here a scheme of processing a periodic signal $f(x)$ belonging to C^p. The commonly encountered situation is when the array of samples is available: $\vec{f^j} = \{f_k^j\} = f(p/2N + k/N)\}_k^j$. The goal of the processing is to transform the original data array into a more informative array. We will process the signal by means of spline wavelets of order p.

First we establish a quadrature formula. Denote

$$F_k^\nu = \int_0^{p/N} f(x - k/N) M^\nu(x) \, dx. \tag{5.4.1}$$

Theorem 12. *If $f \in C^p$ and $p < N/2$ then*

$$F_k^j = 2^{-j} \sum_l f((l + p/2 - k)/N) M^j((l + p/2)/N) + {}_pG^j,$$

where ${}_pG^j = O(N^{-p})$ as p is an even number and ${}_pG^j = o(N^{-p})$ as p is an odd number.

Proof: Without loss of generality, assume that $k = 0$. Provided $p < N/2$, inside the interval $[-1/2, 1/2]$ the periodic $\mathcal{B}$-spline ${}_pM^j(x)$ coincides with $\mathcal{B}$-spline ${}_pB^j(x)$. Therefore, the Proposition 1 is valid for ${}_pM^j$ as well as for the cardinal $\mathcal{B}$-splines ${}_pB_j$. Namely, $\forall\, t \in [0, 1]$

$$\frac{1}{N} \sum_l [(t + l - p/2)/N]^s M^j((t + l)/N) = \mu_s(t),$$

$$\mu_s(t) \;=\; M_s = \int_0^{p/N} (x - p/2N)^s M^j(x) \, dx \qquad \text{if } s < p,$$

$$\mu_p(t) \;=\; N^{-p}(-1)^{p-1}\beta_p(t) + M_p,$$

where $\beta_p(t)$ is the Bernoulli polynomial. If $f \in C^p$ we may write

$$
\begin{aligned}
F_0^j &= \int_0^{p/N} f(x) M^j(x) \, dx \\
&= \sum_{s=0}^p \frac{f^{(s)}(p/2N)}{s!} \int_0^{p/N} (x - p/2N)^s M^j(x) \, dx + o(N^{-p}) \\
&= \sum_{s=0}^p \frac{f^{(s)}(p/2N)}{s!} M_s + o(N^{-p}).
\end{aligned}
$$

Now distinguish two cases.

1. The number p is even. Then

$$\frac{1}{N}\sum_l^j f((l+p/2)/N)M^j((l+p/2)/N) = \frac{1}{N}\sum_l^j f(l/N)M^j(l/N)$$

$$= \frac{1}{N}\sum_l^j M^j(l/N)\sum_{s=0}^p \frac{f^{(s)}(p/2N)}{s!}((l-p/2)/N)^s + o(N^{-p})$$

$$= \sum_{s=0}^p \frac{f^{(s)}(p/2N)}{s!}\mu_s(0) + o(N^{-p})$$

$$= \sum_{s=0}^p \frac{f^{(s)}(p/2N)}{s!}M_s + \frac{N^{-p}\beta_p(0)}{p!}f^{(p)}(p/N) + o(N^{-p})$$

$$= F_0^j + \frac{N^{-p}\beta_p(0)}{p!}f^{(p)}(p/N) + o(N^{-p}).$$

2. The number p is odd. Then

$$\frac{1}{N}\sum_l^j f((l+p/2)/N)M^j((l+p/2)/N)$$

$$= \frac{1}{N}\sum_l^j f((l+1/2)/N)M^j((l+1/2)/N)$$

$$= \frac{1}{N}\sum_l^j M^j((l+1/2)/N)\sum_{s=0}^p \frac{f^{(s)}(p/2N)}{s!}((l+1/2-p/2)/N)^s$$
$$+o(N^{-p})$$

$$= \sum_{s=0}^p \frac{f^{(s)}(p/2N)}{s!}\mu_s(1/2) + o(N^{-p})$$

$$= \sum_{s=0}^p \frac{f^{(s)}(p/2N)}{s!}M_s + \frac{N^{-p}\beta_p(1/2)}{p!}f^{(p)}(p/N) + o(N^{-p})$$

$$= F_0^j + \frac{N^{-p}\beta_p(1/2)}{p!}f^{(p)}(p/N) + o(N^{-p}).$$

If p is an odd number then $\beta_p(1/2) = 0$. Hence, it follows that for the odd p

$$F_k^j = 2^{-j}\sum_k^j f((l+p/2-k)/N)M^j((l+p/2)/N) + o(N^{-p}). \quad \blacksquare$$

Corollary 5. *If* $f \in C^p$ *then*

$$T_r^j(\vec{F^j}) = T_r^j(\vec{f^j})_p u_r^j + {}_p g^j,$$

where $\vec{F^j} = \{F_k^j\}_k^j$ *and* ${}_p g^j = O(N^{-p})$ *as* p *is an even number and* ${}_p g^j = o(N^{-p})$ *as* p *is an odd number.*

The assertion becomes apparent if we note that the expression $2^{-j} \sum_k^j f((l + p/2 - k)/N) M^j((l + p/2)/N)$ is a discrete convolution.

Theorem 13. *Suppose* $f(x)$ *is a 1-periodic, integrable signal and* $\psi^\nu(x) \in \mathcal{W}^\nu$ *is the* $\mathcal{B}$-*wavelet. Let* F_k^ν *be defined as in (5.4.1) and*

$$\Phi_k^\nu =: \int_{-p/2^\nu}^{(p-1)/2^\nu} f(x - k/2^\nu)\psi^\nu(x)\, dx.$$

Then the following relations hold

$$T_r^{j-1}(\vec{F}^{j-1}) = T_r^j(\vec{F^j})\overline{b_r^j} + T_{r+N/2}^j \overline{b_{r+N/2}^j}, \qquad (5.4.2)$$

$$F_k^{j-1} = \sum_r^{j-1} \omega^{2kr} T_r^{j-1}(\vec{F}^{j-1}) = \sum_r^j \omega^{2kr} \overline{b_r^j} T_r^j(\vec{F_j}),$$

$$T_r^{j-1}(\vec{\Phi}^{j-1}) = T_r^j(\vec{F^j})\,\overline{a_r^j} + T_{r+N/2}^j(\vec{F^j})\,\overline{a_{r+N/2}^j}, \qquad (5.4.3)$$

$$\Phi_k^{j-1} = \sum_r^{j-1} \omega^{2kr} T_r^{j-1}(\vec{F}^{j-1}) = \sum_r^j \omega^{2kr}\, \overline{a_r^j}\, T_r^j(\vec{F^j}).$$

Proof: Let $R^\nu(x)$ be the FW dual to the $\mathcal{B}$-spline $M^\nu(x)$ and $\Psi^\nu(x)$ be the MW dual to the $\mathcal{B}$-wavelet $\psi^\nu(x)$. If the spline $S^j(f, x)$ is the orthogonal projection of a signal f onto the spline space ${}_p \mathcal{V}^j$ then

$$S^j(f, x) = \sum_k^j F_k^j\, R(x - k/N) = \sum_r^j \xi_r^j m_r^j(x),$$

$$\xi_r^j = \frac{1}{u_r^j} T_r^j(\vec{F^j}), \qquad T_r^j(\vec{F^j}) = \xi_r^j\, u_r^j.$$

Similarly, if projections of the signal f onto the spaces ${}_p \mathcal{V}^{j-1}$, ${}_p \mathcal{W}^{j-1}$ are:

$$S^{j-1}(f, x) = \sum_r^{j-1} \xi_r^{j-1} m_r^{j-1}(x),$$

$$W^{j-1}(f, x) = \sum_r^{j-1} \eta_r^{j-1} w_r^{j-1}(x),$$

then

$$T_r^{j-1}(\vec{F}^{j-1}) = \xi_r^{j-1}\, u_r^{j-1}, \qquad T_r^{j-1}(\vec{\Phi}^{j-1}) = \eta_r^{j-1}\, v_r^{j-1}.$$

Now we see that (5.4.2) and (5.4.3) are immediate consequences of (4.2.3) and (4.3.4), respectively. ∎

Remark 10. If $f \in C^p$ then it is natural to employ Corollary 5.

As a result of the first step of decomposition we have derived the set $\{\Phi_k^{j-1}\}_k^{j-1}$ from the array $\{F_k^j\}_k^j$. We stress that the value Φ_k^{j-1} carries information on the behavior of the signal f in the frequency strips $[-N/2, -N/4]$, $[N/4, N/2]$ and in the spatial interval $[\frac{2(k-p)}{N}, \frac{2(k+p-1)}{N}]$. By a similar means we acquire the values $\{\Phi_k^{j-\nu}\}_k^{j-\nu}$, $\nu = 2, ..., m$.

Remark 11. We have described transformation of the original array $\{f_k^j\}$ into the array $\mathcal{D}^j$ associated with $\mathcal{B}$-splines and $\mathcal{B}$-wavelets. The elements of this array usually appear as the most informative ones. However, for some special purposes, arrays associated with other FW-MW bases could be of use. The algorithms established in the paper allow performance of straightforward corresponding transformations as well as transformations to arrays allied with wavelet packets.

5.5 Reconstruction of a signal

We now dwell on the situation that is reciprocal to the situation considered in the previous section. We want to reconstruct a signal from the array $\mathcal{D}^j$. The case in point is an approximate reconstruction, of course.

Consider first a single step of the reconstruction.

Problem. *The arrays* $\{F_k^{j-1}\}$, $\{\Phi_k^{j-1}\}$ *are available, where*

$$F_k^{j-1} = \int_0^{p/2^{j-1}} f(x - k/2^{j-1}) M^{j-1}(x)\, dx,$$

$$\Phi_k^{j-1} = \int_{-p/2^{j-1}}^{p/2^{j-1}} f(x - k/2^{j-1}) \psi^{j-1}(x)\, dx,$$

and $f(x)$ *is any 1-periodic integrable signal. The coefficients* $\{q_k^j\}_k^j$ *are wanted of the spline*

$$S^j(f, x) = 2^{-j} \sum_k^j q_k^j M^j(x - k/2^j) \tag{5.5.1}$$

which is an orthogonal projection of the signal f *onto the space* $_p\mathcal{V}^j$.

Emphasize that as written in (5.5.1), a spline $S^j(f, x)$ can be computed and, if need be, displayed graphically immediately.

Solution to the Problem

Carrying out the fast Fourier transform, we obtain the arrays $\{T_r^{j-1}(\vec{F}^{j-1})\}$ and $\{T_r^{j-1}(\vec{\Phi}^{j-1})\}$. Then, using the line of reasoning similar to that of Subsection 5.4, we can maintain that, if the splines

$$S^{j-1}(f,x) = \sum_k^{j-1} \xi_{Sr}^{j-1} m_r^{j-1},$$

$$W^{j-1}(f,x) = \sum_r^{j-1} \eta_r^{j-1} w_r^{j-1}(x),$$

are orthogonal projections of the signal f onto the spaces $_p\mathcal{V}^{j-1}$, $_p\mathcal{W}^{j-1}$ correspondingly, then

$$\xi_r^{j-1} = T_r^{j-1}(\vec{F}^{j-1})/u_r^{j-1},$$
$$\eta_r^{j-1} = T_r^{j-1}(\vec{\Phi}^{j-1})/v_r^{j-1}.$$

Now Theorem 6 enables us to write desired coefficients

$$q_k^j = \sum_r^j \omega^{kr}(b_r^j \xi_r^{j-1} + a_r^j \eta_r^{j-1}).$$

In this manner, given the arrays $\{\Phi_k^{j-\nu}\}$, $\nu = 1, ..., m$, $\{F_k^{j-m}\}$ we are able to reconstruct the spline $S^j(f,x)$, which is an orthogonal projection of the signal f onto $_p\mathcal{V}^j$. In a similar way, the spline $S^j(f,x)$ can be reconstructed when arrays associated with wavelet packets are available.

In conclusion, we point out that the algorithms we have suggested can be straightforwardly extended to the multidimensional case.

Acknowledgments. I thank the organizers of the Neaman Workshop, especially Prof. J. Zeevi and Mrs. D. Maoz, for the invitation and the support of my participation in the Workshop. I am indebted to Professor C. K. Chui for useful discussions.

This work has been supported by the Russian Foundation for Basic Research under research grant No. 93-012-49.

References

[1] Battle, G., A block spin construction of ondelettes. Part I. Lemari'e functions, *Comm. Math. Phys.* **110** (1987), 601–615.

[2] Chui, C. K. and J. Z. Wang, A cardinal spline approach to wavelets, *Proc. Amer. Math. Soc.* **113** (1991), 785–793.

[3] Chui, C. K. and J. Z. Wang, On compactly supported spline wavelets and a duality principle, *Trans. Amer. Math. Soc.* **330** (1992), 903–915.

[4] Chui, C. K. and J. Z. Wang, A general framework of compactly supported splines and wavelets, *J. Appr. Th.* **71** (1992), 263–304.

[5] Chui, C. K., *An Introduction to Wavelets*, Academic Press, Boston, 1992.

[6] Chui, C. K. and J. Z. Wang, Computational and algorithmic aspects of cardinal spline-wavelets, *Appr. Th. Appl.* **9, no. 1** (1993).

[7] Chui, C. K. and Chun Li, Nonorthogonal wavelet packets, *SIAM J. Math. Anal.* **24** (1993), 712–738.

[8] Cohen, A., I. Daubechies, and J.-C. Feauveau, Biorthogonal bases of compactly supported wavelets, *Comm. Pure Appl. Math.* **45** (1992), 485–560.

[9] Daubechies, I., Orthonormal bases of compactly supported wavelets, *Comm. Pure Appl. Math.* **41** (1988), 909–996.

[10] Lemarié, P. G., Ondelettes à localization exponentielle, *J. de Math. Pure et Appl.* **67** (1988), 227–236.

[11] Meyer, Y., Ondelettes, fonctions splines et analyses graduées, Rapport Ceremade No.8703, Université Paris Dauphin, 1987.

[12] Meyer, Y., *Wavelets and Applications*, SIAM, Philadelphia, 1993.

[13] Schoenberg, I. J., Contribution to the problem of approximation of equidistant data by analytic functions, *Quart.Appl. Math.* **4** (1946), 45–99, 112–141.

[14] Schoenberg, I. J., *Cardinal Spline Interpolation*, CBMS #12, SIAM, Philadelphia, 1973.

[15] Subbotin, Yu. N., On the relation between finite differences and the corresponding derivatives, *Proc. Steklov Inst. Math.* **78** (1965), 24–42.

[16] Zheludev, V. A., Asymptotic formulas for local spline approximation on a uniform mesh., *Soviet. Math. Dokl.* **27** (1983), 415–419.

[17] Zheludev, V. A., Local spline approximation on a uniform grid, *Comput. Math. and Math. Phys.* **5** (1989).

[18] Zheludev, V. A., An Operational calculus connected with periodic splines, *Soviet. Math. Dokl.* **42** (1991), 162–167.

[19] Zheludev, V. A., Spline-operational calculus and inverse problem for heat equation, in *Colloq. Math. Soc. J. Bolyai*, **58**, *Approximation Theory*, J. Szabados, K. Tandoi (eds.), 1991, 763–783.

[20] Zheludev, V. A., Periodic splines and the fast Fourier transform, *Comput. Math. and Math. Phys.* **32** (1991), 149–165.

[21] Zheludev, V. A., Spline-operational calculus and numerical solving convolution integral equations of the first kind, *Differ. Equations* **28** (1992), 269–280.

[22] Zheludev, V. A., Wavelets based on periodic splines, *Russian Acad. Sci. Doklady. Mathematics* **49** (1994), 216–222.

[23] Zheludev, V. A., Spline harmonic analysis and wavelet bases, in *Mathematics of Computation 1943-1993: A Half-Century of Computational Mathematics Proc. Sympos. Appl. Math.* **48**, W. Gautcshi (ed.), Amer. Math. Soc., Providence, RI, 1994, pp. 415–419.

[24] Zheludev, V. A., Periodic splines and wavelets, in *Contemporary Mathematics,* **190**, *Mathematical Analysis, Wavelets and Signal Processing*, M. E. H. Ismail, M. Z. Nashed, A. I. Zayed, A. F. Ghaleb (eds.), Amer. Math. Soc., Providence, 1995, pp. 339–354.

Valery A. Zheludev
School of Mathematical Sciences
Tel Aviv University
69978 Tel Aviv, Israel
zhel@math.tau.ac.il

VI.
Filter Banks and Image Coding

A Density Theorem for Time-Continuous Filter Banks

A. J. E. M. Janssen

Abstract. We show in this paper the following result. When $a > 0$ and $g_{nm} = g_m(\cdot - na)$, $n, m \in \mathbf{Z}$, is a frame for $L^2(\mathbb{R})$, where each $g_m \in L^2(\mathbb{R})$ is localized in the frequency-domain around a point $b_m \in \mathbb{R}$, then $(a\bar{b})^{-1} \geq 1$. Here $\bar{b}^{-1}$ is the asymptotic lower bound of the average number of b_m in a symmetric interval around 0 as the interval length tends to ∞. As a particular case it is found that a multi-window Gabor-type system $g^p(t - na) \exp(2\pi imbt)$ with $n, m \in \mathbf{Z}$, $p = 0, ..., P - 1$, can generate a frame for $L^2(\mathbb{R})$ only if $P(ab)^{-1} \geq 1$. The main result of this paper is based on the Ron-Shen criterion in the frequency-domain for duality of two shift-invariant systems of functions, combined with an amplification of Janssen's elementary proof of the fact that a Weyl-Heisenberg system $g(t - na) \exp(2\pi imbt)$ can generate a frame for $L^2(\mathbb{R})$ only if $(ab)^{-1} \geq 1$.

§1 Introduction and results

We consider in this paper function systems of the form

$$g_{nm} = g_m(\cdot - na) \,, \qquad n, m \in \mathbf{Z} \,, \tag{1.1}$$

where $a > 0$ and $g_m \in L^2(\mathbb{R})$, $m \in \mathbf{Z}$. Systems as in (1.1) generate, what is called in [6], shift-invariant spaces. One can also think of the g_m as impulse responses of synthesis filters of a filter bank for time-continuous, finite-energy signals. In either view, one is interested in representing arbitrary finite-energy signals f in the form

$$if = \sum_{n,m} a_{nm} \, g_{nm} \,, \tag{1.2}$$

where the coefficients a_{nm} have the form

$$a_{nm} = (f, \gamma_{nm}) \,; \qquad \gamma_{nm} = \gamma_m(\cdot - na) \,, \qquad n, m \in \mathbf{Z} \,, \tag{1.3}$$

Signal and Image Representation in Combined Spaces
Y. Y. Zeevi and R. R. Coifman (Eds.), pp. 513–523.

with $\gamma_m \in L^2(\mathbb{R})$, $m \in \mathbb{Z}$. In the filter bank approach, one would interpret the γ_m as impulse responses of the analysis filters of the bank.

Representations (1.2) of arbitrary $f \in L^2(\mathbb{R})$ are possible in a stable manner when g_{nm}, $n, m \in \mathbb{Z}$, is a frame. By this we mean that there are numbers $C_g > 0$, $D_g < \infty$, the lower and upper frame bound of g_{nm}, $n, m \in \mathbb{Z}$, such that

$$C_g \|f\|^2 \le \sum_{n,m} |(f, g_{nm})|^2 \le D_g \|f\|^2, \qquad f \in L^2(\mathbb{R}). \qquad (1.4)$$

Equivalently, the frame operator T_g, defined by

$$T_g f = \sum_{n,m} (f, g_{nm}) g_{nm}, \qquad f \in L^2(\mathbb{R}), \qquad (1.5)$$

is a bounded, positive definite operator with

$$C_g I \le T_g \le D_g I, \qquad (1.6)$$

where I is the identity operator of $L^2(\mathbb{R})$. When g_{nm}, $n, m \in \mathbb{Z}$, is a frame with frame bounds $C_g > 0$, $D_g < \infty$, then so is the system $^\circ\gamma_{nm}$, $n, m \in \mathbb{Z}$, where

$$^\circ\gamma_m = T_g^{-1} g_m, \qquad m \in \mathbb{Z}, \qquad (1.7)$$

with frame bounds D_g^{-1}, C_g^{-1}. And any $f \in L^2(\mathbb{R})$ has then the $L^2(\mathbb{R})$-convergent expansion

$$f = \sum_{n,m} (f, {}^\circ\gamma_{nm}) g_{nm}. \qquad (1.8)$$

This expansion is minimal in the sense that for all $\underline{\alpha} \in l^2(\mathbb{Z}^2)$ with

$$f = \sum_{n,m} \alpha_{nm} g_{nm} \qquad (1.9)$$

there holds

$$\sum_{n,m} |(f, {}^\circ\gamma_{nm})|^2 \le \sum_{n,m} |\alpha_{nm}|^2. \qquad (1.10)$$

As an application of (1.10), in which one considers the "trivial" expansion $g_{m'} = \sum_{n,m} \delta_{no} \delta_{mm'} g_{nm}$, one has that

$$|(g_{m'}, {}^\circ\gamma_{m'})| \le 1, \qquad m' \in \mathbb{Z}. \qquad (1.11)$$

We refer for generalities about frames to [3, Sec. 3.2] and [2, Sec. II], and about time-continuous filter banks (shift-invariant systems) to [6, Sec. 1.3].

The question thus arises when the system g_{nm}, $n, m \in \mathbf{Z}$, is a frame for $L^2(\mathbb{R})$. This difficult question has not been answered completely yet, not even in the case of Weyl-Heisenberg systems where one has g_m of the form

$$g_m(t) = e^{2\pi i m b t} g(t) , \qquad m \in \mathbf{Z} , \qquad (1.12)$$

with $b > 0$. For the latter case, a necessary condition that g_{nm}, $n, m \in \mathbf{Z}$, is a frame is that

$$\frac{1}{ab} \geq 1 , \qquad (1.13)$$

see [3, Sec. 4.1] and [2, p. 978], for historical comments and proofs concerning this result. An elementary proof of the result was found by Janssen in [4]; it uses what is now well known as the Wexler-Raz biorthogonality condition [7]. That is, two Weyl-Heisenberg systems g_{nm}, $n, m \in \mathbf{Z}$, and γ_{nm}, $n, m \in \mathbf{Z}$, both having a finite upper frame bound, are dual for the parameters a, b if and only if they are biorthogonal for the parameters b^{-1}, a^{-1}. Here duality means that any $f \in L^2(\mathbb{R})$ has the $L^2(\mathbb{R})$-convergent expansion

$$f = \sum_{n,m} (f, \gamma_{nm}) \, g_{nm} , \qquad (1.14)$$

and biorthogonality means that

$$\int\limits_{-\infty}^{\infty} e^{2\pi i l t / a} \, g(t - k/b) \, \gamma^*(t) \, dt = ab \, \delta_{ko} \, \delta_{lo} , \qquad k, l \in \mathbf{Z} , \qquad (1.15)$$

see [4, Prop. A and Sec. 2] for the formulation and proof of the Wexler-Raz result as used in the present paper. Applying this Wexler-Raz condition with $k = l = 0$ together with (1.11), $m' = 0$, immediately yields $ab \leq 1$, i.e. (1.13).

It is the purpose of the present paper to extend the result (1.13) to more general systems as in (1.1) with g_m that are localized in the frequency-domain as follows.

Theorem 1. *Assume that $g_{nm} = g_m(\cdot - na)$, $n, m \in \mathbf{Z}$, is a frame for $L^2(\mathbb{R})$, that there is a $\delta > 0$ such that $\|g_m\|^2 > \delta$, $m \in \mathbf{Z}$, and that there is a $g \in L^2(\mathbb{R})$ and numbers $b_m \in \mathbb{R}$, $m \in \mathbf{Z}$, such that*

$$|\hat{g}_m(\nu)| \leq |\hat{g}(\nu - b_m)| , \qquad m \in \mathbf{Z}, \quad a.e. \ \nu \in \mathbb{R} . \qquad (1.16)$$

Here $\hat{h}$ denotes the Fourier transform

$$\hat{h}(\nu) = \int\limits_{-\infty}^{\infty} e^{-2\pi i \nu t} \, h(t) \, dt , \qquad a.e. \ \nu \in \mathbb{R} , \qquad (1.17)$$

of $h \in L^2(\mathbb{R})$. Then there holds

$$\frac{1}{ab} \geq 1 , \tag{1.18}$$

where

$$\bar{b}^{-1} = \lim_{W \to \infty} \inf \frac{1}{2W} \#_m(b_m \in [-W, W]) , \tag{1.19}$$

the asymptotic lower bound of the density of the b_m's.

A simple application of Theorem 1 yields the following result.

Theorem 2. *Assume that we have* $g_m^p \in L^2(\mathbb{R})$, $m \in \mathbb{Z}$, $p = 0, ..., P-1$, *and that the system*

$$g_m^p(\cdot - na^p) , \qquad n, m \in \mathbb{Z}, \quad p = 0, ..., P-1 , \tag{1.20}$$

is a frame for $L^2(\mathbb{R})$. *Here the* $a^p > 0$ *are integer multiples of an* $\alpha > 0$,

$$a^p = N^p \alpha ; \qquad N^p \in \mathbb{N}, \quad p = 0, ..., P-1 . \tag{1.21}$$

Assume furthermore that there is a $\delta > 0$ *such that* $\|g_m^p\|^2 > \delta$, $m \in \mathbb{Z}$, $p = 0, ..., P-1$ *and that there is a* $g \in L^2(\mathbb{R})$ *and numbers* $b_m^p \in \mathbb{R}$, $m \in \mathbb{Z}$, $p = 0, ..., P-1$, *such that*

$$|\hat{g}_m^p(\nu)| \leq |\hat{g}(\nu - b_m^p)| , \qquad m \in \mathbb{Z}, \quad p = 0, ..., P-1 , \quad a.e. \ \nu \in \mathbb{R} . \tag{1.22}$$

Then there holds

$$\lim_{W \to \infty} \inf \sum_{p=0}^{P-1} \frac{1}{a^p \, \bar{b}^p(W)} \geq 1 , \tag{1.23}$$

where

$$\left(\bar{b}^p(W)\right)^{-1} = \frac{1}{2W} \#_m(b_m^p \in [-W, W]) , \qquad p = 0, ..., P-1 , \quad W > 0 . \tag{1.24}$$

As a particular case of Theorem 2 we have that multi-window Gabor-systems (with a^p as in (1.21))

$$g_{nm}^p = e^{2\pi i m b^p t} \, g^p(t - na^p) , \qquad n, m \in \mathbb{Z}, \quad p = 0, ..., P-1 , \tag{1.25}$$

as they were considered in [8, pp. 7,23,24], can only be a frame for $L^2(\mathbb{R})$ when

$$\sum_{p=0}^{P-1} \frac{1}{a^p \, b^p} \geq 1 . \tag{1.26}$$

Indeed, this follows from Theorem 2 by taking $b_m^p = m\,b^p$, $\delta = \min\limits_p \|g^p\|^2$, $\hat{g} = \max\limits_p |\hat{g}^p|$.

We also note that an inequality (1.26) has been given in [1] for the harmonic means of lattice areas associated with complete sets of time-frequency translates of the Gaussian $\exp(-\pi t^2)$. In [1] there is, however, no restriction to rectangular lattices, nor do the lattice parameters have to be rationally related.

As a further comment we note that Theorem 1 (and, accordingly, Theorem 2) also holds when the inequality (1.16) is replaced by

$$|\hat{g}_m(\nu)| \leq |\hat{g}(\nu - b_m)| + |\hat{g}(\nu + b_m)|\,, \qquad m \in \mathbf{Z}, \ \text{a.e. } \nu \in \mathbb{R}\,. \qquad (1.27)$$

The proof of this result requires only modest alterations of the proof of Theorem 1 as given in the next section. Hence, our results also apply for function systems of the Wilson type and of the cosine-modulated type.

We present the proofs of the two theorems in the next section. The proof of Theorem 1 consists of amplifying the elementary argument used in [4] for proving (1.13). For this we need a substitute for the Wexler-Raz biorthogonality condition (1.15) for Weyl-Heisenberg systems, together with inequality (1.11) and some hard but elementary analysis. A substitute, adequate for the present purposes, for the Wexler-Raz condition for general systems (1.1) was found by Ron and Shen in [6, Sec. 4]. This is also presented in the next section, together with some further elaborations of it found by the author in [5, Appendix A].

§2   Proof of the main results

We first present a result that gives, in the frequency-domain, a criterion for a system (1.1) to be a frame together with a criterion for two systems (1.1) to be dual and a method for computing minimal duals $°\gamma_m$, $m \in \mathbf{Z}$, for a system (1.1) that is a frame. The first two parts of the result below are due to Ron and Shen and can be found in [6], also see [5, Appendix A]; the third part is due to the author and can be found in [5, Appendix A]. This third part is not strictly required for the proof of the main result, but we have included it since we have not been able to find such a result in this general form in the existing literature.

Proposition 1. *Assume that* $g_{nm} = g_m(\cdot - na)$, $n, m \in \mathbf{Z}$, *has a finite upper frame bound* D_g. *Then*

$$\sum_m |\hat{g}_m(\nu)|^2 \leq a\,D_g\,, \qquad a.e. \ \nu \in \mathbb{R}\,, \qquad (2.1)$$

and the matrix

$$H_g(\nu) := (\hat{g}_m(\nu - k/a))_{k \in \mathbf{Z},\, m \in \mathbf{Z}} \qquad (2.2)$$

defines for a.e. $\nu \in \mathbb{R}$ a bounded linear operator of $l^2(\mathbf{Z})$ with operator norm $\leq (a\,D_g)^{1/2}$. Moreover, for any $g_{nm} = g_m(\cdot - na)$, $n, m \in \mathbf{Z}$, and any $C > 0$, $D < \infty$ there holds

$$g_{nm}, \ n, m \in \mathbf{Z} \text{ is a frame for } L^2(\mathbb{R}) \text{ with frame bounds } C,\ D$$
$$\Leftrightarrow \qquad\qquad\qquad\qquad\qquad\qquad\qquad\qquad\qquad\qquad (2.3)$$
$$a\,C_g\,I \leq H_g(\nu)\,H_g^*(\nu) \leq a\,D_g\,I\ , \qquad a.e.\ \nu \in \mathbb{R}\ ,$$

with I the identity operator of $l^2(\mathbf{Z})$. Next, two systems g_{nm}, γ_{nm}, $n, m \in \mathbf{Z}$, as in (1.1) with a finite upper frame bound are dual if and only if

$$\sum_m \hat{g}_m(\nu - k/a)\,\hat{\gamma}_m^*(\nu) = a\,\delta_{ko}\ , \qquad a.e.\ \nu \in \mathbb{R},\ \ k \in \mathbf{Z}\ . \qquad (2.4)$$

Finally, denote for a.e. $\nu \in \mathbb{R}$ by $\underline{c}(\nu) = (c_m(\nu))_{m \in \mathbf{Z}}$ the least-squares solution

$$\underline{c}(\nu) = a\,H_g^*(\nu)\,\big(H_g(\nu)\,H_g^*(\nu)\big)^{-1}\,\underline{e}\ ; \qquad \underline{e} = (\delta_{ko})_{k \in \mathbf{Z}}\ , \qquad (2.5)$$

of the linear system

$$\sum_{m=-\infty}^{\infty} \hat{g}_m(\nu - k/a)\,c_m = a\,\delta_{ko}\ , \qquad k \in \mathbf{Z}\ . \qquad (2.6)$$

Then we have

$$^\circ\hat{\gamma}_m(\nu) = c_m^*(\nu)\ , \qquad m \in \mathbf{Z},\ \ a.e.\ \nu \in \mathbb{R}\ . \qquad (2.7)$$

Proof of Theorem 1: We shall first establish an upper bound for the number of b_m in an interval. Let $D_g < \infty$ be an upper frame bound for the g_{nm}, $n, m \in \mathbf{Z}$. Then by (2.1) we have for $\nu_1 < \nu_2$, $A > 0$

$$(2A + \nu_2 - \nu_1)\,a\,D_g \ \geq\ \int_{\nu_1 - A}^{\nu_2 + A} \sum_m |\hat{g}_m(\nu)|^2\,d\nu$$

$$\geq\ \sum_{m\,;\,b_m \in [\nu_1, \nu_2]} \int_{-A - (b_m - \nu_1)}^{A + (\nu_2 - b_m)} |\hat{g}_m(\nu + b_m)|^2\,d\nu \quad (2.8)$$

$$\geq\ \#m(b_m \in [\nu_1, \nu_2]) \cdot \left(\delta - \int_{|\nu| \geq A} |\hat{g}(\nu)|^2\,d\nu \right)\ .$$

Hence, when A is sufficiently large,

$$\#(b_m \in [\nu_1, \nu_2]) \leq (2A + \nu_2 - \nu_1)\, a\, D_g \Bigg/ \left(\delta - \int_{|\nu| \geq A} |\hat{g}(\nu)|^2\, d\nu \right) , \quad (2.9)$$

and it follows that

$$\sup_{\nu \in \mathbf{R},\, W \geq 1} \frac{1}{2W} \#_m (b_m \in [-W - \nu, W - \nu]) < \infty . \quad (2.10)$$

Next, let $^\circ \gamma_m = T_g^{-1} g_m$ be the minimal duals for g_{nm}, $n, m \in \mathbf{Z}$, and let $C_g > 0$ be a lower frame bound for the g_{nm}, $n, m \in \mathbf{Z}$. Then

$$D_g^{-1} \delta \leq D_g^{-1} \|g_m\| \leq \|T_g^{-1} g_m\| = \|^\circ \gamma_m\| \leq C_g^{-1} \|g_m\| \leq C_g^{-1} \Delta \quad (2.11)$$

with $\Delta = \|g\|$. Moreover, we have by (1.11) and (2.4) with $k = 0$ that

$$|(g_m, {}^\circ \gamma_m)| = |(\hat{g}_m, {}^\circ \hat{\gamma}_m)| \leq 1 , \qquad m \in \mathbf{Z} , \quad (2.12)$$

and

$$\sum_m \hat{g}_m(\nu)\, {}^\circ \hat{\gamma}_m^*(\nu) = a , \qquad \text{a.e. } \nu \in \mathbb{R} . \quad (2.13)$$

Now for $\varepsilon > 0$, $W > 0$ and $V = (1 - \varepsilon)\, W$, $U = (1 + \varepsilon)\, W$ we have

$$
\begin{aligned}
a &= \frac{1}{2W} \int_{-W}^{W} \sum_m \hat{g}_m(\nu)\, {}^\circ \hat{\gamma}_m^*(\nu)\, d\nu \\
&= \frac{1}{2W} \sum_{m\,;\,|b_m| \leq V} (g_m, {}^\circ \gamma_m) + E_1 + E_2 + E_3 , \quad (2.14)
\end{aligned}
$$

where

$$E_1 = \frac{1}{2W} \int_{-W}^{W} \sum_{m\,;\,|b_m| \geq U} \hat{g}_m(\nu)\, {}^\circ \hat{\gamma}_m^*(\nu)\, d\nu , \quad (2.15)$$

$$E_2 = \frac{1}{2W} \sum_{m\,;\,|b_m| \leq V} \left\{ \int_{-W}^{W} \hat{g}_m(\nu)\, {}^\circ \hat{\gamma}_m^*(\nu)\, d\nu - (\hat{g}_m, {}^\circ \hat{\gamma}_m) \right\} , \quad (2.16)$$

$$E_3 = \frac{1}{2W} \sum_{m\,;\,V < |b_m| < U} \int_{-W}^{W} \hat{g}_m(\nu)\, {}^\circ \hat{\gamma}_m^*(\nu)\, d\nu . \quad (2.17)$$

We shall show that $E_i \to 0$ as $W \to \infty$, $\varepsilon \downarrow 0$, $\varepsilon W \to \infty$ for $i = 1, 2, 3$, and it thus follows from (2.12) and (2.14) that

$$\liminf_{W \to \infty} \frac{1}{2W} \#m(|b_m| \le W) \ge a \,, \tag{2.18}$$

as required.

We shall first estimate E_1. By using (2.1) with $^\circ\gamma_m$, C_g^{-1} instead of g_m, D_g we get by the Cauchy-Schwarz inequality

$$\begin{aligned}
|E_1| &\le \frac{1}{2W} \int_{-W}^{W} \left(\sum_{m \,;\, |b_m| \ge U} |\hat{g}_m(\nu)|^2 \right)^{1/2} \left(\sum_{m \,;\, |b_m| \ge U} |^\circ\hat{\gamma}_m(\nu)|^2 \right)^{1/2} d\nu \\
&\le (a\,C_g^{-1})^{1/2} \frac{1}{2W} \int_{-W}^{W} \left(\sum_{m \,;\, |b_m| \ge U} |\hat{g}_m(\nu)|^2 \right)^{1/2} d\nu \qquad (2.19) \\
&\le (a\,C_g^{-1})^{1/2} \left(\frac{1}{2W} \int_{-W}^{W} \sum_{m \,;\, |b_m| \ge U} |\hat{g}_m(\nu)|^2 \, d\nu \right)^{1/2} \,.
\end{aligned}$$

Next we have

$$\begin{aligned}
&\frac{1}{2W} \int_{-W}^{W} \sum_{m \,;\, |b_m| \ge U} |\hat{g}_m(\nu)|^2 \, d\nu \\
&\le \frac{1}{2W} \sum_{m \,;\, |b_m| \ge U} \int_{-W-b_m}^{W-b_m} |\hat{g}(\nu)|^2 \, d\nu \qquad (2.20) \\
&= \int_{-\infty}^{\infty} |\hat{g}(\nu)|^2 \frac{1}{2W} \sum_{m \,;\, |b_m| \ge U} \chi_{[-W-b_m, W-b_m]}(\nu) \, d\nu \,.
\end{aligned}$$

Then using that

$$\sum_{m \,;\, |b_m| \ge U} \chi_{[-W-b_m, W-b_m]}(\nu)$$
$$\le \begin{cases} \#m(b_m \in [-W-\nu, W-\nu]) \,, & |\nu| \ge U - W \,, \\ 0 & , \quad |\nu| < U - W \,, \end{cases} \tag{2.21}$$

we obtain

$$\frac{1}{2W} \int_{-W}^{W} \sum_{m \,;\, |b_m| \ge U} |\hat{g}_m(\nu)|^2 \, d\nu$$

$$\leq \quad \frac{1}{2W} \sup_{\nu \in \mathbf{R}} \#_m (b_m \in [-W - \nu, W - \nu]) \cdot \int\limits_{|\nu| \geq U - W} |\hat{g}(\nu)|^2 \, d\nu \quad (2.22)$$

Hence, by (2.10) and the fact that $U - W = \varepsilon W$, we see that $E_1 \to 0$ as $W \to \infty$, $\varepsilon \downarrow 0$, $\varepsilon W \to \infty$.

We next consider E_2. There holds

$$\int\limits_{-W}^{W} \hat{g}_m(\nu) \, {}^\circ \hat{\gamma}_m^*(\nu) \, d\nu - (\hat{g}_m, {}^\circ \hat{\gamma}_m) = - \int\limits_{|\nu| \geq W} \hat{g}_m(\nu) \, {}^\circ \hat{\gamma}_m^*(\nu) \, d\nu \; , \quad (2.23)$$

and

$$\left| \int\limits_{|\nu| \geq W} \hat{g}_m(\nu) \, {}^\circ \hat{\gamma}_m^*(\nu) \, d\nu \right| \leq \|{}^\circ \gamma_m\| \left(\int\limits_{|\nu| \geq W - |b_m|} |\hat{g}(\nu)|^2 \, d\nu \right)^{1/2}$$

$$\leq \|{}^\circ \gamma_m\| \left(\int\limits_{|\nu| \geq W - V} |\hat{g}(\nu)|^2 \, d\nu \right)^{1/2} \quad (2.24)$$

when $|b_m| \leq V$. Hence,

$$|E_2| \leq \frac{1}{2W} \sum_{m \,;\, |b_m| \leq V} \|{}^\circ \gamma_m\| \left(\int\limits_{|\nu| \geq W - V} |\hat{g}(\nu)|^2 \, d\nu \right)^{1/2} . \quad (2.25)$$

Therefore, by (2.10–11) and the fact that $V = (1 - \varepsilon) W$, $W - V = \varepsilon W$, we see that $E_2 \to 0$ as $W \to \infty$, $\varepsilon \downarrow 0$, $\varepsilon W \to \infty$.

We finally have for E_3

$$|E_3| \leq \frac{1}{2W} \sup_m \|g_m\| \, \|{}^\circ \gamma_m\| \cdot \#_m \, (|b_m| \in (V, U)) \; . \quad (2.26)$$

Hence, by (2.10–11) and the fact that $U - V = 2\varepsilon W$, we see that $E_3 \to 0$ as $W \to \infty$, $\varepsilon \downarrow 0$, $\varepsilon W \to \infty$. This completes the proof. ∎

Proof of Theorem 2: Let

$$N = \text{l.c.m}(N^0, ..., N^{P-1}) = n^p N^p \; , \qquad p = 0, ..., P - 1 \; , \quad (2.27)$$

with $n^p \in \mathbf{N}$, and let

$$M := n^0 + ... + n^{P-1} \; . \quad (2.28)$$

Any $r \in \mathbf{Z}$ can be written uniquely as

$$r = mM + n^0 + ... + n^{p-1} + n \qquad (2.29)$$

with

$$m \in \mathbf{Z} , \qquad p = 0, ..., P - 1 , \qquad n = 0, ..., n^p - 1 . \qquad (2.30)$$

Set for $r \in \mathbf{Z}$

$$h_r = g_m^p(\cdot - na^p) ; \qquad c_r = b_m^p , \qquad (2.31)$$

with m, p, n as in (2.29–30). When now $a = N\alpha$, the set of all $h_r(\cdot - ka)$ with $r, k \in \mathbf{Z}$ and the set of all $g_m^p(\cdot - na^p)$ with $n, m \in \mathbf{Z}$ and $p = 0, ..., P-1$ are the same and

$$\|h_r\|^2 > \delta , \qquad |\hat{h}_r(\nu)| \leq |\hat{g}(\nu - c_r)| , \qquad \text{a.e. } \nu \in \mathbb{R} , \qquad (2.32)$$

when $r \in \mathbf{Z}$. Since (1.20), and therefore h_{rk}, $r, k \in \mathbf{Z}$, is a frame we get by Theorem 1 that

$$1 \leq \liminf_{W \to \infty} \frac{1}{2Wa} (\#_r \, c_r \in [-W, W])$$

$$= \liminf_{W \to \infty} \sum_{p=0}^{P-1} \frac{n^p}{2Wa} \#_m (b_m^p \in [-W, W]) . \qquad (2.33)$$

And then using that

$$\frac{n^p}{a} = \frac{n^p}{N\alpha} = \frac{1}{N^p \alpha} = \frac{1}{a^p} , \qquad p = 0, ..., P - 1 , \qquad (2.34)$$

we get the result. ■

References

[1] Dana, I., Composite von Neumann lattice, *Phys. Rev.* **A28** (1983), 2594–2596.

[2] Daubechies, I., The wavelet transform, time-frequency localization and signal analysis, *IEEE Trans. Inform. Theory* **36** (1990), 961–1005.

[3] Daubechies, I., *Ten Lectures on Wavelets*, CBMS-NSF Regional Conference Series in Applied Mathematics, Vol. 61, SIAM, Philadelphia, 1992.

[4] Janssen, A. J. E. M., Signal analytic proofs of two basic results on lattice expansions, *J. Appl. Comp. Harmonic Anal.* **1** (1994), 350–354.

[5] Janssen, A. J. E. M., *Density Theorems for Filter Banks*, Nat. Lab. Report 6858, PRL Eindhoven, 5656 AA Eindhoven, The Netherlands, June 1995.

[6] Ron, A. and Z. Shen, Frames and stable bases for shift-invariant subspaces of $L_2(\mathbb{R}^d)$, preprint.

[7] Wexler, J. and S. Raz, Discrete Gabor expansions, *Signal Processing* **21** (1990), 207–220.

[8] Zibulski, M. and Y. Y. Zeevi, *Analysis of Multi-window Gabor-type Schemes by Frame Methods*, CC Pub #101, Technion, Haifa 32000, Israel, April 1995.

A. J. E. M. Janssen
Philips Research Laboratories, WL-01
5656 AA Eindhoven
The Netherlands

Sampling and Interpolation for Functions with Multi-Band Spectrum: The Mean Periodic Continuation Method

Victor E. Katsnelson

Dedicated to the memory of Samson Sas

Abstract. Functions f belonging to $L^2(\mathbf{R})$ are considered in which spectrum is contained in a 'multi-band' set E, *i.e.* in a subset of the real axis, which is the union of finitely many intervals. For such functions, a generalization of the Whittaker-Shannon-Kotelnikov sampling formula is done. The considered problem is also related to Riesz bases of exponentials in $L^2(E)$. No additional restrictions concerning the form of the set E are imposed. We reduce the problem in question to the problem of invertibility of a certain operator. This operator is the composition of two operators. The first one is the operator which realizes a mean periodic continuation of a function from $L^2([0, \text{mes } E])$ to a function defined on $L^2(\mathbf{R})$. (This operator acts from $L^2([0, \text{mes } E])$ into $L^2_{loc}(\mathbf{R})$. It could be constructed as the solution of the Cauchy problem for an appropriate difference equation with the initial data on $([0, \text{mes } E])$.) The second one is the truncation operator from $L^2_{loc}(\mathbf{R})$ onto $L^2(E)$. The sampling sequence appears as the zero set of some almost periodic exponential polynomial, which could be constructed effectively from the set E. The sampling series could also be constructed effectively.

§1 Introduction

Each function from $L^2(\mathbf{R})\,(= L^2(\mathbf{R}, dt))$ is representable in the form

$$f(t) = \int_{\mathbf{R}} \varphi(\omega)\, e^{-it\omega} d\omega \quad (t \in \mathbf{R}) . \tag{1.1}$$

The Fourier-Plancherel transform φ of f:

$$\varphi(\omega) = \frac{1}{2\pi} \int_{\mathbf{R}} f(t)\, e^{it\omega} dt \quad (\omega \in \mathbf{R}) \tag{1.2}$$

Signal and Image Representation in Combined Spaces
Y. Y. Zeevi and R. R. Coifman (Eds.), pp. 525-553.

also belongs to $L^2(\mathbb{R})\,(= L^2(\mathbb{R}, d\omega))$ and the Parseval identity

$$\int\limits_{\mathbb{R}} |\varphi(\omega)|^2 \, d\omega = \frac{1}{2\pi} \int\limits_{\mathbb{R}} |f(t)|^2 \, dt \tag{1.3}$$

holds.

The Paley-Wiener Theorem states that a function $f \in L^2(\mathbb{R})$ is the restriction on the real axis of an entire function of exponential type not exceeding Ω if and only if the condition

$$\varphi(\omega) = 0 \quad \text{for all} \quad \omega \notin [-\Omega, -\Omega] \tag{1.4}$$

holds. This condition can be formulated in terms of the notion 'spectrum of the function.' Let φ be a measurable function.

By definition, *a real point ω_0 belongs to the support of the function φ* if for any $\varepsilon > 0$ the set $(\omega_0 - \varepsilon, \omega_0 - \varepsilon) \cap \{\omega : \varphi(\omega) \neq 0\}$ has positive Lebesgue measure.

Thus, the condition (1.4) can be expressed in such a way: the support of the function φ is contained in the interval $[-\Omega, \Omega]$.

Definition 1. *The spectrum σ_f of the function f belonging to $L^2(\mathbb{R})$ is the support of its Fourier transform φ.*

If we interpret the variable t as a time, it is natural to consider the number ω in the exponent $e^{i\omega t}$ as frequency. In this connection it is natural to call the spectrum of the function f its frequency spectrum.

Definition 2. *Let E be a closed subset of the real axis (of the variable ω). Denote by $L^2_E\,(= L^2_E(\mathbb{R}) = L^2_E(\mathbb{R}, dt))$ the subspace of all the space $L^2(\mathbb{R})$ which consists of all the functions $f \in L^2(\mathbb{R})$ such that $\sigma_f \subseteq E$.*

If the set E is not only closed but also compact, then each function f from L^2_E is the restriction on $\mathbb{R}$ of an entire function of exponential type (we denote this entire function also by f). In particular, in this case the evaluation $f(t)$ is meaningful for each value $t \in \mathbb{C}$. In what follows, we consider only the case of compact E, $E \subseteq \mathbb{R}$. The following result is classical now.

Theorem 1. (Whittaker-Shannon-Kotel'nikov) *Let $f \in L^2_{[-\Omega,\Omega]}(\mathbb{R})$.*

i) *Then the function f could be uniquely determined from its values $f\left(\frac{k\pi}{\Omega}\right)$ at the sequence $\left\{\frac{k\pi}{\Omega}\right\}_{k \in \mathbb{Z}}$. The sequence $\left\{f\left(\frac{k\pi}{\Omega}\right)\right\}_{k \in \mathbb{Z}}$ belongs to l^2, moreover, the identity*

$$\frac{\pi}{\Omega} \sum_{k \in \mathbb{Z}} \left| f\left(\frac{k\pi}{\Omega}\right) \right|^2 = \int\limits_{\mathbb{R}} |f(t)|^2 \, dt \tag{1.5}$$

holds. The function f can be reconstructed from the sequence $\left\{f\left(\frac{k\pi}{\Omega}\right)\right\}_{k\in\mathbf{Z}}$ by the interpolation series

$$f(t) = \sum_{k\in\mathbf{Z}} f\left(\frac{k\pi}{\Omega}\right) \frac{\sin\Omega\left(t - \frac{k\pi}{\Omega}\right)}{\Omega\left(t - \frac{k\pi}{\Omega}\right)} \quad (t\in\mathbb{C}). \tag{1.6}$$

The series converges both in $L^2(\mathbb{R})$ and locally uniformly in $\mathbb{C}$.

ii) The interpolation (1.6) is 'free', i.e. the sequence $\left\{f\left(\frac{k\pi}{\Omega}\right)\right\}_{k\in\mathbf{Z}}$ may be an arbitrary sequence $C = \{c_k\}_{k\in\mathbf{Z}} \in l_2$. For each sequence $C = \{c_k\}_{k\in\mathbf{Z}} \in l^2$ the series

$$f_C(t) = \sum_{k\in\mathbf{Z}} c_k \frac{\sin\Omega\left(t - \frac{k\pi}{\Omega}\right)}{\Omega\left(t - \frac{k\pi}{\Omega}\right)} \quad (t\in\mathbb{C}) \tag{1.7}$$

converges both in $L^2(\mathbb{R})$ and locally uniformly in $\mathbb{C}$ and determines an entire function f_C from L_E^2. The function f_C satisfies the interpolation conditions

$$f_C\left(\frac{k\pi}{\Omega}\right) = c_k \quad (k\in\mathbf{Z}). \tag{1.8}$$

Expansions of entire functions via Lagrange interpolation series (1.6) were considered by E. T. Whittaker as long ago as 1915 [48]. (See also J. M. Whittaker's book [49]). The papers by Hardy [13] and by Ogura [37] should also be mentioned. In 1933 Kotel'nikov paid attention to the fundamental significance of the expansion (1.6) for the transmission information theory. He has formulated the following fundamental proposition:

Let a signal be described by a function f with a bounded frequency spectrum which is contained in $[-\Omega, \Omega]$ (*i.e.* this signal contains no frequencies higher than $\frac{\Omega}{2\pi}$ cycles per second). Let us have a certain communication channel. For the possibility of recovering this signal at the output of this communication channel, it is sufficient to transmit only the values $f(k\Delta)$ of this signal at the sampling points Δ apart, over this channel, where $\Delta = \frac{\pi}{\Omega}$.

Equation (1.6) and the above statement present the content of two fundamental theorems of Kotel'nikov which are contained in his work [23]. In the Russian scientific literature these statements are combined and called "*Theorem of Kotelnikov.*" In the western scientific literature these statements are called "*The Sampling Theorem.*" Shannon [47] has used the sampling theorem in his work in communication theory. Nyquist [36] has pointed out the significance of the interval $\Delta = \frac{\pi}{\Omega}$ for telegraphy. Shannon called this interval the Nyquist interval corresponding to the frequency band $[-\Omega, \Omega]$.

Various proofs of the Sampling Theorem as well as its generalizations can be found in [10, 2.9], [12, 2], [51], [35]. See also overviews [5, 14, 17, 22].

In this paper we deal with sampling and interpolation for functions from L_E^2 for compact $E, E \subset \mathbb{R}$. The paper is organized as follows. Section 2 contains the formulation of the problem concerning sampling and interpolation sequences for the space L_E^2, its relation to the basity of exponentials in $L^2(E)$, and short overview of (nonnumerous up to now) results for multi-band sets E. Section 3 contains discussions of results related to one-band E ($E = [-\Omega, \Omega]$) and a sampling sequence which is the zero set of an entire function of a special form (so-called sine-type function). In Section 4 we discuss the notions of a mean periodic function and a mean periodic continuation of a function from an interval. In Section 5 we formulate and prove a general theorem which gives a criterion for such a mean periodic continuation to lead to sequences T, which are both sampling and interpolating for L_E^2 (or, what is the same, for a sequence $\{e^{i\omega t_k}\}_{t_k \in T}$ to be a Riesz basis in $L^2(E)$). In Section 6 we reduce the σ-m.-p continuation problem to the Cauchy problem for a linear difference equation on the real axis. Finally, in Section 7 we discuss how to construct such a difference equation for a multi-band set.

Detailed considerations of a general multi-band set E (in the planned way) will be pursued in series of papers which we hope to publish in ZAA (Zeitschrift für Analysis und Anwendungen), ISSN 0232-2064.

This paper is dedicated to the memory of Samson Sas who sent me the multi-band sampling problem in 1988. Samson Sas was an extremely talented engineer in the field of communication theory. He lived in the (now Ukranian) town Lwov (=Lemberg). He passed away prematurely in 1990.

§2   Sampling and interpolation problem for the space L_E^2 in the Riesz basis and frame setting

This section deals with vector systems in a Hilbert space, especially with Riesz bases and frames. Details on these topics can be found in [11, 50, 6]. Let $T = \{t_k\}, t_k \in \mathbb{C}$ be a sequence. In what follows we assume that the sequence T satisfies the conditions

$$\sup_{t_k \in T} |\Im t_k| < \infty \tag{2.1}$$

and

$$\inf_{t_j, t_k \in T, t_j \neq t_k} |t_j - t_k| > 0. \tag{2.2}$$

Definition 3. *Let $T = \{t_k\}_{k \in \mathbb{Z}}$ be a discrete sequence of real points (or, more generally, satisfy the condition (2.2). The density $d(T)$ of this se-*

quence is defined as

$$d(\Lambda) = \overline{\lim_{r \to \infty}} \frac{\#\{(-r, r) \cap \Re T\}}{2r} , \qquad (2.3)$$

where $\Re T = \{\Re t_k, \quad t_k \in T\}$ and $\#M$ as usual denotes the cardinality of the set M.

Let $E, E \subset \mathbb{R}$ be a compact set, and $L_E^2(= L_E^2(dt)$ be the space introduced in Definition 2. We associate the sequence $\{f(t_k)\}_{t_k \in T}$ with each $f \in L_E^2$. It is well known that if the sequence T satisfies the conditions (2.1) and (2.2) then inequality

$$m^2 \sum_{t_k \in T} |f(t_k)|^2 \leq \int_{\mathbb{R}} |f(t)|^2 \, dt \quad (\forall \, f \in L_E^2) . \qquad (2.4)$$

This is the so-called Plancherel-Polya Inequalty, see [40]. The paper [20] contains some references related to the Plancherel-Polya Inequalty. This inequality is relatively easy. The opposite inequality

$$\int_{\mathbb{R}} |f(t)|^2 \, dt \leq M^2 \sum_{t_k \in T} |f(t_k)|^2 \quad (\forall \, f \in L_E^2) \qquad (2.5)$$

(with a constant $M < \infty$ not depending on f) is much more delicate. It holds only under some refined conditions on E and T. The combined inequality

$$m^2 \sum_{t_k \in T} |f(t_k)|^2 \leq \int_{\mathbb{R}} |f(t)|^2 \, dt \leq M^2 \sum_{t_k \in T} |f(t_k)|^2 \quad (\forall \, f \in L_E^2) \qquad (2.6)$$

is equivalent to the inequality

$$2\pi m^2 \sum_{t_k \in T} |\langle \varphi, e_{t_k} \rangle|^2 \leq \int_E |\varphi(\omega)|^2 \, d\omega \leq 2\pi M^2 \sum_{t_k \in T} |\langle \varphi, e_{t_k \in T} \rangle|^2 \quad (\forall \, \varphi \in L_E^2), \qquad (2.7)$$

where

$$e_{t_k} \stackrel{\text{def}}{=} e^{it_k \omega} \in L^2(E, d\omega) \quad (t_k \in T) \qquad (2.8)$$

and

$$\langle u, v \rangle \stackrel{\text{def}}{=} \int_E u(t) \overline{v(t)} \, dt \qquad (2.9)$$

is the standard scalar product in $L^2(E)$. The combined inequality (2.7) means that the sequence (2.8) is a frame in the Hilbert space $L^2(E)$.

From general theory of frames (see [6, 3.2]) it follows that under the conditions (2.7) there exists such a sequence $\{\varphi_k\}$

$$\varphi_{t_k} \in L^2(E), \quad \sup_k \|\varphi_{t_k}\|_{L^2(E)} \leq \infty \tag{2.10}$$

that for an arbitrary function φ from $L^2(E)$ the expansion

$$\varphi(\omega) = \sum_k \langle \varphi, e_{t_k} \rangle \varphi_{t_k}(\omega) \quad (\omega \in E) \tag{2.11}$$

holds. We stress that *if the system* $\{e_{t_k}\}_{t_k \in T}$ *is not minimal in* $L^2(E)$, *such a system* $\{\varphi_{t_k}\}$ *is not unique.* Multiplying the latter equality on $e^{-it\omega}$ and integrating with respect to $d\omega$ over E, we obtain (see (1.1)) that

$$f(t) = \sum_{t_k \in T} f(t_k) S_{t_k}(t) \quad (\forall f \in L^2_E, \quad \forall t \in \mathbb{C}) \tag{2.12}$$

where

$$S_{t_k}(t) \overset{\text{def}}{=} \int_E \varphi_{t_k}(\omega) e^{-it\omega} d\omega \quad (\forall t \in \mathbb{C}). \tag{2.13}$$

Every function S_{t_k} belongs to L^2_E, and the series (2.13) converges both in $L^2(\mathbb{R})$ and locally uniformly in $\mathbb{C}$.

The series (2.13) (which exists under the condition (2.6)) allows recovery of a function $f \in L^2_E$ from its sampling sequence$\{f(t_k)\}$.

Thus, the series (2.13) may be considered as a generalization of Whittaker's cardinal series (1.6), and the functions S_{t_k} may be considered as a generalization of Whittaker's cardinal functions $\left(\sin \Omega \left(t - \frac{k\pi}{\Omega}\right)\right) / \left(\Omega \left(t - \frac{k\pi}{\Omega}\right)\right)$.

All of the above motivates the following:

Definition 4. *A sequence* $T = \{t_k\}, t_k \in \mathbb{C}$ *is said to be a a sampling sequence for the space* $L^2(E)$ *if the condition (2.6) is satisfied (with* $m > 0, M < \infty$ *not depending on* f). *The series (2.12) is said to be a sampling series.*

If a sequence T is a sampling sequence for the class L^2_E, then the set T is a uniqueness set for the class L^2_E. It follows from (2.5).

We stress that under condition (2.6), only the interpolation (2.12) is *not free* in general: not every sequence from l^2 is a sequence of the form $\{f(t_k)\}$. (A frame in a Hilbert space is not a minimal system in general.)

Definition 5. *The sequence T is an interpolating sequence for the space L_E^2 if each sequence $\{c_k\}$ may appear as a sequence of the form $\{f(t_k)\}_{k \in L^2}$. In other words, the sequence T is an interpolating sequence for the space L_E^2, if the interpolation problem (with respect to a function $f \in L_E^2$)*

$$f(t_k) = c_k \quad (\forall \; k) \tag{2.14}$$

is solvable for each sequence $\{c_k\} \in l^2$.

If a sequence T is interpolating for the space L_E^2, then the system (2.8) of $\{e_{t_k}\}$ is minimal in $L^2(E)$.

Indeed, let $f_j \in L_E^2$ be any solution of the interpolation problem $f_j(t_k) = \delta_{jk}, f_j(t) = \int_E \varphi_{t_j}(t) e^{-it\omega} d\omega$. Then, the relation $\langle e_{t_k}, \varphi_{t_j} \rangle = \delta_{jk} \; (\forall \; j, k)$ holds.

The following statement is well known:

A system $\{e_k\}$ of elements of a Hilbert space is a Riesz basis if and only if it is both a frame and a minimal system.

From the above discussion follows:

Theorem 2. *Let $T = \{t_k\}_{t_k \in \mathbb{C}}$ be a sequence, and $E \subset \mathbb{R}$ be a compact, mes $E > 0$. A system of exponentials $\left\{e^{it_k\omega}\right\}_{t_k \in T}$ is a Riesz basis in $L^2(E)$ if and only if this sequence is both sampling and interpolating sequence for the space L_E^2.*

Remark. If a sequence $T = \{t_k\}$, $t_k \in \mathbb{C}$ is both a sampling and an interpolating sequence, then the sampling functions S_{t_j} are determined uniquely. They are representable by (2.13) where $\{\varphi_{t_j}\}$ is a system, biorthogonal to the Riesz basis $\left\{e^{it_k\omega}\right\}_{t_k \in T}$. The function S_{t_j} is (unique in the class L_E^2) a solution of the interpolation problem $S_{t_j}(t_k) = \delta_{jk}$.

Whereas plenty of results are known for the case a compact E is an interval, there are only a few results concerning more general E. The classical interpolation machinery such as infinite products and interpolation series fails to work in this situation. One of the main difficulties in the problem in question is connected with too small density of sampling and interpolating sequences. In the remarkable papers [25] and [26] by Landau, it has been shown that if E is a union of an arbitrary finite number of intervals, than the condition

$$d(T) = \frac{\text{mes } E}{2\pi} \tag{2.15}$$

must satisfy for any sampling and interpolating sequence T. However, the density $d(\mathcal{Z}_S)$ of the zero-set $\mathcal{Z}_S$ of any function $S \in L^2(E), S \not\equiv 0$ must satisfy the condition

$$d(\mathcal{Z}_S) = \frac{\text{mes } I_E}{2\pi} \tag{2.16}$$

where I_E is the smallest interval containing E (the so-called supporting interval of the set E).

It is pertinent to note that in his paper [24], Landau constructed an example of a sequence $T \subset \mathbb{R}$, $d(T) = 1$ and a multi-band set E such that this set is a set of uniqueness for the class L_E^2 and mes E can be arbitrarily large. (In this example, the set E can be in particular a finite union of regularly positioned intervals of the same length.) Thus, the condition $d(T) \geq$ mes $I_E / 2\pi$ is not necessary for the sequence T to be a set of uniqueness for the class L_E^2.

In papers [2, 4, 9, 8, 15] another construction is used. Let E be a closed subset of $\mathbb{R}$, and let ρ be a positive number which satisfy the condition

$$\text{mes}\,((E + 2\pi k\rho) \cap E) = 0 \qquad (\forall\, k \in \mathbf{Z} \setminus \{0\}) \qquad (2.17)$$

where $E + 2\pi k\rho \stackrel{\text{def}}{=} \{\omega + 2\pi k\rho : \omega \in E\}$. Then the formula

$$f(t) = \sum_{k \in \mathbf{Z}} f\left(\frac{k}{\rho}\right) S_E\left(t - \frac{k}{\rho}\right) \qquad (2.18)$$

holds, where

$$S_E(t) \stackrel{\text{def}}{=} \frac{1}{\text{mes}\,E} \int_E e^{itl} dl \qquad (t \in \mathbb{R}). \qquad (2.19)$$

The sequence T is a sampling, but not an interpolating one (except in trivial cases). Various aspects of the representation (2.18)–(2.19) (under the condition (2.17)) are discussed in [2, 4, 9, 8, 15].

In Goldman's book [12] functions $f \in L^2(\mathbb{R})$ are considered whose spectrum is contained in a set of the form $E = (-\omega_2, -\omega_1) \cup (\omega_1, \omega_2)$ where $0 < \omega_1 < \omega_2 < \infty$. In [12, Section 2.3], an assertion on such functions has been proved called "Sampling Theorem." However, in Goldman's interpolation formula there appear not only the values of a function f itself, but also the values of its Hilbert transform. Furthermore, Goldman's sampling formula has been established only for functions with *two*-band spectrum (with bands of equal length). It is not clear to us how to extend this result of Goldman to a more general case.

The case E consisting of two intervals of the same length was also considered in the paper [21] by Kholenberg. The methods of this article allow one to directly construct a sampling and interpolating sequence for any set E which is a union of an arbitrary finite number of intervals having commensurable lengths. The corresponding sampling and interpolating sequences appear to be unions of a relevant number of shifted arithmetic progressions. Using another approach, this result was independently obtained in our paper [19], which also contains (for this case) the explicit formulas for recovery of a multi-band signal. In particular, developing the methods of [19] and [21] one can see that if $E = [0, \alpha_1] \cup [2\pi - \alpha_2, 2\pi]$, where $\frac{\alpha_1}{\alpha_2}$ is rational, then it is possible to construct a sampling and interpolating sequence of the form $\mathcal{T} = 2\pi\alpha_1^{-1}\mathbf{Z} \cup (\gamma + 2\pi\alpha_2^{-1}\mathbf{Z})$, where γ is a real number. Thus, in this case Λ may be chosen as a union of two real arithmetic progressions corresponding to the relevant parts of E.

See also [16] where various problems of sampling theory are considered from the viewpoint of Riesz basis setting.

The article [46] gives an elegant approach to the problem which yields the results that are free of arithmetic restrictions on E. In particular, this approach leads one to constructing a real sampling and interpolating sequence for an arbitrary set E consisting of two intervals. In the case of a larger number of intervals, the approach suggested in this article requires some quantitative (not arithmetic!) restrictions on relative lengths of the intervals from E and the gaps between them. The results of [46] are by their nature existence theorems. They are based on Avdonon's result [1] on exponential bases $\left\{e^{it_k\omega}\right\}_{t_k \in \mathcal{T}}$ in $L^2(I)$ (I is an interval) where the sequence $\mathcal{T}$ is a 'perturbation' of the zero-set of some sine-type function. (See appropriate definitions and the prehistory in the next section.) In its turn, the result of [1] is a generalization of our result [19] and is based on the method of [19].

In the paper [34] the problem of constructing a sampling and interpolating sequence $\mathcal{T}$ for L_E^2 (E is the union of finite many intervals) was reduced to a problem of invertibility of some convolution operators acting in the space $L^2(I_E \setminus E)$, where, as before, I_E is a supporting interval of the set E. Sampling and interpolating sequences constructed in [34] are located in a horizontal strip along the real axis. In [34], the question of the sampling of a signal with two band spectrum via its values at the union of two arithmetic progression is also studied. The last mentioned result is based on the invertibility criteria for the above-mentioned operator in the case when its symbol is an almost periodic function. (See [18] concerning this almost periodic machinery.)

The main goal of this paper is to advance a new method of attack to the problem of construction of sequences $\mathcal{T}$ which are both sampling and interpolating for the space L_E^2 ($E \in \mathbb{R}$ is a given compact).

We call this method *the method of mean-periodic continuation* because this method was inspired by the theory of mean-periodic functions originated by Delsart and Schwartz. (Of course, we deal with very special cases of such functions.)

§3   Sine-type functions and exponential bases

Starting from the pioneer work of Paley and Wiener [38], nonharmonic Fourier series, *i.e.* the series of the form

$$\sum c_k \, e^{i\omega t_k}$$

were considered. Most of these considerations deal with the completeness, minimality, or basicity of exponential system

$$\left\{ e^{it_k\omega} \right\}_{t_k \in T} \tag{3.1}$$

on an appropriate interval. Now it is traditional to associate with this exponential system the generating entire function S:

$$S(t) = \lim_{R \to \infty} \prod_{|t_k| < R} \left(1 - \frac{t}{t_k} \right), \tag{3.2}$$

and to express properties of the system (3.1) in terms of the function f. (See [38, 50], and [31, Appendix III].)

Let us recall the notion of a sine-type function introduced by Levin [30, 32].

Definition 6. *Let $a, b \in \mathbb{R}$, $a < b$ be given. An entire function $S(t)$ is called a sine-type function with respect to the interval $[a, b]$, if its zero set $\{t_k\} = T(S)$ satisfies the conditions (2.1) and (2.2) and also the estimates*

$$|S(\xi + i\eta)| \asymp e^{-b\eta}, \quad \eta > 2H; \qquad |S(\xi - i\eta)| \asymp e^{a\eta}, \quad \eta < -2H \tag{3.3}$$

are valid for $H > 0$ are big enough. Here and in what follows, the sign $\asymp$ means that the ratio of the left-hand and right-hand sides lie between two positive constants.

Theorem 3. (Levin [30, 32]) *If $T(S)$ is the zero set of a sine-type function S with respect to the interval $[a, b]$, then it is a sampling and interpolating sequence for $L^2([a, b])$. The exponential system $\left\{ e^{it_k\omega} \right\}_{t_k \in T(S)}$ forms a Riesz basis in the space $L^2([a, b])$.*

Notice that the biorthogonal system of functions $\{\varphi_{t_k}(\omega)\}$ also forms a Riesz basis in the space $L^2([a, b])$. The elements of this system satisfy the relations

$$\frac{S(t)}{S'(t_k)(t - t_k)} = \int_a^b e^{-i\omega t} \varphi_k(\omega) \, d\omega \quad (\forall \, k). \tag{3.4}$$

An elementary proof of Levin's Theorem can also be found in [19, Lemma 7]. In [39] Levin's Theorem is derived from the interpolation theorem due to Shapiro and Shields. (Paper [39] (see also [22]) also contains a far-reaching generalization of Levin's Theorem).

The case of the generating entire function of the special form

$$S(t) = \int_a^b e^{-it\omega} \, d\sigma(\omega) \tag{3.5}$$

has been of our main interest here. ($\sigma(\omega)$ is a bounded variation function on $[a, b]$.) It is clear that the conditions (3.3) are satisfied if and only if both points a and b are jump points for the function σ:

$$\sigma(a) \neq \sigma(a+0), \quad \sigma(b-0) \neq \sigma(b). \tag{3.6}$$

The condition (2.1) follows from (3.3) automatically.

The following theorem is a direct consequence of Levin's Theorem and above-mentioned facts on functions of the form (3.5):

Theorem 4. *Let $S(t)$ be an entire function of the form (3.5) where $\sigma(\omega)$ is a bounded variation function on $[a, b]$ satisfying the condition (3.6), and let the zero set $\mathcal{T}(S) = \{t_k\}$ of the function S satisfy the separability condition (2.2).*

i) *Then the sequence $\{t_k\}$ is a sampling and interpolating sequence for $L^2_{[a,b]}$. The exponential system $\{e^{it_k\omega}\}_{t_k \in \mathcal{T}(S)}$ forms a Riesz basis in the space $L^2([a, b])$.*

ii) *The biorthogonal system $\{\varphi_{t_k}\}$ can be represented in the form*

$$\varphi_{t_k}(\omega) = e^{it_k\omega} \, \psi_{t_k}(\omega) \quad (\omega \in [a, b]) \tag{3.7}$$

where

$$\psi_{t_k}(\omega) = \frac{1}{s_{t_k}} \left(\int_\omega^b e^{-it_k\xi} \, d\sigma(\xi) - \int_a^\omega e^{-it_k\xi} \, d\sigma(\xi) \right) \quad (\omega \in [a, b]), \tag{3.8}$$

with

$$s_{t_k} = 2 \int_a^b \xi \, e^{-it_k\xi} \, d\xi. \tag{3.9}$$

A normalizing sequence $\{s_{t_k}\}$ satisfying the condition $\{s_{t_k}\} \asymp 1$.

Proof: The assertion i) of the theorem is a direct consequence of Levin's Theorem and above-mentioned facts on functions of the form (3.5).

Of course, in the case at hand the biorthogonal system could be constructed using the standard 'infinite product machinery' by the formula (3.4). However, if the generating entire function S is initially given in the form (3.5), the function φ_{t_k} of the biorthogonal system can be expressed more directly, in terms of the function σ and the point t_k (with the same k only). Let us integrate by parts in the expression

$$S(t) = \int_a^b e^{-i(t-t_k)\omega} \, d\left(\int_a^\omega e^{-it_k\xi} \, d\sigma(\xi) \right).$$

In view of

$$\int_a^b e^{-it_k\xi} \, d\sigma(\xi) = 0 \quad (\text{since } S(t_k) = 0) \tag{3.10}$$

the terms outside the integral vanish, and we obtain

$$S(t) = i(t - t_k) \int_a^b e^{-it\omega} \left(e^{it_k\omega} \int_a^\omega e^{-it_k\xi} \, d\sigma(\xi) \right) d\omega. \tag{3.11}$$

Because of (3.10)

$$\int_a^\omega e^{-it_k\xi} \, d\sigma(\xi) = - \int_\omega^b e^{-it_k\xi} \, d\sigma(\xi)$$

$$= \frac{1}{2} \left(\int_a^\omega e^{-it_k\xi} \, d\sigma(\xi) - \int_\omega^b e^{-it_k\xi} \, d\sigma(\xi) \right). \tag{3.12}$$

Taking into account (3.4) and combining (3.11) and (3.12) with

$$S'(t_k) = -i \int_a^b \omega \, e^{-it_k\omega} \, d\sigma(\omega), \tag{3.13}$$

we obtain (3.7)–(3.9). The assertion ii) of the theorem is proved.

Assume that points t_k from the zero-set $T(S)$ of the function S are 'fastened' to the points of the arithmetic progression of an appropriate density, i.e. the conditions

$$\left| \Re t_k - \frac{2\pi}{b-a} k \right| < (1 - \delta) \frac{\pi}{2(b-a)} \quad (\delta > 0 \text{ does not depend on } k)$$

$$\tag{3.14}$$

and sup $|\Im t_k| < \infty$ are fulfilled. Then the points t_k are representable in the form

$$t_k = \frac{2\pi}{(b-a)} k + \Delta(k) \quad (\forall\, k \in \mathbf{Z}) \tag{3.15}$$

where

$$\sup |\Re\Delta(k)| \le (1-\delta)\frac{\pi}{2(b-a)}, \ \sup |\Im\Delta(k)| < \infty. \tag{3.16}$$

Remark. The following result by Krein and Levin, [31, Appendix VI, Section 2], may be interesting in connection with the problem at hand: Let the function σ be a jump function on $[a,b]$, *i.e.* the function S is a sum of an absolute convergent series $S(t) = \sum_l \alpha_l\, e^{i\omega_l t}$ ($\omega_l \in [a,\, b]$ are jump points of the function σ). Assume that condition (3.6) is satisfied, *i.e.* the points a, b are among the points ω_k. Moreover, assume that points t_k from the zero-set $T(S)$ of the function S are fastened to the points of the arithmetic progression of an appropriate density, *i.e.* they are representable in the form (3.15)–(3.16). Then the sequence $\{\Delta(k)\}$ is an almost periodic sequence whose Fourier series converges absolutely and whose exponents are contained in the $\mathbf{Z}$-modulus generated by exponents $\{\omega_k\}$ from the Fourier series representing the function S.

Example: Now we consider the simple example in which we are able to carry all the calculations to completion. Let $[a,b]$ be a zero-centered interval:

$$[a,b] = [-\Omega,\Omega], \quad (\Omega > 0 \text{ is a given number}).$$

Let the entire generating entire function be an exponential polynomial of the special form:

$$S(t) = -2i\left(\sin\Omega t + \sum_{1 \le l \le m} c_k \sin h_l t\right) \tag{3.17}$$

with

$$0 < h_1 < h_2 < h_m < \Omega \quad (h_0 \stackrel{\text{def}}{=} 0,\ h_{m+1} \stackrel{\text{def}}{=} \Omega) \tag{3.18}$$

and

$$c_l = \overline{c_l} \quad (1 \le l \le m)\ (c_{m+1} \stackrel{\text{def}}{=} 1). \tag{3.19}$$

If the coefficients $\{c_l\}$ satisfy the condition

$$\sum_{1 \le l \le m} |c_l| < 1, \tag{3.20}$$

then the zero set $\mathcal{T}(S) = \{t_k\}$ of the function S is real and separable, *i.e.* satisfying (2.2). Moreover, under conditions (3.19)–(3.20) the points t_k are fastened to the points $\frac{\pi}{\Omega} k$:

$$t_k = \frac{\pi}{\Omega} k + \Delta(k) \quad (k \in \mathbf{Z}) \tag{3.21}$$

where $\{\Delta(k)\}$ is an odd real sequence[1]:

$$\Delta(k) = -\Delta(-k) \in \mathbb{R} \quad (\forall\, k \in \mathbf{Z}) \tag{3.22}$$

and

$$\sup |\Delta(k)| < \frac{\pi}{2\Omega}. \tag{3.23}$$

The function S from (3.17) is representable in the form (3.5) with a piece-wise constant function σ. This function jumps only at the points ω_l :

$$\omega_l = -(\operatorname{sign} l)\, h_{|l|} \quad (l = \pm 1, \pm 2, \pm(m+1)). \tag{3.24}$$

The corresponding jump values are

$$\sigma(\omega_l + 0) - \sigma(\omega_l - 0) = c_{|l|} \quad (l = \pm 1, \pm 2, \pm(m+1)). \tag{3.25}$$

According to Theorem 4, the system $\left\{e^{it_k\omega}\right\}_{t_k \in \mathcal{T}(S)}$ forms a Riesz basis in the space $L^2([-\Omega, \Omega])$, with biorthogonal system $\{\varphi_{t_k}\}$ [2] of the form (3.7), where now $\{\psi_{t_k}\}$ is an even real-valued function on $[-\Omega, \Omega]$:

$$\psi_{t_k}(\omega) = \overline{\psi_{t_k}(\omega)} = -\psi_{t_k}(-\omega) \quad (\omega \in [-\Omega, \Omega]) \tag{3.26}$$

which also is a piecewise constant function with jumps only at the points ω_l (3.24):

$$\psi_{t_k}(\omega) = \frac{\cos t_k\Omega + \displaystyle\sum_{l < r \le m} c_r \cos t_k h_r}{2\Omega\, d_k} \quad (\omega \in [\,h_l, h_{l+1}), \, l = 0, 1, \ldots, m) \tag{3.27}$$

with

$$d_{t_k} = \cos t_k\Omega + \sum_{1 \le r \le m} \frac{h_r c_r}{\Omega} \cos t_k h_r. \tag{3.28}$$

We would like to stress that almost all in the above-given construction was absolutely explicit. The calculation of the corrections $\Delta(k)$ (see (3.21)) is the only one which is not quite explicit. The condition (3.20) allows us to construct these 'corrections' using a 'perturbations machinery.'

---

[1]According to the theorem of Krein and Levin, this sequence $\{\Delta(k)\}$ is an almost periodic sequence whose Fourier series converges absolutely.

[2]Which is a Riesz basis in the space $L^2([-\Omega, \Omega])$ as well.

§4 Mean-periodic functions; Mean-periodic extensions

The following perennial question arises naturally: Why (and to whom) this nonharmonic series[3] may be beneficial? We will show below that this nonharmonic Fourier series could be useful in multi-band sampling theory.

Let $E \in \mathbb{R}$ be a compact, mes $E > 0$. If the set E is an interval, then in $L^2(E)$ there exists an exponential basis, namely the classic harmonic basis

$$\left\{ e^{i\,(2\pi/\text{mes } E)\,k\,\omega} \right\}_{k\in\mathbf{Z}}. \tag{4.1}$$

However, in the case the set E is not an interval, there is an evident obstacle for the set E to be basis (even to be complete) in $L^2(E)$. It is the periodicity of the exponents (4.1) which is the obstacle. Indeed, if $t' \in E$, $t'' \in E$ and $t' = t''$ (mod mes E), then the equality $x(t') = x(t'')$ must be satisfied for each function $x \in L^2(E)$ which belongs to the closed (in $L^2(E)$) linear hall of the exponents (4.1). Thus, if the condition

$$\text{mes}\left((E + k\,\text{mes } E) \cap E\right) = 0 \qquad (\forall\,k \in \mathbf{Z} \setminus \{0\}) \tag{4.2}$$

is violated, the system (4.1) is certainly non-complete in $L^2(E)$. A modified notion of periodicity will serve our purposes. It is the *mean-periodicity* (in the sense of Delsarte [7] and Schwartz [42]) which will work to our benefit.

Definition 7. *Let $\sigma(\omega)$ be a nonconstant bounded variation function defined on the interval $[a, b]$, $[a, b] \subset \mathbb{R}$. A function[4] $x(\omega) \in L^2_{loc}(\mathbb{R}, d\omega)$ is said to be a mean-periodic (with respect to σ) if*

$$\int_{[a,b]} x(\omega - \xi)\,d\sigma(\xi) \equiv 0 \quad (\omega \in \mathbb{R}). \tag{4.3}$$

Remark. The integral in (4.3) could be considered as an integral of the form $\int_{[a,b]} (T_\xi(x))\,d\sigma(\xi)$ where $T_\xi(x) \stackrel{\text{def}}{=} x(\omega - \xi)$ is L^2_{loc}- valued continuous function of variable ξ.

The following result is of fundamental importance.

Theorem 5. (The Schwartz Uniqueness Theorem [42]) *Let $x \in L^2_{loc}(\mathbb{R})$ be a mean-periodic function (with respect to a bounded variation nonconstant function σ supported on $[a, b]$).*

If the function x vanishes on the interval[5] $[a, b]$: $x(\omega) \equiv 0 \ \forall\,\omega \in [a, b]$, then the function x vanishes identically: $x(\omega) \equiv 0 \ \forall\,\omega \in \mathbb{R}$.

---

[3]It seems probable that Paley and Wiener [38] developed the nonharmonic Fourier series theory to demonstrate the power of complex analysis methods in harmonic analysis.

[4]Delsart and Schwartz considered the case of $x \in C(\mathbb{R})$ but not $x \in L^2_{loc}(\mathbb{R})$. However, this difference in functional classes is not essential.

[5]Or on any interval whose length is equal $(b - a)$

Exponential mean-periodic functions, *i.e.* functions of the form

$$e_t(\omega) \stackrel{\text{def}}{=} e^{it\omega} \quad (\omega \in \mathbb{R}) \tag{4.4}$$

are of interest. The following statement is clear.

Statement. An exponential function e_t of the form (4.4) is a mean periodic with respect to σ if and only if the number t belongs to the zero-set $T(S_\sigma)$ of the characteristic (entire) function

$$S_\sigma(t) \stackrel{\text{def}}{=} \int\limits_{[a,b]} e^{-it\omega}\, d\sigma(\omega). \tag{4.5}$$

Fourier (or Dirichlet) series are one of the powerful tools in the theory of mean-periodic functions. If x is a m.-p. function (with respect to σ), we associate with x a formal series. In the case all roots of the characteristic equation $S(\sigma) = 0$ are simple, this series has the form

$$x(\omega) \sim \sum_{t \in T(S_\sigma)} c_t\, e^{it\omega} \quad (\omega \in \mathbb{R}) \tag{4.6}$$

with

$$c_t = \int\limits_{[a,b]} x(\omega)\, \overline{\varphi_t(\omega)}\, d\omega, \quad (t \in T(S_\sigma)) \tag{4.7}$$

where $\varphi(\omega) = e^{it\omega}\, \psi(\omega)$ are functions constructed in the same way as the functions (3.7)–(3.8). Modifications needed in the general case of (possible) multiple roots are clear.

Theorem 6. (The Summability Theorem) *The series* (4.6) *is Abel-summable to* $x(\omega)$ *in* $L^2_{loc}(\mathbb{R})$, *(not only in* $L^2([a,b])$*).*

If $x(\omega) = 0$ on $[a,b]$, then, according to (4.7), $c_t = 0\ \forall\, t \in T(S_\sigma)$, hence, $x(\omega) \equiv 0\ \forall\, \omega \in \mathbb{R}$. Thus, the uniqueness theorem is the consequence of the summability theorem.

The uniqueness theorem was proved by Schwartz ([41, 42]) for the case where all the roots of the characteristic equation $\S_\sigma(t) = 0$ are real. (The paper [41] is difficult to obtain, but it is reproduced word by word in [43, Chapter 3].) The extension to the general case (without any restrictions on location of roots of characteristic equation) was done by Leont'ev [27], and [28, Chapter 6, Sections 4 and 5]. The mentioned uniqueness and summability results were proved using a hard analysis entire function technique.

Our main interest here is in the special case: σ is a piecewise constant function. In this case, the uniqueness theorem could be proved elementarily, without complex analysis methods.

By virtue of the uniqueness theorem, the following definition is meaningful:

Definition 8. *Let σ be a given bounded variation function on $[a, b]$, ($[a, b] \subset \mathbb{R}$), , $\sigma \not\equiv$ const. Let $\varphi \in L^2([a, b]), d\omega)$. A function $\Phi \in L^2_{loc}(\mathbb{R})$ is said to be σ-mean-periodic continuation of the function φ if the following two conditions are fulfilled:*

   i) *The function Φ is mean-periodic with respect to σ.*

   ii) $\mathbf{Rstr}_{[a,b]} \Phi = \varphi$.

Notation. Here and in the following, $\mathbf{Rstr}_M f$ stands for the restriction of a function f onto a set M.

In the general case, for given σ not every function $\varphi \in L^2([a, b])$ admits m.-p. continuation onto $\mathbb{R}$ (see an example in [29]: Chapter 5, Sec.1).

Let us define the mean-periodic continuation operator C_σ.

Definition 9. *Let σ be a bounded variation nonconstant function on $[a, b]$. By definition*

   i) *A function $\varphi \in L^2([a, b]$ belongs to the domain $\mathcal{D}_{C_\sigma}$ of the operator C_σ, if the function φ admits σ-m.-p. continuation.*

   ii) *For $\varphi \in \mathcal{D}_{C_\sigma}$, $(C_\sigma \varphi)(\omega) \stackrel{\text{def}}{=} \Phi(\omega)$, $(\omega \in \mathbb{R})$ where Φ is (necessarily unique) σ-m.-p. continuation of the function φ.*

   The linear operator C_σ is said to be an operator of σ-m.-p. continuation (from $[a, b]$ onto $\mathbb{R}$).

Remark. If

$$\varphi(\omega) = \mathbf{Rstr}_{[a,b]} e^{it\omega} \quad (\omega \in [a, b]) \tag{4.8}$$

with $t \in T(S_\sigma)$, then

$$\varphi \in \mathcal{D}_{C_\sigma} \quad \text{and} \quad C_\sigma \varphi = e^{it\omega} \quad (\omega \in \mathbb{R}). \tag{4.9}$$

Theorem 7. (Sedleckiĭ, [44]) *Under condition (3.6) $\mathcal{D}_{C_\sigma} = L^2([a,b])$, and the operator C_σ acts continuously from $L^2([a,b])$ into $L^2_{loc}(\mathbb{R})$.*

The last theorem is an elementary fact. From [45, Theorem 2.12] it follows that the conditions (3.6) are not necessary for the operator C_σ to be continuous.

§5 Mean-periodic continuation and Riesz bases

The following operator is one of the main objects in our considerations.

Definition 10. *Let σ be a bounded variation nonconstant function on $[a,b]$,*

$[a,b] \subset \mathbb{R}$, and let E, $E \subset \mathbb{R}$ be a compact, (mes $E > 0$). By definition,

$$T_{\sigma,E} : L^2([a,b]) \to L^2(E), \quad \mathcal{D}_{T_{\sigma,E}} = \mathcal{D}_{C_\sigma}, \quad T_{\sigma,E} \overset{\text{def}}{=} \text{Rstr}_E \, C_\sigma \ . \quad (5.1)$$

The operator $T_{\sigma,E}$ is said to be a σ-mean-periodic transfer operator (from $[a,b]$ onto E).

As direct consequence of Theorem 7, we obtain:

Theorem 8. *If the function σ satisfies the conditions (3.6), then $\mathcal{D}_{T_{\sigma,E}} = L^2_{[a,b]}$, and the operator $T_{\sigma,E}$ acts continuously from $L^2([a,b])$ into $L^2(E)$.*

Remark. If $t \in T(\mathcal{S}_\sigma)$ ($T(\mathcal{S}_\sigma)$ is a zero set of the function $\mathcal{S}_\sigma$), then

$$T_{\sigma,E} \, \text{Rstr}_{[a,b]}(e^{it\omega}) = \text{Rstr}_E(e^{it\omega}) \ . \quad (5.2)$$

The following result is (formally) the main result of this paper.

Main Theorem. *Let σ be a bounded variation function on $[a,b]$, satisfying the conditions (3.6), and let the zero set $T(\mathcal{S}_\sigma)$ of the function $\mathcal{S}_\sigma$ (4.5) be separable (i.e. the condition (2.2) is satisfied).*

  i) *For the system of exponentials $\left\{\text{Rstr}_E(e^{it\omega})\right\}_{t \in T(\mathcal{S}_\sigma)}$ to be a Riesz basis in the Hilbert space $L^2(E)$, it is necessary and sufficient that the transfer operator $T_{\sigma,E} : L^2([a,b]) \to L^2(E)$ be invertible.*

  ii) *Let the operator $T_{\sigma,E}$ be invertible, and let $\left\{\varphi_{[a,b];\,t_k}\right\}_{t \in T(\mathcal{S}_\sigma)}$ be a Riesz basis biorthogonal to the Riesz basis $\left\{\text{Rstr}_{[a,b]}(e^{it\omega})\right\}_{t \in T(\mathcal{S}_\sigma)}$. Then the system $\left\{\varphi_{E;\,t_k}\right\}_{t \in T(\mathcal{S}_\sigma)}$ with*

$$\varphi_{E;\,t_k} \overset{\text{def}}{=} \left(T_{\sigma,E}{}^*\right)^{-1} \varphi_{[a,b];\,t_k}$$

is a Riesz basis biorthogonal to the Riesz basis $\left\{\mathbf{Rstr}_E(e^{it\omega})\right\}_{t\in T(S_\sigma)}$.

Proof. First of all, we remark that, according to (5.2), for each finite linear combination $\sum c_t e^{it\omega}$ with $t \in T(S_\sigma)$, the equality

$$T_{\sigma,E}\left(\mathbf{Rstr}_{[a,b]}\left(\sum_t c_t e^{it\omega}\right)\right) = \mathbf{Rstr}_E\left(\sum_t c_t e^{it\omega}\right) \qquad (5.3)$$

holds.

a) Assume that the operator $T_{\sigma,E}$ is invertible. Since, from Theorem 4, the system $\left\{\mathbf{Rstr}_{[a,b]}(e^{it\omega})\right\}_{t\in T(S_\sigma)}$ forms a Riesz basis in the Hilbert space $L^2([a,b])$, its image – the system $\left\{\mathbf{Rstr}_E(e^{it\omega})\right\}_{t\in T(S_\sigma)}$ – forms a Riesz basis in the Hilbert space $L^2(E)$.

b) Assume that the system $\left\{\mathbf{Rstr}_E(e^{it\omega})\right\}_{t\in T(S_\sigma)}$ forms a Riesz basis in $L^2(E)$. Then this system is complete in $L^2(E)$, and frame inequalities

$$m_E^2 \sum_{t\in T(S_\sigma)} |c_t|^2 \le \left\|\sum_{t\in T(S_\sigma)} \mathbf{Rstr}_E\left(\sum_{t\in T(S_\sigma)} c_t e^{it\omega}\right)\right\|^2_{L^2(E)}$$

$$\le M_E^2 \sum_{t\in T(S_\sigma)} |c_t|^2$$

hold for each finite linear combination $\sum c_t e^{it\omega}$ with some frame constants $m_E > 0$, $M_E < \infty$. An analogous inequality holds also in $L^2([a,b])$ (with some frame constants $m_{[a,b]} > 0$, $M_{[a,b]} < \infty$. Combining these frame inequalities, we obtain a double inequality

$$\left(m_{[a,b]}/M_E\right)^2 \left\|\sum_{t\in T(S_\sigma)} \mathbf{Rstr}_E\left(\sum_{t\in T(S_\sigma)} c_t e^{it\omega}\right)\right\|^2_{L^2(E)}$$

$$\le \left\|\sum_{t\in T(S_\sigma)} \mathbf{Rstr}_{[a,b]}\left(\sum_{t\in T(S_\sigma)} c_t e^{it\omega}\right)\right\|^2_{L^2([a,b])}$$

$$\le \left(m_E/M_{[a,b]}\right)^2 \left\|\sum_{t\in T(S_\sigma)} \mathbf{Rstr}_E\left(\sum_{t\in T(S_\sigma)} c_t e^{it\omega}\right)\right\|^2_{L^2(E)}$$

$$(5.4)$$

which holds for any finite linear combination $\sum c_t e^{it\omega}$. According to (5.2), the inequality (5.4) means, that

$$\left(m_E / M_{[a,b]}\right)^2 \left\| \sum_{t \in T(S_\sigma)} \mathbf{Rstr}_{[a,b]} \left(\sum_{t \in T(S_\sigma)} c_t e^{it\omega}\right)\right\|^2_{L^2([a,b])}$$

$$\leq \left\| T_{\sigma,E}\left(\sum_{t \in T(S_\sigma)} \mathbf{Rstr}_{[a,b]} \left(\sum_{t \in T(S_\sigma)} c_t e^{it\omega}\right)\right)\right\|^2_{L^2(E)}$$

$$\leq \left(M_E / m_{[a,b]}\right)^2 \left\| \sum_{t \in T(S_\sigma)} \mathbf{Rstr}_{[a,b]} \left(\sum_{t \in T(S_\sigma)} c_t e^{it\omega}\right)\right\|^2_{L^2([a,b])}. \tag{5.5}$$

Linear manifolds $\mathbf{Rstr}_{[a,b]}\left(\sum_{t \in T(S_\sigma)} c_t e^{it\omega}\right)$ and $\mathbf{Rstr}_E\left(\sum_{t \in T(S_\sigma)} c_t e^{it\omega}\right)$ are dense in $L^2([a,b])$ and $L^2(E)$ respectively. Hence, the operator $T_{\sigma,E}$ is invertible, and

$$\left\|\left(T_{\sigma,E}\right)^{-1}\right\| \leq \left(M_{[a,b]} / m_E\right)^2, \qquad \left\|T_{\sigma,E}\right\| \leq \left(M_E / m_{[a,b]}\right)^2. \tag{5.6}$$

The theorem is proved.   ∎

Remark.   Let E be a finite union of intervals. If the sequence $\left\{\mathbf{Rstr}_E(e^{it\omega})\right\}_{t \in T}$ is a Riesz basis in $L^2(E)$, then, according to Landau [25, 26], $d(T) = (\mathrm{mes}\, E)/2\pi$ (see (2.15)). If the sequence $\left\{\mathbf{Rstr}_{[a,b]}(e^{it\omega})\right\}_{t \in T}$ (with the same T) is also a Riesz basis in $L^2([a,b])$, then $d(T) = (b-a)/2\pi$. (Also according to Landau's Theorem. Of course, for the case E is an interval, this result was obtained by means of entire functional methods much earlier [38].) Comparing, we obtain:

Proposition. *If σ is such that the system $\left\{\mathbf{Rstr}_{[a,b]}(e^{it\omega})\right\}_{t \in T}$ be a Riesz basis in $L^2([a,b])$ (for example, if the conditions (3.6) are satisfied), and if the transfer operator $T_{\sigma,E}$ is invertible, then*

$$\mathrm{mes}\, E = b - a \ \ (= \mathrm{mes}\,[a,b]). \tag{5.7}$$

However, this reasoning fails to work for sets E of more general structure than finite union of intervals.

§6   A mean-periodic transfer operator and the Cauchy problem for difference equation

Our Main Theorem reduces the problem of construction of an exponential Riesz basis in $L^2(E)$ to the problem of construction of a bounded variation function σ on $[a, b]$ $(b - a = \text{mes } E)$ such that the mean-periodic transfer operator $T_{\sigma, E}$ is continuous and continuously invertible. (And also it is desirable that the conditions (2.1) and (2.2) be satisfied.) Now the question arises: How to construct such a σ? The theorem of Section 5 does not give any answer to this question. Until now our 'Main Theorem' is only 'an existence theorem'. As long as we have no other way for m.-p. transfer except for the exponential expansion machinery, this theorem looks tautological.

However, there is an alternative approach for m.-p. continuation of a function from $[a, b]$ onto $\mathbb{R}$. This approach is based on a direct consideration of the convolution equation (4.3). We explain this approach in a simple situation.

Let σ be a piecewise constant function on $[a, b]$ satisfying (3.1), and let ω_l, $1 \leq l \leq m$ be jump points of σ:

$$a = \omega_0 < \omega_1 < \omega_2 < \ldots < \omega_m < \omega_{m+1} = b. \qquad (6.1)$$

We assume (without loss of generality) that

$$\sigma(a + 0) - \sigma(a) = 1, \quad \sigma(b - 0) - \sigma(b) = 1.$$

Let us denote $\sigma_p = \sigma(\omega_p + 0) - \sigma(\omega_p - 0)$ $(1 \leq p \leq m)$. Then (4.3) can be rewritten in the form

$$x(\omega - b) = x(\omega - a) + \sum_{1 \leq p \leq m} \sigma_p x(\omega - \omega_p) \quad (\omega \in \mathbb{R}). \qquad (6.2)$$

In the difference equation language the σ-m.-p. continuation problem is:

Given an initial data: a function $\varphi \in L^2([a, b])$, we seek a function x defined on $\mathbb{R}$, satisfying (6.2) on $\mathbb{R}$ and the initial condition

$$x(\omega) = \varphi(\omega) \quad (\omega \in [a, b]). \qquad (6.3)$$

This is nothing more than the Cauchy problem for a linear difference equation. (In the considered case both deviations of argument and coefficients are constant.)

There are (at least) two ways to consider this Cauchy problem. The first way already discussed: This is a version of the Fourier method. First we seek elementary solutions of (6.2). (In our case these solutions have the form $e^{it_k \omega}$ with an appropriate t_k.) Then we expand the initial data via the series

$$\varphi(\omega) = \sum_{t_k} c_{t_k} e^{it_k \omega} \quad (\omega \in [a, b]).$$

Finally, the sum of this series (considered for all $\omega \in \mathbb{R}$, not only for $\omega \in [a, b]$) provides the desired m.-p. continuation. It is the way our Main Theorem was proved.

However, to use the Main Theorem effectively, we need to combine the first and the second way. The second way deals with direct consideration of (6.2). Since the function φ is defined on $[a, b]$, the function $\varphi(\omega - b) - \sum_p \sigma_p \varphi(\omega - \omega_p)$ is defined on $[b + a, b + \omega_1]$. If a defined on $[a, b]$ function x satisfies the condition (6.3), then the equality

$$x(\omega - b) - \sum_{1 \leq p \leq m} \sigma_p \, x(\omega - \omega_p) = \varphi(\omega - b) - \sum_{1 \leq p \leq m} \sigma_p \, \varphi(\omega - \omega_p)$$

$(\omega \in [b + a, b + \omega_1])$ holds. If, moreover, the function x satisfies (6.2), then the equality

$$x(\omega - a) = \varphi(\omega - b) - \sum_{1 \leq p \leq m} \sigma_p \, \varphi(\omega - \omega_p)$$

must be fulfilled for all $\omega \in [b + a, b + \omega_1]$. Therefore, to solve the Cauchy problem (6.2)-(6.3), we are forced to define

$$x(\omega) \overset{\text{def}}{=} \varphi(\omega - (b - a)) - \sum_{1 \leq p \leq m} \sigma_p \, \varphi(\omega - (\omega_p - a)) \quad (\forall \, \omega \in (b, b + \Delta^{(r)}],$$

$$\tag{6.4}$$

where

$$\Delta^{(r)} = \omega_1 - a \tag{6.5}$$

is a step to the right of a difference scheme. Thus, in the first step of our construction we 'propagate' the initial data onto the interval $(b, b + \Delta^{(r)}]$: the function $x(\omega)$, defined originally on interval $[a, b]$ by the initial condition (6.3), is defined already on the larger interval $[a, b + \Delta^{(r)}] = [a, b] \cup (b, b + \Delta^{(r)}]$. This function satisfies both initial conditions (6.3) and (6.2) for $\omega \in (a + b, a + b + \Delta^{(r)}]$. In the same way, in the second step of our construction we 'propagate' the initial data onto the interval $(b, b + 2\Delta^{(r)}] = (b, b + \Delta^{(r)}] \cup (b + \Delta^{(r)}], b + 2\Delta^{(r)}]$. As a result, we obtain a function x defined already on the interval $[a, b + 2\Delta^{(r)}]$, satisfying both initial conditions (6.3) and (6.2) for $\omega \in (a + b, a + b + 2\Delta^{(r)}]$. Acting in this way, we propagate (or continue) the initial data onto the right half axis (b, ∞). Consequently, we obtain a function defined on $[a, \infty)$ satisfying both initial conditions (6.3) and (6.2) for $\omega \in (a + b, \infty)$. Analogously, we can propagate the initial data step-by-step onto the left-half axis $(-\infty, a]$. Now

$$\Delta^{(l)} = b - \omega_m \tag{6.6}$$

is a step to the left of a difference scheme. As result, we obtain a function x defined already on the interval $(-\infty, b]$, satisfying both the initial condition (6.3) and (6.2) for $\omega \in (-\infty, a + b]$. 'Glueing' these left and right continuations, we obtain a function defined on $\mathbb{R}$ and satisfying both (6.2) on $\mathbb{R}$ and the initial condition (6.3) on $[a, b]$.

Thus, the difference equation (6.2) provides us with a construction of the continuation operator C_σ (as well as the transfer operator $T_{\sigma,E}$).

From this construction it is clear that a solution of the Cauchy problem (6.2)–(6.3) is unique, and that the continuation operator $C_\sigma \colon L^2([a, b]) \to L^2_{loc}(\mathbb{R})$ is continuous.

Remark. Since the set E is compact, in order to construct the transfer operator $T_{\sigma,E}$ we need only a finite number of steps to the right and to the left.

§7 The case of multi-band sets in more detail

Now we consider the case of E which is a finite union of intervals:

$$E = \bigcup_{1 \le p \le n} I_p, \quad I_p = [a_p, b_p], \tag{7.1}$$

where

$$a_1 < b_1 < a_2 < b_2 < \ldots < a_n < b_n. \tag{7.2}$$

Denote

$$l_p = |I_p| \; (= b_p - a_p), \quad p = 1, 2, \ldots, n. \tag{7.3}$$

Let us try to construct the transfer operator using the considerations of Section 6. First, we should choose the 'initial' interval $[a, b]$. The only essential value is its length $(= b - a)$. Without loss of generality, we can set

$$a = 0, \quad b = \mathrm{mes}\; E. \tag{7.4}$$

Now we are to construct the difference equation (6.2). We are to choose shift values h_p and a coefficient c_p. The most difficult problem is: *How to choose shift values h_r ?*

Hint. The image of the transfer operator $T_{\sigma,E}$ must coincide with all the space $L^2(E)$. In view of (5.1), *no constraints must be imposed on values of a solution of the difference equation* (6.2) *at points of the set E.*

It means that for each ω, all the points $\omega, \omega + h_1, \omega + h_2, \ldots, \omega + h_m, \omega + \mathrm{mes}\, E$ could not belong to the set E simultaneously. In other words, the condition

$$\mathrm{mes}\left(\bigcap_{0 \le p \le (m+1)} (E - h_r)\right) = 0, \qquad (h_0 \stackrel{\mathrm{def}}{=} 0, \; h_{m+1} \stackrel{\mathrm{def}}{=} \mathrm{mes}\; E) \tag{7.5}$$

must be satisfied. (Moreover, $0 < h_1 < h_2 < h_m < \mathrm{mes}\; E$.)

Definition 11. *Given the set E of the form (7.1)–(7.2), with l_p defined by (7.3), let us define*

$$h_0 = 0, \ h_1 = l_1, \ h_2 = l_1 + l_2, \ h_3 = l_1 + l_2 + l_3, \ \ldots,$$
$$h_{n-1} = l_1 + l_2 + \ldots + l_{n-1}, \ h_n = l_1 + l_2 + \ldots + l_{n-1} + l_n \ (= \text{mes } E). \tag{7.6}$$

Lemma 1. *Let E be a multi-band set of the form (7.1)–(7.2), with l_p defined by (7.3). Shift values h_p, $(p = 0, 1, \ldots, n-1, n)$ are defined by (7.6). Then the condition*

$$\text{mes} \left(\bigcap_{0 \leq p \leq n} (E - h_p) \right) = 0 \tag{7.7}$$

is fulfilled.

According to this lemma, a natural candidate for the difference equation (6.2) is the equation

$$x(\omega + h_n) = x(\omega) + \sum_{1 \leq p \leq n-1} c_p \, x(\omega + h_p).$$

The last equation is not quite satisfactory: we can not guarantee that all the roots of the characteristic equation are real. However, this drawback could be completely overcome. We need only to 'duplicate' the amount of shift parameters.

Definition 12. *Given the set E of the form (7.1)–(7.2), with l_p defined by (7.3), let us define*

$$h_0^* = l_n + l_{n-1} + \ldots + l_2 + l_1 \ (= \text{mes } E),$$
$$h_1^* = l_n + l_{n-1} + \ldots + l_2, \ \ldots, \ h_{n-2}^* = l_n + l_{n-1}, \ h_{n-1}^* = l_n, \ h_n^* = 0. \tag{7.8}$$

In other words,

$$h_p^* = \text{mes } E - h_p \qquad (p = 0, 1, 2, \ldots, n). \tag{7.9}$$

Main Definition. *Let E be a multi-band set of the form (7.1)–(7.2), with l_p defined by (7.3). Let shift values h_p, h_p^* $(p = 0, 1, \ldots, n-1, n)$ be defined by (7.6) and (7.8). Let $c_1, c_2, \ldots, c_{n-1}$ be complex numbers. The difference equation*

$$x(\omega - h_0^*) - x(\omega - h_0) - \sum_{1 \leq p \leq (n-1)} \left(\overline{c_p} \, x(\omega - h_p^*) - c_p \, x(\omega - h_p) \right) = 0 \quad (\omega \in \mathbb{R})$$

$$\tag{7.10}$$

is said to be a difference equation associated with the multi-band set E.

The condition

$$\sum_{1 \le p \le (n-1)} |c_p| < 1 \tag{7.11}$$

guarantees that all the roots of the characteristic equation are real, and that these roots are separated. Moreover, these roots $\{t_k\}$ are fastened to the points $\{(2\pi / \text{mes } E) \, k\}$.

Main Principle. *Let E be a multi-band set of the form (7.1)–(7.2), with l_p defined by (7.3). Let shift values h_p, h_p^* ($p = 0, 1, \ldots, n-1, n$) be defined by (7.6) and (7.8). Let (7.10) be a difference equation associated with the multi-band set E. Then for all values of coefficients $c_1, c_2, \ldots, c_{n-1}$, satisfying the condition (7.11), except a thin set only the appropriate transfer operator $T_{\sigma,E} : L^2([0, \text{mes } E]) \to L^2(E)$ is invertible.*

We proved the principle for an arbitrary two-band set. In this case, we also are able to construct the inverse operator $(T_{\sigma,E})^{-1}$ effectively. We also confirmed this principle in many other cases.

References

[1] Avdonin, S. A., On the question of Riesz bases of complex exponential functions in L^2, *Vestn. Leningrad s. Univ. Ser. Mat.* **13** (1974). English transl. in: *Leningrad. Univ. Math. J.* **7** (1979), 203–211.

[2] Beaty, M. G. and M. M. Dodson, The distribution of sampling rates for signals with equally wide, equally spaced spectral bands, *SIAM J. Appl. Math.* **53** (1993), 893–906.

[3] Bezuglaya, L. and V. Katsnelson, The sampling theorem for functions with limited multi-band spectrum. I, *Zeitschr. für Anal. und ihre Anwendungen* **12** (1993), 511– 534.

[4] Brown, J. L., Sampling expansions for multi-band signals, *IEEE Trans. Acoustics, Speech and Signal Processing* **33** (1985), 312–315.

[5] Butzer, P. L., W. Splettstößer, and R. L. Stens, The sampling theorem and linear prediction in signal analysis, *Jahresber. Dt. Math.-Verein.* **90** (1988), 1–70.

[6] Daubechies, I., *Ten Lectures on Wavelets*, SIAM, CMBS-NSF Regional Conference Series in Applied Math., # 61, Philadelphia, 1992.

[7] Delsarte, J., Les fonctions moyenne-pèriodiques, *Journal de Mathématiques pures et appliquès* **14** (1935), 403–453.

[8] Dodson, M. M. and A. M. Silva, An algorithm for optimal sampling rates for multi-band signals, *Signal Processing* **17** (1989), 169–174.

[9] Dodson, M. M., Diophantine inequalities and sampling rates for multi-band signals, in: *Recent Advances in Fourier Analysis and its Applications*, J. S. Byrnes and J. F. Byrnes (eds.), Kluwer Academic Publishers, Dordrecht · Boston · London (1990), pp. 483—498.

[10] Dym, H. and H. P. McKean, *Fourier Series and Integrals*, Academic Press, New York · London, 1972.

[11] Gohberg, I. C. and M. G. Krein, *Introduction to the Theory of Linear Nonselfadjoint Operators in Hilbert Space* (in Russian), Nauka, Moscow, 1965. Engl. transl.: Amer. Math. Soc., Providence, RI, 1969.

[12] Goldman, S., *Information Theory*, Prentice-Hall, New York, 1953.

[13] Hardy, G. H., Notes on special systems of orthogonal functions, IV: The orthogonal functions of Whittaker's cardinal series, *Proc. Camb. Phil. Soc.* **37** (1941), 331–348. Reprinted in: *Collected Papers of G. H.Hardy. Vol.3.*, The Clarendon Press, Oxford, 1969, pp. 466–483.

[14] Higgins, J. R., Five short stories about the cardinal series, *Bull. Amer. Math. Soc.* **12** (1985), 45–89.

[15] Higgins, J. R., Some gap sampling series for multi-band signals, *Signal Processing* **12** (1987), 313–319.

[16] Higgins, J. R., Sampling theory for Paley-Wiener spaces in the Riesz basis setting, *Proc. Royal Irish Academy*, to appear.

[17] Jerry, A. J., The Shannon sampling theorem – Its various extensions and applications: A tutorial review, *Proc. IEEE* **65** (1977), 1565–1596.

[18] Karlovich, Yu. I. and I. M. Spitkovsky, Factorization of almost periodic matrix functions and the Noether theory for certain classes of equations of convolution type (in Russian), *Izvestija. Akad. Nauk SSSR, Ser. Mat.* **53** (1989), no.2, 276–308 (in Russian). English translation in: *Mathematics of the USSR, Izvestiya* **34** (1990), 281—316.

[19] Katsnelson, V. E., Exponential bases in L^2,(in Russian) *Funkcional'niĭ analiz i ego priloženija* **5** (1971), 1, 37–47. English transl. in: *Funct. Anal. Appl.* **5** (1971), 31–38.

[20] Katsnelson, V. E., An embedding theorem for functions, whose Fourier transforms are weight square summable. *Zeitschr. für Anal. und ihre Anwendungen* **13**:4 (1994), 387– 403.

[21] Kholenberg, A., Exact interpolation of band-limited functions, *J. Appl. Phys.* **24** (1953), 1432–1436.

[22] Khrushchev, S. V., N. K. Nikolskii, and B. S. Pavlov, Unconditional bases of exponentials and reproducing kernels, in *Complex Analysis and Spectral Theory*, V. P. Havin and N. K. Nikolskiĭ (eds.), Lecture Notes in Math. **864**, Springer-Verlag, Berlin Heidelberg 1981, pp. 214–335.

[23] Kotel'nikov, V. A., On the code capacity of 'ether' and wire in the electrical communication (in Russian), in *Vsesojuznniĭ energetičeskiĭ komitet*. Materiali k I Vsesojuznomu sjezdu po voprosam tehničeskoĭ rekonstrukcii dela svjazi i razvitija slabotočnoĭ promišlennosti, Izd. upravlenija svjazi RKK, Moscow, 1933.

[24] Landau, H. J., A sparse regular sequence of exponentials closed on large sets, *Bull. Amer. Math. Soc.* **70** (1964), 566–569.

[25] Landau, H. J., Sampling, data transmission, and the Nyquist rate, *Proc. IEEE* **55** (1967), 1701–1706.

[26] Landau, H. J., Necessary density conditions for sampling and interpolation of certain entire functions, *Acta Math.* **117** (1967), 37–52.

[27] Leont'ev, A. F. On properties of sequences of Dirichlet's polynomials convergent on a subinterval of the imaginary axis (in Russian), *Izvestija Akad. Nauk SSSR, Ser. Math.* **29** (1965), 269–328.

[28] Leont'ev, A. F., *Riady Eksponent* (in Russian), Nauka, Moskow, 1976.

[29] Leont'ev, A. F., *Posledovatel'nosti Polinomov iz Eksponent* (in Russian), Nauka, Moskow, 1980.

[30] Levin, B. Ya., On bases of exponential functions in $L^2(-\pi, \pi)$, *Khar'kov. Gos. Univ. Uchen. Zap* **115** = *Zap. Mat. Otdel. Fiz.-Mat. Fak. i Khar'kov Mat. Obsch.* ser.4, **27** (1961), 39–48 (in Russian).

[31] Levin, B. Ya., *Distribution of Zeros of Entire Functions* (in Russian), GTTI, Moscow, 1956. English translation in *Translations of Mathematical Monographs Ser.*, Vol. **5**, Amer. Math. Soc., Providence R.I., 1964.

[32] Levin, B. Ya., Interpolation by entire functions of exponential type (in Russian), in: *Mat. Fiz. i Funktsional. Anal.*, 1, Marchenko V. A. (ed.), Fiz.-Tehn. Inst. Nizkih Temperatur Akad. Nauk Ukrain. SSR., Naukova Dumka, Kharkov (1972), pp. 136–146.

[33] Levin, B. Ya. and I. V. Ostrovskiĭ, On small perturbations of the set of zeros of funcions of sine type (in Russian), *Izv. Akad. Nauk SSSR* **43** (1979), 89–110. English transl. in: *Math. USSR. Izv.* **14** (1980), 79–101.

[34] Lyubarskiĭ, Yu. and I. Spitkovsky, Sampling and interpolation for a lacunary spectrum, *Proc. of the Royal Edinburgh Society, Section A, Mathematics*, to appear.

[35] Marks II, R. I., *Advances Topics on Shannon Sampling and Interpolation Theory*, Springer-Verlag, New-York, 1992.

[36] Nyquist, H., Certain topics in telegraph transmission theory, *Trans. Amer. Inst. Electr. Eng.* **47** (1928), 617–644.

[37] Ogura, K., On a certain transcendental integral function in the theory of interpolation, *Tôhoku Math. J.* **17**, 1920, 64–72.

[38] Paley, R. and N. Wiener, *Fourier Transform in the Complex Domain*, AMS, New-York, 1934.

[39] Pavlov, B. S., Basicity of an exponential system and Muckenhoupt's condition (in Russian) *Dokl. Akad. Nauk SSSR* **244** (1979), 37–40. English transl. in: *Sov. Math. Dokl.* **20** (1979), 655–659.

[40] Plancherel, M. and G. Polya, Fonction entières et intégrales de Fourier multiples (Seconde partie). *Comment. Math. Helvetici.* **10** (1938), 110–163. Reprinted in: *Pólya, George. Collected papers. Vol. 1.*, The MIT Press, Cambrige MA · London, 1973, 644–697.

[41] Schwartz, L., Approximation d'unu fonction quelconque par les sommes d'exponentielles imaginaires. *Annales de la Faculté des Sciences de l'Université de Toulouse*, ser. 4, **6** (1943), 111–174.

[42] Schwartz, L., Theorie generale des fonctions moyenne-periodiques, *Ann. of Math* **48**:4 (1947), 857–918.

[43] Schwartz, L., *Etude des sommes d'exponentielles (deuxième èdition)*, Hermann, Paris, 1959.

[44] Sedleckiĭ, A. M., A periodic in the mean extension and bases of exponential functions in $L^p(-\pi, \pi)$ (in Russian), *Mat. Zametki* **12**:1 (1972), 37-42. English transl. in: *Math. Notes* **12** (1972), 455–458.

[45] Sedleckiĭ, A. M., Biorthogonal expansions of functions into series of exponents on intervals of the real axis (in Russian), *Uspehi Matem. Nauk* **37** (1982), 51–95. English transl. in: *Russ. Math. Surv.* **37**:5 (1982), 57–108.

[46] Seip, K., A simple construction of exponential bases in L^2 of the union of several intervales, *Proc. of the Royal Edinburgh Society, Section A, Mathematics* **38** (1995), 171–176.

[47] Shannon, C., Communication in the presence of noise, *Proc. IRE* **37** (1949), 10–21.

[48] Whittaker, E. T., On the functions, which are represented by expansions of the interpolation theory, *Proc. Roy. Soc. Edinburgh, Section A* **35** (1915), 181–194.

[49] Whittaker, J. M., *Interpolatory Function Theory*, (Cambridge Tracts in Mathematics and Math. Physics, No. 33). Cambridge Univ. Press, Cambridge, 1935.

[50] Young, R. M., *An Introduction to Nonharmonic Fourier Series*, Academic Press, New York· London, 1980.

[51] Zayed, A. I., *Advances in Shannon's Sampling Theory*, CRC Press, Boca Raton, 1993.

Victor E. Katsnelson
Department of Theoretical Mathematics
The Weizmann Institute of Science
Rehovot 76100
Israel
katze@wisdom.weizmann.ac.il

Group Theoretical Transforms, Statistical Properties of Image Spaces and Image Coding

Reiner Lenz and Jonas Svanberg

Abstract. In the first, theoretical, part of this paper we describe briefly some basic facts about group theoretical generalizations of the discrete Fourier transform and its connection to the Karhunen Loéve transform. We show that the computation of the eigenvectors and eigenvalues of a covariance matrix can be essentially simplified if the underlying stochastic process possesses a group-theoretically defined symmetry.

In the second part of the paper we investigate some properties of a large databases of image patches collected from a standard TV-channel and compare the quality of different approximations of the Karhunen-Loéve transform when they are applied to pixel data from the database. We then illustrate the use of non-standard tilings of the image in image coding. The application of DCT-based algorithms is no longer efficient for these tilings, whereas approximations based on dihedral transforms are still possible.

§1 Introduction

The Fourier transform (and its relatives like the Short-Time Fourier transform) is certainly one of the most important tools in signal processing. One mathematical explanation of its success is its connection with the time-shift operators. If $s(t)$ is a signal and $T_\sigma s(t) = s^\sigma(t) = s(t - \sigma)$ is a time-shifted version of $s(t)$ then the Fourier transforms of s and s^σ are identical up to a complex factor of magnitude one. If we think of s and s^σ as the same signal described in two different coordinate systems, then the Fourier transform gives a description of these signals that is (up to a multiplicative complex constant) independent of the selected coordinate system. Another application in which the connection between the Fourier transform and the time-shift operators is useful is the singular value decomposition of operators that commute with all the time-shift operators. It can be shown that the requirement that an operator commutes with all time-shift operators simplifies the eigenanalysis of such operators considerably.

This relation between the Fourier transform and the time-shift operators has a group theoretical generalization in which the time-shift operators are

Signal and Image Representation in Combined Spaces
Y. Y. Zeevi and R. R. Coifman (Eds.), pp. 555–575.

replaced by a more general class of operators that are connected via a group. For a large collection of groups it is then possible to construct a Fourier-like transform that possesses many of the properties of the Fourier transform. In this context we have a group G, a class of signals $s(x)$, and for each group element $g \in G$ an operator T_g that maps the signal $s(x)$ to a new signal $(T_g s)(x) = s^g(x)$. Again, we can investigate the singular value decomposition of operators that commute with all the operators $T_g, g \in G$. We call such an operator an intertwining operator. The description of this mathematical background is summarized in the first part of this paper.

In the second part of this paper we concentrate on one special class of intertwining operators: correlation and covariance operators of stochastic processes. If the correlation (or covariance) operator of a process is an intertwining operator for a group G of operators, then we will say that the process has the operator group as a symmetry group or that the process is G-symmetric. The eigenvalue analysis of covariance operators is known as the Karhunen Loéve transform (KLT) and it is one of the most important techniques in signal analysis. The results derived in the first part of this paper show that the KLT can be simplified if it is known that the process has a symmetry group. This theoretical result leads then to the following empirical problems:

1. For a given stochastic process and a group of operators G we have to decide if the process is G-symmetric.

2. For a given problem we might have reason to believe that the underlying stochastic process is approximately G-symmetric for several groups G. Each of the symmetry groups will then lead to an approximation of the Karhunen-Loéve transform of this process. The problem is then to compute the quality of these different approximations.

In the second part of the paper we will apply the theoretical results and investigate the properties of different KLT-approximations for large databases of pixel data from TV-images. The results of these experiments will provide us with some insight about the structure of the image spaces considered. Information about such structures is of fundamental importance in all applications where the efficient processing of large amounts of image data is required. A very important technological example of this kind of problem is transform image coding where image data is compressed by removing redundant information from it [20]. Another area where statistical properties of images are of utmost importance are certain models of mechanisms in the early vision processing of animals and human beings [4, 2, 1, 3, 27, 8, 28]. These models are based on the hypothesis that the cells that process the sensor data collected by the retina act as coding devices. Based on this assumption, these theories try to explain the different response patterns found in biological vision systems. In computer vision and pattern recognition, finally, there are numerous algorithms like edge-detectors that compute basic features from an image. Many of these methods are, implicitly or explicitly, based on assumptions about the statistical properties of the intensity distributions in images [18, 22, 23, 25].

Despite its importance there are only a few published investigations in which basic statistical parameters are estimated. In [10] Field uses 6 images of size 256^2 and in [11] the same author uses 55 images of size 512^2. In both cases only natural scenes are considered since the goal of the study is the investigation of mammalian receptive fields. Similar investigations are described in [31]. In many studies it is furthermore assumed that the distribution can be characterized by its power spectrum, an assumption that is never checked.

In our experiments we make no simplifying assumptions about the nature of the data. Therefore, we use large sets of randomly selected regions in a large number of images. We are also primarily interested in image coding problems and therefore we do not select special images, but we collect them with the help of a frame-grabber from a TV-channel over a period of several days. From these datasets we compute the eigenvector system of the correlation matrix of the data and several of its approximations. Then, we test the quality of these approximations by computing the approximation errors for a random selection of patches from the database. Finally, we illustrate the application of non-standard tiling schemes for image coding applications.

§2 Group theoretical transforms

We recall that a group is a set of elements together with a group operation which is associative, which has a neutral element and where each group element has an inverse. For a set X we say that a group G operates on X if every element $g \in G$ corresponds to a transformation $T(g) : X \to X$ such that $T(g_1 g_2) = T(g_1)T(g_2)$. In our applications, X is a part of $\mathbb{R}^2$ and G is a group of geometrical transformations that map X onto itself. Sometimes we will write: gx instead of $T(g)x$. Typical examples are:

1. X = real line and $G = \mathbb{R}$ the group of shift-operators: $T(\sigma)s(x) = s(x - \sigma)$.

2. X = unit circle (or some disc in $\mathbb{R}^2$) and $G = SO(2)$ the group of two-dimensional rotations.

In most applications of group theoretical methods in image processing or vision research, the group G is a Lie-group (examples are [17, 23, 24,

25, 21, 5, 9, 13, 30, 32, 33] and the articles in [19]). This approach has the advantage that the powerful apparatus of Lie-theory can be applied, but it has the drawback that the link between the continuous theory and the discrete data has to be established. Usually this is done via some type of sampling argument. Here we will avoid this difficulty by considering finite groups only. This reduces the whole theory to a part of linear algebra. We will only summarize those results of the general theory that are necessary to get an overview over those aspects of the theory that are of interest in the current application. For more information about the group theoretical background, the reader may consult some of the many books on the subject. More information about group representations in general and finite groups in particular can be found, for example, in: [7, 16, 34, 15, 6, 29] and descriptions of image-processing-oriented applications are described in [21] and [25].

2.1   Representations and intertwining operators

In the rest of this paper we will concentrate on the *dihedral groups* $D(n)$ which are defined as the group of geometry-preserving mappings of the plane that map the regular n-sided polygon onto itself. The set X is a collection of grid points in the plane that are invariant under the dihedral group. Often X will be the set of pixels in a square window.

The most important properties of the dihedral groups are summarized in the following theorem:

Theorem 1.   (1) *The dihedral group $D(n)$ consists of n rotations with rotation angles $\frac{2\pi k}{n}, k = 0 \ldots n - 1$ and the products of these angles $2\pi k/n, k = 0 \ldots n - 1$ and the products of these rotations with reflections about any of the symmetry axes of the polygon.*

   (2) *$D(n)$ has $2n$ elements.*

   (3) *Elements of $D(n)$ can be written as pairs $\left(\rho^k, \sigma^l\right)$ where ρ is the rotation with angle $\frac{2\pi}{n}, \sigma$ is a fixed reflection and l is either equal to zero or one. The multiplication between elements in $D(n)$ is given by:*

$$\left(\rho^k, \sigma^l\right) \left(\rho^m, \sigma^n\right) = \begin{cases} \left(\rho^{k+m}, \sigma^{l+n}\right) & \text{if } l = 0 \\ \left(\rho^{k-m}, \sigma^{l+n}\right) & \text{if } l = 1 \end{cases}$$

   This shows that the dihedral groups are semi-direct products of two cyclic groups with two and n elements. Another example of a semi-direct product is semi-direct product of the rotation group and the group of translations. For details see [15].

Now assume that X is a finite set and G a group operating on X. Assume further that f is a function from X into the complex numbers. Given a group element $g \in G$ we define a new function f^g on X as: $f^g(x) = f(g^{-1}x)$. If the set X consists of N elements then we can identify the space of all complex-valued functions on X with the N-dimensional space $\mathbb{C}^N$ and the mapping $f \mapsto f^g$ defines a linear mapping $\mathbb{C}^N \to \mathbb{C}^N$. Selecting a fixed basis in $\mathbb{C}^N$ we have thus for each $g \in G$ an $N \times N$ matrix $T(g)$ such that $f^g = T(g)f$. This mapping $g \mapsto T(g)$ has furthermore the property that $T(g_1 g_2) = T(g_1)T(g_2)$ for all $g_1, g_2 \in G$. Changing the basis in $\mathbb{C}^N$ with the help of a matrix B, maps the matrices $T(g)$ to $BT(g)B^{-1}$. This leads to the following definition:

Definition 1. (1) *An N-dimensional matrix representation of a group G is a mapping: T from the group G into the space of $N \times N$ complex matrices such that $T(g_1 g_2) = T(g_1)T(g_2)$ for all $g_1, g_2 \in G$.*

(2) *Two matrix representations T_1, T_2 are equivalent if there is a matrix B such that: $T_1(g) = BT_2(g)B^{-1}$ for all $g \in G$.*

(3) *The equation $T_1(g) = BT_2(g)B^{-1}$ is equivalent to $T_1(g)B = BT_2(g)$ which is also meaningful when B is not invertible. We say that a matrix B is an intertwining matrix for the representation T if $T(g)B = BT(g)$ holds for all $g \in G$.*

Since equivalent matrix representations describe the same object in different coordinate systems, we can try to find a system in which a given representation has the simplest description. This leads to the following definition:

Definition 2. (1) *A matrix representation T is reducible if there is a matrix B such that all matrices $BT(g)B^{-1}$ have the form $\begin{pmatrix} T_1(g) & T_{12}(g) \\ 0 & T_2(g) \end{pmatrix}$ with square matrices $T_1(g)$ and $T_2(g)$.*

(2) *A matrix representation T is irreducible if it is not reducible.*

(3) *A matrix representation T is completely reducible if there is a matrix B such that all matrices $BT(g)B^{-1}$ have the form*

$$\begin{pmatrix} T_1(g) & 0 & \cdots & 0 \\ 0 & T_2(g) & \cdots & 0 \\ 0 & 0 & \cdots & 0 \\ 0 & 0 & \cdots & T_K(g) \end{pmatrix}$$

with irreducible components T_k.

For finite groups, the representation theory is fairly simple: There are only finitely many different irreducible representations and all representations are completely reducible. Furthermore, it can be shown that the irreducible representations of finite groups are all finite dimensional. For the dihedral groups, this result is summarized in the following theorem:

Theorem 2. Denote by ρ the rotation with rotation angle $\frac{2\pi}{n}$ and by σ the reflection on a symmetry axis as before. All elements in $D(n)$ are then of the form $\sigma^l \rho^k$ with $(l = 0, 1; k = 0, \ldots n - 1)$ and if we set $\omega = e^{2\pi i/n}$ then the irreducible representations of $D(n)$ are given as follows:

(1) If n is even then there are four one-dimensional irreducible representations:

$$T_1^1(\sigma^l \rho^k) \ = \ 1 \qquad\qquad T_2^1(\sigma^l \rho^k) \ = \ (-1)^l$$

$$T_3^1(\sigma^l \rho^k) \ = \ (-1)^k \qquad\qquad T_4^1(\sigma^l \rho^k) \ = \ (-1)^{k+l}$$

and there are $\frac{n}{2} - 1$ two-dimensional irreducible representations:

$$T_j^2(\rho^k) = \begin{pmatrix} \omega^{jk} & 0 \\ 0 & \omega^{-jk} \end{pmatrix} \qquad\qquad T_j^2(\sigma\rho^k) = \begin{pmatrix} 0 & \omega^{jk} \\ \omega^{-jk} & 0 \end{pmatrix}$$

with $j = 1, \ldots \frac{n}{2} - 1$.

(2) If n is odd then there are two one-dimensional irreducible representations:

$$T_1^1(\sigma^l \rho^k) = 1 \qquad T_2^1(\sigma^l \rho^k) = (-1)^l$$

and there are $\frac{n-1}{2}$ two-dimensional irreducible representations:

$$T_j^2(\rho^k) = \begin{pmatrix} \omega^{jk} & 0 \\ 0 & \omega^{-jk} \end{pmatrix} \qquad\qquad T_j^2(\sigma\rho^k) = \begin{pmatrix} 0 & \omega^{jk} \\ \omega^{-jk} & 0 \end{pmatrix}$$

with $j = 1, \ldots \frac{n-1}{2}$.

In our investigation of singular-value decompositions of G-symmetric processes, we will mainly use the following theorem that describes the structure of intertwining operators of irreducible representations. This theorem is known as Schur's Lemma:

Theorem 3. (1) *Assume T is an irreducible representation and A an intertwining operator. Then $A = c \cdot \text{id}$ where c is a constant and id is the identity operator.*

(2) *Irreducible representations of commutative groups are one-dimensional.*

This can be generalized to intertwining operators A of two irreducible representations S, T of the same group on two different spaces for which $AT(g) = S(g)A$ for all g. Then it can be shown that A is either 0 or an isomorphism.

We know that the finite representations of the dihedral groups are completely reducible and we can thus assume that we have a coordinate system in which the matrices $T(g)$ have the block-diagonal form

$$
T(g) = \begin{pmatrix} T_1(g) & 0 & \cdots & 0 \\ 0 & T_2(g) & \cdots & 0 \\ 0 & 0 & \cdots & 0 \\ 0 & 0 & \cdots & T_M(g) \end{pmatrix}
$$

where the T_m are irreducible representations of $D(n)$. Now we consider an intertwining operator C and write C in the corresponding block diagonal form:

$$
C = \begin{pmatrix} C_{11} & C_{12} & \cdots & C_{1M} \\ C_{21} & C_{22} & \cdots & C_{2M} \\ \cdot & \cdot & \cdots & \cdot \\ C_{M1} & C_{M2} & \cdots & C_{MM} \end{pmatrix}.
$$

Putting this into the intertwining equation $CT(g) = T(g)C$ gives:

$$
T(g)C = \begin{pmatrix} T_1(g)C_{11} & T_1(g)C_{12} & \cdots & T_1(g)C_{1M} \\ T_2(g)C_{21} & T_2(g)C_{22} & \cdots & T_2(g)C_{2M} \\ \cdot & \cdot & \cdots & \cdot \\ T_M(g)C_{M1} & T_M(g)C_{M2} & \cdots & T_M(g)C_{MM} \end{pmatrix}
$$

$$
= \begin{pmatrix} C_{11}T_1(g) & C_{12}T_2(g) & \cdots & C_{1M}T_M(g) \\ C_{21}T_1(g) & C_{22}T_2(g) & \cdots & C_{2M}T_M(g) \\ \cdot & \cdot & \cdots & \cdot \\ C_{M1}T_1(g) & C_{M2}T_2(g) & \cdots & C_{MM}T_M(g) \end{pmatrix} = CT(g).
$$

For $C_{i,j}$ this give the matrix equation $T_i(g)C_{ij} = T_j(g)C_{ij}$ and C_{ij} is thus an intertwining operator for the irreducible representations T_i and T_j. From Schur's Lemma follows that the matrices C_{ij} are either multiples of the identity or zero:

$$
C = \begin{pmatrix} \alpha_{11}E & \alpha_{12}E & \cdots & 0 \\ \alpha_{21}E & \alpha_{22}E & \cdots & 0 \\ \cdot & \cdot & \cdots & \cdot \\ 0 & 0 & \cdots & \alpha_{MM}E \end{pmatrix}
$$

(where the E matrices are identity matrices of different sizes).

Rearranging the basis elements we find that the intertwining operators C all have the block-diagonal form:

$$C = \begin{pmatrix} C_1(g) & 0 & \cdots & 0 \\ 0 & C_2(g) & \cdots & 0 \\ 0 & 0 & \cdots & 0 \\ 0 & 0 & \cdots & C_M(g) \end{pmatrix}.$$

2.2   Stochastic processes with symmetries

We will now apply the results from the previous section in the investigation
of stochastic processes. We assume that we have a stochastic process with
values in a vector space V of dimension N. If we select an orthonormal
basis $\{b_l, l = 1 \ldots N\}$ in V we can write every element $f \in V$ as a sum $f = \sum_{l=1}^{N} \beta_l b_l$ with constants β_l. For a given constant $L < N$ we now try to
find a basis $\{b_l, l = 1 \ldots N\}$ in V which minimizes the average L^2 truncation
error

$$\mathrm{E}\left(\sum_{n=L+1}^{N} |\beta_n|^2 \right)$$

where E is the expectation operator over all elements in the input space.
For simplicity we assume in the following that elements in V are centered,
i.e.: $\mathrm{E}(f) = 0$.

It is a standard result from signal processing [12] that this basis is
given by the largest eigenvectors of the covariance matrix $C = \mathrm{E}(ff')$ of
the input process. We order the eigenvectors to get decreasing eigenvalues
so that: $Cb_l = \lambda_l b_l$ and $\lambda_1 \geq \lambda_2 \geq \ldots \lambda_N$.

In the general case where no additional information is available, we must
first calculate the covariance matrix from the input data. Then we can numerically compute its eigenvectors. In many cases this is computationally
expansive and it can also be numerically unstable.

We now describe how a priori knowledge of a group theoretical structure
of the input space can be used. The elements in this space are functions
of the spatial variable x and the stochastic variable ω : $f(x, \omega)$. If we
assume that the functions are centered then the covariance function $C(x, y)$
is defined as

$$C(x, y) = \int f(x, \omega) f(y, \omega) \, d\omega.$$

In the case where the random variations of the process are identical with
the transformations of the group elements and if each such transformation
occurs with the same probability, then we get for the covariance function:

$$C(x, y) = \int f(x, \omega) f(y, \omega) \, d\omega = \sum_G f^g(x) f^g(y) = \sum_G f(g^{-1}x) f(g^{-1}y).$$

From this it follows easily that the covariance function is invariant under the group transformations: $C(gx, gy) = C(x, y)$. The same argument holds for the case where the probability measure $d\omega$ is the product of the (invariant) Haar measure of the group and an additional measure $d\omega'$, *i.e.* : $\int f d\omega = \int f dg d\omega$. (For details on the Haar measure see any text on integration theory on groups.)

Now assume that the elements f are vectors p and the transformation $p \mapsto p^g$ is described by the matrix multiplication: $p^g = T(g)p$. The invariance of the covariance function under group transformations implies that the covariance matrix commutes with all representation matrices $T(g)$ and the matrix C defines thus an intertwining operator of the representation T.

From the previous results on intertwining operators we know that the covariance matrix is block-diagonal in the special coordinate system in which the representation matrices are block-diagonal with irreducible representations in the diagonal. This simplifies the computation of the Karhunen-Loéve transform considerably since we only have to compute the Karhunen-Loéve transform for the smaller matrices along the diagonal. Finally, we remark that similar results hold for the correlation matrices which will be mainly used in our experiments.

§3   Experiments

In the previous sections we saw that the *a priori* knowledge of a group theoretical symmetry of a stochastic process can be used to simplify the computation of the Karhunen-Loéve transform. If we want to use this in a practical application we have to find out first if the process at hand has such a symmetry, and in the cases where there are several relevant groups, we have to find out which groups are most effective in the application. This is not a mathematical problem but an empirical question. Such investigations are important in the study of biological systems, as we mentioned in the introduction.

In the current context we are more interested in technical applications and we choose, therefore, TV images as our stochastic process. In our experiments we used a commercial frame-grabber to digitize images from a Swedish Television channel. In each image we selected 10 randomly located blocks of a given size and we stored them in a database. When the processing of a frame was completed we picked up the next frame. The database used here consisted of 100 000 blocks of size 8×8 pixels with three channels (RGB) for each pixel. These blocks were collected from 10 000 images. Each day we processed 1000 images, *i.e.* the blocks were collected over a period of 10 days. A similar database with blocks of size 16×16 pixels led to similar results. In the next processing stage the original, 3-

dimensional, RGB-data was transformed to a 1-dimensional signal using a Karhunen-Loéve transform based on the eigensystem of the 3×3 RGB correlation matrix. The new value corresponds roughly to the intensity value.

In the following experiments we tested the performance of different approximations of the Karhunen-Loéve transform. Since we used the database with patches of size 8×8 we get a correlation matrix of size 64^2. In all experiments we computed first a system of 64 vectors $\{b_1, ..., b_{64}\}$ which approximates the Karhunen-Loéve transform. Then we expanded 1000 random patches from the database in this system and computed the average L^2 error. We did this for all truncation orders 1 to 64.

In the experiments we used the following methods to compute the systems b_i :

Karhunen-Loéve transform: The b_i are the eigenvectors sorted according to the eigenvalue.

Stationarity: If X_{ij} is the pixel value at location (i, j) then we enforce the equality $E(X_{ij}X_{kl}) = C_S(k-i, l-j)$ by estimating $C_S(a, b)$ from the true correlations and then using these values in the correlation matrix. The system is the eigenvector system of the correlation matrix.

Radial Symmetry: In the same way we enforce $E(X_{ij}X_{kl}) = C_R(r)$ with $r^2 = (i-k)^2 + (j-l)^2$ and use the corresponding eigensystem.

Toeplitz: First we compute the correlation matrix using the stationary approach and then we replace $C_S(a, b)$ by $C_S(a, 0) * C_S(0, b)$. We use again the eigenvector system.

Dihedral: We compute first the correlation matrix $\tilde{C}$ in the transform domain. If the process is completely invariant under all $D(4)$ transformations then $\tilde{C}$ is a block diagonal matrix. In the case of 8×8 pixel arrays the block-diagonal submatrices are of size: (10, 6, 6, 10, 16, 16). In the approximation we set all entries in $\tilde{C}$ which are outside these blocks to zero. This gives an approximation $\tilde{C}_d$ to $\tilde{C}$ which is then transformed back to the original domain. We use then the eigenvectors of this new correlation matrix.

DCT: The standard (or DCT-II) discrete cosine transform of an $N \times N$ matrix with entries f_{jk} is defined as the mapping:

$$F_{mn} = \frac{2\gamma(m)\gamma(n)}{N} \sum_{i=0}^{N-1} \sum_{j=0}^{N-1} f_{ij} \cos\left(\frac{(2i+1)m\pi}{2N}\right) \cos\left(\frac{(2j+1)n\pi}{2N}\right)$$

$$(3.1)$$

with $\gamma(0) = \frac{1}{\sqrt{2}}$, and $\gamma(k) = 1$ elsewhere. The transform vectors b_l are computed in two steps. First we transform the original correlation matrix C with the DCT to the new matrix $C_f = FCF'$. If the gray-value distribution was perfectly shift invariant then C_f would be diagonal and the values of the diagonal elements would specify the importance of the corresponding Fourier components. Therefore, we order the diagonal elements in C_f and reordered the columns of the DCT-transformation matrix according to the result of the sorting process. The basis vectors are the elements of the reordered DCT basis.

Figure 1 shows the reconstruction errors for the different transforms. Only the approximation orders with the largest differences between the methods are shown. The figure shows that the stationarity assumption is clearly valid in this case. We conclude also that the dihedral approximation

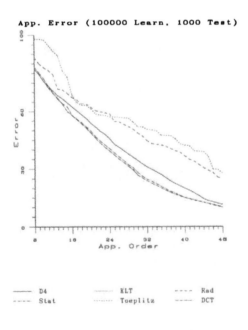

Figure 1. Reconstruction error for the 8 × 8 database.

fits the data reasonably well, whereas the Toeplitz and the radial approximations are of significantly lower quality. In Figure 2 we show the result of an experiment in which we compared the variations within the database with the differences between the different approximation methods. The plots show the error from the Karhunen-Loéve transform expansion plus and minus the standard deviation of this error and the reconstruction errors for the different approximation methods.

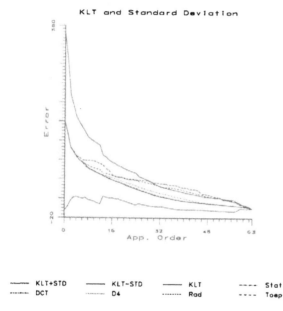

Figure 2. Error for different approximations compared with the standard deviation in the Karhunen-Loéve transform error.

The previous results show that the shift-invariance assumption is clearly valid for the investigated database. The DCT-approximation (which is also the basis of the JPEG coding standard) had therefore the best performance of the investigated approximation methods. When high compression rates must be achieved then it is known that this coding scheme leads to clearly visible block boundaries. It has been argued [14] that this effect can be reduced when other, non-rectangular windows are used in the coding process. The DCT is no longer computationally efficient for most of such image-tilings. One advantage of the $D(4)$-transform over the DCT is that it is not restricted to rectangular blocks. The selection of the form of the window is only restricted by the requirement that the window must be invariant under certain rotations and reflections. In all these cases the $D(4)$-transform is applicable. A different tiling consists of rectangular tiles which are rotated $45°$. Such simple variations do not however improve the quality significantly. Therefore, we did some preliminary experiments with more complicated tilings such as the one shown in Figure 3. In this experiment the "round" tiles consisted of 80 pixels and the "cross-like" tiles of 64 pixels. Using this tiling of the image requires the computation of two basis systems, one for each type of tiles. The dihedral transform is applicable in both cases since both of them are symmetrical with respect to the dihedral group. In the next set of images (4, 5, 6, and 7) we illustrate one of the experiments. Figure 4 is a part of the original image. In Figures 5 and 6 the image was coded using the non-standard tiling shown in Figure 3. In Figure 5 we used the true eigenvectors of the correlation matrix of the image and in Figure 6 its dihedral approximation. The standard JPEG coding is shown in Figure 7. All codings were done with the same rate. The coding effects are especially noticeable in the homogeneous regions of the image where the JPEG coding produces long, straight borders, whereas the coding based on the non-standard tiling breaks up these structures.

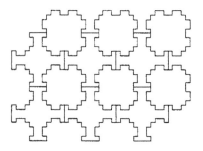

Figure 3. Non-standard tiling scheme.

§4 Discussion and conclusions

We showed that for each finite group there is a DFT-like transform and that the application of this transform block-diagonalizes operators with group theoretical symmetries. In cases where the operators do not possess the full symmetry we get approximations to the KLT. We concentrated mainly on the description of the dihedral transform and compared it with several other popular approximations to the KLT with the help of a large database

Figure 4. Original image.

Figure 5. Karhunen-Loéve transform coding.

Figure 6. Dihedral transform.

Figure 7. JPEG coding.

of image blocks. The experiments showed that the symmetry assumptions are largely valid and that the Karhunen-Loéve transform, the DCT, and the approximation based on the stationary assumption gave the best results in the general case.

An important drawback of the standard transform coding methods (like JPEG) is the visibility of block artifacts when the images are coded with a very low bit-rate. In the image-coding literature it is sometimes suggested that the human visual system is especially sensible for horizontal and vertical structures and that this leads to better recognition of these block boundaries. We therefore tested some coding methods based on non-rectangular tilings and illustrated the results with an example. For such tiling schemes the DCT is no longer efficient, whereas the $D(4)$-transform is still applicable.

We did not discuss efficient implementations of these transforms in detail but it can be shown that the transforms based on the $D(4)$ symmetry group involve basically only sets of four and eight points all connected via rotations and reflections (for details see [26]). This makes them more flexible than the FFT and it also allows very simple implementations.

In our discussion we restricted ourselves to the dihedral group $D(4)$, but by using a hexagonal tiling of the image, we would also allow the application of transforms to be connected to the group $D(6)$. These and other variations of the same technique should provide interesting applications of the basic ideas based on dihedral symmetries.

Acknowledgment. The authors were supported by NUTEK grant P1839-1.

References

[1] Atick, J. J., Could information theory provide an ecological theory of sensory processing? *Network* **3**(2) (1992), 213–251.

[2] Atick, J. J. and N. Redlich, Towards a theory of early visual processing, *Neural Computation* **4** (1990), 196–210.

[3] Atick, J. J. and N. Redlich, What does the retina know about natural scenes, *Neural Computation* **4** (1992), 449–572.

[4] Barlow, H. B., The coding of sensory messages, in *Current Problems in Animal Behaviour*, W. H. Thorpe and O. L. Zangwill (eds.), Cambridge University Press, 1961, pp. 331–360.

[5] Blatt, N. and J. Rubinstein, The canonical coordinates method for pattern recognition-ii: Isomorphisms with affine transformations, *Pattern Recognition* **27**(1) (1994), 99–107.

[6] Burrow, M., *Representation Theory of Finite Groups*, Dover, Mineola, 1965.

[7] Curtis, C. W. and I. Reiner, *Representation Theory of Finite Groups and Associative Algebras*, Wiley Classics Library. Wiley, New York, 1988.

[8] Daugman, J., Complete discrete 2-D Gabor transforms by neural networks for image analysis and compression. *IEEE Trans. Acoustics, Speech and Signal Processing* **36**(7) (1988), 1169–1179.

[9] Eagleson, R., Measurement of the 2-D affine Lie group parameters for visual motion analysis, *Spatial vision* **6**(3) (1992), 183–198.

[10] Field, D. J., Relations between the statistics of natural images and the response properties of cortical cells, *J. Optical Soc. Amer.* **4**(12) (1987), 2379–2394.

[11] Field, D. J., Scale-invariance and self-similar 'wavelet' transforms: an analysis of natural scenes and mammalian visual systems, in *Wavelets, Fractals and Fourier Transforms*, M. Farge, J. C. R. Hunt, and J. C. Vassilicos (eds.), Claredon Press, 1993, pp. 151–193.

[12] Fukunaga, K., *Introduction to Statistical Pattern Recognition*, Academic Press, Boston, 1990.

[13] Gertner, I. and R. Tolimieri, A group theoretic approach to image representation, *J. Visual Comm. Image Representation* **1**(1) (1990), 67–82.

[14] Gray, R. M., P. C. Cosman, and K. L. Oehler, Incorporating visual factors into vector quantiziers for image compression, in *Digital Images and Human Vision*, A. B. Watson (ed.), MIT PRESS, Cambridge, MA, 1993, pp. 35–52.

[15] Gurarie, D., *Symmetries and Laplacians, Introduction to Harmonic Analysis, Group Representations and Applications*, North Holland, Amsterdam, 1992.

[16] Hamermesh, M., *Group Theory and Its Applications to Physical Problems*, Addison-Wesley, 1962.

[17] Hoffman, W. C., The Lie algebra of visual perception, *J. Math. Psychology* **4** (1966), 65–98.

[18] Hummel, R. A., Feature detection using basis functions, *Computer Graphics and Image Processing* **9** (1979), 40–55.

[19] Humphrey, G. K., Invariance, recognition, and perception: A special issue in honour of P. C. Dodwell. *Spatial Vision* **8**(1) (1994), 1–166.

[20] Jayiant, N. S. and P. Noll, *Digital Coding of Waveforms*, Prentice Hall, Englewood Cliffs, NJ, 1984.

[21] Kanatani, K., *Group Theoretical Methods in Image Understanding*, Springer Verlag, Heidelburg, 1990.

[22] Koenderink, J. J. and A. J. van Doorn, Generic neighborhood operators, *IEEE Trans. Pattern Anal. Machine* **14**(6) (1992), 597–605.

[23] Lenz, R., A group theoretical model of feature extraction, *J. Optical Soc. Amer.* **6**(6) (1989), 827–834.

[24] Lenz, R., Group-invariant pattern recognition, *Pattern Recognition* **23** (1990), 199–218.

[25] Lenz, R., *Group Theoretical Methods in Image Processing*, Lecture Notes in Computer Science (Vol. 413), Springer Verlag, Heidelburg, 1990.

[26] Lenz, R., Investigation of receptive fields using representations of dihedral groups, *J. Visual Comm. Image Representation*, **6**(3) (1995), 209–227.

[27] Li, Z. and J. J. Atick, Toward a theory of the striate cortex, *Neural Computation* **6** (1993), 127–146.

[28] Linsker, R., Self-organization in a perceptual network, *IEEE Computer* **21**(3) (1988), 105–117.

[29] Lomont, J.S., *Applications of Finite Groups*, Dover, Mineola, 1959.

[30] Rubinstein, J., J. Segman, and Y. Zeevi, Recognition of distorted patterns by invariance kernels, *Pattern Recognition* **24**(10) (1991), 959–967.

[31] Ruderman, D. L., *Natural Ensembles and Sensory Signal Processing*, PhD thesis, University of California, Berkeley, 1993.

[32] Segman J., J. Rubinstein, and Y. Zeevi, The canonical coordinates method for pattern deformation: Theoretical and computational considerations. *IEEE Trans. Pattern Anal. Machine* **14**(12) (1992) 1171–1183.

[33] Segman, J. and Y. Zeevi, Image analysis by wavelet-type transforms: Group theoretic approach, *J. Math. Imaging and Vision* **3** (1993), 51–77.

[34] Vinberg, E., *Linear Representation of Groups*, Birkhäuser Verlag, Basel, 1989.

Reiner Lenz
Department of Electrical Engineering
Linköping University
S-58183 Linköping
Sweden
reiner@isy.liu.se

Jonas Svanberg
Department of Electrical Engineering
Linköping University
S-58183 Linköping
Sweden
svan@isy.liu.se

Subject Index

WAVELET ANALYSIS AND ITS APPLICATIONS

CHARLES K. CHUI, SERIES EDITOR

Printed and bound by CPI Group (UK) Ltd, Croydon, CR0 4YY

03/10/2024

01040425-0018